FOUNDATIONS
OF DATA
ORGANIZATION

FOUNDATIONS OF DATA ORGANIZATION

Edited by
Sakti P. Ghosh
IBM Almaden Research Center
San Jose, California

Yahiko Kambayashi
Kyushu University
Fukuoka, Japan

and

Katsumi Tanaka
Kobe University
Kobe, Japan

PLENUM PRESS • NEW YORK AND LONDON

Library of Congress Cataloging in Publication Data

International Conference on Foundations of Data Organization (1985: Kyoto, Japan)
  Foundations of data organization.

  "Proceedings of the International Conference on Foundations of Data Organization, held
May 22–24, 1985, in Kyoto, Japan."—T.p. verso.
  Includes bibliographies and indexes.
  1. Data base management—Congresses. 2. File organization (Computer science)—Con-
gresses. I. Ghosh, Sakti P. II. Kambayashi, Y. III. Tanaka, Katsumi, 1951–   IV. Title.
QA76.9.D3I5585   1985   005.74   87-2440
ISBN 0-306-42567-X

Proceedings of the International Conference on Foundations of Data Organization,
held May 22–24, 1985, in Kyoto, Japan

© 1987 Plenum Press, New York
A Division of Plenum Publishing Corporation
233 Spring Street, New York, N.Y. 10013

Printed in the United States of America

IN MEMORIAM

WITOLD LIPSKI, JR.

1949-1985

This book is dedicated to the memory of our friend Witold Lipski, Jr.,
who died May 30, 1985 in Nantes, France of a terminal illness. Witold
was a program committee chairperson of the International Conference on
Foundations of Data Organizations on May 21-24, 1985 in Kyoto, Japan and
unfortunately, he could not see the success of this conference and this
book.

Witold was an exceptionally talented researcher in computer science.
He published over 70 papers and 3 books in the wide areas of computer
science:theory of databases, algorithms, data structures, computational
geometry, VLSI design and combinatorics. Especially, his fundamental
contribution in the theory of incomplete information database was widely
recognized.

We will remember him as a person of very warm and modest personality
as well as his academic talent. His memory will remain a constant
inspiration for all of us.

PREFACE

Foundations of data organization is a relatively new field of research in comparison to, other branches of science. It is close to twenty years old. In this short life span of this branch of computer science, it has spread to all corners of the world, which is reflected in this book. This book covers new database application areas (databases for advanced applications and CAD/VLSI databases), computational geometry, file allocation & distributed databases, database models (including non-traditional database models), database machines, query processing & physical structures for relational databases, besides traditional file organization (hashing, index file organization, mathematical file organization and consecutive retrieval property), in order to identify new trends of database research.

The papers in this book originally represent talks given at the International Conference on Foundations of Data Organization, which was held on May 21-24, 1985, in Kyoto, Japan. This conference was held at Kyoto University, and sponsored by the organizing committee of the International Conference on Foundations of Data Organization and the Japan Society for the Promotion of Science. The conference was in cooperation with: ACM SIGMOD, IEEE Computer Society, Information Processing Society of Japan, IBM Research, Kyushu University, Kobe University, IBM Japan, Kyoto Sangyo University and Polish Academy of Sciences. This Conference was the follow-up of the first conference, which was hosted by the Polish Academy of Sciences and held at Warsaw in 1981. The Warsaw conference focused mainly on consecutive retrieval property and it's applications. The technical contents of this conference has been expanded to include many new and challenging fields of foundations of data organizations.

In the last five years database processing has expanded from logical processing of information to the field of artificial intelligence, statistical processing of information, and distributed databases. The database structures needed to support these new advances are reflect in papers presented by researchers from Japan, USA, Canada, Europe, and China. Advances in consecutive retrieval property are being made by researchers in Japan and China, which are covered in this book also. New results in classical topics of data organization such as: hashing, mathematical file organization using computational geometry, database theory, database models, structures and performances of physical database models, are being continually discovered. Some new results in these topics from researchers in Japan, China, USA, and Europe are covered in this book. Applications of databases in the field of engineering is an immersing new area of database research. Papers on CAD/VLSI databases

are being presented by researchers from USA and Europe. Database
machines have been the dream of engineers, since databases were invented.
Researchers from Japan and Europe are presenting some of their research
results in this book. In a multiuser environment, simultaneously
processing of multiple queries directed at the same database, is always a
challenge to database researchers. New results in query processing and
physical structures for relational databases are presented by researchers
from Japan, USA and Europe.

After the conference, the authors carefully revised their papers for
publishing this book. Overall, the papers of this conference reflect a
snap shot of the major advances around the globe in the important topics
of foundations of data organization. We truly hope this book will be a
start of a new database research and of a new communication among world-
wide database people. Also, we hope that the readers will find this book
to be of value for their future activities.

Finally, we would like to express our sincere appreciation to
Professor Sumiyasu Yamamoto, the honorary chairperson of the
International Conference on Foundations of Data Organization and many
people who organized, worked, or helped us for the conference.

Sakti P. Ghosh
(IBM Almaden Research Center, U.S.A.)

Yahiko Kambayashi
(Kyushu University, Japan)

Katsumi Tanaka
(Kobe University, Japan)

CONTENTS

DATABASES FOR ADVANCED APPLICATIONS

HASHING

CONSECUTIVE RETRIEVAL PROPERTY

FILE ALLOCATION AND DISTRIBUTED DATABASES

MATHEMATICAL FILE ORGANIZATION AND COMPUTATIONAL GEOMETRY

DATABASE MACHINES

DATABASE MODELS

STRUCTURES AND PERFORMANCE OF PHYSICAL DATABASE MODELS

CAD/VLSI DATABASES

QUERY PROCESSING AND PHYSICAL STRUCTURES FOR
RELATIONAL DATABASES

DATABASE THEORY

DATABASE RESEARCH AND DEVELOPMENT
IN THE PACIFIC AREA COUNTRIES

Databases for
Advanced Applications

STATISTICS METADATA: LINEAR REGRESSION ANALYSIS

Sakti P. Ghosh

Computer Science Department
IBM Almaden Research Center
San Jose, California

ABSTRACT

This paper deals with statistics metadata in the context of estimating the parameters and analysis of data, using statistical linear regression analysis techniques. Statistics metadata consists of complex data objects and the rules for manipulating them. The paper includes a brief outline of statistical linear estimation and establishes the important role that statistics metadata can play in real time linear regression analysis. The paper also discusses the structure of the statistics metadata for range-select-query, link-query, insert-query, delete-query and update-query. Statistics metadata locking is also discussed.

1. INTRODUCTION

Researchers in database management have felt for a long time that the data collected from the real world and stored in the computer are not adequate for complex manipulating of information and abstracting all the implicit knowledge in the database. Thus they have invented different semantic and physical models for data processing (Date (1977), Ghosh (1977), Ullman (1982)). Most of these models were invented for facilitating processing of the logical information contained in the data. Statisticians on the other hand, have been concerned with the processing of statistical information contents of data, for more than a century (Kendall & Stuart (1958)). In the last five years the statisticians and the database management researchers have been working together to combine the logical and statistical processing of data (Wong (1982a, 1982b), Hammond & McCarthy (1983), Annual conferences on Computer Science and Statistics). Different researchers have taken different approaches in these *meetings of the two disciplines*. The University of Florida school (see Su (1983)) have concentrated on extending the present database management models to accommodate corporate and scientific/statistical data modelling. Many schools, *e.g.*, Lawrence Berkeley Laboratory (Wong (1982b)), have contributed significantly in this approach. Other schools have taken the statisticians approach to data management. The University of Wisconsin (Madison) (Boral, DeWitt & Bates (1982)) is notable for this approach.

The concept of *metadata*, which has been around for many years (Teitel (1977)) has received a new life in the hands of the statistical database management system (SDBMS) researchers (McCarthy (1982)). Many early researchers in Information Retrieval and Information Management Systems (Baxendale (1966)) have discussed in details that derived/descriptive data should be treated as metadata structures. Many derived/descriptive data like, data dictionary and data directory, have been used in Database Management Systems (DBMS) for many years (Lefkovits (1977), Allen, Loomis & Mannino (1982)) without being referred to as meatadata. Many researchers in SDBMS have rediscovered that metadata can play a vital role not only in data description but also in enhancing statistical processing of the data (Bishop & Freedman (1983)). Researchers in metadata have a good idea about some of the properties of a metadata, but a no formal definition has been established up to now, because it is too new a subject and researchers are not sure that all the necessary properties have been identified. Metadata has two broad set of properties: functional (based on usage) and operational (based on the objects actually used and stored). The functional properties are: storage and retrieval of data, presentation and display of data, analysis of data, and physical manipulation of the data file. The operational properties of metadata consists of those information most commonly included in various data dictionary and directories, *e.g.*, data structure, long name, abbreviated names, title, column headers, footnotes, data types, summary statistics, *etc*. Most of these metadata have played some role in SDBMS but still have not succeeded in reducing the time needed for most of the common types of statistical analysis. Some attempts have been made in this area very recently (Ghosh (1983, 1984)). In this paper we shall discuss in details, some statistical metadata and their relation to statistical estimation of parameters and analysis of data in the context of linear regression analysis.

This paper is written for readers who are not experts in statistical *linear regression analysis*, hence we shall cover some fundamental concepts of linear regression analysis for completeness of the paper. Readers who would like to be more familiar with the subject should refer to the cited materials (Kendall & Stuart (1961), Scheffe (1959)). Suppose we have n observations of n random variables $y_1, y_2, \ldots\ldots, y_n$, which are constituted as linear combinations of p unknown quantities $\beta_1, \beta_2, \ldots\ldots, \beta_p$ plus errors $e_1, e_2, \ldots\ldots, e_n$, *i.e.*,

$$y_i = x_{1i}\beta_1 + x_{2i}\beta_2 + \ldots\ldots + x_{pi}\beta_p + e_i \qquad (i = 1, 2, \ldots\ldots, n) , \qquad (1.1)$$

where the x_{ji}s are known constant coefficients. The β_js are numeric parameters, which idealizes some aspects of the physical world that the model is trying to represent. The purpose of the gathering of statistical data is to estimate the β_js, which are also referred to as the *regression coefficients* of the y_is on the x_{ji}s. The values of the x_{ji}s depend on the statistical environment, under which the data has been collected. In a statistical randomized block design experiment (also referred to as one-way layout) the n observations are divided into subgroups and for the *j*th subgroup $x_{ji} = 1$ for a fixed j and all other x_{ji}s are zeros. The statistical design of experiment determines the pattern of the x_{ji}s. In a multivariate environment the x_{ji}s are random variables (correlated with the y_is. In this paper we shall discuss statistics metadata for estimating the β_js, thus our results will be applicable for any linear regression analysis, including design of experiments, multivariate analysis, statistical applications in manufacturing, *etc*.

Most of these statistical estimates of the parameters are mathematical functions of all the observed data. Thus in a large database, if the data are stored as raw data, then calculating the estimates of the parameters can be a very time consuming effort. This is the major complaint of most of the statisticians in using a DBMS. The SDBMS that are commercially available, do not solve this problem. The statistics metadata that researchers have discussed up to now (McCarthy (1982)) do not solve this problem. Solving the efficiency problem in statistics computation using certain types of metadata is the theme of this paper.

A statistics metadata contains some numerical functions (could be vector valued) of observed data, which can be used to compute other statistics. Thus statistics metadata may contain partially computed sum of observations, or partial sum of squares of observations or partial sum of products of observations, *etc.* Some database management researchers would like to label statistical metadata as *abstract data types*, but this is inaccurate because of the unique generic nature of statistics metadata and the processing rules associate with them. Some of the advantages of the use of metadata have been shown by this author in manufacturing test environment (Ghosh (1983) and in an interactive information display system (Ghosh (1984)). Some times we may include certain types of computational logic as a part of our statistics metadata. When computational logic is included in the statistics metadata it will provide modularity and transportability to them. In this paper we shall show that certain types of *statistics metadata* can be used to calculate the estimates of the parameters and statistical data analysis (including analysis of variance, analysis of data from a design of experiment, multiple regression and correlation) much faster than conventional statistical formulas.

In the next section we shall provide some background on linear regression analysis for those, who are not familiar with the subject. In section 3 we will discuss statistics metadata in detail. In section 3 we will also present structures of statistics metadata and their properties associated with some typical DBMS queries. In section 4 we will discuss locking mechanism in statistics metadata. In section 5 we will compare the computation time associated with different statistics when they are computed from statistics metadata versus usual formula. In section 6 we shall discuss some mathematical properties of statistics metadata. We will include concluding remarks in section 7.

2. LINEAR REGRESSION ANALYSIS

Regression analysis refers to studying functional dependencies between different variables. In eq. (1.1) y_is are the values of a dependent variable and the x_{ji}s are the values of p independent variables. The dependencies between the values of the dependent and independent variables are not exact functional dependencies. The dependencies are *statistical dependencies*. The e_is represent the statistical errors associated with the values of the dependent variable, *i.e.*, y_is. Usually, in a regression analysis the independent variables are assumed to be non-statistical variables. This is not true in multiple regression analysis. In general, the functional form of the dependency can be complex, but in many practical situations the form is assumed to be linear, as in eq. (1.1). The functional form is also known as the model for representing the data. The determination of the functional form is outside the scope of this paper. All the analysis of variance techniques (see Scheffe (1959)) are based on the linear form of the functional dependency. Every linear form has some unknown parameters; in the eq. (1.1) the β_js are the unknown parameters. These unknown parameters are statistically estimated from the data (involves the values of both the dependent and independent variables). There are many statistical estimation methods. The most common method is known as, the *Method of Least Squares*. In the least square estimation method, the unknown parameters, *i.e.*, β_js are estimated in such a manner that the sum of squares of the errors, *i.e.*, Σe_i^2 , is minimum.

In estimating the parameters of eq. (1.1) it is assumed that the expected value (sum of the product of the probabilities and the values of the random variable) of the errors is zero, *i.e.*,

$$E(e_i) = 0 \qquad (i = 1, 2,\ldots\ldots, n) \; . \tag{2.1}$$

This assumption is justified because, if there is a constant term in the model it can be accommodated by assuming that $x_{ji} = 1$ for a specific j. The e_is are assumed to be statistically independent with a constant variance, *i.e.*,

$$E(e_i e_j) = \sigma^2 \delta_{ij} \; , \tag{2.2}$$

5

where σ^2 is the unknown constant variance and δ_{ij} is 0 or 1, according as $i \neq j$ or $i = j$, respectively.

Example 2.1

Suppose $x_{1_i} = 1$ and $x_{2_i} = x_{3_i} = \ldots \ldots \ldots x_{p_i} = 0$, *for all i*, then eq. (1.1) becomes

$$y_i = \beta_1 + e_i .$$

This model is referred to as the one parameter model. The estimate of β_1 is the mean value of the observations.

If $x_{1_i} - x_{2_i} - 1$ and $x_{3_i} = x_{4_i} = \ldots \ldots \ldots x_{p_i} = 0$, for all i, then eq. (1.1) becomes

$$y_i = \beta_1 + \beta_2 + e_i .$$

This model is referred to as the two-parameter model. It is easy to see that, it is difficult to estimate β_1 and β_2 without some more information. The theory of linear estimation establishes the conditions under which the different $\beta_i s$ can be estimated. Sometimes the estimation conditions are enforced by capturing the data, according to the different design of experiment schemes.

Using eq. (1.1) the error sums of squares can be expressed as

$$\sum_{i=1}^{n} \left(y_i - x_{1i}\beta_1 - x_{2i}\beta_2 - \ldots \ldots - x_{pi}\beta_p \right)^2 . \tag{2.3}$$

Equation (2.3) has to be minimized with respect to the $\beta_i s$. If eq. (2.3) is differentiated with respect to the $\beta_i s$ and equated to zero then, the set of p equations thus obtained are called the *Normal Equations* and is given by

$$\sum_{i=1}^{n} y_i x_{ji} = \beta_1 \sum_{i=1}^{n} x_{1i} x_{ji} + \beta_2 \sum_{i=1}^{n} x_{2i} x_{ji} + \ldots \ldots \ldots + \beta_p \sum_{i=1}^{n} x_{pi} x_{ji} .$$

for $j = 1, 2, \ldots \ldots p$. \tag{2.4}

The least squares estimates of the $\beta_i s$ are obtained from the solutions of the eq. (2.4). The computation of the solutions involve two steps, (i) calculating the $\Sigma y_i x_{ji}$ and $\Sigma x_{k_i} x_{ji}$, and (ii) solving the equations in eq. (2.4). The computational complexity of (i) is of the order n and the computational complexity of (ii) is of the order p^2. In any practical statistical database n is very very large, hence computations of the $\Sigma y_i x_{ji}$ and $\Sigma x_{k_i} x_{ji}$, take most of the time. Moreover the set of observations from which these sums of products and squares have to be calculated, also depends on the actual query. These sums are the foundations of multivariate analysis and analysis of variance, which play a very important role in statistical analysis. Some discussions of multivariate analysis and analysis of variance are covered in the next few paragraphs. One of the goals of this paper is to show, how to reduce the computation time for these sums of products and squares, and thus solve some major problems of real time statistical analysis.

In multivariate analysis $(y_i, x_{1_i}, \ldots \ldots, x_{p_i})$ is assumed to be an observation on $(p+1)$ random variables, say, $(Y, X_1, X_2, \ldots \ldots, X_p)$. All these random variables are correlated. There is no special role for the variable Y, thus in most multivariate analysis the random variable Y is omitted. The eq. (1.1) is referred to as the multiple regression equation and the $\beta_i s$ are referred to as the multiple regression coefficients (see Kendall & Stuart (1961)). The multiple regression coefficients are estimated using the least square estimation techniques and can be expressed in terms of covariances, variances and correlations coefficients of the $X_i s$. The covariance between X_j and X_k is given by

$$\mu_{jk} = \sum_{i=1}^{n} x_{ji} x_{ki}/n - \left(\sum_{i=1}^{n} x_{ji} \right) \left(\sum_{i=1}^{n} x_{ki} \right)/n^2 \; . \tag{2.5}$$

The variance of the random variable X_j is given by

$$\sigma_j^2 = \mu_{jj} \; . \tag{2.6}$$

The correlation coefficient between the random variables X_j and X_k is given by

$$\rho_{jk} = \frac{\mu_{jk}}{\sigma_j \sigma_k} \; . \tag{2.7}$$

The *regression coefficients, multiple correlations coefficients, partial correlation coefficients*, all can be expressed as recursive expressions in terms of the $\rho_{jk}s$ and the σ_js (see Kendall & Stuart (1961)). Thus creating statistics metadata, which enables fast computation of the sums of squares and products from the records relevant to the different queries in a DBMS, will enable real time statistical multivariate analysis.

In order to perform the *analysis of variance* on a set of data, the values of the random variable Y, have to be observed according to certain schemes conforming to the underlying statistical model. This is needed for estimability of the parameters of the model. Usually the underlying model is assumed to be linear. In a one-way layout the observations are grouped into I groups of sizes $n_i(i = 1,2,........,I)$ and the *jth* observation in the *ith* group, in the model, is given by

$$y_{ij} = \mu + \alpha_i + e_{ij} \qquad i = 1,2,......,I; \quad j = 1,2,......,n_i \tag{2.8}$$

and the e_{ij} are statistically independently Gaussian distributed with mean 0 and variance σ^2. Estimation of the $\alpha_i s$ and performing the analysis of variances, need the computation of the sums, Σy_i and Σy_i^2 (see Scheffe (1959)). If the readers are interested in more complex design of experiments, they may read Scheffe (1959).

3. STATISTICS METADATA

McCarthy's (1982) paper provided an excellent overview of metadata, including statistical metadata, as it is known to the researchers in DBMS. According to McCarthy, metadata is data about data, *i.e.*, systematic descriptive information about data content and organization that can be retrieved, manipulated, and displayed in various ways. His definition of statistical metadata includes default specifications at the database level, error expressions, weighting expression, aggregate or disaggregate expressions, suppression expression, *etc.* Most of the researchers of DBMS will agree with these definitions and at the same time complain that statistical processing of very large statistical databases are extremely time consuming. Hence, it appears that these researchers have taken the first step in this new field of SDBMS but, statistical metadata have to be examined in more detail from both statistics and DBMS point of views, to achieve real time statistical analysis.

Most of the statistical metadata, provided by different DBMS, contain rules for processing raw data. As the volume of raw data is very large in most of the statistical databases hence, providing only the rules for processing raw data, will not achieve real time statistical analysis. *It is important that statistical metadata contain some complex type of data and the rules for processing them.* In this paper these new objects will be referred to as *statistics metadata.* The rules for processing the complex data objects in the statistics metadata are different from the rules for processing or representing or retrieving (information) of the raw data. The statistics metadata will contain not only computation rules for the complex data objects but also logical manipulation rules for these complex data objects. In this section, we shall discuss in details some desired global rules for manipulating the complex data objects of the statistics metadata. We shall also discuss some of the structures of the complex data objects, their properties and how they can be used to achieve real time statistical analysis in a SDBMS environment.

Consider the random variables: $X_1 =$ Age, $X_2 =$ Height and $X_3 =$ Weight of individuals. If we are interested in the average height of individuals, whose age $= 20$ years, then we have to select the records of all individuals whose age $= 20$ years. This set will be an aggregation of individuals of all weights, so long as their age is 20 years. If we had constructed some statistics metadata for the heights of the individuals then a desirable property for these statistics metadata, for expediting computations, should be that *they be additive under aggregation* of the values of the weights. This additive property would enable the derivation of the statistics metadata needed for computation of the average height much faster than, if the average height had to be calculated from the observed values of the heights. It is easy to see that if the average heights of all individuals were stored as the statistics metadata for every pair of values of age and weight, they would not have the additive property needed to compute the average height of individuals whose age are 20 years. The same problem would arise if we were interested in the standard deviation of the heights of individuals, whose age are 20 years, and the statistics metadata were standard deviations of heights for every combination of values of age and weight. If instead, we had stored as statistics metadata: the sum of the heights, the sums of squares of the heights, the number of individuals for each pair of values of age and weight, and the rules for computing the means and standard deviations under aggregation; then the average height and standard deviation for individuals whose age are 20 years could be calculated very rapidly (from these statistics metadata).

The author (Ghosh (1984)) has provided the constructions of certain complex data objects needed for statistics metadata, which have additive properties under categories consolidations and statistical tables consolidations. In that paper, the author has shown that partial sums, partial sums of squares, sums of products, frequency distributions, *etc.*, all preserve additive properties under categories and statistical tables consolidations. These complex data objects will be components of our statistics metadata. We will also discuss the rules needed for manipulating these complex data objects, their limitations and efficiencies, *etc.*

The least square solution to the linear regression is given by the eq. (2.4). Both the left side and the right side of those p equations contain sums of squares and products. Suppose the sums of squares and the products are computed for each atomic category. Let us denote the p Normal equations for the *uth* atomic category by

$$E_{1u}, \quad E_{2u}, \quad \ldots\ldots\ldots, \quad E_{pu} . \tag{3.1}$$

Let $\Sigma^{\#}_{u} E_{iu}$ represent the *equation sum*, *i.e.*, the left side terms and the right side terms of equations are summed separately over the categories represented by the range of the subscript u. Suppose terms in E_{iu} are represented by

$$t_{lu} = t_{r1u} + t_{r2u} + \ldots\ldots\ldots + t_{rpu} \tag{3.2}$$

then $\Sigma^{\#}_{u} E_{iu}$ would represent the following equation

$$\sum_{u} t_{lu} = \sum_{u} t_{r1u} + \sum_{u} t_{r2u} + \ldots\ldots\ldots + \sum_{u} t_{rpu} . \tag{3.3}$$

Suppose S denotes the set of atomic categories contained in a query, then the Normal equations for solving the regression coefficients for that query will be given by

$$\sum_{u \in S}^{\#} E_{1u}, \quad \sum_{u \in S}^{\#} E_{2u}, \quad \ldots\ldots\ldots\ldots, \quad \sum_{u \in S}^{\#} E_{pu} \ . \tag{3.4}$$

Thus, $\sum^{\#}_u$ would be included as a processing rule in the statistics metadata for solving Normal equations of linear regression analysis. These statistics metadata would also include for each atomic category, the complex data objects $\sum y_j x_{ji}$ and $\sum x_{k_i} x_{ji}$. These statistics metadata will reduce considerably, the computation time for estimating the regression coefficients for a statistical query. If the number of atomic categories associated with the statistical query is reasonably small, then the computations can be performed in real time. It is to be noted that, we have assumed that the regression equation is of the same form in all the atomic categories (only the estimates of the regression coefficients, which are data dependent, will be different) of the statistical query. If this assumption is not valid, then the rules for combining the partial sums of squares and products will be different (any student of linear algebra can figure the rules).

Statistics is the science of averages hence consolidation of atomic categories is very common in statistical queries. In general, *if the complex data objects in the statistics metadata are sums of raw data then they will have the additive property under consolidation of atomic categories. Thus, for expediting computation of statistics in statistical queries, the formula for computing statistics have to be decomposed into components, which are sums of raw data; these sums and the rules for computing the statistics from the sums, have to be stored in the statistics metadata.*

In this paper we shall discuss only additive properties of the complex data objects and not multiplicative property because, (i) multiplicative operations can be transformed into additive operations by using the logarithmic transformations, (ii) in statistics computation most of the multiplicative operations between variables, are within atomic categories, which we are preserving.

The formula for the covariance is given by eq. (2.5), but if the complex data objects in the atomic categories are stored as eq. (2.5), then they will not have the additive property under aggregation of the atomic categories. If instead, the statistics metadata for the atomic categories contain the following complex data objects, then additive property under aggregation is preserved.

$$\sum_{i=1}^{n} x_{ji} x_{ki} \ , \quad \sum_{i=1}^{n} x_{ji} \ , \quad \sum_{i=1}^{n} x_{ki} \ , \quad n \ . \tag{3.5}$$

Let us denote first three objects by O_{2jk}, O_{1j}, O_{1k}, then the following rule for computing the covariance from these complex data objects has to be included in the statistics metadata

$$\mu_{jk} = O_{2jk}/n - O_{1j} O_{1k}/n^2 \ . \tag{3.6}$$

It is easy to show that *the partial sums: O_{2jk}, O_{1j}, O_{1k}, are the largest complex data objects that preserves the additive property under atomic categories consolidations in statistical queries, which contain covariance, variance and correlation computations.*

Let us consider the analysis of variance of a two-way layout with interactions and equal number of observations per cell (see Scheffe (1959)). The linear model is given by

$$y_{ijk} = \mu + \alpha_i + \beta_j + \gamma_{ij} + e_{ijk}$$

$$i = 1,2,\ldots\ldots,I \ ; \quad j = 1,2,\ldots\ldots,J \ ; \quad k = 1,2,\ldots\ldots,K \ ; \tag{3.7}$$

where $\alpha_i s$ denote the effects of the first factor, $\beta_j s$ denote the effects of the second factor and γ_{ij} denotes the interaction effect between the *ith* level of the α factor and the *jth* level of the β factor. The e_{ijk} denotes the statistical error associated with y_{ijk}. μ the mean effect. I is the

9

number of levels of α factor, J is the number of levels of β factor and K is the number of observations per cell. The formula for the sums of squares of the main effect of the α factor is given by

$$SS_\alpha = \sum_i \left(\sum_j \sum_k y_{ijk} \right)^2 /JK - CF ,$$

where

$$CF = \left(\sum_i \sum_j \sum_k y_{ijk} \right)^2 /IJK . \tag{3.8}$$

The formula for the sums of squares of the main effect of the β factor is given by

$$SS_\beta = \sum_j \left(\sum_i \sum_k y_{ijk} \right)^2 /IK - CF . \tag{3.9}$$

The formula for the sums of squares for the interaction effect between the α and the β factors is given by

$$SS_{\alpha\beta} = \sum_i \sum_j \left(\sum_k y_{ijk} \right)^2 /K - CF - SS_\alpha - SS_\beta . \tag{3.10}$$

The formula for the total sums of squares of all the data is given by

$$SS_{tot} = \sum_i \sum_j \sum_k y_{ijk}^2 - CF . \tag{3.11}$$

The error sums of squares, SS_e, is obtained by subtracting (3.8), (3.9) and (3.10) from (3.11). In order to preserve the additive property under aggregation of the atomic categories for statistical queries containing analysis of variance for the two-way layout, the statistics metadata should contain the following complex data objects for each of the atomic categories

$$\sum_i \sum_j \sum_k y_{ijk}^2 , \ \left(\sum_i \sum_j \sum_k y_{ijk} \right)^2 , \ \sum_i \left(\sum_j \sum_k y_{ijk} \right)^2 , \ \sum_j \left(\sum_i \sum_k y_{ijk} \right)^2 ,$$

$$\sum_i \sum_j \left(\sum_k y_{ijk} \right)^2 , \ I , \ J , \ K . \tag{3.12}$$

The statistics metadata will also contain the rules for computing the sums of squares from these complex data objects for the different statistical queries.

The *Granularity of Statistics Metadata* plays a very important role in SDBMS. In general, the complex data objects are values of set functions of raw data, hence there is a many (individual) to one mapping involved. Thus the identity of the original individuals associated with the raw data are lost in the statistics metadata. The operations associated with the statistics metadata allows the creation of more complex aggregate data objects but the reverse is not possible. The advantage associated with the creating of the complex data objects in the statistics metadata is to reduce the statistics computation time. The granularity selection of the statistics metadata is a trade-off between (i) the query capability of the SDBMS and (ii) the desired real time response of query transactions. The choice of the granularities associated with the different statistics metadata has to be decided during the design of the SDBMS. The granularity design of the statistics metadata is not the scope of this paper and is being omitted. In the remaining of this section, we shall discuss structures of statistics metadata and their properties associated with some typical DBMS queries.

Range-Select Query (RSQ)

A query which specifies a range or a set of values of the qualification attribute (or attributes) is referred to as the range select query. An example of a RSQ would be: 'Find the average height of people between the ages 20 to 30 years.' The record set relevant to a RSQ is the union of the records relevant to the single value select queries (SVSQ) for the specified range. In the example, the SVSQ refer to: 'Average height of people for a specified age.' For statistical queries in a SDBMS, the category attributes of the atomic categories may be different from the attributes specified in the requested statistics. In the previous RSQ, the category attribute was 'age of people' and the statistics attribute was 'height of people.' Sometimes the category attributes are non-numeric attributes, *e.g.*, names; and hence cannot be statistics attributes.

In some statistical queries the category attribute and the statistics attribute may be the same. Consider a frequency distribution of people by height with the following frequency class ranges (in inches): 61-65, 66-70, 71-75, 76-80. The atomic categories are these classes. The statistical query: 'Find the standard deviation of height of people, who are 61 to 65 inches tall,' will be considered as a SVSQ. Each of these atomic categories can have statistics metadata associated with them. We can also consider RSQ such as: 'Find the skewness in the heights of people, whose heights are between 71 and 80 inches.' Based on these discussions, we can state the following properties of complex data objects in statistics metadata associated with RSQ.

If the classification attributes of the atomic categories are different from the statistics attributes and the complex data objects of the single-valued-select-queries have additive properties then, the complex data objects in the statistics metadata of the range-values-queries (on the category attributes) can be obtained by using the additive rules.

Thus if, the category attribute is 'age,' the atomic categories are specified by 'year of age' and the statistics attribute is 'height' and the complex data objects in the statistics metadata are: 'sums of squares of height, sums of height, and number of observations,' then complex data objects needed to answer the following RSQ can be obtained by adding the corresponding components in the atomic categories: 'Calculate the standard deviation of the heights of people, who are between the ages 25 and 30 years.'

The properties of statistics metadata, discussed in the previous paragraph, will also hold for the case when the category attributes are the same as the statistics attribute. There could be some exceptions but the author has not been able to find one.

When a range select statistics query involves linear regression estimation, then the normal equations (expressed in terms of the normal equations of the atomic categories) will be of the form (3.4). Thus, all the advantages of real time solutions of these equations will also be valid for RSQ.

Link-Queries (LQ)

Suppose we have records of people, which contain many fields, including 'name of individual' and 'name of parent.' Consider the query 'List the name and average income of the different people along with the standard deviation of the age of their children.' To answer this query a record has to be selected and the name field has to be compared with the 'name of parent' field of all the other records to locate the records of the children of that individual. As the answer to this query involves linking together different records it is referred to as a link-query. In relational model for DBMS, this link operation is also referred to as a *join operation*. This query is a *statistics-link-query (SLQ)*) because it involves the computation of some statistical information in addition to the linking operation. This query requests two statistics information: (i) average income of people, and (ii) the standard deviation of the age of the children. The average income is a field value of the original record and remains invariant under the link operation. The standard deviation of the age of the children has to be computed from the ages of the children, after the link operation identifies the records of the children of an individual. The link operation does not change the original ages values of the children. If the original records had complex data objects then, they would also remain invariant under the link operation. It is possible (as in this example) that the answer to the SLQ may involve some more statistical computation on the complex data objects. Thus in order to handle the SLQ some additional computational rules may have to be added to the statistics metadata of the atomic categories. We can generalize these results as follows:

If two or more atomic categories are joined together in a link-query, then the complex data objects remain invariant provided the complex data objects do not contain the linking attributes. If the join is an equi-join on a statistics attribute then also the complex data objects remain invariant.

Statistics metadata for linear regression analysis remains invariant under link-query. Additional computational rules may have to be augmented to the statistics metadata if the statistics-link-queries contain more statistical analysis.

If the links are non-equi-joins on statistics attributes, the results are difficult to predict and are not covered in this paper.

Insert Queries (IQ)

The process of inserting records into an existing database is referred to as insert query. If the new records are of existing types then the process of updating the statistics metadata involves updating the complex data objects with the new values of the statistics attributes. As we have discussed, only complex data objects which are based on additive properties of raw data, hence the process of updating will involve adding new values (or some functions of them) to the existing complex data objects. The rules for manipulating the complex data objects remain invariant because no new record types are added by the insert query.

If the insert query involves inserting new record types into the statistics database then the new complex data objects and the rules for manipulating them have to be added to the statistics metadata. The new complex data objects and their manipulating rules will dependent on the statistical queries, which are directed at the new objects.

Delete Queries (DQ)

If the delete process involves the deletion of existing atomic categories then the process is simple. It will involve the deletion of the complex data objects associated with the deleted atomic categories. This process may effect some rules for manipulating the complex data objects (*e.g.*, the handling of null values), if so, the rules in the statistics metadata have to be modified accordingly.

If the delete query involves the deletion of some individuals only of an atomic category, then the successful execution of the process will depend on whether the raw data associated with the individuals are accessible or not. If they are accessible, then the process will involve the deletion of the raw data associated with the individuals, modifying the complex data objects and the rules of the statistics metadata to reflect the deletion process. If the raw data associated with the deleted individuals are not accessible, then some type of statistical approximation have to be used.

Replace Queries

A replace query is a combination of a delete query and a insert query with the restriction that the format of the deleted information is the same as that of the inserted information. Usually, the inserted information is physically written in the same space where the deleted information was stored. Thus the behavior of the statistics metadata for a replace query can be derived by combining the behavior under a delete query, followed by an insert query.

4. STATISTICS METADATA LOCKING

In a multi-user database, instances of data segments need to be locked during dynamic updating to prevent ambiguity. In most databases, limited access privileges (who can see what data, who can modify what data, *etc.*) are provided to different users. The granularity of the locking is determined by the nature of the updating transaction, which varies from one transaction to another. The granularity of the access privileges or the sensitivity of the data segments are statics in nature and are determined from security considerations. The complex data objects in the statistics metadata are usually set functions of raw data, hence they do not compromise any security at the individual level. Some times it is possible to direct different statistical queries at the same database and penetrate some individual security (Denning & Schlorer (1983)). *As complex data objects are set functions of raw data hence, a set of access privileges on complex data objects provide an higher level of security than the same access privileges on the raw data.*

If transactions are directed at the statistics metadata only, then very often it is possible to achieve desired security, by restricting the execution authority of the rules in the statistics metadata, rather than restricting access privileges to complex data objects. Suppose the system would like to restrict the disclosure of salary information of individuals and at the same time would permit statistical queries directed at atomic categories. If the number of individuals in a category is two, then by requesting the mean salary and the standard deviation of salaries, it is possible to deduce the salaries of the individuals. Suppose, the rules of the statistics metadata are modified to include the following rule: "If the number of individuals in an atomic category is equal to n, then do not disclose to the same userid more than n-1 statistics calculated from the same statistics attribute." This would prevent access to individual salary information by an userid. Under this rule, it would still be possible for two userids to deduce the individual salary information. This can also be prevented by tightening the rule, by never disclosing to the user community more than n-1 statistics on the same statistics attribute, until the data has been updated. This security rule can be even relaxed by restricting the constraints to 'n-1 independent statistics,' but it would be very time consuming to enforce this rule in real time.

Locking of data segments for updating is performed at different hierarchical levels of the data structure. Thus, the locked structures may or may not, coincide with the complex data objects of a statistics metadata. Locking complex data objects may some times provide the same effect as locking an hierarchical data structure. Sometimes, locking complex data objects may provide the desired effects with more flexibility than hierarchical structure locking. Suppose we have an hierarchical data structure of a department along with its employees. The records contain many attributes along with salary information of employees. The statistics metadata contains complex data objects for all statistics attributes. All statistical queries are answered from the statistics metadata. Suppose some employees are to be given a salary raise. During this updating process it is sufficient to lock only the salary complex data object and not the other complex data objects in the statistics metadata. Transactions can be directed to the other complex data objects while the lock is enforced on the employee salary complex data object. Instead, if the hierarchical data structure of the department (along with it's children segments) was locked, then it would have not been possible to direct any statistical queries at the particular department during the lock enforcement. Thus, complex data object locks can be much more flexible than hierarchical structure locks. It is also possible to enforce the locks through the rules of the statistics metadata, but they are more complex.

5. PERFORMANCE OF STATISTICS METADATA

All the statistics metadata, discussed in this paper, contain partial sums of certain functions of raw data. These partial sums are components of the statistics computation, hence any statistics computation based on these statistics metadata will require much less computation time than direct computing the statistics from the raw data. In general, the computational complexity of the statistics, from the statistics metadata, is of the order $O(m)$, where m is the number of operations. On the other hand, the computational complexity of the same statistics from the raw data would be of order $O(n + m)$, where n is the number of observations. As n>>m, hence the improvement in efficiency of using statistics metadata *versus* computing from raw data (for any statistics computation) is of the order $O(n)$. Thus statistics metadata can be very cost efficient for statistics computation in very large databases.

In all database management techniques (access methods, indexing, search algorithms, *etc.*) applications, there is an initial set up cost (*e.g.*, sorting file, index creation, *etc.*), which is amortized over repetitive use of the particular program. Thus, most of these techniques are not efficient if the program is used only once, but the techniques and algorithms pay off because of the repetitive use of the programs. In the case of statistics metadata the pay off is at using the statistics metadata only once. The reason for this is that, the statistics metadata contains partial sums, which are parts of the raw data computational formula. Thus, the cost of statistics metadata generation is amortized over one usage. From the second usage, the statistics metadata is free (no computational cycles are needed). The only cost involved is the cost of storage, like any other database management algorithm or technique.

6. MATHEMATICAL PROPERTIES

It is important to investigate, if the statistics metadata satisfy the the three fundamental laws of mathematical operators, namely, *(i) Associative law, (ii) Commutative law*, and *(iii) Distributive law* We shall discuss these three laws for the statistics metadata associated with eq. (3.4). The verification of these laws for other statistics metadata is left to the reader. The complex data objects associated with eq. (3.4) are $\Sigma y_j x_{ji}$ and $\Sigma x_{ki} x_{ji}$. The rules, for processing the complex data objects associated with this statistics metadata is:
$$\sum_{u \in S}^{\#}.$$

Two more rules are added to this metadata.

(i) $\underset{s_1 + s_2}{\cup}$ → union over data sets S_1 and S_2.

(ii) $\underset{s_1 - s_2}{\Delta}$ → deletion of data set S_2 from S_1.

Consider three data sets S_1, S_2 and S_3.

$\Sigma y_i x_{ji}$ satisfies the associative law because

$$\sum_{i \epsilon S_1} y_i x_{ji} \ \underset{S_1 + (S_2 + S_3)}{\cup} \left(\sum_{i \epsilon S_2} y_i x_{ji} \ \underset{S_2 + S_3}{\cup} \ \sum_{i \epsilon S_3} y_i x_{ji} \right)$$

$$= \left(\sum_{i \epsilon S_1} y_i x_{ji} \ \underset{S_1 + S_2}{\cup} \ \sum_{i \epsilon S_2} y_i x_{ji} \right) \ \underset{(S_1 + S_2) + S_3}{\cup} \ \sum_{i \epsilon S_3} y_i x_{ji} \ . \tag{6.1}$$

If the data sets satisfy the condition: $S_1 \supseteq S_2 \supseteq S_3$ then the associative law holds for Δ, *i.e.*,

$$\sum_{i \epsilon S_1} y_i x_{ji} \ \underset{S_1 - (S_2 - S_3)}{\Delta} \left(\sum_{i \epsilon S_2} y_i x_{ji} \ \underset{S_2 - S_3}{\Delta} \ \sum_{i \epsilon S_3} y_i x_{ji} \right)$$

$$= \left(\sum_{i \epsilon S_1} y_i x_{ji} \ \underset{S_1 - S_2}{\Delta} \ \sum_{i \epsilon S_2} y_i x_{ji} \right) \underset{(S_1 - S_2) - S_3}{\Delta} \ \sum_{i \epsilon S_3} y_i x_{ji} \ . \tag{6.2}$$

It is easy to show that the same associative laws also hold for $\Sigma x_{ki} x_{ji}$ and $\underset{u \epsilon S}{\Sigma^\# E_{iu}}$. The laws are similar to eqs. (6.1) and (6.2). The commutative laws also hold for $\cup$ because

$$\sum_{i \epsilon S_1} y_i x_{ji} \ \underset{S_1 + S_2}{\cup} \ \sum_{i \epsilon S_2} y_i x_{ji} = \sum_{i \epsilon S_2} y_i x_{ji} \ \underset{S_1 + S_2}{\cup} \ \sum_{i \epsilon S_1} y_i x_{ji} \ . \tag{6.3}$$

It is easy to show that the same commutative laws also hold for $\Sigma x_{ki} x_{ji}$ and $\underset{u \epsilon S}{\Sigma^\# E_{iu}}$. The laws are similar to eq. (6.3). The commutative laws do not hold for Δ When the data sets satisfy the condition: $S_1 \supseteq S_2 \supseteq S_3$ when the distributive laws also hold, between $\cup$ and Δ because

$$\sum_{i \epsilon S_1} y_i x_{ji} \ \underset{S_1 + (S_2 - S_3)}{\cup} \left(\sum_{i \epsilon S_2} y_i x_{ji} \ \underset{S_2 - S_3}{\Delta} \ \sum_{i \epsilon S_3} y_i x_{ji} \right)$$

$$= \left(\sum_{i \epsilon S_1} y_i x_{ji} \ \underset{S_1 - S_2}{\Delta} \ \sum_{i \epsilon S_2} y_i x_{ji} \right) \ \underset{(S_1 - S_2) + S_3}{\cup} \ \sum_{i \epsilon S_3} y_i x_{ji} \tag{6.4}$$

also

$$\sum_{i \epsilon S_1} y_i x_{ji} \ \underset{S_1 - (S_2 + S_3)}{\Delta} \left(\sum_{i \epsilon S_2} y_i x_{ji} \ \underset{S_2 + S_3}{\cup} \ \sum_{i \epsilon S_3} y_i x_{ji} \right)$$

$$= \left(\sum_{i \epsilon S_1} y_i x_{ji} \ \underset{S_1 - S_2}{\Delta} \ \sum_{i \epsilon S_2} y_i x_{ji} \right) \ \underset{(S_1 - S_2) - S_3}{\Delta} \ \sum_{i \epsilon S_3} y_i x_{ji} \ . \tag{6.5}$$

It is easy to show that the same distributive laws between $\cup$ and Δ hold for $\Sigma x_{ki} x_{ji}$ and $\underset{u \epsilon S}{\Sigma^\# E_{iu}}$. The laws are similar to eq. (6.4) and eq. (6.5).

7. CONCLUDING REMARKS

The main goal of this paper was to introduce a new type of complex data objects and their computational rules to expedite the computation of statistics. These objects have been called statistics metadata. We have introduced statistics metadata for linear regression analysis, which is widely used in many common statistical analysis. In the paper, we have introduced statistics metadata for handling normal equations obtained from least square estimation methods. We have discussed statistics metadata for the one-factor and the two-factors linear models in details. We have discussed statistics metadata for analysis of variance techniques. The statistics metadata structures for range-select queries, link queries, insert queries, delete queries and replace queries have been discussed. Locking of statistics metadata have also been covered. It has been shown that statistics metadata reduces the computational time of statistics computation by a factor of n, where n is the number of records. Amortization of statistics metadata creation has also been discussed.

This paper discusses only the statistics metadata for linear regression analysis and there are other fundamental classes of statistical analysis; the structure of statistics metadata should be investigated for these also. It is possible that statistics metadata may turn out to be very fundamental in real time statistical analysis for very large statistical data bases. The answer to this conjecture is left to future researchers.

8. REFERENCES

Allen, Frank W., Loomis, Mary E. S., and Mannino, Michael V., 1982, The Integrated Dictionary/Directory System, "ACM Computing Surveys," Vol. 14, No. 2, pp. 245-286.

Baxendale, Phylis, 1966, Content Analysis, Specification, and Control, "Annual Review of Information Science and Technology," C. A. Cuadra, ed., Vol. 1, John Wiley Publication, New York.

Bishop, Yvonne, M. and Freedman, Stanley R., 1983, Classification of Metadata, "Proc. of Second Int. Workshop on Stat. Database Management," Los Altos, California, September 27-29, 1983, pp. 230-234.

Boral, H., DeWitt, D. and Bates, D., 1982, A Framework for Research in Database Management for Statistical Analysis, "Proc. of ACM/SIGMOD 1982 Conference," June 1982.

Date, Chris J., 1977, "An Introduction to Database Systems," Addison-Wesley, Reading, MA, USA.

Denning, Dorothy E. and Schlorer, J., 1983, Inference Controls for Statistics Databases, *IEEE Computer*, July 1983.

Ghosh, Sakti P., 1977, "Data Base Organization for Data Management," Academic Press, New York, USA, Chapter 1.

Ghosh, Sakti P., 1983, An Application of Statistical Databases in Manufacturing Testing, "Proc. of COMPDEC (1984)," Vol. 1, pp. 96-103. Also published as an *IBM Research Report No. RJ 4055*.

Ghosh, Sakti P., 1984, Statistical Relational Tables for Statistical Database Management, *IBM Research Report No. RJ 4394*.

Hammond, R. and McCarthy, John L., edited 1983, "Proceedings of the Second International Workshop on Statistical Database Management," Los Altos, California, September 27-29, 1983.

Kendall, Maurice G. and Stuart, Alan, 1958, "The Advanced Theory of Statistics," Vol. 1, Charles Griffin & Company, London, Chapters 1-6.

Kendall, Maurice G. and Stuart, Alan, 1961, "The Advanced Theory of Statistics," Vol. 2, Hafner Publishing Co., New York, UDA, Chapters 17-20.

Lefkovits, H., 1977, Data Dictionary Systems, *Information Sciences*, 1977.

McCarthy, John L., 1982, Metadata Management for Large Statistical Databases, "Proc. of Eighth International Conference on Very Large Databases," Mexico City, Mexico, September 8-10, pp. 234-243.

Rao, C. R., 1952, "Advanced Statistical Methods in Biometric Research," Wiley Publications in Statistics, New York.

Scheffe, Henry, 1959, "The Analysis of Variance," Wiley Publication in Mathematical Statistics, New York.

Su, Stanley Y. W., 1983, A Semantic Association Model for Corporate and Scientific/Statistical Database, *Journal of Inf. Sc.*, Vol. 26, No. 3.

Teitel, R. F., 1977, Relational Database Models and Social Science Computing, "Proc. of Computer Science and Statistics: Tenth Annual Symposium on the Interface," Gaithersburg, MD, April 1977, pp. 165-177.

Ullman, Jeff D., 1982, "Principles of Database Systems," Computer Science Press, Potomac, MD, USA.

Wong, Harry K. T., 1982a, "Proceedings of the First LBL Workshop on Statistical Database Management," Menlo Park, California, December 2-4, 1981.

Wong, Harry K. T., edited 1982b, "A LBL Perspective on Statistical Database Management," Lawrence Berkeley Laboratory of University of California, Berkeley.

DYNAMIC PATRICIA

T. H. Merrett and Brenda Fayerman

School of Computer Science
McGill University
Montreal, PQ Canada H3A 2K6

ABSTRACT

We add dynamic capabilities to PATRICIA, the only known sublinear
searching algorithm for text databases on secondary storage. This we do
by giving constant-time insertion and deletion algorithms for the PATRICIA
index and by storing the text on conventional database structures, namely
B-trees. We discuss the application of dynamic PATRICIA to an editor for
large texts on secondary storage.

INTRODUCTION

Morrison's [1968] Practical Algorithm to Retrieve Information Coded in
Alphanumeric (PATRICIA) is an elegant strategy for searching text data in
time independent of the amount of data stored. It has been described by
Knuth [1973] and so is reasonably well-known, but it has not been much exploited. We propose its use in a text editor, for which it needs to be
extended to handle dynamically changing data. Morrison originally proposed
insertion algorithms for adding pieces of text larger than a certain size,
and Bays [1974] published a deletion algorithm, but one which does not use
Morrison's data structures, and is not useful for text editing.

Knuth, Morris and Pratt [1970] offered competition with a linear-time
pattern matching algorithm, based essentially on a finite automaton created
from the pattern to be matched. Aho and Corasick [1975] extended this to
apply to arbitrary patterns obeying a general syntax, and Boyer and Moore
[1977] gave a faster, but still linear, algorithm to do the original job.
These algorithms are more powerful than PATRICIA, which must predefine the
substrings that can be matched and which extends in only a limited, and
predefined, way to patterns more general than explicit substrings. But they
are impossibly slow for a text editor operating on sizeable documents on
secondary storage.

Because PATRICIA seems to be inadequately used or understood, we give
a tutorial introduction to it in Section 2. We hope this will make Morrison's
paper more accessible. Sections 1 and 3 respectively outline the problem of
representing text on secondary storage and the connection between this representation and PATRICIA. Then Section 4 presents the major new contribution of this paper, the dynamic extensions to PATRICIA. We adhere to Morrison's original data structures and offer dynamic extensions which are useful

19

for text editing: insertions into and deletions from the middle of a text, and changes to individual words.

1. Text on Seconday Storage

The work reported in this paper is part of the construction of an editor for texts which are assumed to be too large to fit entirely into RAM. We require a data structure which provides fast access to arbitrary sections of text and which allows them to be changed, extended or deleted.

A *document* contains both text and a hierarchical structure of sentences, paragraphs, sections, chapters, etc., as well as provision for lists, quotations, etc. We have built both structure and text editors, but this paper is concerned only with the latter.

We consider text to be a sequence of words. The data structure for a text editor must allow scanning the text in the order of this sequence as well as insertions into and deletions from this order. It must also allow searches by substrings or phrases of the text. The sequence-order operations require a capacity for high-activity*, one-dimensional†, dynamic processing, such as provided by B-trees. The substring searches must be done in sublinear time, and PATRICIA is the only algorithm we know for doing this. It became necessary for us to extend PATRICIA from searches on static data by adding a dynamic capability for the changes that an editor makes to text. This extension is the subject of this paper.

In addition to the above considerations, we have discovered that it is valuable to represent text on secondary storage as relations. Merrett [1984a] discusses some of the text operations, such as indexing and statistical analysis of text, that can be handled easily by the relational algebra as a result of the right relational representation of text. This representation simply adds a sequence number attribute to the set of words in the text:

TEXT	(WORD	SEQ)
	Old	0
	King	1
	Cole	2
	was	3
	a	4
	merry	5
	old	6
	soul	7
	,	8
	And	9
	:	:

This representation leads to some considerations for the editor and for the use of PATRICIA, to which we will return after introducing PATRICIA. (Note that this *representation* of text does not dictate how it is *stored*, a subject discussed in the rest of this paper.)

---

*The *activity* of a search on a data file is a measure of the proportion of the file that must be retrieved to satisfy the search. Activity of over a few percent is generally considered high. Secondary indexing techniques are inadequate for high activity searches. See Merrett[1984].
†The *dimensionality* of an access mechanism is the number of data attributes for which efficient independent access can be provided. One-dimensional access structures are well documented, but high-activity ones can at best be logarithmic. See Merrett[1984].

2. A Tutorial on PATRICIA

We start by considering the trie representation of four binary strings, ABABB, ABBABABB, BABABBA and BABBA. (We follow Morrison [1968] in using a binary alphabet, {A,B}, because it stands out more in the examples than {0,1}). The trie is shown in Figure 1. Searching is easy:

> *Starting at the root of the trie and the first position of the search string, move left if the search string contains A and right if it contains B; increment the level of the trie and the position in the search string and repeat.*

> *The search fails if a node does not have a descendent corresponding to the current position of the search string. Otherwise it succeeds when the search string is exhausted.*

Since this search will find any initial substring of the strings in the trie, its success will yield two or more matching strings if the trie has branches below the point at which the search string is exhausted.

For example, ABABBA is not found, BB is not found, ABAB is found once and AB is found twice in the trie of Figure 1. Notice that only initial substrings can be found: the search must start at the root.

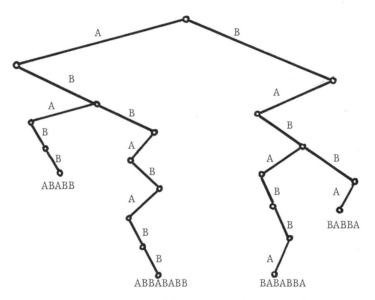

Figure 1. Trie Holding Four Binary Strings

Text, however, is not a set of independent strings such as that considered above. The strings to be searched are embedded in a single sequence of symbols. The four strings above might appear as follows in a text,

```
(ABB(ABABB)(BA(BABBA)
 1    4      9 11
```

where the open parentheses are additional symbols used to mark the *start* of each of the four strings (the start position in the text is also given)

and the close parentheses mark the *stops*. Notice that strings can be completely contained in others as long as they share the same stop. Morrison
defines a *book* to be a stop, its associated starts, and all the text between.

Because the full text must be stored somewhere in its proper sequential
order, storing all of its searchable strings in a trie such as Figure 1 is
redundant. We can take advantage of this situation by using the trie as an
index to the text and by trimming from it all nodes which do not lead to a
decision, either to go left or right or to terminate a search. Thus we are
left with *branches* (where we must choose left or right) and *ends* (which are
the terminal nodes of the trie). Figure 2 shows the result: since we have
left out intermediate parts of the trie, we must store the *height* of each
node, which is the length of the substring prior to the node, so that we
know where in the search string to match the node. The *start* positions in
the text are also given.

A search for the string ABAB now takes the form:

> *Start at the root and find its height,* h = 0.
> *Since position* h + 1 *of the search string is* "A",
> *go left and find the height,* h = 2, *of the next node
> found. Since position* h + 1 *of the search string
> is again* "A", *go left again and find* h = 5. *This
> exceeds the length of the search string, so look up
> the text at the corresponding start, position* 4.

The search has been done without reference to the text until this last step,
when checking the text is necessary because of the gaps in the trie. The
cost of the trie-index search does not exceed ℓ comparisons, where ℓ is the
length of the search string, and the search of the text requires the inspection of only ℓ symbols. This is true even when search string has multiple
occurences within the text.

We can formalize this search procedure further by uniquely labelling
the ends and branches. Then with one additional mechanism we can do the
index search by consulting only some arrays, instead of the informal

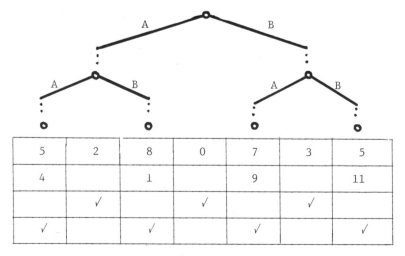

5	2	8	0	7	3	5	height
4		1		9		11	start
	√		√		√		branch
√		√		√		√	end

Figure 2. The Trie as an Index

diagrams we have so far used. Since the alphabet is binary, we have a
binary trie, and there is exactly one more end than branches. So Morrison
proposed using the odd integers to label ends and using the corresponding
even integers to identify branches. Ends can be labelled in any order,
since unique identification is all we need. Branches, on the other hand,
must be labelled to show their positions in the trie, and so they are
assigned an (even) integer which is one less than the label of the descendent end chosen by the following rule:

> *To find branch* (n), *move up the trie from end* (n+1) *until
> you find the branch which end* (n+1) *has in common with
> end* (m) *for some* m < n.

5	2	8	0	7	3	5	height
4		1		9		11	start
	2		4		6		branch
3		1		5		7	end
2	4	3	1	6	5	7	twin

branch, end grouped as } chain

Figure 3.   PATRICIA in Array Form

Since ends and branches are distinguishable by their labels being odd
and even, respectively, and since these labels are used in the same ways in
the search algorithm to be described, Morrison placed them in one array,
called *chain*. This is shown in Figure 3, along with *twin*, the final array,
which we now describe.

We need a mechanism which tells us, when we are at a node, which chain
is its left and which its right descendent. The *twin* index tells us this:
for each node, it stores the branch number of the parent if the node is a
left child, or 1 plus the branch number of the parent if the node is a right
child. The root node has a twin value of 1.

The PATRICIA search algorithm now takes the form:

> *Starting at the root node* (twin = 1), *find the
> corresponding chain, c. If the search string is "A"
> at the position* (height + 1), *look up* twin = c, *find
> the new corresponding chain and repeat. Otherwise
> look up* twin = c + 1 *and do the same. The algorithm
> terminates successfully or unsuccessfully as before.*

The only thing we have not done is to display the arrays in a form
which makes the lookups quick. We will expect both text and indexes to
exceed RAM capacity and require secondary storage. Figure 4 shows the
form Morrison gives. Figures 3 and 4 are equivalent, apart from efficiency
of lookups.

All of these methods can be extended to alphabets of three or more
symbols, but the discussion is more complicated [Dhillon, 1984].

The strength of PATRICIA is that it has an access cost which is pro-

t	1	2	3	4	5	6	7
TC(t)	4	3	1	2	6	5	7

a) Twin Index: TC(t) = the chain which contains twin (t).

c	1	2	3	4	5	6	7
H(c)	8	2	5	0	7	3	5

b) Height Index: H(c) = length (in bits) of longest member in chain (c)

c	1	2	3	4	5	6	7
S(c)	1	4	4	9	9	11	11

c) Start Index: S(c) = position in text of first symbol of end (c) or

of end (c+1)

Figure 4.   Search Indexes for PATRICIA

portional to the length of the search string and independent of the size of the text being searched.* Its weakness is that starts and stops must be pre-defined, so that it cannot be used to search arbitrary substrings, let alone general patterns. Further, because of the trie construction, we have the restriction:

Rule 1. No end may be identical to the initial portion of another end.

Thus a start at position 12 in our example would be illegal. This can be circumvented by the artifice of padding the text before each stop with a unique ending:

Rule 2. Each book ends uniquely.

3. PATRICIA and the Relational Representation of Text

We can avoid the significant weaknesses of PATRICIA by placing a start before each word in the TEXT relation and a stop after each sentence (punc-tuated by ? ! and .). We can append to each sentence its unique sequence number to implement Rule 2.

Then the TEXT relation of Section 1 can be stored as a B-tree in se-quence number order, with a PATRICIA index for substring searches. The only thing we may want to do, but cannot, is to search for a part of a word other than its beginning, such as "body" in "everybody". Also, we cannot search for a sequence of words which extends past a sentence stop. The absence of blanks between words makes further difficulty in distinguishing between "token" and "literal" modes of search.†

---

*Neglecting the fact that, strictly, the size of the integer indices in the tables is proportional to log n where n is the size of the file.
†The "*token*" mode of search matches only words; searching for "the" finds only occurrences of "the" in the text, not "there", "these", "other", etc., all of which contain the literals "the". The *literal* mode of search will, by contrast, retrieve any word containing the sequence specified. The li-teral mode imposed by PATRICIA will also not distinguish "bookends" from "book ends".

More compact ways of storing the text, which eliminate the sequence numbers for instance, are possible, but a way which allows a sublinear search for a position (as B-trees do) will need some other organizing principle, such as by chapter and section.

4. Inserting and Deleting Ends

Morrison gives an algorithm for inserting ends to which we here add the converse, an algorithm to delete ends.

Suppose first that we wish to insert the end AABABAB. In the trie of Figure 1, this would create a leftmost branch, as shown in Figure 5. In the index trie, we would have a new end, 9, of height 7, and a new branch, 8, of height 1, because AABABAB differs from the existing ends after the 1st position. All this is shown in Figure 6. In addition, we need new entries for the twin index. The twin of end 9 is easily seen to be 8 by the rule for twins, because end 9 is the direct descendent of branch 8. The twin of branch 8, however, requires us to look into the existing trie: its parent is branch 4, whose former descendent, branch 2, it has now replaced. Thus the twin of branch 8 is 4 while that of branch 2 must be changed to 9. See Figure 6.

The algorithm follows. (It is Morrison's algorithm ADDp.)

Algorithm ENDI   Adds end e to existing index.

1. (Check Rule 1) Use the search procedure to find if e or its initial part is already there. If so, terminate because Rule 1 has been violated. (This will not occur in our application because ends terminate uniquely, including new ones - Rule 2.).

2. (Identify location in index) The above search will find the longest initial part of e already present - call it q. This tells us two things: the length, $\ell(q)$, is the height of the new branch, and the shortest chain containing q has a twin which will be the twin of the new branch.

3. (Update PATRICIA) The new branch is chain $(n + 1)$ and the new end is chain $(n + 2)$ where n was the former size of PATRICIA. The twin of chain $(n + 2)$ is $n + 1$ (or $n + 2$, depending on the bit in e following q). The twin of chain $(n + 1)$ is the twin found in step 2. The twin of the chain containing q, found in step 2, is changed to $n + 2$ (or $n + 1$). The height of the new branch is $\ell(q)$ and the height of the new end is $\ell(e)$. Increment $n \leftarrow n + 2$.

The start of the new end is wherever we wanted the end inserted in the text: PATRICIA is not concerned, apart from recording the position (say, 8.1 in the example). ▯

Deleting an end is just the converse of inserting, except that we would like to compact the index if the end deleted was not the last one inserted. (In this case, the deletions will be from the middle of the twin, height and start indexes in Figure 4.) Compacting turns out to be easy because the ends are essentially unordered: we can just re-label the last end and insert with the chain numbers of the deleted end.

Suppose we want to delete BABBA from the index we have just extended (Figures 5 and 6). This requires removal of the rightmost branch of the trie in Figure 5, i.e., removal of the rightmost columns of Figure 6. We show this in Figure 7. In addition, we must make one change to the twin

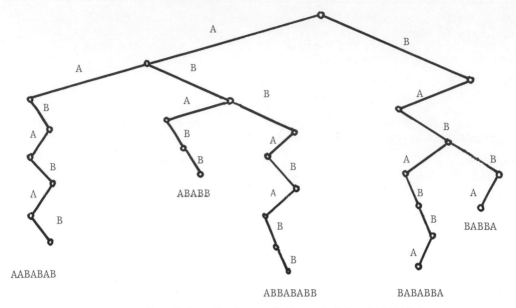

Figure 5. Trie of Figure 1 Extended by AABABAB

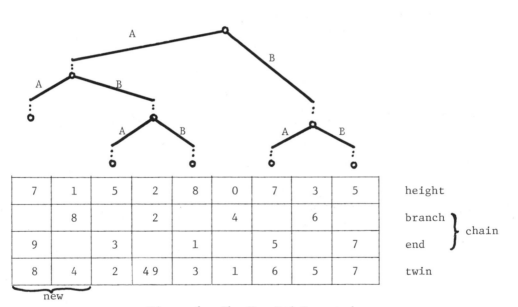

7	1	5	2	8	0	7	3	5	height
	8		2		4		6		branch ⎫
9		3		1		5		7	end ⎭ chain
8	4	2	4 9	3	1	6	5	7	twin

new

Figure 6. The New End Inserted

index, namely to replace the twin of the end which was not removed by the twin of its former twin (the deleted branch).

We see that the indexes that we would generate from Figure 7 (e.g., the twin, height and start indexes of Figure 4) will have gaps because chains 6 and 7 and twins 6 and 7 are missing. To compact these indexes, we relabel chains 8 and 9, the last two to be added. The result is Figure 8.

Algorithm ENDD   Deletes end e from existing index.

1. (Identify location in index) Search for e. It will be found, but we need to know t, the twin of the end e and, t', the twin of its immediate ancestor branch (i.e., respectively the last and penultimate twins found in the search).

2. (Update PATRICIA) Find the other immediate descendent (branch or end) of the penultimate chain found (i.e., the one with twin = $t - 1$ if t is odd, or with twin = $t + 1$ if t is even) and change its twin to t'. Delete the two columns of the PATRICIA arrays corresponding to the branch to be removed.

3. (Compact) Replace n and n − 1 everywhere in PATRICIA with the chain numbers respectively of the end and branch just deleted. Decrement n ← n − 2. (This last step requires ability to access the twin index (Figure 4a) by chain number: instead of storing an inverted twin index, which must be maintained both on deletion and insertion, we recommend that the search procedure be used by retrieving the appropriate end from the text and searching for it until the desired chain number is reached. The cost is less than incurred by most inversion methods and storage is not required.) ☐

Both ENDI and ENDD clearly run in $O(\ell(e))$ time.

5. Insertion, Deletion and Text Editing

In our context, a PATRICIA Book is a sentence, and one of the most frequent insertions (or deletions) is likely to be a Book. The algorithms are easy: just repeatedly insert (or delete) each end in the book; i.e., for each word in the sentence execute ENDI (ENDD) for the end beginning at that word.

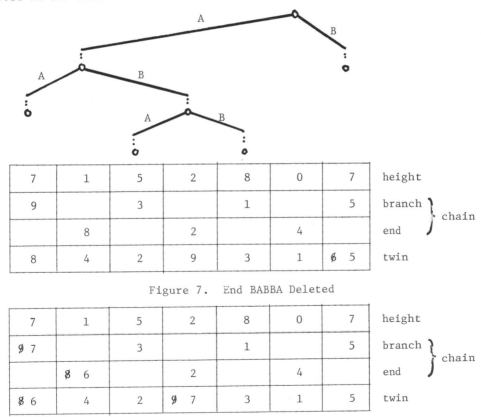

7	1	5	2	8	0	7	height
9		3		1		5	branch ⎫ chain
	8		2		4		end ⎭
8	4	2	9	3	1	~~6~~ 5	twin

Figure 7.   End BABBA Deleted

7	1	5	2	8	0	7	height
~~9~~ 7		3		1		5	branch ⎫ chain
	~~8~~ 6		2		4		end ⎭
~~8~~ 6	4	2	~~9~~ 7	3	1	5	twin

Figure 8.   PATRICIA Compacted after Deletion

Algorithm <u>BOOKI</u> (Insert a book: Morrison's BUILDINDEX)

1. Insert the book into the text (e.g., conventional insertion in sequence
    number order in a B-tree).

2. Proceeding from left to right in the book, for each end e (i.e., from
    each start to the stop) call ENDI to insert e into the index. □

Algorithm <u>BOOKD</u> (Delete a book)

1. Proceeding from right to left in the book, for each end e call ENDD to
    delete e from the index.

2. Delete the book from the text. □

A frequent change in the process of text editing is to individual
words, as in spelling typographical corrections. PATRICIA can handle
this type of change, but in a complicated way. Usually, these errors are
detected by the typist before the word - or at least the sentence - is complete, so it is useful to use BOOKI to enter the whole sentence and provide
a cursor-driven editor to correct it before entry. We give the algorithms
for modifying a word, for the infrequent occasion on which we must change
a word in a sentence which has already been inserted.

> Consider again the original text
>
>   (ABB(ABABB)(BA(BABBA)
>   1   4   9 11

and the requirement of inserting B between positions 4 and 5. We must modify
all ends starting before this insertion, i.e., those at positions 1 and 4.
A worse example is the insertion of BB before position 4 but after the start:
the end at position 4 must be replaced entirely. Insertion of this sort is
most expensive just after the last start of a book. In the general case, it
is sufficient - and not far from necessary - to delete the old ends and re-
insert the new, using ENDD and ENDI repeatedly. Deletion from a word is
essentially the same as insertion in this approach.

Algorithm <u>BITSI (BITSD)</u> (Insert (Delete) part of an end)

1. (Delete ends) Proceeding left from the insertion (deletion) site, use
    ENDD to delete the end beginning at each start encountered in the work.

2. (Change bits) Insert (delete) bits at site in text.

3. (Insert ends) Proceeding right to the insertion (deletion) site, use
    ENDI to insert the end beginning at each start in the modified book. □

To insert or delete a whole word, we run BITSI or BITSD, but with an
unequal number of calls to ENDD or ENDI: one more call to ENDI in the case
of BITSI and one less in the case of BITSD.

To split one word into two, we call ENDI once to create the end starting
at the new second word. To merge two words, call ENDD once to delete the
end.

6. Conclusions

We have reviewed Morrison's PATRICIA in a tutorial way which we hope
will make it more widely used for text databases and processing. We have
contributed deletion procedures for "ends" and "books", and insertion and

deletion procedures for words and parts of words in our representation of text using PATRICIA. These dynamic extensions of PATRICIA make it useful for text editing as well as for storage of static documents for retrieval purposes only.

Our representation of text using relations is briefly described. Its advantages for processing of text, in applications more general than editing and formatting, have been presented elsewhere. The combination of relations, conventional database file structures and PATRICIA is noticeably fruitful for the significant operations on text.

A weakness of PATRICIA, but one which we do not believe will seriously affect our use of it, is its need to search from predefined start points in the text. This weakness is more than compensated by the fact that PATRICIA is the only sublinear alogrithm for finding substrings of a text.

Further work which can be done is to study the class of pattern-matching operations (more general than just substring finding) that is implemented by extending PATRICIA. This would make PATRICIA still more competitive with linear-time algorithms such as that of Aho and Corasick.

Acknowledgements

This work has been supported by the Natural Science and Engineering Research Council of Canada (Grant NSERC-A 4356) and by the Québec Fonds formation des chercheurs et actions concertées (Grant FCAC-EQ 2561). I thank Vera Gorni for her painstaking typing of this paper.

References

A. V. Aho & M. J. Corasick, 1975. Efficient string matching, CACM 18, 333-40.

J. C. Bays, 1974. The compleat PATRICIA, Ph.D. dissertation, U. of Oklahoma.

R. S. Boyer & J. S. Moore, 1977. A fast string searching algorithm, CACM 20, 10, 762-72.

R. Dhillon, 1984. Private communication.

D. E. Knuth, 1973. The Art of Computer Programming, vol. III, Sorting and Searching, Addison-Wesley, Reading, Mass.

D. E. Knuth, J. H. Morris & V. A. Pratt, 1970. Fast pattern searching in strings, SIAM J. Comput. 6, 2, 323-50.

T. H. Merrett, 1984. Relational Information Systems, Reston Publishing Co. Inc., Reston, Virginia.

T. H. Merrett, 1984a. First steps to algebraic processing of text in G. Cardarin & E. Gelenbe eds., New Applicationf of Databases, Acad. Press, 109-128.

D. R. Morrison, 1968. PATRICIA: practical algorithm to retrieve information coded in alphanumeric, JACM 15, 514-34.

DESIGN OF AN INTEGRATED DBMS

TO SUPPORT ADVANCED APPLICATIONS

V. Lum*, P. Dadam, R. Erbe, J. Guenauer, P. Pistor,
G. Walch, H. Werner, and J. Woodfill

IBM Scientific Center
Heidelberg, W. Germany

Abstract

New applications of DBMS's in areas of sciences, engineering and offices have produced new requirements that are not satisfied in current DBMS's. Included among these requirements are support for both the normalized and non-normalized models directly at the system interface level, support for text processing, and support of the temporal domain. To provide these supports, one can try to build additional functions on top of or into an existing DBMS. This approach has been deemed to be inefficient. It is believed that much can be gained by designing a new system to satisfy the new requirements more directly.

This paper describes the specific design of a DBMS directed to satisfy the three requirements just cited. The authors first discuss the overall architecture before proceeding to discuss some aspects of its internal data management. Included here is the internal data structure and how they are used to provide the necessary supports. Summarized in the conclusion are the features of the system that are not available generally in current DBMS's. In addition, the status and planned enhancements are also outlined.

1. Introduction

In recent years the use of DBMS's has moved from the traditional area of commercial applications such as inventory control and banking to include many new areas such as engineering (e.g., CAD/CAM) and office applications (e.g., Di84, GP83, HL82, HR82, KL82, Lo82, LP83, Lue83, Lu84, SP82, St82, St83). As a result, new requirements for DBMS's have been discovered and current DBMS's have been found to be lacking in support of these new applications.

The Advanced Information Management (AIM) project in the IBM Heidelberg Scientific Center, established several years ago in anticipation of some of these advanced

A version similar to this one has been published in the Proceedings of the GI Special Conference on "Datenbanksysteme in Buero, Technik und Wissenschaft", Karlsruhe, March 20 - 22, 1985, Informatik-Fachberichte No. 94, Springer, Berlin, Heidelberg, New York, Tokyo.

*Current affiliation and address: Computer Science Dept., Naval Postgraduate School, Monterey, CA 93943

31

applications, originally saw the need to incorporate textual retrieval into the normal data management facility in a DBMS. The need for such a capability has not only been established in office document retrieval, but has also been found in engineering applications as well (IBM82). Since then, other basic facilities necessary to support the divergent applications have been uncovered.

Our initial thought has been to build a system on top of an existing DBMS. Our analysis soon showed that this is not a good approach as the constraints to fit into a given system are too restrictive, the modifications required to achieve our objectives are too much, and that the resulting system will likely perform unsatisfactory (see Lu84). These observations and analyses led us to the decision that a completely new design is needed, with all constraints removed. This paper describes the design of such a system to meet the objectives discussed below. Moreover, while some of the objectives is similar to those in the paper by Deppisch (De85), these two papers are different in the following way: Our paper describes a system intended to have normal level DB interface, the other paper emphasizes on the construction of a low level kernel data management system that can be integrated into the file and data management portion of an operating system.

One objective in the design of our system is to support non-atomic data types and relations with complex - (i.e., non-flat or non-normalized) - tuples in addition to normalized relations and simple data types. Recent applications of DBMS have shown that a non-normalized form of relations (also referred to as hierarchical structure) can be used to advantage (e.g., GP83, HL82, KL82, KTT83, Lo82, Lue83, Lu84, SP82, St82, St83). Some researchers feel that a hierarchical structure built with normalized relations would be sufficient. However, we think there is considerable merit in implementing a system which directly supports normalized and non-normalized relations at the same time.

Part of the attractiveness of normalized relations lies in their mathematical soundness and tractability. The non-normalized form has similar appeal, as recent theoretical work (ja84, JS82, SS84) has indicated that algebra for non-normalized relations, which include normalized relational algebra as a special case, is now available. This provides us with mathematical soundness as a foundation on which to construct such a system. Further, Shu and Housel (SHL75, HS76) and Pistor (PHH83) have demonstrated that high level languages similar to the relational languages for normalized forms can be defined. The question, then, is how to construct a system to have the flexibility and efficiency similar to relational systems based on the normalized form.

Another objective is to have a database system for supporting history data. Bubenko (Bu77), Clifford (CW83), among others (e.g., Gu83, Ki83, KL83, MSW83) have stated the desirability of having the temporal domain information. For example, temporal information has always been used for tracking and control purposes in office applications. One can, however, easily see applications beyond the office for such a facility. Thus, the second major objective is to incorporate time information or history data in the databases of the system.

Generally, when the temporal aspect of the information is needed and when this function is not available in the DBMS as in current systems, the users themselves explicitly manage this part by introducing an additional "time" attribute as part of their data. We believe, as do Clifford (CW83) and others, that the time aspect should be included directly in the system right from the beginning. This approach would make the system easier to use, more efficient, more uniform and more consistent. Further, when history data is kept, the need to have temporal information for the system's catalog can also be established. This cannot be easily done as catalog information management is deeply embedded into the internals of any DBMS. The papers by Lum (lu84) and Dadam (DLW84) discussed this aspect sufficiently and it will not be reiterated here.

While other requirements have been uncovered, we have decided to concentrate on the above ones. The decision is mostly pragmatic, because we think that these are basic functions needed in many applications, and because we are limited by resources and other constraints. Further, it is not clear to us whether one can in fact build an efficient system to support too many requirements. Frequently systems die from their own weight when too much is to be included.

Thus the three main objectives for us are: to integrate textual retrieval support into a database management system, to support normalized and non-normalized relations concurrently, and to provide history data support. In the following sections, we will first discuss the overall architecture of the system; then we will describe the internal data structures, and how they are used to provide support for the defined objectives.

2. Architecture Overview

In pursuit of our goals as given in the previous section, we have decided to consider normalized relations as a special case of non-normalized (NF2) forms. As already said, the theoretical results for non-normalized relational algebras allow one to use this approach with a solid foundation. Designing a system to support this approach with good performance is a different matter. We should define the system with operations such that

- the relational operations are a subset of the whole set of operations,

- relations can be manipulated to form non-normalized table and vice versa,

- users of either model would be minimally, if at all, penalized with respect
  to the performance of their tasks.

In an attempt to localize the different problems to a limited number of modules in the system, the architecture of our system has been laid out as shown in the schematic diagram (Figure 1). The Buffer Manager, the Segment Manager, the Catalog Manager, the Session Manager, the Supervisor, and the Communication Manager perform the usual functions. We shall not go into further details about them. The other components, however, contain functions that are not so obvious and we shall now briefly describe them.

Although the architecture of the Query Processor has been defined to allow a number of operations to be executing at the same time as in a multi-tasking mode, the discussion of this aspect is beyond the scope of this paper. However, in the next section, we will give an overall description of the Query Processor operations, as it is believed that such a description provides our readers with a better understanding of the system.

The Index Manager has been designed to provide threefold functions: First, it provides the conventional access support as secondary indexes. Second, it provides the indexing support for textual data search, using the special search technique called Fragment String Index as described in Sc78 and KW81. Although this indexing technique is different from the normal secondary index organization, the same index access mechanism can be used as the basis for constructing the Fragment String Index support. Finally, when no primary key has been defined for a table, the system will create an index structure using the Index Manager to support sequential scan.

As will be seen later, a unit of data access is a subtuple, which can be an instance of a relational (normalized) tuple, an instance of an element in a repeating group, or a structural unit. (A complex or NF2 tuple is composed of subtuples of both kinds, structural and data.) These data units are managed by the Subtuple Manager, which delivers on demand either a subtuple as a whole or only selected attribute values. It will qualify data at the subtuple level for a certain class of predicates.

Another important task of the Subtuple Manager is to create and maintain history data. All time information management will be done by this component. In fact, the system is so designed that the catalog itself is just one of the tables in the system. Thus, it will have history data, too. In this way, the system will allow structural changes of data in an incremental manner after the creation of a database without even the need to update the tuples, a feature deemed necessary when history data is kept in the databases.

The Subtuple Manager also provides some support for concurrency control and recovery for transactions as well as index maintenance, the latter by giving the modified attribute values to the Index Manager. The Complex Tuple Manager component works closely with the Subtuple Manager and the Query Processor to manage complex tuples. Its responsibility to the Query Processor includes relieving the Query Processor's concern about the physical aspect of data placement and retrieval. It determines how a piece of data required by the Query Processor is to be realized and invokes the proper components to write a selected part of this data on secondary storage when space is needed, and to retrieve it when that data is needed again. Naturally, it must provide suitable catalog entry information to the Catalog manager during this process.

In cooperation with the Subtuple Manager, the Complex Tuple Manager places the data on disk in a way that improves performance. For example, as the subtuples in a complex tuple have a strong inter-relationship and they are likely to be processed together, the Complex Tuple manager will try to place this data close together if possible (physical clustering). Since a complex tuple may be very large, storing the subtuples contiguously may be impossible. If so, this component must make the decision which subtuples should be placed together to maximize the chance of achieving good performance.

3. Internal Data Structure and Management

3.1 System Interface

In an earlier paper Pistor (PHH83) has demonstrated that a high level language similar in structure to SQL (Ch76) can be constructed to operate on the non-normalized model. In even earlier work (HS76, SHL75), a language of such a level has been shown to be specifiable. Since the emphasis of this paper does not include the discussion of language aspects, we shall not go deeper into this topic. It suffices to say here that a high level system interface can be and has been constructed on which applications can be built.

It is worthwhile, however, to mention that users of a NF2 system may process their data using views which are different from the structure of the stored data. In particular, a user may see the data as relational tables whereas the data is stored in NF2 forms. In this case, the necessary mapping can be described in the NF2 operations. The problem of lossless transformations in the operations has been investigated (Sc82) and useful results have been obtained.

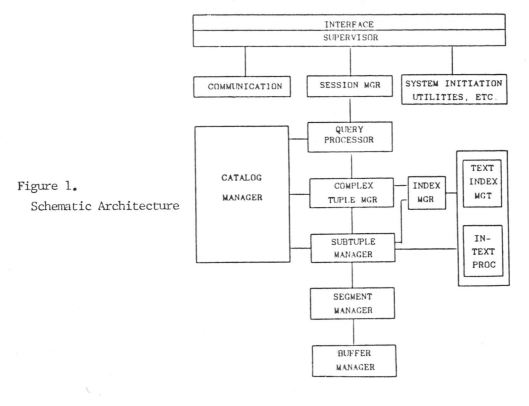

Figure 1.

Schematic Architecture

| Project # | Employee | | Equipment | |
	Empl #	Empl Name	Equ #	Equ Descr
1	1701	Mueller	1716	158
	1705	Major	1221	Series/1
	1712	Edmonds		

(a): Complex Tuple of NF2 Relation Projects

Figure 2. Basic
Structure for Storing
a Complex Tuple

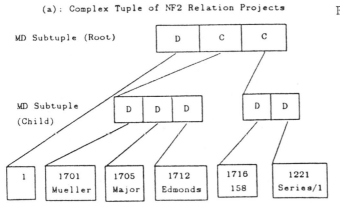

(b): Access Path to the Data in the Complex Tuple
(MD = Mini Directory)

3.2 Internal Data Management

3.2.1 Basic Structure of Complex Tuples

Although the mathematical analysis indicates that the relational model is a special case of the NF2 relational model, it is not a simple matter at all to find a design which satisfies our various goals, particularly with respect to performance. Conceptually one may store NF2 tables with NF2 tuples as one stores relational tables. However, such a design makes processing of complex tuples very difficult. In a paging environment, we see that the separation of the structural information and data can give us complete flexibility without performance degradation. Thus, we have pursued this approach in the design of our system.

The basic structure for storing a complex tuple is shown in Fig. 2 (a more detailed analysis of this structure and an alternative method is given in DGW85). It is composed of two classes of records, data records (or data "subtuples") and structural records (or "Mini-Directory subtuples"). The data subtuples correspond to tuples of flat relations. The mini-directory subtuples (or MD subtuples for short) are essentially pointer lists which combine data subtuples to form complex objects. In essence. MD subtuples consist of pointers to data subtuples ("D" in Fig. 2b), and of pointers to other MD subtuples ("C"), each representing a repeating group (e.g., "EQUIPMENT"   Fig. 2). Except for the root level, a MD subtuple contains as many "DC...C" sequences as there exist child instances under a given parent data subtuple.

As can be seen from Fig. 2, MD subtuples representing tuples of a "flat" repeating group (e.g., "EQUIPMENT") will be composed entirely of "D" pointers. On the other hand, a MD subtuple can have all "C"s, indicating that there is no attribute with atomic values at that level. In the case that some parent instances have no child data for that NF2 tuple (empty repeating group), the corresponding C's will be set to null.

Though not shown in Fig. 2,.each subtuple contains a pointer to its parent MD subtuple and a pointer to the root subtuple (naturally, these two pointers will be absent for the root subtuple). Thus a mini-directory is an access path structure for an instance of a NF2 table, i.e., for a complex tuple of that table. It describes the hierarchical structure of this given instance, and it supports navigational hierarchical access to all subparts of it, as well as direct access to instances of specific repeating groups. By following the "D" pointers, we obtain the user data for a complex or NF2 tuple. By following the "C" pointers, we are traversing the hierarchies without directly touching the data.

The mini-directory even supports ordered repeating groups, since the MD subtuples can be used to specify the sequence in which the instances are to be accessed. The data subtuple instances are physically stored in manners that contain no information on the logical sequence.

As already indicated above, the data of a NF2 tuple is stored such that all atomic attributes of an instance in a repeating group are put together as an addressable unit (the data subtuple). For example (see Fig. 2), PROJECT data and EQUIPMENT data are put into different subtuples. This schema allows access to specific parts of one complex tuple without reading it as a whole. Secondary indexes will directly point to the single subtuples and to the root of an NF2 tuple.

For addressing a data subtuple we use the concept of tuple identifiers (TID's) as in System R (AS76). A TID is a stable address which consists of a logical page-id and a slot-id. The slot-id points to a slot in a page where the actual position of the subtuple in that page is stored. This allows the shifting of a subtuple within a page without changing the TID. In fact, a TID is not changed even when a page overflow occurs.

The MD subtuples can be accessed in the same way as data subtuples via their TID's. Thus, both the "D" and the "C" entries of a MD subtuple are essentially TID's.

Conceptually, sharing of subtrees by several NF2 tuples can be implemented by sharing MD subtuples. In fact one can use this mechanism for defining several structures on the same set of dat. It will be very useful for supporting access to data being differently viewed in different applications. However, in these cases the semantics of operations must be made clear. This topic is being explored and will be reported later.

In the case that tables are normalized relations, the mini-directories are not needed. The system can see this from the catalog information and it will then perform just like an ordinary relational system.

3.2.2 History Data Management

As mentioned before, history data management is completely done inside the Subtuple Manager. It maintains history for both structural subtuples and user data subtuples. Every subtuple along with all its predecessors (i.e. history data of the subtuple) is associated with a unique tuple-id (TID). The addressing of a particular subtuple has to be done via this TID and the appropriate time (though not included explicitly in Fig. 2, all subtuples have timestamps stored in them). Current and 'old' subtuples (history data) are addressed in the same way with the default being current.

Current data and history data are stored in physically separate areas to enhance the performance of accessing the current data (Fig. 3). However, when dealing with history data, one cannot completely neglect the problem of storage space. Therefore, we have chosen the approach of recording only those attribute values which have changed. If 'in-field' operations (e.g., deletion within a field of variable length, or partial update of a field) are used to manipulate an attribute value, even less than an attribute value may be stored in many cases. In addition, we allow the users to explicitly define the attributes for which history information is to be kept. In so doing, the amount of 'useless' history data is under the control of users (or database administrators). The history information is stored as 'undo' information, i.e., the system stores all the changes required to generate from a subtuple value the next older one. All changes to a subtuple performed in one transaction are collected in one history subtuple. All history subtuples belonging to the same tuple-id are chained in descending order of time and the anchor to this chain is contained in the current subtuple (see Fig. 3). Several alternatives have been analyzed and this one was chosen because we think it is reasonable for one to pay a higher price for older history information and pay no penalty for current data.

When a subtuple, for which history is kept, is to be deleted, the subtuple is transformed into a history subtuple (with complete attribute values) and only a small 'deletion mark' with a pointer to the deleted tuple in the history area remains in the current database (subtuples 2 and 5 in Figure 3). a detailed description of the way history data is maintained is given in a paper by Dadam (DLW84).

Separation of current and history data alone can not satisfy the required performance. To allow fast processing of history data, index support must be provided not only for current data but also for history data. Our solution to this problem is illustrated in Fig. 4.

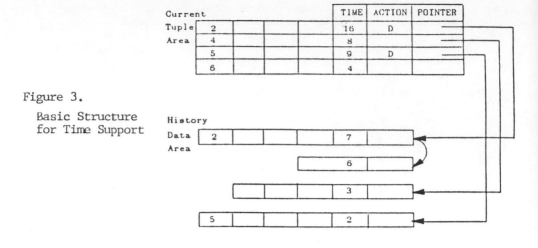

Figure 3.

Basic Structure
for Time Support

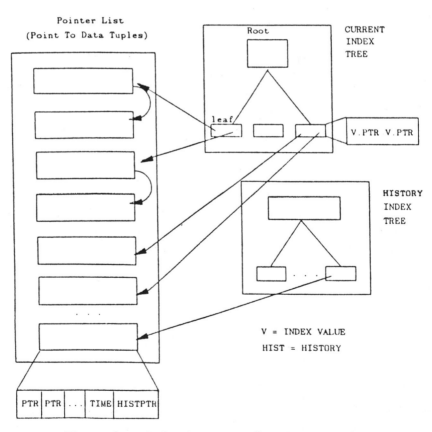

Figure 4.   Index Structure for History Paths

38

As shown in the diagram, the 'current' index contains all keys which occur in the current data, and the 'history' index contains all the keys that existed in the past but exist no longer in any current tuple in the corresponding NF2 table. Hence, history queries using an index may have to look up both index trees, while queries addressing the actual state need only to look up to "current" index tree. The pointer lists associated with the index tree are maintained in versions and contain pointers to the tuples in the NF2 table, whether current or historical.

Briefly stated, the system accesses history data as follows: It first searches the appropriate version of the catalog to see whether any index on the qualifying attributes of a query exists. If so, then the current index tree and the history index tree are searched, and the corresponding pointer lists are fetched to retrieve (in case of current state queries) or to generate (in case of history queries) the pointer lists matching the time period defined in the query. Once the appropriate index pointer is available, the system can retrieve the associated data subtuples, either in their current or in some historical state.

Simple algorithms have been developed to maintain the current and the history indexes and the pointer lists. The maintenance of the indexes can be done in a manner similar to the maintenance of normal secondary indexes, and the maintenance of the pointer lists is similar to the maintenance of an NF2 table (Lu84). Thus, the method used in our system will process queries of current data with practically no performance penalty and queries of history data with some small penalty.

3.2.3 Support for Textual Data

Handling textual data means allowing tuples in table to become large. Although a long text can be structured in a hierarchical fashion (e.g., section, chapter, paragraph etc.), this does not solve all the problems. For example, one must provide mechanisms for handling a long text even in cases where a user buffer is not large enough to hold the entire text. The Subtuple Manager has been designed to allow the system's users to retrieve any part of an attribute value. Further, when text data is not formatted, processing any textual query requires a search in the entire text data to determine which tuples satisfy the predicate in a query, when no indexing terms are provided. This kind of search would be prohibitive in performance.

To support text data processing, we are incorporating the Fragment String Index technique into the system (Sc78), KW81). This method uses a static set of index terms or keys consisting of words and word fragments. This key is determined from a sample (e.g., a subset of the stored data) which exhibits the same statistical properties as the text entities for which an index is being built. The key set is constructed by a heuristic algorithm which is biased towards medium frequency long words or fragments. If a word or a word fragment is selected as a key, its frequency is used to adjust the frequency of potential shorter keys that are substrings of keys already selected. These candidates will be adopted only if the adjusted frequency exceeds a preset threshold. For instance, assume that 'warm' and 'arm' have the frequencies 50 and 70, resp.. Then, the adjusted frequency means, that 'ARM' can be found 20 times outside the context of 'WARM'. In case of frequencies 70/70, 'ARM' would not add any useful information to 'WARM' and therefore not accepted as index key. The heuristic also suppresses stopwords (e.g., 'AND', 'OR', 'THE'). Provisions are made, however, that any string with 2 letters or more remains searchable.

Using such a key set, an index is built by associating each key with identifiers pointing to the tuples where this key occurs at least once. Redundant references are suppressed. E.g., with 'WARM' and 'ARM' as keys, the item 'WARM WATER' would be referred to solely by 'WARM', while 'ARM IN WARM WATER' would be referred to by both keys.

To resolve a query, one needs to identify index keys covering the search argument. The associated tuple identifier lists are merged according to the given query, thus resulting in a pointer list identifying a superset of records matching the search argument. Non-matching candidates remain to be discarded in a subsequent verification step.

Based on the Fragment String Index technique, the text processing component has efficient capabilities to respond to search arguments with a large variety of match patterns (e.g., within some word, adjacent, before, after, within the same sentence, distance, etc.). For example, one can ask for '@comput@' meaning that one would accept mini-computer, microcomputer, computerization, computer, etc. Or one can ask for 'control ADJ system' meaning that the two words must be adjacent to each other and in the order specified.

The text processing component works closely with the Subtuple Manager and the Index Manager to provide the support needed in the system.

3.2.4 Query Processor

The emphases in determining the architecture and design of the Query Processor are flexibility and extendibility. As a result, the Query Processor not only supports processing of high level queries like the ones specifiable in relational languages, for example, but also lower level operations for the traversal of a hierarchical data instance as in IBM IMS's DL/1 (Da81). This is done for the following reasons: (1) We wish to avoid getting into a constraint where we may find difficulty in accommodating certain applications not expected at this time, (2) we believe that host language processing of hierarchical data is simplified in certain cases, (3) the delivery of data in small pieces is more manageable than the delivery in units of complex tuples, and (4) this provision can avoid excessive buffer size requirements.

The Query Processor is composed mainly of four major components: the Parser, the Query Tree Optimizer, the Query Tree Evaluator, and the Walk Manager. We shall now briefly discuss them in the following. First, however, for easier understanding, we should briefly state our notion of a query tree.

A query tree is a tree representation of a query where each node contains information defining operations to be done at that node. Operations at each node may be

- getting/putting a value from/into the current data subtuple,

- looping over a repeating group, whose data is contained in the subtuples, to perform certain tasks,

- evaluating basic expressions (Boolean, arithmetic, comparison).

A query tree is evaluated in the order of top to bottom and left to right. As a rule, it is advantageous to perform the most discriminatory evaluation at the earliest suitable time.

The Parser accepts queries on NF2 tables in linear syntax. Basically it does preliminary name resolution and semantic checks, and converts queries into preliminary query trees in internal representations for the next step in the query tree optimizer. A preliminary query tree obtained in this step reflects only the syntactical structure of a query and is incomplete in its information content, because the Parser does not use the catalog information in deriving its result (Fig. 5(a)).

The Query Tree Optimizer's first task is to obtain information from the system catalog tables to complete the information needed for specifying a query tree. This tree can be used to evaluate the tuples to deliver the output as specified. However, as this query tree reflects merely the syntactic structure of the query, we see that we can get performance improvement in most cases by restructuring the query tree. With the use of the information in the catalog, the query tree in Figure 5(a) can be modified as shown in Figure 5(b).

The optimization process, in general, takes into account data structures and the specification of the predicates in queries to do the transformation in this process. The changes may be

- loop reordering and predicate shifting for join optimization,

- separating predicates that are more efficiently evaluated by other components in the system (e.g., evaluation by the Subtuple Manager),

- operation specific optimization.

In some cases, the preliminary query trees can be quite close to the transformed tree; in other cases, they are quite different.

The details of the algorithm is beyond the scope of this paper. However, one can see intuitively in this example that the evaluation of a predicate defined for the data at a level closer to the root of the data structure of the complex tuples is more efficient. After a query tree has been transformed, the Query Tree Evaluator, an interpretive component, is invoked to process the tree recursively.

The Walk Manager is a collection of functions providing access to hierarchically structured tuples. Its main purpose is to package Complex Tuple Manager services in a way tailored to the Query Processor's needs. Some of the functions for traversing the hierarchical data as mentioned at the beginning of this section, are almost directly exposed at the Query Processor interface.

3.2.5 Basic Transaction Support

In addition to the functions mentioned above, the Subtuple Manager component is also designed to provide some basic transaction support like transaction undo (rollback) and transaction redo in case of a crash. The method to do this is outlined in the following:

Every completed update transaction consists of a work phase (insert, update, delete operations), a commit phase, and a write phase (propagation phase). At the beginning of the work phase a private workspace, called TWS (transaction's workspace) is assigned to it. During the work phase, all updates and deletions of a transaction are reflected only in its TWS, while inserts of new subtuples into a database are performed directly in the database - but not necessarily forced to disk immediately - to get their database TID's. (These database TID's are needed at insertion time to update or construct the corresponding mini-directory subtuples). In addition, the TWS collects all of the update and delete information necessary to compute the history data when the write phase is to be performed.

When a transaction reaches its COMMIT phase, the TWS is forced to an associated disk segment, the newly inserted tuples are forced to the database disks, and an appropriate COMMIT entry is put into the log file. By doing so, the subsequent write phase is made repeatable. That is, the transaction is no longer vulnerable against system crashes (the TWS serves from now on as a transaction oriented redo logfile). If the write phase fails due to a crash it is simply repeated by restoring the TWS from the associated disk segment.

To avoid I/O, the pages are not forced to disk during the write phase. As in System R (As76), pages are forced to disk only when performing a checkpoint but not during the write phase - thus reducing I/O. For this reason a TWS will not be immediately reused after the completion of the write phase but will be kept until the next checkpoint.

If a transaction has to be rolled back due to an ABORT command or due to a crash which has occurred before the completion of the commit phase, only the inserted tuples have to be removed from the database using undo log information.

The approach described above offers the potential for some additional functions with only some relatively small extensions. For example, user defined save points can be provided rather easily. This can be done by simply copying the TWS at a given point in time to disk (and by refetching it if necessary). Further, the approach of separating commit and write phases fits well into a distributed environment where a two-phase commit protocol has to be supported. In the case of a distributed database, updates may be prepared at a place different from the one where the update data is finally stored.

Another point worthy of note is that this approach is a good starting point for deferred update strategies. Sometimes an update may be a lengthy task in advanced information system applications and the system must adopt this strategy to satisfy the performance requirements. For example, the maintenance of the Fragment String Indexes may be such an example. The decoupling of the index update is a reasonable way from the logical view-point of transaction management.

Finally our approach fits also best for times-tamping the updated tuples with the transaction's commit timestamp. As opposed to other approaches (see DLW84 for an extensive discussion of this point), the timestamping of updated tuples can be done here without any additional overhead. Thus, the write phase can be used to stamp the subtuples 'on the fly' with their transaction's commit timestamp while moving them from the TWS to the database buffer. Only newly inserted subtuples which are no longer available in the database buffer have to be separately refetched from the database to assign the final timestamp to them.

3.2.6. System Generated Tuple-Names

Although the system as described can have tables of tuples that are hierarchically structured, there remain occasions where tables must be connected to form more complex tuples. One can use the 'key' property to join tables as normally done in relational systems. As discovered and discussed in Lo84, for example, in many applications it is better to have the system build and maintain connections between tables. Tuple names are such a mechanism.

Briefly stated, tuple names are system generated keys for each tuple. No two tuple names are the same across the whole database. Tuple names can be used as references to other tables in lieu of explicit keys. However, tuple name values cannot be changed by users although they can be accessed and searched. Internally tuple names are actually a form of

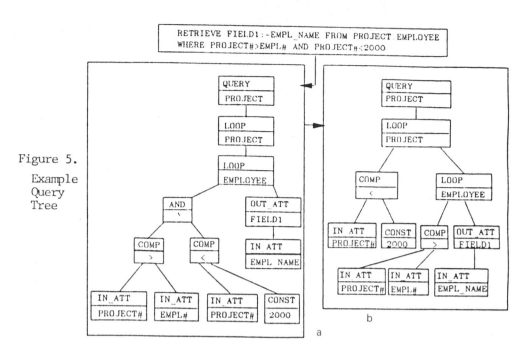

Figure 5.
Example
Query
Tree

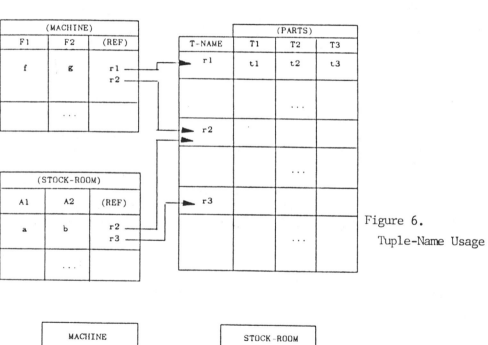

Figure 6.
Tuple-Name Usage

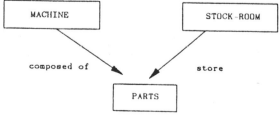

addresses. Thus the main advantage of using tuple names for connection is speed. By using tuple names instead of keys, joins can be performed much faster. Of course, one also has the benefit of not having to maintain the key property. Figure 6 shows an application of using tuple names (T-NAME in PARTS table) for connecting MACHINE and STOCK-ROOM to it. The term REF is a system reserved keyword, indicating that only tuple names can be entered there.

While the users can use tuple names to build more complex objects, the system must maintain additional catalog information on any of these complex structures. In this way, the user' task of database maintenance is reduced.

4. Conclusion and Discussion

4.1. Status Review

The system as described has not only been designed but a version of the prototype, though partially done with only a small subset of the planned functions, already has many features not found in other database systems. Included among these features are:

1. The ability to support both the normalized and non-normalized relational models in such a way that users of the system can change from one to another without penalty. Further, it is expected that as a relational system, there is hardly any penalty paid as a result of the dual model support. The same is true for the non-normalized case.

2. The system can process history data as well as current data. When processing only current data, there is little penalty as compared to normal database systems. Moreover, users of the system can dictate whether history data is to be kept, or on which attributes history data is not to be kept. Thus, additional processing time would be imposed only when history data is needed.

3. The system can process textual data as well as formatted data in that content search of text fields can be handled efficiently through the use of indexing as opposed to brute force search. As a result, queries that require join operation of a substring of text can be handled. For example, suppose one wishes to find, from two, tables which have 'NAME' common to both but stored in different formats (like one with first name then last name and the other the opposite), one can use this system to do the join operation even if the name formats are different.

4. Query operation allows processing of sets (especially: repeating groups) in ways that are not generally available in other DBMS's.

5. With the use of the history data handling capability, changes in data structures in a table can be supported dynamically, within certain constraints.

These are some sophisticated capabilities that are expected to be useful in supporting advanced applications in science, engineering and offices. From various investigations, we have found that these functions are basic and are indeed needed.

There are many other functions that will be needed, but are not yet implemented. Included among these are multi-user management, security control, recovery, complex data types like lists of sets and lists of lists, etc., long fields that run into millions of bytes, abstract data type, and others. Some of these have been planned and others are being studied at this time.

4.2 Desirable Functional Extensions

Even with all the planned functions implemented, we wish to state that the system being constructed is meant to provide a foundation on which further development can be done. For example, objects of certain engineering applications (e.g., car models) need to be maintained in different versions. While versions can be handled in many different ways, we think that a good basis to generate versions is based on time. The system mentioned here does not directly support version management. However, it is believed that with the history data handling capability and the capturing of timestamps, one can provide such capabilities using the functions provided.

The comment on versions also applies to history data management proper, in that our system is capable to provide the basis for elaborate higher level history management concepts (e.g., KL83, GT84), which also take into consideration virtual data and history related consistency constraints. It should be noted, however, that a lot of interesting issues need further be clarified which are directly related to our more basic approach, but are not fully covered by these concepts either. One area has already been mentioned along with physical and logical time. Another area is the proper treatment and generation of null values in the context of structure redefinitions (examples: addition/deletion of attributes, turning on/off history maintenance). The linkage of histories is another interesting issue to be encountered in different contexts. One end of the spectrum is marked by the problem of binding together logically related objects of different structures (e.g., tuples before and after a "vertical" table split). The other end is illustrated by the question how to unify the histories of an employee who worked for a company during several periods.

A design challenge untackled so far is end-user interfaces. At this time we plan to design an application program interface. Later, we will work on an end-user oriented one. Because of limited space we have not discussed the system interface in this paper. However, the papers (SP82, PHH83) provide an idea of the direction we are pursuing. We feel that, because of the diversified possible applications, different interfaces may be needed for different applications, as opposed to the single query interface in today's systems. In fact, it is quite uncertain at this time whether a single database management system can be constructed to satisfy all the diversified applications!

4.3 Performance Related Issues

Another area to be explored relates to the development of techniques to optimize the placement of data. Much work has been done in the database area on the problem of clustering data for better performance. It is hoped that some of these techniques and other new ones may be applicable to enchance the Complex Tuple Manager component for fine tuning the performance of the system.

Still another problem area to be addressed is the execution of joins. Much work has been done for normalized relational joins; hardly anything has been done for joining complex tuples of NF2 tables. Since joins are very complex and time-consuming, extreme care is necessary to achieve the desired performance level. Considering that complex tuples contain repeating groups, that joins should be allowed at any level, and that inversion of hierarchies should also be permitted, intolerable performance problems may arise unless good solutions are found.

In the history data handling facility, further work is needed to find an efficient method for migrating aged history data to off-line devices and to recall off-line data to be available on-line. Again we are looking for a method that not only allows us to do migration in batch mode, but also in an incremental fashion. We believe that, when this can be accomplished, a DBMS of the kind discussed here will become generally applicable in all normal environments.

Acknowledgement

The authors wish to thank their management, specifically Dr. A. Blaser, for the support of this project, and Prof. H. J. Schek who not only started this project but has continued to provide stimulating discussions in formulating the architecture and design of the system.

References

As76

Astrahan, M.M. et al.: System R: Relational Approach to Database Management, ACM Trans. on Database Systems, Vol. 1, No. 2, June 1976. pp. 97-137.

Bu77

Bubenko, J. A.: The Temporal Dimension in Information Processing. Architecture and Models in Database Management, G.M. Nijssen, Ed. North Holland, 1977, pp. 93-118.

Ch76

Chamberlin, D.D. et al: SEQUEL 2: A Unified Approach to Data Definition, Manipulation, and Control, IBM Journal of Research and Development, Nov. 1976. pp. 560-575.

CW83

Clifford, J., Warren, D.S.: Formal Semantics of Time in Databases. ACM TODS, Vol. 8, No. 2, 1983. pp. 214-254.

Da81

Date, C.J.: An Introduction to Database Systems, Addison-Wesley Publishing Company, 1981.

De85

Deppisch, U et. al.: Ein Subsystem zur stabilen Speicherung versionenbehafteter hierarchisch strukturierter Tuple. Proceedings of the German Informatik Conference on DB Systems for Science, Engineering, and Office, March 20-22, 1985, Karlsruhe, W. Germany.

DGW85

Deppisch, U.; Guenauer, J.: Walch, G.: Speicherungsstrukturen and Addressierungstechniken fuer komplexe Objekte des NF2-Relationenmodells. Proceedings of the German Informatik Conference on DB Systems for Science, Engineering, and Office, March 20-22. 1985, Karlsruhe, W. Germany.

Di84

Dittrich, K.R.; Kotz, A.R.; Muelle, J.A.; Lockemann, P.C: Datenbankkonzepte fuer Ingenieur-Anwendungen: Eine Uebersicht ueber den Stand der Entwicklung (Data Base Concepts: A State of the Art Overview). GI Jahrestagung, 1984.

DLW84

Dadam, P.; Lum, V; Werner, H.-D.: Integrating Time Versions into a Relational Data Base System. Proceedings of the 1984 VLDB Conference, August 27-31, Singapore.

GP83

Gruendig, L; Pistor, P.: Landinformations-systeme und ihre Anforderungen an Datenbankschnittstellen (Geographic Information Systems and Their Requirements Regarding DB Interfaces). J.W. Schmidt (ed.): Sprachen fuer Datenbanken (Data Base Languages). Informatik Fachberichte 72, Springer Verlag, Berlin-Heidelberg-New York. 1983.

GT84

Ginsburg, S; Tanaka, K.: Interval Queries on Objects Histories: Extended Abstract. Proceedings of the 1984 VLDB Conference, August 27-31, Singapore.

GU83

Guenther, K.D.: PLOP- A Predicative Programming Language for Office Automation. Proc IEEE Workshop on Languages for Automation, Chicago, Nov. 7-9, IEEE computer Society Press, p. 94-101.

HL82

Haskin, R.L.; Lorie, R.A.: On Extending the Functions of a Relational Database System. Proc. SIGMOD 82, Orlando, June 1982. pp. 207-212.

HR82

Haerder, T.; Reuter, A.: Data Base Systems for Non-Standard Applications. Universitaet Kaiserslautern, Fachbereich Informatik, Interner Bericht 54/82. 1982.

HS76

Housel, B.C.; Shu, N.C.; A High-Level Data Manipulation Language for Hierarchical Data Structures. Proc. Conf. on Data Abstraction, Definition and Structure, Salt Lake City, Utah, March 1976.

IBM82

Das allegemeine Abfrage-System STAIRS/VS-MIKE in der Fertigungsindustrie (The General Purpose Query System STAIRS/VS MIKE in Assembly Industries). IBM Deutschland GmbH, 1982.

Ja84

Jaeschke, G.: Recursive Algebra for Relations with Relation Valued Attributes. HDSC Technical Report 84.01.003, 1984.

JS82

Jaeschke G.; Schek, H.-J.: Remarks on the Algebra of Non First Normal Form Relations. Proceedings of the ACM SIGACT-SIGMOD Symposium of Principles of Database Systems, Los Angeles, March, 1982.

Ki83

Kinzinger, H.: Erweiterungen einer Datenbank-Anfragesprache zur Unterstuetzung des Versionenkonzepts (Query Language Extensions to Support Time Version Concepts). J.W. Schmidt (ed.): Sprachen fuer Datenbanken (Data Base Languages). Informatik Fachberichte 72, Springer Verlag, Berlin-Heidelberg-New York. 1983.

KL82
    Katz. R.H.; Lehman, T.J.: Storage Structures for Versions and Alternatives. Computer Science Report #479, July 1982, University of Wisconsin-Madison.

KL83
    Klopprogge, M.R.; Lockemann, P.C.: Modelling Information Preserving Databases: Consequences of the Concept of Time. Proc. VLDB 1983, Florence, Italy, Oct./Nov. 1983.

KTT83
    Kambayashi, Y.; Tanaka, K; Takeda, K.: Synthesis of Unnormalized Relations Incorporating more Meaning. International Journal of Information Sciences, Special Issue on Databases, 1983.

KW81
    Kropp, D.; Walch, G.: A Graph Structured Text Field Index based on Word Fragments. Information Processing and Management, Vol. 17, No. 6, 1981.

Lo82
    Lorie, R.: Issues in Databases for Design Applications. File Structures and Databases for CAD. (J. Encarnacao and F.-L. Kraus, ed.), North Holland Publishing Company, 1982.

Lo84
    Lorie, R.; Kim, W.; McNabb, D.; Plouffe; Meier, A.: User Interface and Access Techniques for Engineering Databases. IBM Research Technical Report RJ 4155 (45943) January 1984.

LP83
    Lorie, R.; Plouffe, W.: Complex Objects and Their use in Design Transaction. Proc. Engineering Design Applications of ACM-IEEE Database Week, San Jose, Ca., May 23-26, 1983.

Lue83
    Lueke, B.: DANTE: Ein semantisches Datenmodell fuer Anwendungen aus dem Konstruktionsbereich (DANTE: A Semantic Data Model for Mechanical Engineering Applications). Universitaet Karlsruhe, Fakultaet fuer Informatik, Interner Bericht 17/83, 1983.

Lu84
    Lum, V; Dadam, P; Erbe, R.; Guenauer, J.; Pistor, P.; Walch, G; Werner, H.-D.; Woodfill, J.: Designing DBMS Support for the Time Dimension. Proceeding of the 1984 SIGMOD Conference, June 18-21, Boston, Mass.

MSW83
    Mueller, T.; Steinbauer, D.;Wedekind, H.: Control of Versions of Data Base Application. HDSC Technical Report TR 83.09.003, September 1983; J.W. Schmidt (ed.): Sprachen fuer Datenbanken (Data Base Languages). Informatik Fachberichte 72, Springer Verlag, Berlin-Heidelberg-New York. 1983.

PHH83
    Pistor. P.; Hansen. B.: Hansen. M.: Eine sequelartige Schnittstelle fuer das NF2
    Modell (A SEQUEL Like Interface for the NF2 Model). J.W. Schmidt (ed.): Spra-
    chen fuer Datenbanken (Data Base Languages). Informatik Fachberichte 72,
    Springer Verlag, Berlin-Heidelberg-New York, 1983.

Sc78
    Scheck, H.-J.: The Reference String Indexing Method. Information System Metho-
    dology (Editors: G. Bracchi & P. Lockemann), Springer Verlag, LNCS 65, 1978.

Sc82
    Scholl, M.: Algebraische Frageoptimierung in Datenbanksystemen mit nichttrivialen
    Abbildungen zwischen konzeptuellem und internem Datenmodell (Algebraic Query
    Optimization in Data Base Systems with Non-Trivial Transformation Between Con-
    ceptual and Internal Data Model). University of Darmstadt Master thesis, August
    1982.

SHL75
    Shu, N.; Housel, B.C.; Lum, V.: Convert: A High Level Translation Definition
    Language for Data Conversion. Comm. of ACM, Oct., 1975. Vol. 18, No. 10.

SP82
    Schek, H.J.; Pistor, P.: Data Structures for an Integrated DB Management and Infor-
    mation Retrieval System. Proceedings of the VLDB Conference 1982, Mexico City,
    September 8-10, 1982.

SS84
    Schek, H.-J.: Scholl, M.: An Algebra for the Relational Model with Relation-Valued
    Attributes, Tech. University Darmstadt Technical Report DVSI-1984-T1.

St82
    Stonebraker, M. et. al.: Support for Document Processing in a Relational Database
    System. Electronics Research Laboratory, Memo M82/15, March 1982.

St83
    Stonebraker, M. et. al.: Application of Abstract Data Types and Abstract Indices to
    CAD Databases. Proc. Engineering Design Applications of ACM-1EEE DataBase
    Week, San Jose, Ca, May 1983.

USE OF THE RELATIONAL MODEL FOR DATA REPRESENTATION

IN A DEDUCTIVELY AUGMENTED DATABASE MANAGEMENT SYSTEM

Kioumars Yazdanian

ONERA - CERT / DERI (Computer Science Department)
2 Avenue Edouard Belin
31055 - Toulouse
France
(uucp mail: mcvax!inria!tls-cs!yaz)

ABSTRACT

Database Management Systems deal usually with explicit data i.e. data explicitely introduced in the Database (DB). Deductive or inferential DB uses general rules to derive implicit data i.e. data not explicitely stored in the DB. The Codd's relational model introduced as a formalism for information representation can be used to represent internal (system specific) data needed for the implementation of a deductive process over a relational DBMS. This representation remains user transparent.

INTRODUCTION

In the DataBase Management Systems (DBMS) context, data usually concerned are explicit data i.e. data explicitely introduced in the DataBase (DB). To improve DB capabilities, research work has been done to manage also general information represented by general rules and used either as integrity rules to be enforced, or as deduction rules to derive implicit data i.e. data not explicitely stored in the DB. The basis of this research has been the relational model described by Codd (1970) the concepts of which are similar to those of the first order logic (Gallaire, Minker and Nicolas, 1984), commonly used for general rules expression and automatic deduction. Some implementation attemps have been done to validate concepts, formalisms and methods (Nicolas and Yazdanian, 1983). Any system implementation needs internal data structures to represent its specific data which will be accessed through specific algorithms.

In the context of relational DB with deductive capabilities, this paper presents an attempt to use the relational model itself to represent internal (system specific) data needed for the implementation of a

Acknowledgment to European Economic Community ESPRIT project under grant of which this work has been done

deductive process over a relational DBMS. This representation remains at the conceptual level of the DB and once excluded from the external user views, it will remain transparent to all common users but the data stored in it are available for the system and eventually the Database Administrator usage through the classical functionalities and Data Management Language of the relational DB.

In the following, the first paragraph is a brief review of relational concepts used for this presentation, the second one describes succintly the deductive processes used for DB applications, the third paragraph presents the relational representation of the general rules and additional data needed to perform deduction in a relational database, and the fourth one drawback and advantages.

The reader is supposed to be more or less familiar to the concepts and languages of the relational model and to the first order logic and notions of unification, bottom up and top down search strategy used in automatic problem solvers. Nevertheless, the first two paragraphs could be considered as an informal introduction sufficient for a good comprehension of the sequel.

I - THE RELATIONAL MODEL

The concept of relation is defined on a set of domains which are not necessarily distinct. Given domains D1, D2, ... , Dn, a relation of degree i (or i-ary relation), is a subset of the cartesian product D1 x D2 ... x Di. Each subset corresponding to a relation is also called a relation extension and will be denoted by a set of tuples (inside parenthesis), components of which are elements of the domains, and preceded by the relation name.

Information is represented as relationships between objects or elements belonging to domains.

Follow some elementary information expressed in english:

- Peter lives in Tokyo
- Mary lives in New-York
- John lives in Paris
- John teaches physics in the classroom n° 2 to students of the 3rd level
- Factory F2 makes green tables
- Supplier S1 supplies part P3 to project J2 in quantity 4
- The motor n° 3 is 20 kg weight and its power is 2 kw
- The motor n° 3 drives wheel W1
- The axis n° 5 connects wheels W1 and W2.
- Wheel W3 gears wheels W4 and W5

Assuming different domains containing names, town names, subset of integers, course names, factory names, color names, object names, supplier identifiers, part numbers, project identifiers, ... , the above information will be expressed using relation names (in upper case letters) by the relational extensions below:

            LIVE   ("John","Paris")
                    ("Mary","New-York")
                    ("Peter","Tokyo")

            TEACH   ("John","physics","2","3")

            MAKE   ("F2","green","tables")

```
SUPPLY   ("S1","P3","J2","4")

MOTOR    ("3","20","2")

DRIVE    ("3","W1")

CONNECT  ("5","W1","W2")
         ("9","W2","W3")
         ("7","W5","W6")

GEAR     ("W3","W4") .
         ("W3","W5")
```

All elements of domains (constants values) which could be so will be written double quoted in order to distinguish them from variables (one alphabetic letter) and argument names.

To use a relational model for information representation, a relational schema is first defined, which gives the relation names which will be used and for each relation, its arity or degree (its number of arguments) and as many different argument names as it has arguments in order to avoid the reference to an argument by its position inside the relation. An exemple of relational schema is given below for the data presented above.

```
LIVE     (person,town)
TEACH    (person,course,classroom,level)
MAKE     (factory,color,object)
SUPPLY   (supplier,part,project,quantity)
MOTOR    (motor_identifier,weight,power)
CONNECT  (axis,wheel1,wheel2)
GEAR     (wheel1,wheel2)
DRIVE    (motor_identifier,wheel)
```

II - THE DEDUCTIVE PROCESS

A classical notation will be used for the logical expressions, using the symbols : ^, &, v, -->, ¥, respectively for the logical operators ´not´, ´and´, ´or´, ´implication´, and the quantifier ´for all´. The equality relation will have the special notation $(x=y)$ for better readability.

A litteral, positive or negative, is a predicate symbol (corresponding to a relation name), preceded or not by the ´not´. More details about correspondence between mathematical logic and relational databases are given by Nicolas and Gallaire (1978).

Two litterals are unifiable if they have the same predicate or relation name, and if their corresponding arguments are either both identical constants, or at least one of them is a variable (the other being a constant or a variable); in this case, the variable has to be substituted by the constant in all of its occurrences before going on the unification process. If the litterals are unifiable, a unification substitution (eventually empty) is defined at the end.

A deduction general rule is a logical expression under the clausal form (prenex conjunctive normal form) containing one and only one positive litteral, or under an equivalent form such as:

<pre>
              premise   -->   conclusion
          (left hand)   (right hand)
</pre>

where the premise is a conjunction of positive litterals, and the conclusion is a positive litteral.

Another constraint is that all variables of the conclusion (or in the positive litteral of the clause) must be present in the premise (or the remainder of the clause).

Two main methods exist to perform deduction in a DB: the derivation method corresponding to a top down strategy, and the generation method corresponding to a bottom up strategy.

Let us give an example.

Some general rules are:

- If two wheels are connected and a motor drives one of them, then it drives the other.
- If two wheels are geared and a motor drives one of them, then it drives the other.
- Every axis which connects one wheel to another connects also the second one to the first (commutativity of the "CONNECT" property with respect to a given axis).
- If one wheel gears another, the second one gears also the first (commutativity of the "GEAR" property).
- If one wheel gears a second one which in its turn gears a third wheel then the first one gears the third one (transitivity of the "GEAR" property).

They will be expressed in a first order logic syntax by:

(1) $\forall x \, \forall y \, \forall z \, \forall u$ (CONNECT (u,x,y) & DRIVE (z,x) --> DRIVE (z,y))

(2) $\forall x \, \forall y \, \forall z$ (GEAR (x,y) & DRIVE (z,x) --> DRIVE (z,y))

(3) $\forall x \, \forall y \, \forall z$ (CONNECT (x,y,z) & $^\wedge(y=z)$ --> CONNECT (x,z,y))

(4) $\forall x \, \forall y$ (GEAR (x,y) --> GEAR (y,x))

(5) $\forall x \, \forall y \, \forall z$ (GEAR (x,y) & GEAR (y,z) & $^\wedge(x=z)$ --> GEAR (x,z))

In the deduction by derivation method, each time a query is asked, the extensions of the relations contained in the query are evaluated on the database. This evaluation includes two stages:

- First looking for the explicit part of the relation i.e. facts explicitely introduced in the DB. This is done by a classical data retrieval depending on the data structure used.

- Second looking for the implicit part of the relation i.e. the facts which can be derived from other explicit or implicit facts of the DB and the general rules used for deduction. This involves a top-down or problem decomposition process commonly used in theorem provers and Artificial Intelligence processes.

For instance, the answer to the query "what are the wheels driven by the motor 3" will be computed first by looking for all wheels associated with "3" in the extension of the relation DRIVE. Then have to be added all implicit data consisting on wheels connected to or geared by the first ones. To find these wheels, the two steps are necessary again: first to find all wheels connected to the previous ones, as "W2" which is connected to "W1" i.e. the couple "W1","W2" is present in that order in the explicit extension of the relation CONNECT, and second to find all implicit data indicating a link between two wheels through deduction rules such as (3), (4) or (5).

In the deduction using the generation method, the implicit data are retrieved by the uptdate time, when a new data is introduced in or deleted from the DB. To use the same terms of "data introduction" both for storing a t-uple in a relation and for deleting a t-uple from a relation, we will consider in the following that a data could be positive or negative; to introduce a positive data in a relation corresponds to a storing operation and to introduce a negative data corresponds to a deletion operation. The process introduces first the explicit data into the corresponding relation, and then looks for new implicit data which can be derived given this data and the existing general rules. These new implicit data are then physically introduced into the DB in order to limit further retrieval to the processing of the explicit information only.

For instance by the introduction of the data "motor 3 drives wheel W1", and given the set of general rules above, all wheels which are driven by motor 3 depending on different links existing between wheels, are found and explicitely introduced (if not already present) into the explicit extension of the relation DRIVE; another example is the introduction of a data corresponding to a link between two wheels, either by an axis connection or a gearing system. All consequences of this link, for the CONNECT, GEAR or DRIVE relation will be computed and introduced into the DB. The deletion of a data will also involve the deletion of other data which are no longer true.

In the following are considered different steps involved in the deduction process using general rules. The management of the elementary data can be done by the classical methods and data structures (the physical level of the DBMS). What is interesting is to show how the DBMS can be used to manage also data needed for the deduction process essentially to store and fetch the general rules on user request, and to select and instanciate corresponding general rules for a given operation in the process.

In the derivation method, what is needed is to find every deduction rule the conclusion of which is unifiable with one of the relations contained in the query.

In the generation method, what is needed is to find all deduction rules which contains in their premise one litteral unifiable with the relation on which update is performed. In the special case where a deleting operation concerns a data having implicit occurrences, the deduction rules unifiable on their conclusion part with the involved relation are also requested in order to determine the minimal sets of purely explicit facts the deletion of which will suppress all implicit occurrences of the data to delete (Nicolas and Yazdanian, 1983).

Whether in a derivation method or in a generation method, the deductive process needs to select the set of rules involved in a deductive path; For performance improvement, this set has to be reduced as much as possible before putting in work the deduction process which will act either using a Modus Ponens like inference rule (generate the conclusion in every case the premise is true) or using a refutation like process: if a positive litteral of a rule unifies a negative one, then use the remainder of the rule (subproblems or subqueries), after appropriate substitution, in place of the negative one (initial problem or query).

III - REPRESENTATION OF THE GENERAL RULES WITH A RELATIONAL MODEL

In the following we will consider general rules and a representation of them in the relational model.

Related to relation definition, new domains are requested:

- The domain DR of rules identifiers, each element of which will denote one and only one deduction rule.

- The domain DRN of relation names.
- The domain DV of variable names.
- A two elements domain DS to characterize the premise side or the conclusion side of a rule (subproblem/problem side for a clause). The elements of DS will be for instance "p" and "c".

Each deduction rule will be denoted by a unique identifier associated to it by the time of its introduction into the Deductive DB.

For the first step of the deduction process, the general rules containing a given relation in their premise (or conclusion) side have to be found. Two relations could be used to store these information. Let us call them

> LEFT (rule_id, relation_name) and
> RIGHT (rule_id, relation_name)

An alternative structure could be a ternary relation:

> RULE (rule_id, left_rel_name, right_rel_name)

but in this case the right_rel_name will have as many occurrences as there is left_rel_name in the rule.

The second step in the deduction process is to find possible unification between the external relation (the relation involved in the update or in the query or subquery) and litterals of deduction rules. For each relation present in a deduction rule, (and more generally for each relation R of the DB) there is a corresponding relation R′ containing information about general rules (rule_id) in which the relation R occurs and also about its arguments (which ones are constants and which others are variables). This set of relations will be used to find potential unifiers for the external relation. In such a relation R′, distinction is done between variable and constant, the last ones should match the corresponding constant of the external relation, but a variable could be unified with any constant of the external relation.

Two possible ways to define R′ relations are:

- If R is of degree n, over domains D1, D2, ... , Dn let us define R′ as a relation of degree n+2 over the domains:

> D1 U DV, D2 U DV, ... , Dn U DV, DR, DS.
> (U denotes the union operator)

But in this case, no variable name could be identical to an element of a domain Di, for otherwise ambiguity should occur.

- Let us take a domain DT for the argument types, instead of the variable names domain. This domain contains only two elements, corresponding to the variable argument and the constant argument type (for instance "1" and "0"). Thus R´ could be defined in the same way over the domains:

$$(D1 \ U \ DV) \ x \ DT, \ (D2 \ U \ DV) \ x \ DT, \ \dots \ , \ (Dn \ U \ DV) \ x \ DT, \ DR, \ DS$$

An equivalent definition for R´ would be a relation of degree 2n+2 over the domains:

$$D1 \ U \ DV, \ D2 \ U \ DV, \ \dots \ , \ Dn \ U \ DV, \ DT, \ DT, \ \dots \ , \ DT, \ DR, \ DS$$

Let us notice that with the set of R´ like relations, the relations LEFT and RIGHT are no more necessary for the deduction process usage. But they may be helpful for other functionalities of the deductive DBMS such as listing all the deduction rules_identifiers of rules containing a given relation in the premise (or in the conclusion). Without the relation LEFT, the deductive DBMS would have to evaluate as many queries as there are relations R´ (i.e. same number as the number of relations in the DB). For storing, retrieving, listing the rule expressions, an EXTERN relation can be used with two arguments: rule_id and rule_expr containing the expression of deduction rules as written by the user (and eventually in a normalized form). The domains of this arguments are DR and DE (domain of deduction rules expressions).

As example, let us consider a DB with relations P(x,y), Q(x,y,z) and R(x,y,z).

Explicit extensions of these relations will be in a given state of the DB:

```
P("el","e2")
 ("el","e8")
 ("e2","el")

Q("el","e3","e4")
 ("el","e2","e5")

R("e5","e6","e7")
 ("el","e6","e8")
```

Let us consider for this DB the set of following general rules to be used for automatic deduction:

```
(1) P(x,y) --> P(y,x)
(2) Q("el",x,y) & R("el",z,u) --> P("el",u)
(3) P(x,y) --> Q(x,y,"e9")
```

These general rules will be identified by the rule_id corresponding for instance to the rule number.

The other relations (called prime relations) to be added are:

```
P´(arg1,arg2,arg1´,arg2´,rule_id,s)
Q´(arg1,arg2,arg3,arg1´,arg2´,arg3´,rule_id,s)
R´(arg1,arg2,arg3,arg1´,arg2´,arg3´,rule_id,s)
```

The additional arguments arg1´, arg2´, ... indicate respectively if arg1, arg2, ... are variables or not, rule_id indicates in which general rule the litteral is present, the argument s indicates if it is in the premise or in the consequent part, and the other arguments arg1, arg2, ... (corresponding to the arguments of the corresponding relations P, Q, or R) take either constant value, or variable value depending on what the litteral inside the deduction rule contains.

With respect to our example, the extensions of the prime relations are:

```
P´("x","y","1","1","1","p")
  ("y","x","1","1","1","c")
  ("e1","u","0","1","2","c")
  ("x","y","1","1","3","p")

Q´("e1","x","y","0","1","1","2","p")
  ("x","y","e9","1","1","0","3","c")

R´("e1","z","u","0","1","1","2","p")
```

A given external relation can be considered as a member of a special general rule restricted to one relation and therefore the corresponding information represented by the prime relation, where the ´rule_id´ argument takes a special value "E" (corresponding to no rule reference) and the ´s´ argument takes value "p" or "c" depending on the deduction method to be used (respectively "p" for derivation method and "c" for a generation method).

For instance let us consider the external relation Q("e1","e2",x) its representation will be Q´("e1","e2","x","0","0","1","E","p"). Notice that this external relation can only be involved in a derivation method, since it contains variable: in the generation method, with respect to the constraints on general rules structure and variable, only constants should be present).

In the sequel either for a derivation method or for a generation method, only one step of the deduction process is considered i.e. the decomposition of one query to subqueries, or the generation of new data in one application of the inference rule. The whole deduction process is iterative and stops either if no decomposition into subqueries is possible (and in some cases the problem cannot be solved without accessing the data inside the DB), or if all data generated already exist (no new data is generated).

The candidate rules for unification are selected by a relational query such as:

```
retrieve into SELECT
    T1.rule_id
    T1.arg1 T2.arg1 (T1.arg1´ & T2.arg1´)
    T1.arg2 T2.arg2 (T1.arg2´ & T2.arg2´)
    T2.arg3 T1.arg3 (T1.arg3´ & T2.arg3´)

from      T1:Q´
          T2:Q´

where     (T1.arg1="e1" v T1.arg1´="1")
        & (T1.arg2="e2" v T1.arg2´="1")
        & (T1.s ^= T2.s)
        & (T2.rule_id = "E")
```

(Notice that the order of T1.arg T2.arg is reversed for the variable arguments of the external relation, which are not present in the ´where´ part of the query).

The answer to this query will be stored in the relation SELECT (temporary relation) and is corresponding to the selection of general rules having the Q relation as conclusion, and for which the conclusion is unifiable with Q("el","e2",x). The evaluation of query will also provide the set of substitutions used for the unification:

The rule_identifier is the first argument of the answer.
The following arguments are split into subgroups of three elements
For each subgroup, the first element is the element which is to be substituted by the second, and the third one indicates if it is a variable/variable substitution (value "1") or not (value "0").

In our example, the extension of SELECT will be:

     SELECT("3","x","el","0","y","e2","0","v","e9","0")

If the external relation was: P("el","e8")
Then P´ would contain the t-uple:
     ("el","e8","0","0","E","p") for a derivation method, and

     ("el","e8","0","0","E","c") for a generation method.

And the extensions of SELECT would be for the first step:

     SELECT ("1","y","el","0","x","e8","0")
           ("2","el","el","0","u","e8","0")
                              for a derivation method, and

     SELECT ("1","x","el","0","y","e8","0")
           ("3","x","el","0","y","e8","0")
                              for a generation method.

Finally to illustrate the case of an incomplete unification where variables remain in the SELECT relation let us consider the external relation P("el","v") and the derivation method. This leads to:

     P´("el","v","0","1","E","p")

     SELECT ("1","y","el","0","x","v","1")
           ("2","el","el","0","v","u","1")

The unification is then effectively possible if two identical variables are not substituted by different constants. This could happen with a premise or conclusion containing for instance, Q("el",x,x) and with an external relation such as Q("el","e2","e3").

If unification is possible at this level, the expression of selected rules are found through the relation EXTERN, instanciated with the adequate set of substitutions in the corresponding side (conclusion or premise side depending on the deduction method) and evaluated upon the DB to instanciate the eventually remaining variables. In fact, for a given general rule, all the possible instanciations are found through one evaluation of a premise or conclusion.

Then the substitutions, now completely defined, are applied to the other side of the selected rules and lead to new external relations, composing new subproblems (subqueries) in the derivation method, or new data for DB update in the generation method.

And the process iterates.

IV - DRAWBACK AND ADVANTAGES

The representation of general rules in a deductively augmented Relational DBMS involve generally ad-hoc extension of the system. Using the relational model to represent these rules has the advantage of easy implementation since no change has to be done in the system.

The only drawback could be on the performance aspects, but there is in counterpart several advantages :

- insertion, deletion, modification and listing of rules is easily done using DBMS functionalities, including a retrieval of rules with respect to given specification (for instance looking for all rules having a given relation in its right hand side).

- the symbolic storage of the rule is well suited for interfacing Expert Systems or high level programming languages (LISP, PROLOG) commonly used for writing inference engines,

- all possible relevant unifications (possible substitutions of the rule variables) for a rule in a given context (an external query to evaluate or an update to process) can be obtained with a classical access to the database using a selection expression of the DBMS,

- to trace the applied rules, this representation allows a step by step selection of the rules, and determines at the same time the instanciations applied to them.

This logical representation rely upon classical physical structures used in the DBMS and rises no new problems since it uses the underlying DBMS access method for which solutions already exist.

CONCLUSION

The interest of such a data representation for a deductive system is of course not a response time improvement while this method uses a DBMS which in its turn has corresponding internal (physical) data structures for data representation. The aim is to provide a basis to represent and manage easily data needed for a deductive extension of an existing classical DBMS. One application could be storage of general rules of an Expert System.

With a simple extension of the original schema of the DB, main information needed for the deduction process could then be stored, retrieved, and managed, up to a partial unification of variables, and to get profit from the global evaluation upon the data of the database to find with one query all the involved rules, or all the unification possibilities, and at the end all the instances of the premise or consequence of a rule, given an external relation.

BIBLIOGRAPHY

Adiba, M., Bancilhon, F., Delobel, C., Demolombe, R., Gallaire, H., Gardarin, G., Le Bihan, J., Nicolas, J.-M., Scholl, M., 1982, "Bases de Données : nouvelles perspectives", Rapport du groupe BD3, ADI-INRIA, Paris (1982). Marseille (Jun. 1981).

Chang, C.L., 1978, DEDUCE 2. Further Investigations of Deduction in Relational Data Bases, in: "Logic and Data Bases", H. Gallaire and J. Minker, ed., Plenum Publishing Corporation, New York, N.Y., 201:236.

Chang, C.L., 1979, On evaluation of queries containing derived relations in a relational data base, Journées d´études : Bases Formelles pour Bases de Données, Toulouse, France.

Chang, C.L., Lee, R.C.T., 1973, "Symbolic logic and mechanical theorem proving", Academic Press, New York.

Codd, E.F., 1970, A relational model for large shared data banks, Comm. of the ACM, 13(6), 377:397

Date, C.J., 1981, "An introduction to database systems", 3rd edition, Addison Wesley, Reading.

Furukawa, K., 1977, A deductive question answering system on relational databases, Proc. of 5th International Joint Conference on Artificial Intelligence, 1977, Cambridge, Mass., USA, 59:66.

Gallaire, H., and Minker,J. Eds., 1978, "Logic and Data Bases", Plenum Press, New-York.

Gallaire, H., 1981, Impacts of logic on databases, Proc. 7th VLDB Conf., Cannes (Sept. 1981), 248:259.

Gallaire, H., Minker, J., Nicolas J.-M., 1984, Logic and Databases : a deductive approach, Computing Surveys, Vol. 16 n° 2.

Homeier, P.V., 1981, "Simplifying integrity constraints in a relational database : an implementation", Master Thesis, University of California L.A., Los Angeles.

Kellogg, C., Travis, L., 1979, Reasoning with Data in a Deductively Augmented Data Management System, Journées d´Etudes : Bases Formelles pour bases de Données, Toulouse, France.

Lipski, W., Jr., 1979, On semantic issues connected with incomplete information databases, ACM TODS 4,3, 262:296.

Mendelson, E., 1978, "Introduction to mathematical logic", 2nd edition, Van Nostrand, Princeton (1978).

Minker, J., 1978, Search strategy and selection function for an Inferential Relational System, ACM TODS, 3,1, 1:31

Nicolas, J.-M., 1982, Logic for improving integrity checking in relational databases, Acta Informatica 18, 227:253

Nicolas, J.-M., 1979, "Contribution l´étude théorique des bases de
        données - Apports de la logique mathématique". Thèse d´Etat,
        Toulouse déc. 79, n° 902.
Nicolas, J.-M., Gallaire, H., 1978, Data Bases theory vs interpretation,
        in: "Logic and Data Bases", H. Gallaire, J. Minker, ed., Plenum
        Press, New York N.Y., 33:54

Nicolas, J.-M., Yazdanian, K., 1978, Integrity checking in deductive data
        bases, in: "Logic and Data Bases", H. Gallaire, J. Minker,
        eds., Plenum Press New York, N.Y., 325:343

Nicolas, J.-M., Yazdanian, K., 1983, An outline of BDGEN: a deductive
        DBMS, Proc. 9th IFIP Congress, Paris, France, 711:717.

Reiter, R., "Toward a logical reconstruction of relational database
        theory" personal communication.

Ullman, J.D., 1980, "Principles of database systems", Computer Science
        Press, Woodlands Hills.

Yazdanian, K., 1977, "Cohérence des bases de données déductives", thèse
        de 3ème cycle, Toulouse, nov. 77, n° 2065

Yazdanian, K., 1984, Déduction dans les bases de données relationnelles,
        fondements logiques et mise en oeuvre, Journées d´études AFCET:
        "Bases de données, sécurité et intelligence", Paris.

Hashing

HASH-BASED FILE ORGANIZATION UTILIZING LARGE CAPACITY MAIN MEMORY

Isao Kojima[1] and Yahiko Kambayashi[2]

[1]Electrotechnical Laboratory, Tsukuba, Japan
[2]Kyushu University, Fukuoka, Japan

ABSTRACT

Due to the recent development of computer technology, it is possible to realize a reasonable amount of database on main memory. Although extensive use of main memory will produce good search performance and flexibility, conventional file organizations designed for disks seem not to be suitable for this environment. In this paper we first summarize characteristics of main memory compared with disks and problems to be solved in order to develop file organization suitable for main memory. B-trees and dynamic hashing schemes are well-known file organizations for database applications. Problems of direct implementations of these files under this environment are shown. A new file organization called structured variable length hashing scheme is introduced. Its access performance compared with conventional file organizations is presented. This comparison also shows that B-tree(and AVL tree) is preferable to B+tree in main memory environment.

1. INTRODUCTION

By the current development of memory technology, the whole or a part of the directories and data in a database will be stored in main memory for some applications. Dewitt et al. presented B+trees are suitable for main memory environment[5]. Those conventional file organizations designed for disk drives are, however, not suitable for large capacity main memory. In this paper a dynamic hash-based file organization is introduced to solve this problem. Conventional file organizations make use of the following properties of disk devices.

1. The time to retrieve a set of data is mainly determined by the number of direct access, since sequential access is much faster than direct access.

2. The storage utilization factor of disks is not always the most important factor to design a file. In main memory, on the other hand, there is no distinction between direct access and sequential access. Thus the size of pages can be arbitrary and access time is proportional to the page size. We can use pages to simplify access procedures rather than to access a set of data.

We can also conclude that binary trees are suitable for main memory. Files with dynamic reorganization capability usually attain a relatively low utilization factor. We have to achieve high utilization factor for main memory by the following reasons.

1. Main memory is still expensive.

2. If there is a space in main memory, data in disks should be moved to the space, since main memory are much faster than disks.

In Section 2 we summarize characteristics for main memory together with problems to be handled in order to develop file organizations suitable for large capacity main memory. Section 3 discusses problems when we use conventional file organizations. We also show typical dynamic hashing schemes and their importance to database applications. A structured variable length hashing scheme is introduced in Section 4. Its directory has a composite structure of trees and tables so that it can take advantages of dynamic hashing schemes using tables and trees. Its performance evaluations compared with conventional file organizations are shown in Section 5. Section 6 presents remaining problems and conclusions.

2. CHARACTERISTICS OF MAIN MEMORY BASED FILES

As conventional file organizations have been developed for secondary storage devices(disks), we will summarize characteristics of main memory realized by semiconductor memory chips in order to develop file organizations suitable for main memory.

1. Access time is very short: access time for main memory is typically 10 ms to 1 μs, while access time for disk storage is between 10 to 100 ms. disk storage is more than 10 times slower than main memory.

2. Access time for the next data is independent of the address of the currently accessed data: In disk devices, access time varies depending on the current position of the disk head, because of their mechanical implementation. Especially time for direct access and time for sequential access are different very much. File organizations for secondary storage utilize such characteristics.

3. Access time is proportional to the data size: Because of (2), in secondary storage devices data are partitioned into pages, while in main memory access time is determined by the data size.

4. Storage cost is high: Although the price of semiconductor memory chips is decreasing rapidly, it is still more expensive than disk devices to store the same amount of data.

5. Reliability is much lower than that of disks: When main memory is used all its contents may be influenced by power failure. In disk devices only data currently handled by the disk head may be influenced.

6. Volatility: By electric power shutdown, contents of main memory may be lost while in disk storages data will not be lost.

Because of these characteristics the following problems should be handled.

- (P1) Averaging access time gap is required: We usually store data with high usage frequency in main memory. Consider a set of data partitioned into two subsets, one stored in main memory and the other stored in disk storages. In this case, the access time for the data in the second subsets is more than 10 times slower. Usually, there is small gap between the highest frequency for the data in the first subset and the lowest frequency for the data in the second subset, we have to find some method to narrow the access time gap.

- (P2) Locality of data is not required: Many file organizations utilize the property that consecutively stored data can be accessed by one direct access to a disk device. Such data locality is very important for file organizations[7]. In main memory access time is determined by the data size, clustering of data is used in main memory by the following reasons.

    1. To simplify access strategy, since access procedures using individual data items cause complicated data placement problems(such as garbage collection problems).
    2. To realize an interface between main memory and disk storages.

- (P3) Tree structures suitable for main memory are different from those for disk storages: In main memory binary trees are suitable while in disk devices multiway trees are used. Consider the multiway tree shown in Fig.1(a). In secondary storages all data in one node are retrieved by one disk access. If we simulate it using main memory, the number of direct access will be increased. Fig.1(b) shows the case when 4-way tree structure as shown in Fig.1(a) is realized in main memory. In this case, 4 data accesses are required in the maximum. If it is a k-way tree we need maximum k accesses. If a binary tree is used as shown in Fig 1.(c), the maximum number of access is at most log k. Furthermore, in disk storages balancing of trees is very important, but unbalanced trees can be used in main memory without serious problems.

- (P4) Memory utilization factor should be high: One reason for this requirement is the cost of main memory. Also due to the speed gap between memory and disks, it is better to store data as much as possible in main memory to improve average access time. Disk utilization factor of conventional files is usually 50% to 75%.

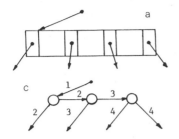
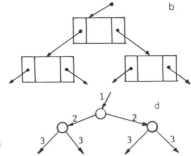

Fig.1. A Multiway Tree and a Binary Tree

- (P5) Memory initialization time must be short: In order to use the file system we have to load data into main memory. Loading time for data of 1 Gygabytes takes typically 20 minutes. Thus we need to shorten the memory initialization time.

- (P6) Transfer of a large amount of data between main memory and disk storage at a time should be avoided: The system will be slowed down during the data transmission between main memory and disk storage. A large amount of data transmission may be required for restructuring of files.

- (P7) Backup process is required: Because of (5) and (6), we have to develop a file organization suitable for realizing backup process easily. Since structures suitable for main memory and disk storage are different, we have to handle interface problems.

3. PROBLEMS OF CONVENTIONAL FILE ORGANIZATIONS

In this section we will discuss problems of conventional file organizations as files for main memory. We will consider only dynamic files, since the number of data changes dynamically in database applications. Typical dynamic file organizations are B-trees and dynamic hashing schemes.

These file organizations consist of stored data and directory structure to access data. A basic method to use such conventional files for main memory is as follows.

1. The directory is stored in the main memory, since it is frequently accessed.

2. If there remains an empty space in the main memory, it is used as a disk cache.

These methods contribute to reduce the access time. In this case,the directory part is duplicated in the main memory and disks for backup purpose. In order to access the required data, first the directory is searched and if the data is not in the cache, it is retrieved from disk. Data in the cache will be swapped out, when (1) data not in the cache is required or (2) the size of the directory part increases. Reorganizations of the directory follow the file organization algorithms. Frequently used data must be stored in the memory location which is hardly replaced by the increase of the directory part. Problems of typical dynamic file organization to be used as files for large capacity main memory are as follows.

3.1. B-trees[3]

B-tree is a dynamic file organization where its directory is realized as a balanced multiway tree. B-trees have following features.

1. Multiway tree structure can reduce the height of the tree.

2. Balanced tree structure can improve the worst case access time.

3. Page splitting algorithm keeps the storage utilization more than 50%.

In B+trees, all data is contained in leaf pages to achieve efficient sequential access and reduction of the height of the tree. When we realize B-trees on main memory environment, following problems arises.

- (B1) Pagewize manipulations are not efficient in main memory(P2). Thus the dynamic page splitting method based on fixed size pages may not suitable for main memory.

- (B2) The multiway structure is not suitable for main memory as shown by (P3).

- (B3) Memory utilization factor is too low to be used in the main memory. Many improvements are applied to basic B-tree in order to achieve good search performances and storage utilizations.

As shown in the Section 5, simple performance evaluation shows following results.

- B-trees have almost the same performance as B+trees, so that The multiway tree implementation is not efficient.

- Memory utilization factor is very much effective to the access performance.

3.2. Dynamic Hashing Schemes[8,9,11,12,6]

Dynamic hashing schemes can solve the hash collision problems by changing hash address space. These hashing schemes allow the file grow and shrink dynamically. Since logical hash function is difficult to change and it requires the reorganization of entire file, some directory structures or overflow procedures are needed. Dynamic hashing schemes can be classified as follows.

- (a) A hash table directory is realized in main memory[6].

- (b) A binary tree(or trie) structure using hash values is created in main memory[12,14].

- (c) By using overflow chaining, high storage utilization factor is obtained without directory structures in main memory[11,9].

Among these hashing schemes, (a) and (b) can read data within one disk access. Case (c) uses overflow chain and more than one disk access are required to read overflow pages. Although these dynamic hashing schemes use main memory efficiently, some problems are still considerable. For example they cannot cover the problem when data in the main memory is overflowed. Other problems of dynamic hashing schemes are as follows.

- (D1) Memory utilization factor is low in case (a) and (b).

- (D2) Pagewize handling is not efficient as shown by the discussion on B-trees.

- (D3) Although (c) can achieve good memory utilization, chaining of overflow pages increases the number of disk accesses. Thus, we can conclude that some directory structure must be maintained in main memory to minimize disk accesses. Two kind of directory structures are considered to be complement. In a tree structure, internal nodes are required but these nodes are not required when hash tables are used. In a tree directory the total number of nodes is limited by disk pages but table structure requires 2 entries when bad hash functions are used.

- (D4) Table structured directory: Even one overflow causes large amount of additional table size. This requires many data to be swapped out. Table may become quite large when data distribution is bad.

- (D5) Tree structured directory: By the increase of files, trees become large and this causes the increase of waste space in the main memory. This also causes the increase of the number of access of internal nodes when bad hash functions are used.

4. STRUCTURED VARIABLES LENGTH HASHING SCHEME

Here we will introduce a new file organization called a structured variable length hashing scheme which is based on dynamic hashing.

4.1. Basic Concepts for Constructing Directory Structure

One of the characteristics of this file organization is directory structure. First, we consider a virtual trie representing hash address space.(Fig.2(a)) This trie grows and shrinks dynamically by insertion and deletion of records as shown in Fig.2.(b). In general, this is an incomplete binary tree. If we implement this trie directly, we can get a trie structured directory. The directory structure of our file is the composite structure of trees and tables. The basic idea of the directory with a composite structure is as follows.

1. Corresponding to a complete binary tree of height h a hash table with 2^h entries are created. If we use table structure we don't need $2^h - 1$ extra internal nodes.

2. An incomplete binary tree (there are less than 2^h leaf nodes if the height is h) is decomposed into complete subtrees. Since this incomplete tree changes by insertions and deletions, we need to develop a decomposing procedure to handle the change of the incomplete tree.

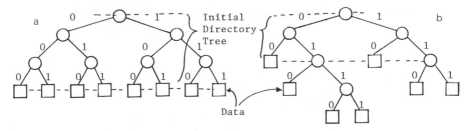

Fig.2. A Trie Index Corresponding to Hash Address Space

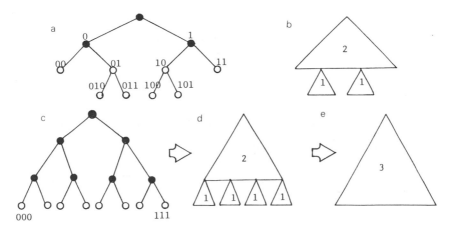

Fig.3. Decomposition of Subtrees

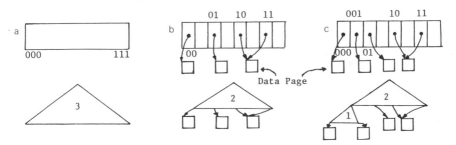

Fig.4. Directory Structure of SVLH

Next, we show an example. Example 1: An incomplete tree in Fig.3(a) can be decomposed into complete subtrees like Fig.3(b). Each triangle shows a complete subtree and the number in it shows the height. If a node 00 is splitted we have a tree as shown in Fig.3(c). We can obtain the decomposition shown in Fig.3(d) at first and then it is replaced by the decomposition shown in Fig.3(e). For this purpose we need to keep the number of complete subtrees which are connected to each complete subtree. In general if the height of a subtree T is i and it has 2^i complete subtrees, T can be replaced by a subtree of height i+1 . In each complete subtree, we can access directly to arbitrary leaf node by using the hash table. Thus the number of accesses is determined by the number of complete subtrees used to access the data from the root. We use a hash function H for such direct accesses. Let d_i be a datum and $H(d_i)$ is the hashed value of d_i . The value is assumed to be represented by a binary value. Let $H_{1,k}(d_i)$ be the first k digits of $H(d_i)$ and $H_{j,k}(d_i)$ be the number starting from j-th digit and ending at k-th digit. How to retrieve data is shown in the following example. Example 2: The labels of nodes in Fig.3(a) show hash values. We assume $H(d_1) = 00101...$, and $H(d_2) = 11010...$, . Since the height of the first complete subtree is 3, we use $H_{1,2}(d_i)$ for direct access. $H_{1,2}(d_1) = 00$ and the leaf node storing the pointer to the page containing d_i is obtained. $H_{1,2}(d_1) = 11$ and after the direct access to the node we find another subtree of height 2. We use $H_{3,3}(d_2) = 0$ and the leaf node for d_2

is retrieved. In this directory structure, total reorganization of a subtree to construct a hash table is required whenever each complete subtree is created. We avoid this problem by constructing a directory structure in which each nodes has predetermined location. We use the memory area partitioned to the database to the directory. First the maximum height h of the tree to be realized by the directory is calculated. The memory area is partitioned into 2^h locations each of which can store a pointer to the data page and some additional information. The initial directory is a complete tree of height k which is smaller than h . Only 2^k locations are used by the initial tree and other areas are used to store data, which may be replaced by directory trees if the number of data increases. Example 3: We assume the maximum height of the directory is 3, and that we have the pointer location shown in Fig.4(a). If the height of initial tree is 2, the correspondences between the location and the tree structure are shown in Fig.4(b) . In the area 00 there are some empty space, but if the page corresponding to 00 is splitted, the assignment shown in Fig.4(c) results. This page splitting method creates new two directory nodes, each of which contains a pointer to the new disk pages. The pointer in the old directory node is replaced by the height h of the subtree created by new directory nodes. The value h is used to calculate $H_{1,i+k}$. These internal nodes are deleted whenever a complete tree is created. This method also guarantee 50

4.2. Memory Overflow Procedure in Structured Variable Length Hashing Scheme

Another characteristics of this file is high main memory utilization factor caused by sending records to the next page. Similar method is used in B*tree[3], and Dense multiway tree[4]. This is realized by changing split keys of the father node of the tree and sending records to brother nodes. There is a trade-off problem between the improved disk access cost by high utilization factor and the memory access cost to send records to brother nodes. In memory environment this method holds high storage utilization since access cost of records is cheap in main memory. In our file organization, memory utilization is improved by the following procedure.

- (a) If a page overflow occurred in main memory, an overflowed record is sent to the next page and then it is examined whether there is free space to store that record.

- (b) If there is free space, store that record and update directory. In this case, we don't use split value but use pointer which points the data area of the next page. This is shown in Fig.5. In this

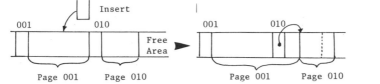

Fig.5. Memory Overflow Procedure

example, area 010 is used to store the overflow record of the page 001. Directory of the location 010 is updated by this overflow.

- (c) If there is no space for the record and the page is within the search range for free space, then, another overflow occurs and return (a). In above example, record overflow at page 001 affects the page 011 when the page 010 is full. In this case, if the search range is 1, there is no space in main memory and goto (d).

- (d) If there is no space within the search range, then overflow records are stored into disk pages. if the disk page is also full, then page split occurs.

This method does not require the change of the split value since variable size pages are used. This method has the following features.

1. If we store split values instead of pointers, we cannot delete that directory node when the node is internal node of the complete tree. This requires total reorganization of a subtree in which the split value is stored in the root.

2. When the page size of a page become quite large and overflow occur at that page, we cannot store the page since disk page size is fixed. In this case, the directory will splitted until the page size of memory page matches the page size of disk storage.

3. Directory area is small since no split values are used.

4. Leaf nodes of the directory tree of this file are used to point disk pages. Thus, this area can be also used to store the pointers of memory pages when no disk overflow occurs.

4.3. Directory Structures and Data Manipulation Algorithms

At each location corresponding to a complete subtree, following information is stored.

1. If it is a leaf node of the directory, pointer to the corresponding data pages in the main memory and disks is stored.

2. If it is a root of a complete subtree, height i of the subtree i is stored to determine the part of sequence of the hashed value.

3. The number n of subtrees connected to the leaf nodes of the tree is used to delete internal nodes. As previously discussed in Fig.3(d) and (e), the subtree of height i is replaced by a subtree of height i+1 by using this information.

Initialization step of the structured variable length hashing scheme is as follows.

- (a) We first determine the maximum size directory which can be stored in the main memory.

- (b) Determine the height t of the initial hash tree. Datum d_i is stored in the page corresponding to the leaf node labeled by $H_{1,t}(d_i)$

- (c) By the addition or deletion of data the structure will change. As the directory increases, data are swapped out to disks. When some incomplete subtrees are replaced by a complete subtree, internal directory nodes are deleted and used to store data.

Next, we present data manipulation algorithm briefly.

- **Retrieval**

  - (a) Hash the key of the record and get initial hash key $H_{1,t}$.
  - (b) Access the directory and get the directory value.
  - (c) If the directory is not a leaf, then $H = H_{1,t} + H_{t+1,t+k}$. Goto (b).
  - (d) If it is a leaf node, access data page using the pointer of the directory.
  - (e) Search within the page.

- **Insert**

  - (a) Access the page as shown in Retrieval step. Set the search range R
  - (b) If there is some empty space, store the record and return.
  - (c) If overflow occurs, Send the record which has the maximum(or minimum) hash value in the page to the adjacent page in main memory.
  - (d) Update directory pointer and decrement R.
  - (e) If R is zero, disk overflow occur. Store the data in the disk page if there is free space in the disk page.
  - (f) If disk overflows, then split the directory and data.
  - (g) If complete subtree is composed by page splitting, Increase the height of the subtree and internal directory area of the subtree is used for data area.

- **Deletion** Deletion is considered to be the inverse of Insertion. However, record deletion in main memory always causes disk access to keep high memory storage utilization factor.

The number of disk accesses for these operations are as follows.

- **Retrieval**: 0 to 1 disk accesses

- **Insertion**: 1 to 3 (Split of disk page) disk accesses

- **Deletion** : 1 to 3 (Page merging) disk accesses

We will discuss how this structure can handle some of the problems in Sections 2 and 3.

- (P1) Averaging access time: We assume that the hash function H is designed to distribute data uniformly so that the frequency of access of each page is almost balanced in the initial state.

Thus, the number of data items stored in main memory for each page is almost the same. If a page is splitted into two pages due to the increase of data, each page has nearly one half of the memory space. Although the number of data has increased, the total frequency to access the both pages is similar to the original frequence. This leads that each splitted page has the half of the frequence of access so that we need the half of the memory space for that page. If two pages are merged, the frequence of accesses into new page will doubled. We can calculate that the memory space is almost proportional to the access frequency and averaging the access time is achieved.

- (P2 and P3) In order to avoid a difficult problem caused by using different data structures used in main memory and disks, we use the separate backup process. In this system, data contained in main memory and disk data which is accessed from directory are disjoint. Thus, both can use different structures appropriate for both devices if required. We use (logically) different disk to backup the main memory which is a full copy of the memory.

- (P3) A binary tree structure is used in the main memory.

- (P4) If there is an empty space in a directory it is used to store data. If there is a page which is smaller than the main memory space which is assigned to the page, we have some space which can be used as an overflow area of other pages.

- (P5) As in the main memory location of directory is fixed, we can load data from backup disk incrementally. If some data is required to be accessed for the first time, the part of the directory corresponding to the data is loaded into the main memory. We do not need to load the whole data into the main memory when the system starts.

- (P6) Data transmission is required when page splitting or page merging occurs. In this structure the data transmission is not large, since directory structure is small so that only overflow records are stored into the disk storages.

- (P7) Backup is realized by different backup process to handle file organization and backup problems separately.

5. PERFORMANCE COMPARISONS

In this section, we evaluate typical file organizations including our method under the assumption that main memory can contain at least half of the entire file. The reason for this assumption is that we omit the discussion on the cases when the index cannot be stored in the main memory. As shown in Section 3, directory structure is fixed on main memory. In tree structure, internal nodes near the root node should be contained in main memory, since these nodes are frequently accessed. This may not be suitable for entire sequential scan, but covers all direct accesses and most of range queries. Our performance evaluation method is generally based on the static evaluation[5] except this assumption. The reason for this is that random replacement algorithms are not suitable for main memory so that unnecessary disk accesses are required. Following two aspects are taking into account in the performance evaluations.

1. Access cost of the file.

2. The amount of storage area required when there is a change on the data distribution.

The following symbols are used.

- Page size of disk = P,

- Average record size = L,

- Key length = K,

- Disk I/O time = DIO, (about 1000 times slower than memory, containing elapsed time for disk accesses)

- Time to compare = COMP,

- Time to Hash = HASH,

- Pointer length = 4,

5.1 B-tree, B+tree,

First, we analyze access cost of B+trees. In B-trees, 69% of the node is filled on the average. Thus, the average number of edges A is,

$$A = \frac{0.69 * P}{K + 4}$$

The number of leaf pages will be about

$$D = \frac{N * L}{0.69 * P}$$

data pages. So, the height of a B+tree index is,

$$H = \frac{logD}{logA}$$

The total number of pages is approximately as follows.

$$P_{total} = \frac{D * A}{A - 1}$$

The number of internal directory pages is

$$P_{internal} = P_{total} - D$$

Since directory is fixed in main memory, probability when required data is not in main memory is given below.

$$P_{fault} = 1 - \frac{M - P_{internal}}{P_{total} - P_{internal}}$$

The total cost of a B+tree access is as follows.

$$Cost_{b+tree} = DIO * P_{fault} + log_2 N * COMP$$

Next we consider conventional B-trees. In this section, dynamic behavior of files is not considered so that binary trees(including AVL trees) are considered to be special cases of B-trees when edges of the node=2. B-trees are usually higher than B+trees since internal nodes contain data and thus they have less pointers. The average number of edges of a B-tree is as follows.

$$A = \frac{0.69 * P}{L + 4}$$

Consequently

$$P_{total} = \frac{N * (L + 4)}{0.69 * P}$$

pages are required to contain N records. Thus, the height of a index tree is,

$$H = \frac{logP_{total}}{logA}$$

The total access cost of a B-tree access is as follows.

$$Cost_{b-tree} = DIO * (1 - \frac{M}{P_{total}}) + H * COMP$$

5.2. Dynamic Hashing Schemes

Here, we cover following two types of dynamic hashing schemes. To simplify the evaluation we consider only the best and the worst cases.

1. **Table Structured Dynamic Hashing Schemes** This method realizes the table(array) for hash space and grows the table when overflow occurred. Data pages are pointed from the table. This method is considered to be good when data distribution is random. Data pages are handled by page splitting. Thus the total number of pages to store N records is,

$$P_{data} = \frac{N * L}{0.69 * P}$$

In the best case, The number of directory pages to point these pages is,

$$P_{directory} = \frac{4 * 2^{log_2 P_{data}}}{P}$$

Then, the total number of pages is,

$$P_{total} = P_{data} + P_{directory}$$

Access cost with the search cost within the page is given as shown below.

$$Cost_{ehash} = HASH + DIO * (1 - \frac{M - P_{directory}}{P_{data}}) + log_2 \frac{0.69 * P}{L} * COMP$$

In the worst case, the main memory is occupied only by the directory. Then,

$$Cost_{ehash} = HASH + DIO + log_2 \frac{0.69 * P}{L} * COMP$$

2. **Tree structured Dynamic Hashing Schemes** This method construct a tree(or trie) structure by hash values. Although the storage area to store directory tree becomes large as data increases, this method gives good storage utilization factor when data distribution is not random. The number of pages required to store a tree structure index is,

$$P_{dir} = \frac{(P_{data} - 1) * (8 + 4)}{P}$$

In the best case, the height of the directory tree is given as follows.

$$H = log_2(P_{data} - 1)$$

Thus, the total cost to access a data is,

$$Cost_{trie} = HASH + DIO * (1 - \frac{M - P_{directory}}{P_{data}}) + H * COMP + log_2 \frac{0.69 * P}{L} * COMP$$

In this method, access cost is proportional to the number of data. It becomes worse than disk access when the number of nodes to access one record is more than DIO and $H = P_{data}$.

5.3. Structured Variable Hashing Scheme

The access cost of this file is calculated in a similar method to the above examples. We assume U_{til} be the storage utilization factor obtained form the overflow handling method of section 4.2[LEUN84]. In this file,

$$D_p = N * L - \frac{M * (P - 4) * U_{til}}{0.69 * P - 4 * U_{til}}$$

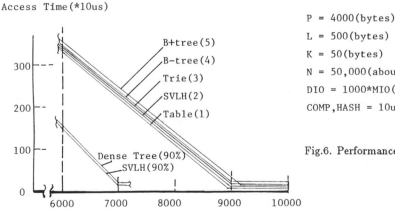

Access Time (*10us)

300

200

100

0

6000   7000   8000   9000   10000

No. of Pages of Main Memory

B+tree(5)
B-tree(4)
Trie(3)
SVLH(2)
Table(1)

Dense Tree(90%)
SVLH(90%)

P = 4000(bytes)
L = 500(bytes)
K = 50(bytes)
N = 50,000(about 25 megabytes)
DIO = 1000*MIO(about 10ms)
COMP,HASH = 10us

Fig.6. Performance Comparisons

pages are stored in disk. The probability to read disk drive is

$$P_{fault} = 1 - \frac{M}{M + D_P}$$

Thus, the access cost is determined as follows.

$$\begin{aligned}
Cost_{svlh} = \ & HASH + D * (HASH + COMP) + DIO * P_{fault} \\
& + (1 - P_{fault}) * log_2(U_{til} * P) * COMP + P_{fault} * log_2(0.69 * P) * COMP
\end{aligned}$$

Fig.6 is an example of the above cost values. The following two results are derived. (This also includes example of the Dense tree)

1. Storage utilization is the most important factor to the average access cost. The effect of the difference of directory structure is small but this becomes large as the DIO decreases.

2. B+tree is not always the best method.

Another cost evaluation is the access cost when the data distribution is changed. This is important when hash-based files are used. Here we only present Fig.7. which shows the amount of required directory structure when the hash functions are changed. This is easily obtained from the shape of the directory tree. This example shows that our method can avoid this problem by using the composite structure.

6. CONCLUSIONS

In this paper we presented some problems of conventional file organizations when large capacity main memory is available. A new file organization called a structured variable length hashing scheme which is based on dynamic hashing scheme is introduced. Major characteristics of this file are as follows.

1. Composite directory structure can reduce the directory area.

2. Overflow record is send to next page in a different manner from B*trees and Dense-trees. This method is suitable for (1) and main memory realization.

One of the problems of this file is that it is not order preserving. Following methods are possible to overcome this problem.

1. Use of Clusters: Instead of hashing all data, first data are classified by the starting character. Then for each subset the hash file structures are created. We only sort data in each subset.

2. Use of quasi order preserving hash functions: To reduce the number of pages to be sorted, we can use several bits preserving the ordering from entire hash function. These bits are not required to keep its order in hashed key. These are used to reduce the number of the pages to be accessed. If there is one bit which keeps key ordering in the hashed key, only half of the entire pages are searched in some queries.

Except the case when there are frequent requirements for sequential access, we believe that this file organization is suitable for large capacity main memory. Another problem is that we can determine the value of search range when memory page overflow occur. In this case, we can calculate this value dynamically to the data distribution. Some algorithms to calculate this range is required.

Height of directory tree

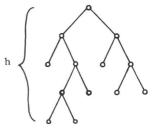

	h	3	4	5	6	7
Tree		15	15	15	15	15
Table		2^3	2^4	2^5	2^6	2^7
SVLH		8	10–12	10–12	12	15

(Number of leaf nodes = 8)

Fig.7. The Number of Nodes of Directory Structure

Acknowledgments This work started when the authors belonged to Kyoto University. We would like to acknowledge the valuable suggestions of Prof. S. Yajima. We are also grateful to Dr. S. Uemura of Electrotechnical Lab. for his helpful discussions and comments.

REFERENCES

1. Arnow,D. and Tenenbaum,A.M. "Emprical Comparison of B-trees, Compact B-trees and Multi-way Trees", SIGMOD,1984.

2. Burkhard,W.A. "Interporation - Based Index Maintenance", 3rd PODS,1983

3. Comer,D. "The Ubiquitous B-trees",ACM Comput. Surv. Vol.11, No.3, June 1979.

4. Culik,K,II. Ottman,Th. and Wood,D. "Dense Multiway Trees" ACM TODS, Vol.6., No.3,September, 1981.

5. Dewitt,D., Katz,R. Olken,F. Shapiro,L. Stonebraker,R. and Wood,D. "Implementation Techniques for Main Memory Database Systems", SIGMOD '84. June, 1984.

6. Fagin,R. Nievergelt,J. Pippenger,N. and Strong,H.R. "Extendible Hashing - A Fast Access Method for Dynamic Files", ACM TODS, Vol.4, No.3, September,1979.

7. Ghosh,S.P. Kambayashi,Y. and Lipski,W.Jr. (Eds.) "Data Base File Organization: Theory and Applications of the Consecutive Retrieval Property", Academic Press, July, 1983.

8. Larson,P.A. "Dynamic Hashing" BIT, Vol.18, 1978.

9. Larson,P.A. "Linear Hashing with Partial Expansions", 6th VLDB, 1980

10. Leung,C.H,C. "Approximate storage utilisation of B-trees: A simple derivation and generalisations", Inf. Pro. Letters, Vol 19, No 2, 1984.

11. Litwin,W. "Linear Hashing: A New Tool for File and Table Addressing", 6th VLDB, 1980.

12. Litwin.W. "Trie Hashing", SIGMOD 1981.

13. Lomet,D.B. "Digital B-trees", 7th VLDB, September, 1981.

14. Lomet,D.B. "A High Performance Universal Key Associative Access Method", SIGMOD 1983.

15. Mullin,J.K. "Unified Dynamic Hashing", 10th VLDB, August, 1984

16. Orenstein,J.A. "A Dynamic Hash File for Random and Sequential Accessing", 9th VLDB. 1983.

17. Scholl,M. "New File Organizations Based on Dynamic Hashing", ACM TODS, Vol.16, No.1, March, 1981.

18. Tamminen,M. "Order Preserving Extendible Hashing and Bucket Tries", BIT, Vol.21, May, 1981.

TRIE HASHING : FURTHER PROPERTIES AND PERFORMANCE

Witold Litwin

INRIA, 78150 Le Chesnay, France

ABSTRACT

Trie hashing is one of the fastest access methods for dynamic and ordered files. We show new properties of this method and, in particular, its performance shown by simulations. The results confirm earlier expectations. Furthermore, they show new aspects of the file behavior, especially under sorted insertions. They also indicate that some refinements that looked promising are not that worthy.

1. INTRODUCTION

In the past few years, a new class of file access methods has emerged, now usually called _dynamic_ or _virtual_ or _extendible hashing_ /LAR78/, /LIT78/, /FAG79/. About forty various algorithms have appeared so far, only some could be included into our references. Like the classical hashing, most of them, create unordered files. Some however are order preserving, like the _trie hashing_ (TH) /LIT81/, the _interpolation based_ hashing (IH) /BUR83/, and the _grid file_ (GF) /NIV84/. TH was designed for single key files, IH and GF – for the multikey ones.

TH showed that one may create large ordered dynamic files where any key search needs at most one disk access. This property rendered TH one of the fastest access methods. Properties and extensions of the method were further studied in /JON81/, /FLA83/, /TOR83/, /KRI84/, /LIT84/ and others.

The description of the method in /LIT81/ contained mainly the basic algorithm with some refinements and the performance analysis. The analysis concerned almost only random insertions and was rather short. Main factors under consideration were the trie size and the load factor. These factors dictated the maximal file size for which the one disk access per search performance could hold.

Below we first present the principles of the method. Afterwards, we discuss the performance as observed through simulations. Finally, we compare TH to other algorithms for dynamic files.

The simulations confirm the initial expectations. They further show that for unexpected sorted insertions, the load factor attains 60-70 %. It is thus significantly higher than the corresponding 50 % of a B-Tree. Finally, the simulations show some subtle properties of the method.

Section 2 discusses the principles of the method. Section 3 presents the performance study. Section 4 compares TH to other methods. Section 5 concludes the paper.

2. TRIE HASHING
2.1 File structure

For TH, a _file_ is a set of records identified by primary keys. Keys consist of digits of some alphabet. The smallest digit of the alphabet called _space_ will be denoted '_' , while ':' will denote the largest digit. The set of all possible keys is called _key space_ and is assumed totally ordered. The part of the record outside the key is irrelevant to access computation. Records are stored in _buckets_ that are units of transfer between the file and the core. Buckets are assumed to be of the same length and capacity. The _capacity_, noted b below, is the ratio of bucket size to record size. One rather chooses larger b values especially for ordered files. $5 \leq b \leq 200$ is about the range of typical choices.

Each bucket has an _address_. Successive addresses are $0, 1, 2, ...$. The _access method_ finds from the key the address where the record should be. It then allows to perform operations on the file, which are basically key search, insertion and deletion. Ordered files also make it possible to process range queries efficiently. Fig 1 shows TH file of 31 most used English words /KNU73/. The file is addressed through the trie in fig 1.c. The trie is basically an M-ary (digital) tree whose nodes correspond to digits. TH rather uses a binary representation of tries, as in the figure. The trie in fig 1.c was dynamically generated by the insertions of words in fig 1.a. The way in which the access method uses the trie will be shown later. Deeper discussion may be found in /LIT81b/ and /LIT85/.

The trie represented as shown consists usually of _internal_ and _external nodes_. It may in particular consist of only one external node, being then called _null trie_. An internal node contains a pair of attributes called _digit value_ and _digit number_, usually noted below (d, i). An external node, called _leaf_, contains a bucket address or a value called _nil_. The latter is called _nil leaf_ or nil node. Nil value indicates that no bucket corresponds to the leaf.

The initial empty file consists of bucket 0 and of the null trie consisting only of leaf 0. Then, insertions dynamically expand the file and the trie and deletions contract them. The corresponding algorithms for operations on the file are as follows.

2.2 Operations on the file
2.2.1 Key search

Any search starts at the root node. It then follows a path determined by a comparison of key digits to trie nodes. As the result of each comparison either the node under the left or the one under the right of the visited node is chosen as the next node to visit. The traversal stops when a leaf is reached. The key address is then the one in the leaf. The traversal principles are as follows :

A1 : key search

Let c be the searched key. Let $(c)_i$ denote the prefix $c_0 c_1 ... c_i$ of c. Let C', C'' be string variables, $C' = C'' = ':'$ initially. Let c'_i denote digit number i of C' ; $i = 0, 1, ...$. Let (d'', i'') be the visited node if the node is internal ; otherwise let A' denote the visited leaf value. Finally, let A denote the address to be calculated for c.

1. While the visited node is an internal node do :

   if $i'' = 0$ then $C'' \leftarrow d''$ else $C'' \leftarrow c'_0 ... c'_i{}''^{-1} d''$;

   if $(c)_i{}'' \leq C''$ then (i) $C' \leftarrow C''$ and (ii) traverse the left edge under (d'', i''), else traverse the right edge.

(a) the, of, and, to, a, in, that, is, i, it, for, as, with, has, his, he, be, not, by, but, have, you, which, are, on, or, her, had, at, from, this.

(b)

are and a	to this the that	or on of not	it is in	by but be	you with which was	i	her he have had	his	at as	from for
0	1	2	3	4	5	6	7	8	9	10

(c)

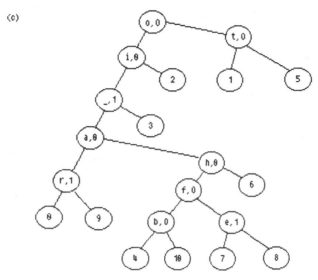

Fig 1 Example file
(a) insertions with collisions underlined
(b) buckets
(c) the trie

2. $A \leftarrow A'$. Return.

For instance, $c = \text{'s'}$ leads through the example trie to $A = 1$. The keys $c = \text{'he'}$ or $c = \text{'gun'}$ leads to $A = 7$. The corresponding final C' is 'he' in both cases. If (A1) ends up with $A = \text{nil}$, then c cannot be in the file as there is no corresponding bucket.

2.2.2 Range queries

Let the digits of the final value of C' be $c'_0 \ldots c'_k$. Let $k'+1$ be the limit on the key length, usually imposed by the file management system. Let C be $(k'+1)$-digit value as follows :

$$C = c'_0 \ldots c'_k \qquad\qquad \text{if} \quad k = k'$$
$$C = c'_0 \ldots c'_k ::\ldots: \qquad \text{if} \quad k < k'$$

C will be called maximal key for the bucket A. The reason is that (A1) implies that C is the largest key value that the bucket A may contain. It may further be observed that when leaves are visited according to the *inorder*, then the successive corresponding values of C are strictly in ascending order. We recall that the inorder traversal of a binary tree is recursively defined as : (1) Traverse the left subtree inorder (2) Process the root node, (3) Traverse the right subtree inorder. In our example, the successive maximal keys and buckets are as follows :

79

$$\mathcal{C} = \text{'ar::::...:'} \qquad \text{for bucket 0}$$
$$= \text{'a::::...:'} \qquad \text{for bucket 9}$$
$$= \text{'b::::...:'} \qquad \text{for bucket 4}$$
$$= \text{'f::::...:'} \qquad \text{for bucket 10}$$
$$= \text{'he::::...:'} \qquad \text{for bucket 7}$$
$$\dots$$
$$= \text{'t::::::...:'} \qquad \text{for bucket 1}$$
$$= \text{':::::::...:'} \qquad \text{for bucket 5}$$

Thus the TH file is ordered, i.e. all keys may be examined in ascending order through only one access to each bucket. <u>Range queries</u> such as "find all records whose keys c are within some interval $[c', c'']$ " $: c' < c''$; may thus be efficiently processed. They need indeed only to access the buckets pointed by the leaf $A(c')$ and the leaves that follow in inorder up to the leaf $A(c'')$. The corresponding sequence of addresses results also from preorder and postorder traversals, as all these traversals visit the leaves in the same order.

Simple algorithms for various traversals are in /KNU73/ (Vol. 1, pp 313-33) and, in particular in Pascal, in /TRE84/ (pp. 346-350). If keys should be examined in descending order, one may use the *converse inorder* (or converse preorder etc.) We recall that such orders result from the interchange of the words "left" and "right" in the corresponding definitions. On the other hand, one may put the corresponding forward and/or backward chains to the buckets. In this case, it suffices to find through (A1) one of the bounds and then to follow the chain until the other bound is attained.

2.2.3 Insertions

If bucket A is not full, then the insertion simply adds the record to the bucket. Otherwise, an empty bucket is appended to the end of the file and the trie is modified. The modification decreases the maximal key $C(A)$ and assigns to the new bucket the keys that were mapped to A. but now exceed $C(A)$. The set of records that were mapped to the bucket A is then split into two parts whose sizes are usually almost equal. One part contains the records whose keys are smaller than any key in the other part. The split is therefore order preserving. The records with smaller keys are reinserted into bucket A. The other records migrate to the new bucket. Further insertions that would enter bucket A are henceforward directed either to bucket A or to the new one, depending on their keys with respect to new $C(A)$.

The splitting algorithm is presented below. Its principle is to insert between leaf A and its parent a new internal node with right child pointing to the new bucket. The value of $C(A)$ decreases because of the contribution of the new node. In some cases, it is however necessary to insert more than one internal node. The algorithm generates then also some nil leaves.

A2 : splitting

Let A be the address of the bucket that overflowed and let C be the corresponding maximal key, both determined through (A1). Let N be the last address in the file. Let p denote the position of a key within a sequence of keys. Let S be the ordered sequence of $b + 1$ keys to be split (where the new key is thus not necessarily the last one). Let <u>middle key</u> denoted c'' be the key in S with the position $p'' = INT (b/2 + 1)$. Let c' be a key in S called <u>split key,</u> whose position will be noted p'. We assume that basically $p' = p''$ that is $c' = c''$. but provided $p' \le b$, any other choice is admissible as well. Finally, let c''' be the last key in S.

1. Find the smallest i for which $(c')_i < (c'')_i$

2. If $i > 0$, then go to step (4).

3. Set $N \leftarrow N + 1$. Replace leaf A with node (c'_i, i). Attach leaf A again, as the left child of (c'_i, i) and attach leaf N as the right child of (c'_i, i). Append bucket N and move to it all keys c in S such that $(c)_i > (c')_i$. Return.

4. Cut from $(c')_i$ the largest $(c')_I$, $I < i$, such that $(c')_I = (C)_I$. If $I = i - 1$, then go to step (3).

5. For each $j = I + 1, \ldots, i - 1$ do :
 - replace leaf A with node (c'_j, j).
 - attach leaf A again as the left child of (c'_j, j) and attach nil leaf as the right child of (c'_j, j).

6. Go to step (3).

   For example, let 'hat' be the key to be inserted. It would lead to the split of bucket 7. Leaf 7 would then be replaced with node (a, 1) with two children : leaf 7 itself on the left and leaf 11 on the right. The order will remain preserved, as all keys pointed henceforth by leaf 7 will be smaller than those of leaf 11.

   The steps (1) to (4) correspond to the case where only one new internal node is created. Step (5) creates several internal nodes and nil leaves. Nil leaf is then replaced with the actual address $N + 1$ when an insertion leads to it. The corresponding bucket is appended and the key inserted. An example dealing with nil leaves is in /LIT81b/.

   It follows from Step (3) that for any split all records with keys $c \leq c'$ stay in bucket A. If, as is usual, $p' < b$, then this may also happen to some records with keys $c > c'$, namely for which $(c)_i = (c')_i$ (thus it cannot happen to the record whose key is the last in S). The number of such records is random between zero and $b - p'$. In this sense, TH split is usually partially random. For $p' = p''$, the split is half random.

   The partial randomness is a property presently unique to TH. It places the method somewhere between B-Trees where splits are totally deterministic and other algorithms for dynamic hashing, where splits are fully random. The property results from the idea in TH "to chop off some aspects of the key and to use this partial information as a basis for searching". This idea is in fact the general one in hashing (see /KNU73/), and that is why we referred to hashing while terming our method. A general consequence is that many of TH characteristics are similar to those of other algorithms for dynamic hashing.

   The partial randomness works in favor of bucket A. This is in the sense that after the split bucket A contains at least p' keys, while the new bucket has at most $(b + 1) - p'$ keys. This phenomenon has consequences on the load factor, discussed in Section 3.

2.2.4 Deletions

   A deletion may simply remove a record. Furthermore, it may test whether the leaf pointing to the bucket has a sibling. This would mean that the other child of the leaf's parent is a leaf as well. The deletion may then merge sibling buckets that contain together at most b records. The merge puts the records of both siblings into the left one, say bucket A'. Next, the records in bucket N (the last one in the file) are moved to the right sibling, say bucket A'', freeing the space at the file end. The leaf N is then updated to the value A''. One way to find this leaf is to recompute (A1) using any key in bucket N. Finally, the parent node and leaf A'' are deleted from the trie. If it happens then that leaf A' gains a sibling which is a nil node, then the merge is repeated, until a leaf that is not nil is attained.

(a)

UP
DU ┊ DN
LP

UP – upper pointer
LP – lower pointer
DU – digit value
DN – digit number

(b)

-4	2	3	-5	5	6	-7	8	9	10
0,0	i,0	_,1	a,0	t,0	h,0	f,0	e,1	r,1	b,0
-1	-2	-3	-8	1	-6	-9	7	0	4

Fig 2 Standard representation
(a) cell structure
(b) example trie

The test of the number of records in the sibling bucket may be done at each deletion. However, if the bucket that underwent the deletion is yet quite full, the chance that the former is empty enough is low. Probably the best practical trade-off between the deletion cost and the value of the load factor is to wait until the bucket undergoing the deletion is about half empty.

If a bucket without the sibling becomes empty, the corresponding leaf is rendered nil. The records of the last bucket in the file are then transferred into the bucket and leaf N is updated as above.

2.3 Trie implementation

The trie may be represented in storage in many manners. Two types of such representations were called respectively standard and sequential representations /LIT81/. Both differ from the "classical" representations in /KNU74/, because of the dynamic nature of TH. Below we will discuss the standard representation only. In this representation, the trie is a linked list whose elements are called cells. Cells are created in the order of splitting, i. e. splits append cells to the list end. Each cell represents one internal node in the trie and the connections to its two sons.

The general structure of a cell is shown in fig 2a. If the left (right) child of the represented node is a leaf, then the corresponding value of LP (UP) is an external pointer to the bucket. The pointer is then stored as a positive value. Otherwise, it is an internal pointer to the cell representing the child, stored as a negative value. Fig 2b shows the standard representation of the example trie. To reformulate the above discussed algorithms with respect to the standard representation is a trivial task (see /LIT81/).

3. PERFORMANCE

The behavior of the method was studied mainly through simulations, under the assumption of a sequence of insertions. Key values in the sequence were assumed to be either (uniformly) random or sorted in ascending order. Sorted insertions were assumed either expected or unexpected. b values were 5, 10, 20, 100. The characteristics under study were as follows :

- the load factor a : $a = x/(b * N)$, where x is the number of keys in the file.
- the evolution of trie size and in particular, of the number of nil nodes.

The latter subject was studied in order to determine the file size for which the trie may fit in the core. This is the normal case the method was designed for.

3.1 Random insertions
3.1.1 Load factor
3.1.1.1 Splitting using the middle key

We first consider the case of $p' = p''$. Fig 3 presents $a(N)$ in $log_2 N$ scale, for different bucket capacities b and $c' = c''$. This presentation gives better picture of a given various b, although it privileges smaller and so less practically important values of N. The following properties appear :

- curves display the existence of a <u>transient state</u> for small N. Then, it becomes more regular, displaying therefore convergence to some <u>stable state.</u> In this state the curves are basically periodical in $log_2 N$. This pattern agrees with the behavior already observed for other methods using dynamic hashing /FAG79/, /LIT79/, /YAO80/, /REG83/. However, TH curves show also a kind of random noise consisting of harmonics of lower frequencies. This seems due to the fact that the randomness of TH split is only partial.

- the average values of a over one or more periods are almost independent of b and about 70 % . This agrees with the expectations in /LIT81/. Table 1 shows the average load, noted a calculated, because of irregularities, over two last periods (75 % of insertions). The table shows in particular that the average load is slightly better for smaller buckets. This pattern was also observed for other algorithms for dynamic hashing, as long as splits are uncontrolled /LIT80/.

- in contrast, oscillation amplitude increases with b. This phenomenon was also observed for the other methods. It results from the tendency of larger buckets to become full more simultaneously.

3.1.1.2 Splitting under the middle key

Let f be the number of keys that stay in bucket A after the split. Partial randomness leads to $t > p'$ on the average. In particular, $p' = p''$ leads to uneven splitting, i.e. to $t/(b+1) > 0.5$ on the average. Uneven splits usually lower a. It may therefore be advantageous to choose the split key under the middle one, that is, to choose $p' < p''$.

Table 1   Random insertions.

b	5	10	20	50	100
a	0.72	0.70	0.69	0.68	0.68
u	0.55	0.52	0.54	0.55	0.57
a_o	0.72	0.70	0.70	0.69	0.69
p'_o	3	5	9	22	45
r'_o	0.5	0.46	0.43	0.43	0.44
$n\%$	0.5	0.5	0.4	0.4	0.4

Table 1 shows the results of simulations corresponding to these expectations. First, it shows the average value of $t/(b+1)$, denoted u, for $p' = p''$. Next, it shows the optimal a, denoted a_o and the corresponding optimal p'_o. Finally, it shows the optimal relative position r'_o : $r'_o = p'_o/(b+1)$. The results confirm on one hand the expectations with respect to u. In contrast, they show that a improves only for larger b, $(b \geq 20)$. Finally, they show that the gain is, after all negligible, $(\underline{a}_0 - \underline{a} \leq 1\%)$. In fact, the curves of $\underline{a}$ for $p' = p'_o$, not shown here, are visually identical to those of Fig 3 for $p' = p''$.

In this analysis, p' was the same for all the splits. In /LIT81/ we also spoke about refinements where, in order to improve $\underline{a}$, p' changes from split to split. Although detailed analysis of these refinements is not done yet, the above results render their practical interest doubtful.

3.1.2 Nil nodes

Finally, Table 1 shows the average percentage of nil nodes denoted $\underline{n\%}$. As it was expected, for random insertions (and alphabetical digits) this percentage is effectively negligible $(\underline{n\%} \leq 0.5\%)$.

3.1.3 Trie and file sizes

The value of $\underline{n\%}$ shows that the trie expands at the rate of practically one cell per split. The number of cells in its representation is thus practically N. Since $x \approx 0.76N$, if I_n is the cell size in bytes and I_t is the trie representation size, then :

$$I_t \approx I_n \, N \approx 1.43 \, I_n \quad x/b$$

For $b = 100$ for instance, quite frequent for B-Trees, and for $I_n = 6$ that should be typical for the standard representation, the file size is :

$$x = 11.6 \, I_t$$

Therefore, a 10 Kbyte buffer for instance suffices for a file growing to about 100 000 records. A 64 Kbyte buffer suffices for more than 750 000 records. The trie may thus usually enter the core, even of a personal computer.

3.1.4 Access performance

Since there is no overflow records, the trie in core provides the successful search cost of only one disk access. An unsuccessful search costs at most one access, since nil nodes avoid accesses. Although this property is theoretically interesting, the simulations show that in practice it has almost no effect. $\underline{n\%}$ values imply indeed the average number of accesses per unsuccessful search not less than 0.995.

An insertion needs : 2 accesses when there is no split and the bucket exists, 1 access when it meets a nil node and 3 accesses when a collision occurs. Since about only 1 insertion per $0.7 b$ collides, the average insertion cost is about 2 accesses.

3.2 Unexpected order

In this case, keys come sorted in ascending order, while the splitting strategy is the one used for the random insertions. Batched updates, joins etc., lead to this case. The unexpected order usually deteriorates the load and/or access performance. In particular, a B-Tree load usually decreases to 50 %.

3.2.1 Load factor

Fig 4 shows $\underline{a}$ for $c' = c''$. Table 2a displays $\underline{a}$ and other corresponding factors. As one may see, there is almost no load deterioration for $b \leq 20$. For larger buckets, the deterioration is quite small, only about 5-10 %. Finally, the curves are generally more flat.

TH file keeps thus its 70 % load for smaller buckets and provides slightly lower, but less oscillating load of 60-65 % for larger ones. This surprisingly good performance is due to the partial randomness. For sorted

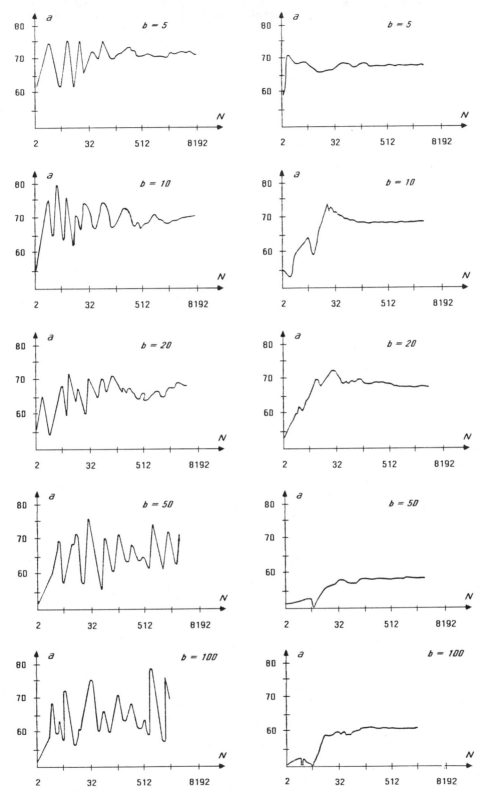

Fig 3 Load factor for random insertions   Fig 4 Load factor for unexpected order

Table 2 Sorted insertions :
(a) - unexpected order
(b) - expected order

b	5	10	20	50	100	
a	0.69	0.70	0.68	0.60	0.62	(a)
$n\%$	1.5	0.7	0.4	0.6	1.2	
a	0.79	0.73	0.63	0.58	0.63	(b)
$n\%$	1.7	2.3	2.0	1.6	4.1	

insertions, it is indeed more likely that keys $c > c'$ start with the same digits. The effect is a higher u . The new bucket usually ends up underloaded. But then new insertions all go to this bucket, until it is full and splits in turn etc. The average load is equal to the average t / b. i. e is about u . For $c' = c"$. this ratio is over 50 %. So is the average load.

On the other hand, if insertions are in descending order, then the partial randomness plays against the good load. However, this case is infrequent. It suffices then to decrease p' . This may be done automatically, through a mechanism similar to the one of the load control /LIT79/ and /LIT80/.

The good load for unexpected order means that with TH, one may use the same splitting strategy for both insertions and the initial loading.

3.2.2 Other factors

Table 2a shows the average ratio of nil nodes $n\%$ Although usually slightly larger than for random insertions, $n\%$ remains negligible, (under 1.5 %). The trie and file sizes, as well as access performance for sorted insertions are thus practically the same as for the random ones.

3.3 Expected order

In this case, since one knows in advance that keys are sorted, one may try to improve the load through higher p' . This is frequently the case of the initial loading of a B-Tree. One may even choose $p' = b$. obtaining then a compact B-Tree where $a = 100 \%$. This possibility reveals however only moderately useful for dynamic files. Just a few insertions decrease the load even under its usual value of 70 %. This is because almost any insertion generates a split since buckets are full, and many buckets become then 50 % full. Probably the best choice is to raise p' only to about $0.7 b$.

For TH, one may also intuitively expect that $p' = b$ leads for sorted insertions, to $a = 100 \%$. As for a compact B-Tree, splits indeed leave the buckets that overflowed completely full. Fig 5 and Table 2b show the corresponding simulation results.

As one may see, this expectation reveals erroneous. a is always quite far from 100 %. In fact, a values are close to those characterizing the

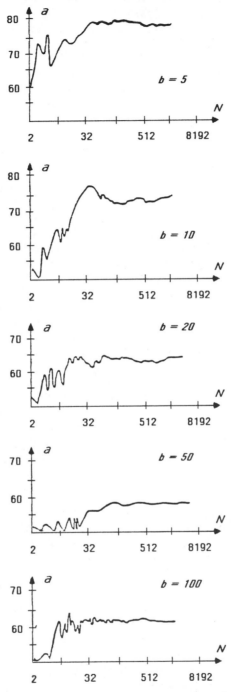

Fig 5 Load factor for expected order

other types of insertions. The usual strategy, i. e. $p' \approx p''$, is thus about equally good.

A closer analysis shows that this surprising property is due to a subtle influence of nil nodes. Apparently, these nodes do not influence a, since no bucket corresponds to them. In fact, for $p' = b$, they act as follows :

- the insertions, sorted or not, generate many nil nodes. The reason is that the $b + 1$ key in the sequence udergoing the split, usually shares many digits with its predecessor. For sorted insertions, a nil node is then usually rapidly selected by an insertion and the corresponding bucket is created. But afterwards, particularly for larger b, such a bucket receives all together less than b sorted insertions on the average. All further insertions go to other buckets. The bucket remains thus underloaded and renders the overall load under 100 %, despite the 100 % load of normally created buckets.

This behavior means in particular that the best load for expected sorted insertions should correspond to some p' such that $(b + 1) / 2 < p' < b$. This optimal p' remains to be studied.

Although, splits create in the discussed case more nil nodes, Table 2b shows that $n\$$ is again negligible (although for large b it reaches 4 %). This is due to the short life time of most of these nodes. The performance factors remain therefore almost the same as in the other cases.

3.4. Overall discussion

The simulations confirm the initial expectations relative to performance of TH. Furthermore, they show a unique performance stability. As long as the trie is in core, performance factors are indeed almost independent of b and of x. Next, they are very similar for both random and sorted insertions. In particular, the same splitting strategy may be used for both normal insertions and the initial loading.

The negligible average ratio of nil nodes $n\$$ suggests that the concept of nil leaves is not that useful. Instead one may simply generate actual empty buckets. TH algorithms then become even simpler, practically without damage to disk usage performance. Note however that we discuss here only the usual alphanumerical digits. Alphabets of smaller size may lead to more nil nodes.

4. COMPARISON TO OTHER METHODS

For the discussed sizes of files, access performance of TH are usually better than those of other currently known methods. This is true for both dynamic and static files. The search is usually 2-4 times faster than the one in a B-Tree and at least two times faster than in GF /NIV84/ (but GF provides multikey access as well). In contrast it may be only slightly faster on average than with IH /BUR83/ for random insertions. It may nevertheless be several times faster for a particular key and should be much faster if the file resulted from nonuniform and, in particular, sorted insertions.

For random insertions, the load factors provided by all the discussed methods are about the same. In contrast, for sorted insertions, TH load is 10-20 % better than the 50 % of a B-Tree. This is due to the flexibility of d resulting from the partial randomness. The poor load of a B-Tree is due to the lack of such flexibility because of deterministic splits. Similar inflexibility of predefined splits leads to even greater difference.

If the trie cannot fit in the core, we presently recommend other methods. However, a way exists to bound the search cost of TH to two accesses for files over 10^8 records /LIT85/. The corresponding variant is under study.

5. CONCLUSION

Work on the trie hashing and other methods that appeared in the meantime, brought to light new properties of the algorithm. Performance analysis confirmed most of earlier expectations. It also showed good behavior for sorted insertions, usually better than that of other known methods. On the other hand, it showed that some refinements do not improve performance sufficiently for being interesting in practice. Finally, some subtle aspects of the algorithm behavior appeared.

The work remains to be done on the analytical study of the performance factors. The results of /FLA83/ should be useful for this obviously complex task. One should also analyze the use of the bounded disorder. Finally, multikey extensions should be investigated.

ACKNOWLEDGMENTS

We would like to thank K. Vidyasankar for helpful comments.

REFERENCES

/BAY72/ Bayer, R., Mc. Creight, E. Organization and maintenance of large ordered indexes. Acta Informatica, 1, 3, (1972), 173–189.

/BAY77/ Bayer, R., Unterauer, K. Prefix B-Trees. ACM TODS, 2, 1,(Mar 1977), 11–26.

/BUR83/ Burkhard, W. Interpolation-Based Index Maintenance. PODS 83.ACM, (March 1983), 76–89.

/DAT86/ Date, C., J. An Introduction to Relational Database Systems. 4-th ed., Addison-Wesley, 1986, 639

/ELL83/ Ellis, C., S. Extendible Hashing for Concurrent Operation and Distributed Data. PODS 83. ACM, (March 1983), 106–116.

/FAG79/ Fagin, R., Nievergelt, J., Pippenger, N., Strong, H.R. Extendible hashing – a fast access method for dynamic files. ACM-TODS, 4, 3, (Sep 1979), 315–344.

/FLA83/ Ph. Flajolet : On the Performance Evaluation of Extendible Hashing and Trie Searching. Acta Informatica, 20, 345–369 (1983).

/KJE84/ Kjelberg, P., Zahle, T., U. Cascade hashing.VLDB-84, Singapore (Aug. 1984), 481–492.

/KNU73/ Knuth, D.E. : The art of computer programming. Vol.3. Addison-Wesley, 1973.

/KRI84/ Krishnamurty, R., Morgan S., P. Query Processing on Personal Computers – A Pragmatic Approach. VLDB-84, Singapore (Aug. 1984), 26–29.

/JON81/ de Jonge, W., Tanenbaum, A., s., van de Riet R. A Fast, Tree-based Access Method for Dynamic Files. Rapp IR-70, Vrije Univ. Amsterdam, (Jul 1981), 20.

/LAR78/ Larson, P., A. Dynamic hashing. BIT 18, (1978), 184–201.

/LAR82/ Larson, P., A. Performance Analysis of Linear Hashing with Partial Expansions. ACM TODS, 7, 4, (Dec 1982), (566–587).

/LAR82a/ Larson, P., A. A single file version of linear hashing with partial expansions. VLDB 82, ACM, (Sep 1982), 300–309.

/LIT78/ Litwin, W. Virtual hashing : a dynamically changing hashing. VLDB 78. ACM, (Sep 1978), 517–523.

/LIT80/ Litwin, W. Linear hashing : A new tool for files and tables addressing. VLDB 80, ACM, (Sep 1980), 212–223.

/LIT81/ Litwin, W. Trie hashing. SIGMOD 81. ACM, (May 1981), 19–29.

/LIT84/ Litwin, W. Data Access Methods and Structures to Enhance Performance. Database performance, State of the Art Report 12:4. Pergamon Infotech, 1984, 93–108.

/LIT85/ Litwin, W. Zegour, D. Multilevel trie hashing. Res. Rep. 1106851043, INRIA-Sesame, 36.

/LIT86/ Litwin, W., Lomet, D. Bounded Disorder Access Method. 2-nd Int. Conf. on Data Eng. IEEE, Los Angeles, (Feb. 1986).

/LOM81/ Lomet, D. Digital B-trees. VLDB 81. ACM, (Sep 1981), 333-344.

/LOM83/ Lomet, D. A high performance universal key associative access method. SIGMOD 83, ACM, (May 1983), 120-133.

/MAR79/ Martin, G. Spiral storage : Incrementally augmentable hash addressed storage. Theory of Computation Rep. 27, Univ. of Warwick, (March 1979).

/MUL84/ Mullin, J., K. Unified Dynamic Hashing. VLDB-84, Singapore (Aug. 1984), 473-480.

/NIE84/ Nievergelt, J., Hinterberger, H., Sevcik, K., C. The Grid File: An Adaptable, Symmetric Multikey File Structure. ACM TODS, (March 1984).

/ORE83/ Orenstein, J. A Dynamic Hash File for Random and Sequential Accessing. VLDB 83, (Nov 1983), 132-141.

/OTO84/ Otoo, E., J. A mapping Function for the Directory of a Multidimensional Extendible Hashing. VLDB-84, Singapore (Aug. 1984), 493-508.

/OUK83/ Ouksel, M. Scheuerman, P. Storage Mapping for Multidimensional Linear Dynamic Hashing. PODS 83. ACM, (March 1983), 90-105.

/RAM82/ Ramamohanarao K. and Lloyd J.K. Dynamic hashing schemes. Computer J., 25, 4, (1982), 478-485.

/RAM83/ Ramamohanarao, K., Lloyd, J., W., Thom, J., A. Partial-Match Retrieval Using Hashing and Descriptors. ACM TODS, 8, 4, (Dec 1983), 552-576.

/RAM84/ Ramamonohanarao, K., Sacks-Davis, R. Recursive Linear Hashing. ACM-TODS, 9, 3, (Sep. 1984).

/REG82/ Regnier, M. Linear hashing with groups of reorganization. An algorithm for files without history. In Sheuermann P. (ed) : Improving Database Usability and Responsiveness, Academic Press, (1982), 257-272.

/SCH81/ Scholl, M. New File Organization Based on Dynamic Hashing. ACM TODS, 6, 1, (March 1981), 194-211.

/TAM82/ Tamminen, M. Extendible hashing with overflow. Inf. Proc. Lett. 15, 5, 1982, 227-232.

/TOR83/ Torenvliet, L., Van Emde Boas, P. The Reconstructive and Optimization of Trie Hashing Functions. VLDB 83, (Nov. 1983), 142-157.

/TOR84/ Torn, A., A. Hashing with overflow indexing. BIT, 24 (1984), 317-332.

/TRE85/ Tremblay, J-P., Sorenson, P., G. An Introduction to Data Structures. 2-nd ed., McGraw-Hill, 1984, 861.

/TRO81/ Tropf, H., Herzog, H. Multidimensional range search in dynamically balanced trees. Agnew. Inf. 2, 71-77.

/YAO80/ Yao, A., C. A note on the analysis of extendible hashing. Inf. Proc. Lett. 11, 2, 1980, 84-86.

THE STUDY OF A LETTER ORIENTED MINIMAL

PERFECT HASHING SCHEME

C. C. Chang

Institute of Applied Mathematics
National Chung Hsing University
Taichung, Taiwan

ABSTRACT

In this paper, a simple method shall be presented to construct minimal perfect hashing functions suitable for letter oriented keys. We applied this minimal perfect hashing method successfully to four non-trivial
key sets: (1) the set of twelve months in English, (2) the set of thirty
four non-printable ASCII identifiers, (3) the set of thirty one most frequently used English words, (4) the set of thirty six PASCAL reserved words.

1.   INTRODUCTION

Hashing is known variously as a fast technique for data storage and
retrieval. With this technique, a key is used to generate the address
where the record of the key is stored. That is why hashing is also called
the address-computation technique or the scattered-storage technique or the
key-to-address transformation. Actually, hashing is a mapping function
which maps a key into a near-random number. A general hashing function h
maps a set of keys $\{k_1,k_2,\ldots,k_n\}$ into the address space $\{0,1,2,\ldots,m-1\}$.

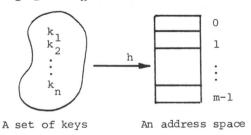

A set of keys   An address space

Fig. 1.1.   The basic model of hashing.

Using hashing may cause the key collision problem. Thus, the user
needs to handle the key collision problem. Some collision resolution
strategies have been studied in [Knuth 1973, Maurer 1968, Maurer and Lewis
1975, Morris 1968, Severance 1974]. One strategy of solving the key collision problem is to establish a perfect hashing function. If a hashing
function can be found that is one-to-one mapping from the set of keys to
the address space, it is a perfect hashing function. In the last decade,
many literatures reported the construction of perfect hashing functions

[Chang 1984a, Chang 1984b, Cichelli 1980, Cook and Oldehoeft 1982, Du, Hsieh, Jea and Shieh 1983, Ghosh 1977, Jaeschke 1981, Sprugnoli 1977, Yang, Du and Tsay 1983]. Some of the methods proposed can construct hashing functions which are not only one to one but also onto. Such hashing functions, which map from the set of keys with size n to an address space of length n, have been called minimal perfect hashing functions [Chang 1984a, Chang 1984b, Ghosh 1977, Jaeschke 1981].

Jaeschke [1981] presented a method for establishing minimal perfect hashing functions for static key sets. Given a finite key set $K=\{k_1,k_2,...,k_n\}$ of positive integers, Jaeschke's method (called reciprocal hashing) attempts to find three integer constants C, D and E, such that the function h defined by $h(k)=\left\lfloor \dfrac{C}{Dk+E} \right\rfloor$ mod n is a minimal perfect hashing function. Jaeschke [1981] showed that the existence of h was guaranteed. He also gave two algorithms, called Algorithm C and Algorithm DE, to find such C and D, E respectively.

Chang [1984a] proposed a method which is similar to Jaeschke's method. He gave a hashing function $h(k_i)=C$ mod $p(k)$, instead of $h(k)=\left\lfloor \dfrac{C}{Dk+E} \right\rfloor$ mod n. Based upon the Chinese Remainder Theorem, he proved that: Given a finite set $K=\{k_1,k_2,...,k_n\}$ of positive integers, there exists an integer C such that $h(k_i)=C$ mod $p(k_i)$ is a minimal perfect hashing function if $p(x)$ is a prime number function on K. In [Chang 1984a], an efficient algorithm to find such a constant C was given. Furthermore, the keys in K can be stored in ascending order by applying $h(k_i)$. Another method based on Euler's Theorem to construct minimal perfect hashing functions is also presented in [Chang 1984b].

In [Cichelli 1980], a simple method is proposed to construct minimal perfect hashing functions. In this method, each character is associated a value. The form of hashing function is defined as h(k)=length of k+associated value of the first character of k+associated value of the last character of k for any key k.

Cichelli [1980] proposed an essentially exhaustive search method for finding the proper values for different characters. He applied his method successfully to a set of PASCAL reserved words and a set of frequently used English words, etc. Later, Jaeschke and Osterburg [1980] pointed out that Cichelli's method was unable to find a minimal perfect hashing function in many cases.

Cook and Oldehoeft [1982] improved Cichelli's method. They developed an essential algorithm that handles more than one word in the search for assigning suitable associated values to characters. In the following sections, we shall propose a new minimal perfect hashing scheme suitable for letter-oriented keys.

2. A MINIMAL PERFECT HASHING SCHEME BASED UPON CHINESE REMAINDER THEOREM

Recently, Chang [1984a] proposed a method for establishing minimal perfect hashing functions. His method is based upon the Chinese Remainder Theorem. Given integers $m_1,m_2,...,m_n$ and $r_1,r_2,...,r_n$, a constant C can be found such that $C \equiv r_1$ (mod m_1), $C \equiv r_2$ (mod m_2),..., $C \equiv r_n$ (mod m_n) if m_i and m_j are relatively prime for all $i \neq j$. To apply the Chinese Remainder Theorem for perfect hashing, let us assume that m_i's are keys and they are relatively prime with one another. In this case, we may let $r_i=i$ and let the hashing function be $r_i=C$ mod m_i. Through this method, m_i will be hashed to i for all i.

Unfortunately, we can not assume that m_i and m_j are relatively prime for all $i \neq j$. Thus, Chang [1984a] proposed a method to transform m_i to $p(m_i)$ such that $p(m_i)$ is a prime number. In [Chang 1984a], it was pointed out that there exist several prime number functions. A function $p(x)$ is called a prime number function for $a \leq x \leq b$, where x, a, b are all positive integers, if $p(x)$ is a prime number for $a \leq x \leq b$ and $p(x_1) > p(x_2)$ if $x_1 > x_2$.

How to find the constant C. Chang [1984a] gave an essential algorithm, called Algorithm A, to find such a C. In the following, we describe Chang's Algorithm A.

Algorithm A [Chang 1984a]

This algorithm finds a smallest positive constant C for constructing a minimal perfect hashing function for a prime number function $p(x)$.

Input : $k_1, k_2, \ldots, k_n$. (Let $k_1 < k_2 < \ldots < k_n$)

Output: A constant C.

Step 1: [Input k_i's].

   Input all keys k_i.

Step 2: [Calculate all $p(k_i)$'s].

   Compute $m_i = p(k_i)$, $1 \leq i \leq n$.

Step 3: [Calculate all M_j's]

   Compute $M_j = \Pi_{i \neq j} m_j$, $1 \leq j \leq n$.

Step 4: [Calculate all b_i's].

   Calculate $M_i' = M_i \bmod m_i$.

   DEND $\leftarrow m_i$,

   DSR $\leftarrow M_i'$.

   $j \leftarrow 0$.
   While RMD$\neq$1 do
            DEND $\leftarrow$ DSR
            DSR $\leftarrow$ RMD
            $j \leftarrow j+1$
            $Q_j \leftarrow \left\lfloor \dfrac{\text{DEND}}{\text{DSR}} \right\rfloor$
            RMD $\leftarrow$ DEND$-Q_j$*DSR
   end.
   $k \leftarrow j$.
   $B_0 \leftarrow 1$.
   $B_{-1} \leftarrow 0$.
   DO $j \leftarrow 0$ to k-1
       $B_{j+1} = -B_j Q_{k-j} + B_{j-1}$
   end.
   $b_i = B_k$.

Step 5: [Calculate C]

   Compute $C = \sum_{i=1}^{n} b_i M_i i \bmod \Pi_{i=1}^{n} m_i$.

Step 6: [Output C]

   Output the constant C.

Example 2.1

Let $k_1=1$, $k_2=3$, $k_3=51$. $p(x)=x+2$ is a prime number function. Using Chang's [1984a] Algorithm A, C will be found to be 427. It is important for the reader to verify that

C mod $p(k_1)$=427 mod (1+2)=427 mod 3=1,

C mod $p(k_2)$=427 mod (3+2)=427 mod 5=2,

C mod $p(k_3)$=427 mod (51+2)=427 mod 53=3.

In this paper, we shall show that the method proposed in [Chang 1984a] can be modified to handle letter oriented keys. By letter oriented keys, we mean keys such as PASCAL reserved words (ARRAY, AND, ..., WHILE), twelve month in English (JANUARY, FEBRUARY, ..., DECEMBER), non-printable ASCII characters (ACK, BEL, ..., VT) and most frequently occurring English words (A, AND, ..., YOU). One straightforward method to handle such kind of keys is to consider these keys as numbers. This poses one problem: the resulting constant C will be too large.

In the next section, we shall show that the method proposed in [Chang 1984a] can be modified to process letter oriented keys.

3. A LETTER ORIENTED MINIMAL PERFECT HASHING SCHEME

We assume that we have a set of keys where each key is a string of characters. Besides, we shall assume that there exist i and j such that the pairs formed by the i-th and the j-th characters of the keys are distinct.

Consider the case of twelve months as shown below:

JANUARY
FEBRUARY
MARCH
APRIL
MAY
JUNE
JULY
AUGUST
SEPTEMBER
OCTOBER
NOVEMBER
DECEMBER

We may pick the second and the third characters. This will produce twelve pairs as shown below:

(A,N)
(E,B)
(A,R)
(P,R)
(A,Y)
(U,N)
(U,L)
(U,G)
(E,P)
(C,T)
(O,V)
(E,C)

The reader can convince himself that the above twelve pairs are all distinct. Of course, if the number of keys is too large, two keys may not be sufficient to distinguish all of the keys. This occurs in the COBOL reserved

94

word case. There are roughly five hundred COBOL reserved words and there do not exist such two characters which will form five hundred distinct pairs.

Let us now illustrate our hashing scheme. Consider the case of twelve months. We may order the distinct pairs extracted from these twelve keys by lexical ordering as shown below:

Group	Location	Extracted Pair	Original Key
1	1	(A,N)	JANUARY
	2	(A,R)	MARCH
	3	(A,Y)	MAY
2	4	(C,T)	OCTOBER
3	5	(E,B)	FEBRUARY
	6	(E,C)	DECEMBER
	7	(E,P)	SEPTEMBER
4	8	(O,V)	NOVEMBER
5	9	(P,R)	APRIL
6	10	(U,G)	AUGUST
	11	(U,L)	JULY
	12	(U,N)	JUNE

There are six groups. The first group starts with A, the second group starts with C and so on. As shown above, we shall hash (A,N) to location 1. (A,R) to 2 and finally, (U,N) to 12.

Our minimal perfect hashing scheme works as follows:

(1) Each distinct pair is denoted as (k_{i1}, k_{i2}).

(2) To each k_{i1}, associated it with a positive integer $d(k_{i1})$ according to the following rule: If k_{i1} first appears in location m, then k_{i1} is associated with m-1.

(3) To each distinct k_{i2}, associate it with a prime number $p(k_{i2})$. Since there are at most 26 distinct k_{i2}'s, we may assign the smallest 26 prime numbers to A,B,...,Z respectively. That is, we may assign 2 to A, 3 to B and finally 101 to Z.

(4) To each k_{i1}, associate it with a constant $C(k_{i1})$ determined by all of the accompanying k_{i2}'s appearing together with k_{i1}. Let $p_1, p_2, \ldots, p_b$ be the prime numbers assigned to $k_{i2}, k_{i+1,2}, \ldots, k_{i+b-1,2}$ respectively. Then $C \equiv 1 \pmod{p_1}$, $C \equiv 2 \pmod{p_2}, \ldots$, $C \equiv b \pmod{p_b}$. C can be found by consulting Algorithm A.

(5) The minimal perfect hashing function is $H(k_{i1}, k_{i2}) = d(k_{i1}) + (C(k_{i1}) \bmod p(k_{i2}))$.

Let us illustrate our hashing function by considering the case of (A,R).

(1) A is associated with 0 because the first A appears in location 1. Thus d(A)=0.

(2) R is associated with prime number 61. Thus p(R)=61.

(3) There are three k_{i2}'s appearing together with A. They are N, R and Y which are associated with prime numbers 43, 61, and 97 respectively.

(4) By using the result in [Chang 1984a], C(A) is found to be 161896.

(5) The minimal perfect hashing function hashes (A,R) as follows:

$$H(A,R) = 0 + (161896 \bmod 61)$$
$$= 0 + 2$$
$$= 2$$

which is correct.

Let us consider another example: (E,C).

(1) d(E) is 4 because E first appears in location 5.

(2) p(E)=11.

(3) B, C and P appear together with E in group 3. Their respective prime numbers are 3, 5, and 53.

(4) By using the result in [Chang 1984a], C(E) is found to be 427.

(5) The minimal perfect hashing function hashes (E,C) to
$$H(E,C) = 4 + (427 \bmod 5)$$
$$= 4 + 2$$
$$= 6$$

which is correct.

It should be easy to see that our hashing scheme is a minimal perfect hashing scheme.

4. SOME EXAMPLES

In this section, we shall present examples to illustrate how we can apply this letter oriented minimal perfect hashing scheme.

Example 1 (The Twelve Months)

For the English words of the twelve months as discussed in Section 3, we select the second and third characters. The extracted pairs can be found in Section 3. Our minimal perfect hashing function can be summarized as follows:

x	A	B	C	D	E	F	G	H	I	J	K	L	M	N	O	P	Q	R	S	T	U	V	W	X	Y	Z
d(x)	0		3		4										7	8					9					
c(x)	161896		1		427												1	1			12989					
p(x)	2	3	5	7	11	13	17	19	23	29	31	37	41	43	47	53	59	61	67	71	73	79	83	89	97	101

Let us consider "APRIL". P and R are selected.

$$H(P,R) = d(p) + (C(p) \bmod p(R))$$
$$= 8 + (1 \bmod 61)$$
$$= 8 + 1$$
$$= 9.$$

Example 2 (Non-Printable ASCII Characters)

There are thirty four non-printable ASCII identifiers {ACK,BEL,BS, CAN,CR,DLE,DCI,DCJ,DEL,DCW,DCZ,ETB,ESC,EM,ENQ,EOT,ETX,FF,FS,GS,HT,LF,NAK, NUL,RS,SUB,SOH,SI,SYN,SO,SP,STX,US,VT}. We choose the first and the third letters, if the identifier contains more than two letters and we choose the first and the second letters if otherwise. Thus we have the following thirty four distinct pairs {AK,BL,BS,CN,CK,DE,DI,DJ,DL,DW,DZ,EB,EC,EM,EQ, ET,EX,FF,FS,GS,HT,LF,NK,NL,RS,SB,SH,SI,SN,SO,SP,SX,US,CT}. Our minimal perfect hashing function is summarized as follows:

x	A	B	C	D	E	F	G	H	I	J	K	L	M	N	O	P	Q	R	S	T	U	V	W	X	Y	Z
d(x)	0	1	3	5	11	17	19	20				21		22			24	25		32	33					
c(x)	1	1074	1893	1553917124	60396652	404	1	1				1		187			1	9818876641		1	1					
p(x)	2	3	5	7	11	13	17	19	23	29	31	37	41	43	47	53	59	61	67	71	73	79	83	89	97	101

Let us now consider ACK. This will be hashed as follows:

$$H(A,K)=d(A)+(C(A) \bmod p(K))$$
$$=0+(1 \bmod 31)$$
$$=0+1$$
$$=1.$$

Similarly, BEL will be hashed as follows:

$$H(B,L)=d(B)+(C(B) \bmod p(L))$$
$$=1+(1074 \bmod 37)$$
$$=1+1$$
$$=2.$$

Example 3 (Frequently Occuring English Words)

This example considers the 31 most frequently used English words. They are A, AND, ARE, AS, AT, BE, BUT, BY, FROM, FOR, HAD, HE, HER, HIS, HAVE, I, IN, IS, IT, NOT, OF, ON, OR, THAT, THE, THIS, TO, WHICH, WAS, WITH and YOU. We choose the first letter and the third letter if the word contains more than two letters and the first and last letters if otherwise. The corresponding pairs are: AA, AD, AE, AS, AT, BE, BT, BY, FO, FR, HD, HE, HR, HS, HV, II, IN, IS, IT, NT, OF, ON, OR, TA, TE, TI, TO, WI, WS, WT, YU.

The following illustrates the hashing function:

x	A	B	C	D	E	F	G	H	I	J	K	L	M	N	O	P	Q	R	S	T	U	V	W	X	Y	Z
d(x)	0	5				8		10	15					19	20					23			27		30	
c(x)	98695	23574				612		10647309	22686914					1	202255					4281			70151		1	
p(x)	2	3	5	7	11	13	17	19	23	29	31	37	41	43	47	53	59	61	67	71	73	79	83	89	97	101

Consider the word YOU which should be hashed to the last location.

$$H(Y,U)=d(Y)+(C(Y) \bmod p(U))$$
$$=30+(1 \bmod 73)$$
$$=30+1$$
$$=31.$$

Example 4 (PASCAL's Reserved Words)

The thirty six reserved words of PASCAL are as follows: ARRAY, AND, BEGIN, CASE, CONST, DOWNTO, DO, DIV, END, ELSE, FUNCTION, FILE, FOR, GOTO, IF, IN, LABEL, MOD, NIL, NOT, OTHERWISE, OF, OR, PROCEDURE, PROGRAM,

PACKED, REPEAT, RECORD, SET, TYPE, THEN, TO, UNTIL, VAR, WITH, WHILE. We choose the first and fourth letters if possible. If a word contains less than four letters, then we choose the first and the last letters. The extracted pairs are as follows: AA, AD, BI, CE, CS, DN, DO, DV, ED, EE, FC, FE, FR, GO, IF, IN, LE, MD, NL, NT, OE, OF, OR, PC, PG, PK, RE, RO, ST, TE, TN, TO, UI, VR, WH, WL. The hashing function is illustrated by the following table:

x	A	B	C	D	E	F	G	H	I	J	K	L	M	N	O	P	Q	R	S	T	U	V	W	X	Y	Z
d(x)	0	2	3	5	8	10	13		14			16	17	18	20	23		26	28	29	32	33	34			
c(x)	9	6	50010 672		57	3236	::		131			1	1	1777	3785	716		331 1	15654	1	1	39				
p(x)	2	3	5	7	11	13	17	19	23	29	31	37	41	43	47	53	59	61	67	71	73	79	83	89	97	101

Consider the reserved word WHILE. Again, this key should be hashed into the last location.

$$H(W,L)=d(W)+(C(W) \bmod p(L))$$
$$=34+(39 \bmod 37)$$
$$=34+2$$
$$=36.$$

5. CONCLUDING REMARKS

In this paper, we proposed a simple minimal perfect hashing scheme to handle letter-oriented keys. Minimal perfect hashing functions for twelve months in English, thirty four non-printable ASCII identifiers, thirty one most frequently used English words and thirty six PASCAL reserved words are successfully presented as example in Section 4.

REFERENCES

1. Chang, C. C., (1984a): The Study of an Ordered Minimal Perfect Hashing Scheme, Communications of the Association for Computing Machinery, Vol. 27, No. 4, April 1984, pp. 384-387.
2. Chang, C. C., (1984b): An Ordered Minimal Perfect Hashing Scheme Based Upon Euler's Theorem, Information Sciences, Vol. 32, No. 3, June 1984, pp. 165-172.
3. Cichelli, R. J., (1980): Minimal Perfect Hash Functions Made Simple, Communications of the Association for Computing Machinery, Vol. 23, No. 1, January 1980, pp. 17-19.
4. Cook, C. R. and Oldehoeft, R. R., (1982): A Letter Oriented Minimal Perfect Hashing Function, Sigplan Notices, Vol. 17, No. 9, September 1982, pp. 18-27.
5. Du, M. W., Hsieh, T. M., Jea, K. F. and Shieh, D. W., (1983): The Study of A New Perfect Hash Scheme, IEEE Transactions on Software Engineering, Vol. SE-9, No. 3, May 1983, pp. 305-313.
6. Ghosh, S. P., (1977): Data Base Organization for Data Management, Academic Press, New York, 1977, pp. 148-151.
7. Jaeschke, G., (1981): Reciprocal Hashing: A Method for Generating Minimal Perfect Hashing Functions, Communications of the Association for Computing Machinery, Vol. 24, No. 12, December 1981, pp. 829-833.
8. Knuth, D. E., (1973): The Art of Computer Programming, Vol. 3, Sorting and Searching, Addison-Wesley, Reading, Mass., U.S.A., 1973.
9. Maurer, W. D., (1968): An Improved Hash Code for Scatter Storage, Communications of the Association for Computing Machinery, Vol. 11, No. 1, January, 1968, pp. 35-37.

10. Maurer, W. D. and Lewis, T. G., (1975): Hash Table Method, Computing Surveys, Vol. 7, No. 1, March 1975, pp. 5-19.
11. Morris, R., (1968): Scatter Storage Techniques, Communications of the Association for Computing Machinery, Vol. 11, No. 1, January 1968, pp. 38-44.
12. Severance, D. G., (1974): Identifier Search Mechanisms: A Survey and Generalized Model, Computing Surveys, Vol. 6, No. 3, September 1974, pp. 175-194.
13. Sprugnoli, R., (1977): Perfect Hashing Functions: A Single-Probe Retrieving Method for Static Sets, Communications of the Association for Computing Machinery, Vol. 20, No. 11, November 1977, pp. 841-850.
14. Yang, W. P., Du, M. W. and Tsay, J. C., (1983): Single-Pass Perfect Hashing for Data Storage and Retrieval, Proceedings 1983 Conference on Information Sciences and Systems, Baltimore, Mayland, May 1983, pp. 470-476.

Consecutive Retrieval Property

ON THE RELAXED CONSECUTIVE RETRIEVAL PROPERTY

IN FILE ORGANIZATION

Hiroshige Inazumi and Shigeichi Hirasawa

School of Science and Engineering
Waseda University
Shinjuku, Tokyo 160, Japan

ABSTRACT

            In the information retrieval systems, the consecutive
retrieval property first defined by S.P. Ghosh is an important relation
between a set of queries and a set of records. Its existence enables the
design of the system with a minimal search time and no redundant storage.
However, the consecutive retrieval property cannot exist between every
arbitrary query set and every record set unless the duplication of
records are allowed. We consider the file organization satisfying the
relaxed consecutive retrieval property which tolerates both the loss of
search time and the redundancy of records, and determine the
relationships between these parameters by rate—distortion theoretic
approach. The result indicates that it is worthwhile to search for
algorithms that will generate the storage locations satisfying the
relaxed consecutive retrieval property, when the system size becomes
sufficiently large.

I. INTRODUCTION

    Recent development on computer hardware and software has led to
growth of the amount of information in data bases. The growth has
brought about the need for flexible intelligent interfaces between the
users and the large scale data base systems so that the vast information
can be effectively utilized. Studies on intelligent interfaces have been
done in the aspect on the efficiency of various filing and retrieval
procedures in computer systems.

    The design of a file organization scheme in computer systems is
concerned with techniques of storing the information of records on a
storage medium in a way that every query of given query set can be
efficiently answered. Especially, the amount of storage and the search
time are most important parameters required for storing and retrieving
those records, which may not be decreased simultaneously.

    In so far as the time required for retrieving each record pertinent
to a query is concerned, an efficient organization of records is to store
them in consecutive storage locations, since storage devices used for
storing records are essentially linear. In this aspect, the concept of
the consecutive retrieval (CR) property has been introduced [1]-[4].

The formal definition of the CR property is as follows [1]: Given a query set and a record set, suppose there exists an organization of the records in the set which satisfies the following properties: (i) the organization contains no records stored more than once, and (ii) the records pertinent to any queries belonging to the set are stored in consecutive storage locations. Then, the query set is said to have the consecutive retrieval property.

Since it is obvious that the CR property cannot exist between every arbitrary query set and every record set, there are several variations of file organization [5]: (a) CR property with duplication of records, (b) CR property with addition of dummy records, (c) CR property with permission of multiple access, (d) CR property with decomposition of a query set, (e) quasi-consecutive retrieval (QCR) property [6], and (f) buffer limited QCR (BL-QCR) property [7].

In this paper, we propose the relaxed CR (RCR) property which tolerates the multiple access and the organization of records stored more than once [2][3], assuming the sufficient buffer size in the main storage. Then, we investigate the relationships between the redundancy of records and the loss of the search time by a rate-distortion theoretic approach. In Section II, we introduce the storage systems based on the CR property, and we relax the CR property to allow for at most r retrievals, which is refered to as the r-relaxed CR (r-RCR) property. Furthermore, with the duplication of records, we consider the file organization which tolerates the r-RCR property, i.e., each query is satisfied 1 retrieval, 2 retrievals, ..., or r retrievals. In Section III, we define the system efficiency from a viewpoint of rate-distortion theory, which clarifies the relationships between the loss of search time and the redundancy of records. In Section IV, we evaluate the file organization with the r-RCR property for moderately large number of records, and we analyze them by using the technique first proposed in the question-answering systems [8]-[14].

II. THE RELAXED CR PROPERTY

We consider information storage and retrieval systems based on the CR property, in which the file organization is located in the cyclic access storage systems, e.g., the magnetic bubble domain storage system and the magnetic disk storage system. In the case of the magnetic disk storage system, track selection will be done by the direct-access procedure [15].

The motivation for the investigation of the CR property is that it corresponds to the least cost or least time retrieval because it provides the most concise dictionary possible: it lists next to each query the start record and the total number of records involved. Also the storage of records themselves involves no pointers. Unfortunately, however, a query set does not generally have the CR property with respect to a record set.

If a storage device is a cyclic access storage tolerating multiple access, the CR file organization enables us to retrieve necessary records with a single run of records without gaps of unneeded records. Let ρ be the record set, and q be the query set such that

$$\rho = \{\rho_i \mid i=1, 2, \ldots, N\}, \tag{1}$$

$$q = \{q_i \mid i=1, 2, \ldots, Q\}, \tag{2}$$

and $B_1(N)$ be the number of the queries on N records that have the CR property. $B_1(N)$ has been shown to be disappointingly small number compared with the size of a query set [2]. Thus, the CR property is relaxed to allow retrievals with upto r accesses. Let the relaxed CR (RCR) property permit both the multiple access and the duplication of records in file organization, and let the r-RCR property tolerate retrievals within r accesses.

Let $B_i(N)$ be the number of queries on ρ with N records that require i distinct runs of records on cyclic access storage, and $C_r(N)$ be the number of queries on ρ with the r-RCR property. Then we have the following equations.

$$C_r(N) = \sum_{i=1}^{r} B_i(N),$$ (3)

$$C_{r_{max}}(N) = \sum_{i=1}^{r_{max}} B_i(N) = Q,$$ (4)

where r_{max} is the maximum number of retrievals.

For example, given N records and 2^N possible query answers associated with them, $B_i(N)$ and $C_r(N)$ are calculated as follows:

$$B_i(N) = \binom{N+1}{2i},$$ (5)

$$C_r(N) = \sum_{i=0}^{r} \binom{N+1}{2i},$$ (6)

$$C_{\lfloor \frac{N+1}{2} \rfloor}(N) = \sum_{i=0}^{\lfloor \frac{N+1}{2} \rfloor} \binom{N+1}{2i},$$ (7)

where $\lfloor x \rfloor$ is the greatest integer not greater than x, and $\lfloor \frac{N+1}{2} \rfloor = r_{max}$.

Furthermore, concerning the duplication of records equiportable in file organization, let $s^{(r)}$ be the storage locations and $N(r)$ be the number of the records on the storage locations which enable the queries to satisfy the r-RCR property, such that

$$s^{(r)} = \{ s_i^{(r)} \mid i=1, 2, \ldots, N(r) \},$$ (8)

where $N(1) \geq \ldots \geq N(r) \geq \ldots \geq N(r_{max}) = N$. Let $B_i\{N(r)\}$ be the number of queries on $N(r)$ records requiring retrieval with i accesses, $C_r\{N(r)\}$, the number of queries on $N(r)$ records requiring the r-RCR property, and $M\{N(r)\}$, the average number of retrievals with the r-RCR property. Assuming that each query is equiprobable, we have the following:

$$\sum_{i=0}^{r} B_i\{N(r)\} = C_r\{N(r)\} = Q,$$ (9)

$$M\{N(r)\}=\sum_{i=0}^{r} iB_i\{N(r)\}/Q. \tag{10}$$

Note that, in the RCR file organization, we need to provide a buffer in the main storage whose size is the same as the largest query size. Therefore, we consider the file organization scheme in the secondary storage, and regard the RCR property as the model for evaluating the performance of the secondary storage.

III. RATE-DISTORTION THEORETIC APPROACH

In this section, we try to evaluate the relationships between the factors of each of the proposed new schemes, and define the system efficiency.

We consider information theory, especially rate-distortion theory, and regard the proposed scheme as the analogy of communication channel. Though this rate-distortion theoretic approach has been analyzed in question-answering systems which involve memory versus error trade-offs [8]-[14], we extend this analysis and clarify the redundancy of storage versus the loss of search time in the the modified version of CR file organization.

According to rate-distortion theory [16][17], given an input alphabet I, $I=\{{}^{i}_{p_i}\}$, an output alphabet J, $J=\{{}^{j}_{p_iP(j|i)}\}^1$, and distortion measure $d(i,j)$, it is not possible with any coding scheme to transmit or store the source information at a rate less than $R(D)$ and with average distortion not exceeding fidelity criterion D. $R(D)$ is called the rate-distortion function and is given by

$$R(D)=\min_{P(j|i)\in P_D} \sum_{i,j} p_iP(j|i)\ln \frac{P(j|i)}{p_i}, \tag{11}$$

where

$$P_D=\{P(j|i) \mid \sum_{i,j} p_iP(j|i)d(i,j)\leq D\}. \tag{12}$$

In the file organization scheme tolerating the RCR property from the viewpoint of rate-distortion theory, the storage locations satisfying the 1-RCR property for any queries can be recognized as the input message, and those satisfying the r-RCR property, output message. Let the source code be the algorithm that will generate the storage locations and will process the consecutive retrievals within r times retrievals. In this respect, a distortion of the input message can be measured by the number of retrievals, i.e., the loss of search time. Unlike ordinary transmission problems, the quality of code is not judged by its ability to faithfully reproduce the input on a symbol by symbol basis, but rather by the efficiency of the retrieval it helps generate.

---

1. The symbol p_i is the apriori probability of the input alphabet i, and $P(j|i)$ is the conditional probability of the output alphabet j assuming the input alphabet i, where $p_iP(j|i)$ is the aposteriori probability of the output alphabet j.

From (9) and (10), assuming the ideal file arrangement of the $N(r)$ records where all queries will have the r-RCR property, we have the following definition.

Definition 1:   The loss of search time as a function of r, $D(r)$ is defined as follows,

$$D(r) = [M\{N(r)\} - M\{N(1)\}] / [M\{N(r_{max})\} - M\{N(1)\}]$$

$$= [\sum_{i=0}^{r} iB_i\{N(r)\} - B_1\{N(1)\}] / [\sum_{i=0}^{r_{max}} iB_i(N) - B_1\{N(1)\}].\qquad(13)$$

where the ideal arrangement will be generated by mapping from the query set to the storage locations. We define the set of assignment $f_{D(r)}$ for given r as follows,

$$f_{D(r)}: \quad q \mapsto s^{(r)}.$$

Considering Shannon's source coding theorem [16], we have the following lemma.

Lemma 1:   For every $D(r)$, admissible computation yields an entropy not smaller than $R\{D(r)\}$. That is

$$H = -\sum_{i=1}^{N(r)} P(s_i^{(r)}) \ln P(s_i^{(r)}) \geq R\{D(r)\},\qquad(14)$$

where $P(.)$ is the probability induced on $s^{(r)}$ by mapping $f_{D(r)}$.

Especially, if the records and queries are independent and equiprobable,

$$H = \ln N(r) \geq R\{D(r)\}.\qquad(15)$$

Therefore, we can define the upper bound on $R\{D(r)\}$, $R_U\{D(r)\}$, such that

$$R_U\{D(r)\} = \ln N(r).\qquad(16)$$

The definition of $R\{D(r)\}$ here takes its operational significance from the negative part of Shannon's source coding theorem, which states that there exists no code for which both the average distortion is less than $D(r)$ and the rate is less than $R\{D(r)\}$. This implies in particular that the filing system with the r-RCR property must be provided with an average storage size of at least $R\{D(r)\}$ nats per query set in order to achieve a mean distortion of at most $D(r)$. This fact leads to the following theorem.

Theorem 1:   From Definition 1 and Lemma 1, in the filing system with the r-RCR property, the redundancy of the records is defined by $R_{rcr}\{D(r)\}$ and the loss of search time, $D(r)$. The trade-offs between these two factors is given in the following parametric representation.

$$R_{rcr}\{D(r)\} = \{\ln N(r) - \ln N\} / \{\ln N(1) - \ln N\},\qquad(17)$$

$$D(r) = [\sum_{i=0}^{r} iB_i\{N(r)\} - B_1\{N(1)\}] / [\sum_{i=0}^{r_{max}} iB_i\{N\} - B_1\{N(1)\}]. \tag{18}$$

where $N(r)$ satisfies the following condition:

$$\sum_{i=0}^{r} B_i\{N(r)\} = Q. \tag{19}$$

Since $R_{rcr}(D)$ is the normalized function of which the property can be equivalent to the rate-distortion function $R(D)$, we can assess $R_{rcr}(D)$ from the source coding theorem. Therefore, treating r as a parameter between 1 and r_{max}, $R_{rcr}(D)$ is regarded as a system efficiency.

IV. PROPERTIES OF THE SYSTEM EFFICIENCY

In this section, we will treat the following typical filing system, where N records and 2^N possible query answers associated with them are given, and evaluate the system efficiency according to the definition of elasticity which has first defined in the analysis of the question-answering systems [8]-[14]. Elasticity means that drastic savings of the amount of storage with a small loss tolerance is feasible, as the system size becomes sufficiently large. In storage systems, when we can define the system efficiency $R*(D)$ which is equivalent to the normalized rate-distortion function $R(D)/R(0)$, we have the following definition.

Definition 3: if $R*(D)$ tends to zero for $\forall D$, $0 < D < 1$ when the number of records $N \to \infty$, then the storage systems will be called elastic.

In the RCR file organization, the number of redundant records $N(r)$ satisfies the following equation,

$$\sum_{i=1}^{r} \binom{N(r)+1}{2i} = 2^N - 1. \tag{20}$$

Therefore, the system efficiency $R_{rcr}\{D(r)\}$ can be described in the following parametric representation.

$$R_{rcr}\{D(r)\} = \{\ln N(r) - \ln N\} / \{\ln N(1) - \ln N\}, \tag{21}$$

$$D(r) = [\sum_{i=1}^{r} i\binom{N(r)+1}{2i} - \binom{N(1)+1}{2}] / [\sum_{i=1}^{r_{max}} i\binom{N+1}{2i} - \binom{N(1)+1}{2}]$$

$$= [\sum_{i=1}^{r} i\binom{N(r)+1}{2i} - (2^N - 1)] / [2^N(N+1)/4 - (2^N - 1)], \tag{22}$$

where $R_{rcr}\{D(r)\}$, $D(r)$, and $R_{rcr}(D)$ are depicted in Fig.1 and Fig.2 for N=10, 50, 100.

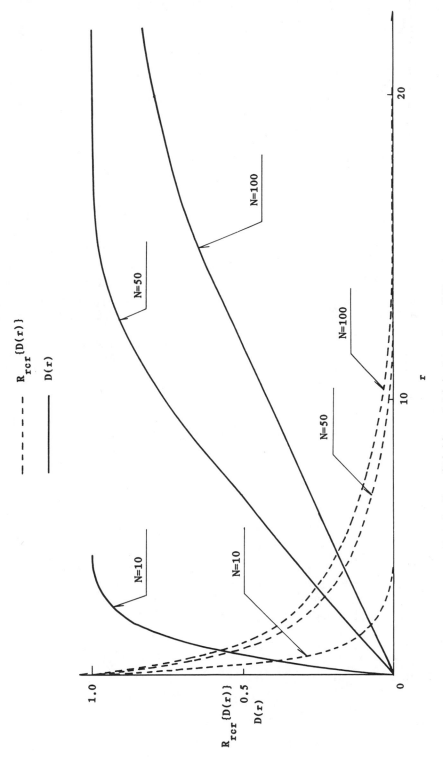

Fig.1. $R_{rcr}\{D(r)\}$ and $D(r)$ for the r-RCR property.

109

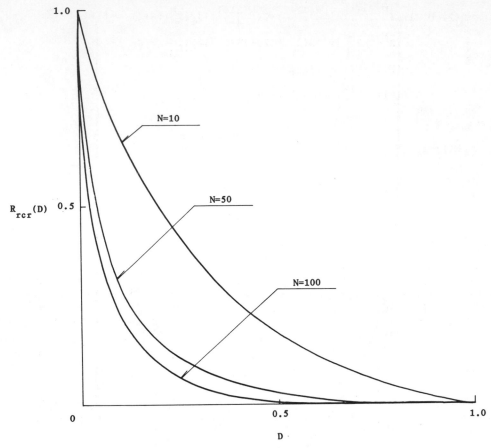

Fig.2. $R_{rcr}(D)$ for the r–RCR property.

Since it is practical to analyze the system efficiency when r is small and N, large, we consider a local property of $R_{rcr}(D)$ for r=1, 2. $R_{rcr}\{D(1)\}-R_{rcr}\{D(2)\}$ and $D(1)-D(2)$ are approximately represented as follows:

$$R_{rcr}\{D(1)\}-R_{rcr}\{D(2)\}=[\ln N(1)-\ln N(2)]/[\ln N(1)-\ln N]$$
$$\sim[(N/2-N/4)\ln 2]/[(N/2)\ln 2-\ln N]$$
$$\sim 1/2, \tag{23}$$

$$D(1)-D(2)=[(\tbinom{N(1)+1}{2})-\{(\tbinom{N(2)+1}{2})+2(\tbinom{N(2)+1}{4})\}]/[2^N\{(N+1)/4-1\}+1]$$

$$=[2^N-1-2^{N+1}+2+N(2)\{N(2)+1\}/2]/[2^N\{(N+1)/4-1\}+1]$$

$$=[N(2)\{N(2)+1\}/2-2^N+1]/[2^N\{(N+1)/4-1\}+1]$$

$$\sim(-1+2^{-N/2})/\{(N+1)/4-1\}, \qquad\qquad\qquad (24)$$

where

$$(\tbinom{N(1)+1}{2})=2^N-1, \quad (\tbinom{N(2)+1}{2})+(\tbinom{N(2)+1}{4})=2^N-1,$$

$$N(1)=0(2^{N/2}), \quad N(2)=0(2^{N/4}).$$

From (34),(35) and Fig.2, as $N\to\infty$, $[R_{rcr}\{D(1)\}-R_{rcr}\{D(2)\}]\to 1/2$, $\{D(1)-D(2)\}\to 0$, and $[R_{rcr}\{D(1)\}-R_{rcr}\{D(2)\}]/\{D(1)-D(2)\}\to -\infty$. For relative small r, it is shown that a slight loss tolerance of search time leads to the possibility of significant savings of records' redundancy, as the number of records N, becomes large, which satisfies the necessary condition of Definition 3. Therefore, this result suggests that it would be worthwhile to search for algorithms that will generate the storage locations satisfying the r-RCR property for relative small r.

V. DISCUSSION

The problem discussed in this paper is almost analogous to that of question-answering systems [8]-[14], whose background and objective is as follows: In tasks involving data retrieval and answering non-numerical questions, error have traditionally been unacceptable. However, when the data size or complexity of problems exceeds the practical limits, error tolerance becomes computationally necessary. Especially, trade-offs between storage and error have been studied as the question-answering systems using rate-distortion theoretic methods for various information systems which also involve information storage and retrieval systems. The main objective for analyzing question-answering systems is to provide simple tests for determining conditions under which a small error tolerance results in drastic savings of the amount of storage.

In many problems which involve trade-offs between the system parameters, e.g., trade-offs between error and complexity, time and error, storage and error, etc., it is important to consider those relationships. Furthermore, it is worthwhile to clarify conditions under which the system shows the existence of any algorithms for realizing the optimal or suboptimal system.

Especially, in the CR file organization, we can apply the parameters dependent on the system redundancy and the loss of search time from the viewpoint of rate-distortion theory. As the result, we can suggest the significance to search for algorithms for realizing these suboptimal conditions. We plan a future paper in which the system efficiency will be strictly evaluated and shown to be elastic.

REFERENCES

[1] S.P.Ghosh, "File organization: The consecutive retrieval property," *Communications of the ACM*, vol.15, no.9, pp.802-808, Sept., 1972.

[2] A.Wakesman and M.W.Green, "On the consecutive retrieval property in file organization," *IEEE Trans. on Computers*, vol.C-23, no.2, pp.173-174, Feb., 1974.

[3] S.P.Ghosh, "Consecutive storage of relevant with redundancy," *Communications of the ACM*, vol.18, no.8, pp.464-471, Aug., 1975.

[4] S.Yamamoto, K.Ushio, S.Tazawa, H.Ikeda, F.Tamari, and N.Hamada, "Partition of query set into minimal number of subsets having consecutive retrieval property," *J. of Statistical Planning and Inference*, no.1, pp.41-51, 1977.

[5] S.P.Ghosh, "Future of the consecutive retrieval property ," in Data Base File Organization. Academic Press, 1983.

[6] K.Tanaka, Y.Kambayashi, and S.Yajima, "Organization of quasi-consecutive retrieval files," *Information Systems*, vol.4, pp.23-33, 1979.

[7] Y.Kambayashi, "The buffer limited quasi-consecutive retrieval file organization and its application to logic circuit layout," in *Data Base File Organization*. Academic Press, 1983.

[8] J.Pearl, "On the storage economy of inferential question-answering systems," *IEEE Trans. on Syst., Man, Cybern.*, vol.SMC-5, no.6, pp.595-602, Nov., 1975.

[9] J.Pearl, "Memory versus error characteristics for inexact representations of linear orders," *IEEE Trans. on Computers*, vol.C-25, no.9, pp.922-928, Sept., 1976

[10] J.Pearl,"On coding precedence relations with a pair-ordering fidelity criterion," *IEEE Trans. on Inform. Theory*, vol.IT-22, no.1, pp.118-120, Jan., 1976.

[11] J.Pearl, "Theoretical bounds on the complexity of inexact representation," *IEEE Trans. on Inform. Theory*, vol.IT-22, no.1, pp.118-120, Jan., 1976.

[12] A.Crolotte and J.Pearl, "Bounds on memory versus error trade-offs in question-answering systems," *IEEE Trans. on Inform. Theory*, vol.IT-25, no.2, pp.193-202, Mar., 1979.

[13] A.Crolotte and J.Pearl, "Elasticity conditions for storage versus error exchange in question-answering systems," *IEEE Trans. on Inform. Theory*, vol.IT-25, no.6, pp.656- 664, Nov., 1979.

[14] H.Inazumi and S.Hirasawa, "Memory versus error trade-offs in binary-valued retrieval problems," *IEEE Trans. on Syst., Man, Cybern.*, vol.SMC-14, no.5, Sept., 1984.

[15] H.S.Stone, ed., *Introduction to computer architecture*. Chicago: Science Research Associates, Inc., 1975.

[16] C.E.Shannon, "Coding theorems for a discrete source with a fidelity criterion," *1959 IRE Nat. Conv. Rec.*, pt.4, pp.142-163.

[17] T.Berger, *Rate Distortion Theory*. Englewood Cliffs, NJ: Prentice-Hall, 1971.

CONSECUTIVE RETRIEVAL ORGANIZATION AS A FILE

ALLOCATION SCHEME ON MULTIPLE DISK SYSTEMS

Chin-Chen Chang and Jaw-Ji Shen

Institute of Applied Mathematics
National Chung Hsing University
Taichung, Taiwan

ABSTRACT

Consecutive retrieval organization can be used to arrange records into buckets. In this paper, we shall propose an important organization, called bucket-oriented consecutive retrieval organization which can be used to arrange buckets into different disks. Some interesting properties of this new organization will be discussed and proved.

Section 1. Introduction

It is surely a great progress in the recent years on the way how to design a most efficient multi-attribute file system for partial match queries [Aho and Ullman 1979, Bentley and Friedman 1979, Bolour 1979, Chang, Du and Lee 1984, Chang and Fu 1981, Chang, Lee and Du 1980, Chang, Lee and Du 1982, Chang and Su 1983(a), Chang and Su 1983(b), Chang and Su 1984, Ghosh 1977, Jakobsson 1980, Lee 1981, Lee and Tseng 1979, Lin, Lee and Du 1979, Liou and Yao 1977, Rivest 1976, Rothnie and Lozano 1974]. The so called multi-attribute file system means that each record in the file contains more than one attribute, and the form of partial match query is:

$q=(A_1=a_1, A_2=a_2,...,A_N=a_N)$, a_i is the i-th attribute value of "*", that is the i-th attribute value is unspecified, where $1 \leq i \leq N$. For instance, in a 3-attribute file system, the query $(A_1=*, A_2=b, A_3=c)$ means to retrieve all the records that satisfy the 2nd attribute value and 3rd attribute value which equal b and c respectively. The above mentioned query can also be written as (*,b,c) for short.

We presume in this paper that each file contains several buckets. And this can be explained by Table 1.1 and Table 1.2 to indicate how to design a multi-attribute file system.

Table 1.1(a). Simple 2-attribute file system.

Bucket 1	Bucket 2	Bucket 3	Bucket 4
(a,a)	(a,b)	(a,c)	(a,d)
(b,b)	(b,c)	(b,d)	(b,a)
(c,c)	(c,d)	(c,a)	(c,b)
(d,d)	(d,a)	(d,b)	(d,a)

Table 1.1(b). The corresponding inverted list of the file shown in Table 1.1(a).

Queries	Buckets to be Examined
$Q_1 = (a, *)$	1,2,3,4
$Q_2 = (b, *)$	1,2,3,4
$Q_3 = (c, *)$	1,2,3,4
$Q_4 = (d, *)$	1,2,3,4
$Q_5 = (*, a)$	1,2,3,4
$Q_6 = (*, b)$	1,2,3,4
$Q_7 = (*, c)$	1,2,3,4
$Q_8 = (*, d)$	1,2,3,4

Table 1.2(a). Simple 2-attribute file system.

Bucket 1	Bucket 2	Bucket 3	Bucket 4
(a,a)	(a,c)	(c,a)	(c,c)
(a,b)	(a,d)	(c,b)	(c,d)
(b,a)	(b,c)	(d,a)	(d,c)
(b,b)	(b,d)	(d,b)	(d,d)

Table 1.2(b). The corresponding inverted list of the file shown in Table 1.2(a).

Queries	Buckets to be Examined
$Q_1 = (a, *)$	1,2
$Q_2 = (b, *)$	1,2
$Q_3 = (c, *)$	3,4
$Q_4 = (d, *)$	3,4
$Q_5 = (*, a)$	1,3
$Q_6 = (*, b)$	1,3
$Q_7 = (*, c)$	2,4
$Q_8 = (*, d)$	2,4

From Table 1.1(b), we can easily calculate the number of examined buckets to be 4 over all possible partial match queries of the file system which was shown in Table 1.1(a); from Table 1.2(b) we can also get average number of examined buckets to be 2 over all possible partial match queries of the file system which was shown in Table 1.2(a).

Therefore, the problem of designing the multi-attribute files for partial match queries can be stated as follows: How to arrange a set of given multi-attribute records into the buckets in such a way that the average number of buckets to be examined, over all possible partial match queries, is minimized.

Suppose there is already a well constructed file system in a single disk system and also suppose it needs to take a time unit to access a bucket from a disk; in such a case we know that assigning different buckets to disks influences the performance.

Imagine that we have more than one disk, and then buckets are assigned

to these disks.   Therefore, to a partial match query, the response time of
this query is no longer equal to the total number of buckets which need to
be examined but equal to the maximum number of buckets which need to be
examined in a certain disk.   Let's review the previous file system that
showed in Table 1.2.   We suppose there are two disks.   If we allocate these
buckets as follows:

Table 1.3(a).   Disks and buckets assigned.

Disk Number	Buckets
0	1,2
1	3,4

Table 1.3(b).   Time required to answer each query for
buckets distribution shown in Table 1.3(a).

Queries	Response time
(a,*)	2
(b,*)	2
(c,*)	2
(d,*)	2
(*,a)	1
(*,b)	1
(*,c)	1
(*,d)	1

In this case, for example, for query (a,*), buckets 1 and 2 are to
be examined.   Since they reside on two disks and they can be accessed in
parallel, only one unit of time is needed.

In fact, the reader should be able to verify by himself that the
average response time is 1.5 for the above allocation.   The time needed to
respond to each possible partial match query is 1.   If we allocate these
buckets as follows:

Table 1.4(a).   Disks and buckets assigned.

Disk Number	Buckets
0	1,4
1	2,3

Table 1.4(b).   Time required to answer each query for
buckets distribution shown in Table 1.4(a).

Queries	Response time
(a,*)	1
(b,*)	1
(c,*)	1
(d,*)	1
(*,a)	1
(*,b)	1
(*,c)	1
(*,d)	1

Thus the performance under this allocation method is superior to that under
previous allocation.

Thereby, we understand that in a multi-disk system, assigning different buckets to disks influences the performance.   The problem of the
arrangement of a file system on multi-disk system can be stated as follows:

Given a multi-attribute file system designed primarily for partial match queries and an ND-disk system (ND means the total number of disks), our job is to find an appropriate allocation method for buckets on ND disks such that the average response time, over all possible partial match queries, is minimized.

Du and Sobolewski [1982] discussed the problem of allocating buckets to different disks. They proposed a heuristic method, called the Disk Modulo (DM) allocation method, for distributing buckets in multi-disk system. In the DM allocation, the concept of Cartesian product file is used. The interested reader is encouraged to read [Du and Sobolewski 1982] for more information.

Section 2. Consecutive Retrieval Organization

Ghosh [1977] was the first one who tried to construct a file with the consecutive retrieval (CR) property with respect to a given set $\{Q_i\}_1^m$, i.e., a set of records answering every query in $\{Q_i\}_1^m$ are contiguous. In most file organizations, a significant portion of the processing time of a query is spent in seek times for record retrievals. By forcing the relevant records be stored in contiguous blocks, an improved performance can be expected. In practice, moreover, the CR property can be extended to fit a set of records together and load them into one physical storage area such that the access time for the queries in $\{Q_i\}_1^m$ is minimized. For instance, a file of nine records may be related to three queries as follows:

	Q_1	Q_2	Q_3
R_1	1	0	0
R_2	0	1	0
R_3	0	1	1
R_4	1	1	0
R_5	0	0	1
R_6	0	0	1
R_7	0	0	1
R_8	1	1	0
R_9	0	0	1

If we store these records sequentially into three buckets and the bucket size is three, the average number of buckets to be examined for $\{Q_i\}_1^3$ is 3. If we rearrange these records as the following, the average number of buckets to be examined for $\{Q_i\}_1^3$ is reduced to approximately 1.66:

		Q_1	Q_2	Q_3
	R_1	1	0	0
Bucket 1	R_4	1	1	0
	R_8	1	1	0
	R_2	0	1	0
Bucket 2	R_3	0	1	1
	R_9	0	0	1

116

	Q_1	Q_2	Q_3
R_6	0	0	1
R_5	0	0	1
R_7	0	0	1

(Bucket 3)

As the reader can see, the records are now stored in such a way that all relevant records are stored in consecutive storage locations.

In this case, the relation between queries and records is expressed by the record-query incidence matrix. This incidence matrix is an $NOR \times NOQ$ 0-1 matrix, where NOR is the total number of records and NOQ is the total number of queries. The entry (i,j) of the matrix contains 1, if R_i is pertinent to $Q_j \in \{Q_i\}_1^m$ and contains zero if otherwise.

The CR organization can be formally stated as:

Definition

An organization of the records set $\{R_i\}_1^n$, in which pertinent to any query in $\{Q_i\}_1^m$ are stored in consecutive storage locations, is defined to be a CR organization of $\{R_i\}_1^n$ for $\{Q_i\}_1^m$.

Definition

A query set $\{Q_i\}_1^m$ is defined to have the consecutive retrieval (CR for short) property with respect to a set of records $\{R_i\}_1^n$ if there exists an organization of the records $\{R_i\}_1^n$ without duplication of any record, such that for every $Q_i \in \{Q_i\}_1^m$ all pertinent records can be stored in consecutive storage locations.

Many results have been published on this field of research [Al-Fedaghi and Chin 1979, Nakano 1973(a), Nakano 1973(b), Ghosh 1972, Ghosh 1973, Ghosh 1977].

Section 3. Bucket-Oriented Consecutive Retrieval Organization

We indicated before that the CR organization can be used to arrange records into buckets. This CR organization can be slightly modified into the bucket-oriented consecutive retrieval organization and then can be used to arrange buckets into disks.

Definition

An organization of the buckets set $\{B_i\}_1^n$ in which buckets pertinent to any query $\{Q_i\}_1^m$ are stored in consecutive storage areas is defined to be a bucket-oriented CR organization $\{B_i\}_1^n$ for $\{Q_i\}_1^m$.

Example 3.1

Let us consider a well constructed file system. The nine buckets are to be examined by queries as shown below:

Buckets	Queries
B_1	Q_1
B_2	Q_2
B_3	Q_2, Q_3
B_4	Q_1, Q_2
B_5	Q_3
B_6	Q_3
B_7	Q_3
B_8	Q_1, Q_2
B_9	Q_3

The bucket-query incidence matrix is given by

$$
M = \begin{array}{c} B_1 \\ B_2 \\ B_3 \\ B_4 \\ B_5 \\ B_6 \\ B_7 \\ B_8 \\ B_9 \end{array}
\begin{array}{ccc} Q_1 & Q_2 & Q_3 \end{array}
\left(
\begin{array}{ccc}
1 & 0 & 0 \\
0 & 1 & 0 \\
0 & 1 & 1 \\
1 & 1 & 0 \\
0 & 0 & 1 \\
0 & 0 & 1 \\
0 & 0 & 1 \\
1 & 1 & 0 \\
0 & 0 & 1
\end{array}
\right)
$$

The bucket-query incidence matrix with buckets permuted as $\{B_1, B_4, B_8,$ $B_2, B_3, B_9, B_6, B_5, B_7\}$ is given by

$$
M' = \begin{array}{c} B_1 \\ B_4 \\ B_8 \\ B_2 \\ B_3 \\ B_9 \\ B_6 \\ B_5 \\ B_7 \end{array}
\begin{array}{ccc} Q_1 & Q_2 & Q_3 \end{array}
\left(
\begin{array}{ccc}
1 & 0 & 0 \\
1 & 1 & 0 \\
1 & 1 & 0 \\
0 & 1 & 0 \\
0 & 1 & 1 \\
0 & 0 & 1 \\
0 & 0 & 1 \\
0 & 0 & 1 \\
0 & 0 & 1
\end{array}
\right)
$$

M' has consecutive 1's in each column; we say the query set $\{Q_1, Q_2, Q_3\}$ has the bucket-oriented CR property with respect to the bucket set $\{B_1, B_2, B_3, B_4, B_5, B_6, B_7, B_8, B_9\}$.

Assuming these nine buckets are assigned into a 3-disk system (disk labelled 0,1,2). For reasons that will become obvious later, a bucket associates with the j-th row of M' may be assigned to disk unit j mod 3. The distribution of assigning all buckets to three disks is shown in Table 3.1(a) and the time that needed to respond to each query is shown in Table 3.1(b).

Table 3.1(a). The distribution of all buckets onto
the 3 disks.

Buckets	Assigned Disk
B_1	1
B_2	1
B_3	2
B_4	2
B_5	2
B_6	1
B_7	0
B_8	0
B_9	0

Table 3.1(b). Buckets to be examined and time required to
answer each query for buckets distribution
shown in Table 3.1(a).

Queries	Buckets to be examined	Response time
Q_1	B_1, B_4, B_8	1
Q_2	B_2, B_3, B_4, B_8	2
Q_3	B_3, B_5, B_6, B_7, B_9	2

The reader can now see that we can use a bucket-oriented CR organization to obtain a distribution of all buckets among multiple disks,
Amazingly, the response time of each query in Table 3.1(b) is the shortest.

Section 4. Bucket-Oriented CR Organization as a File Allocation Scheme

In the previous section, we gave an outline about how bucket-oriented
CR organization can be used as a file allocation scheme to distribute arbitrarily well constructed file system onto multiple disk systems. In this
section, we show the exact algorithm.

We are given a well established file organization F. Let F contain
n buckets $B_1, B_2, \ldots,$ and B_n. Let $Q=\{Q_1, Q_2, \ldots, Q_m\}$ be a set of all possible
partial match queries. How do we know an incidence matrix has a bucket-
oriented CR structure and how can we rearrange an incidence matrix into
one with a bucket-oriented CR structure? The reader is encouraged to read
[Al-Fedaghi and Chin 1979].

Suppose we now know M, a bucket-query incidence matrix, has a bucket-
oriented CR structure, we can apply the following algorithm to arrange all
buckets onto the ND disks.

Algorithm A: The algorithm which arranges all buckets in disks

Input : A bucket-query incidence matrix $M_{n \times m}$ and the total number of disks
(ND).

Output: The distribution of all buckets onto the ND disks.

Step 1: Rearrange $M_{n \times m}$ into $M'_{n \times m}$ (Let B'_i associated to the bucket B_{j_i} be
the i-th row of matrix $M'_{n \times m}$) a bucket-oriented CR organization.

119

Step 2: Assign bucket B_{j_i} to disk unit i mod ND, for all i=1,2,...,n.

Example 4.1:

Consider Example 3.1 again. Since

$$M' = \begin{array}{c} \\ B_1 \\ B_4 \\ B_8 \\ B_2 \\ B_3 \\ B_9 \\ B_6 \\ B_5 \\ B_7 \end{array} \begin{array}{ccc} Q_1 & Q_2 & Q_3 \\ \left(\begin{array}{ccc} 1 & 0 & 0 \\ 1 & 1 & 0 \\ 1 & 1 & 0 \\ 0 & 1 & 0 \\ 0 & 1 & 1 \\ 0 & 0 & 1 \\ 0 & 0 & 1 \\ 0 & 0 & 1 \\ 0 & 0 & 1 \end{array}\right) \end{array}$$

is a bucket-oriented CR organization of the bucket-query incidence matrix, we can use Algorithm A to arrange all buckets as follows:

B_1 is assigned to disk unit 1 mod 3 = 1

B_4 is assigned to disk unit 2 mod 3 = 2

B_8 is assigned to disk unit 3 mod 3 = 0

B_2 is assigned to disk unit 4 mod 3 = 1

B_3 is assigned to disk unit 5 mod 3 = 2

B_9 is assigned to disk unit 6 mod 3 = 0

B_6 is assigned to disk unit 7 mod 3 = 1

B_5 is assigned to disk unit 8 mod 3 = 2

B_7 is assigned to disk unit 9 mod 3 = 0.

Theorem 4.1

Given a well constructed multi-attribute file F and an ND-disk system. Let $\{B_i\}_1^n = \{B_1, B_2, ..., B_n\}$ be a set of buckets and $\{Q_i\}_1^m$ be a set of queries. If there exists a bucket-oriented CR organization of $\{B_i\}_1^n$, then Algorithm A can achieve the optimal response time for all possible partial match queries.

Proof:

Assuming that the bucket set $\{B_1, B_2, ..., B_n\}$ has a bucket oriented CR organization. Let $\{B_1', B_2', ..., B_n'\}$ for $\{Q_i\}_1^m$ be a bucket-oriented CR organization.

For a query Q_j, let the buckets pertinent to Q_j be $\{B_i', B_{i+1}', ..., B_k'\}$. According to Algorithm A, since B_i' is assigned to disk unit i mod ND, for $1 \leq i \leq n$, it is obvious that the response time of Q_j is $\left\lceil \frac{k-i+1}{ND} \right\rceil$ and $\lceil x \rceil$ indicates the smallest integer which is greater than x. Because the number of buckets need to be accessed in response to query Q_j is k-i+1, the time required to respond to Q_j is optimal. Thus, we can say that Algorithm A

120

can achieve the optimal response time for all possible partial match queries in this case.

Q.E.D.

For any given well constructed file if the theorems and lemmas enable one to establish that the bucket set has the bucket-oriented CR organization with respect to the pertinent set of queries, then we can say that the bucket-oriented CR organization is the best for bucket distribution by using Algorithm A, because it provides the minimum access time without any redundancy.

In practice, the query structure is very complex and there may not exist a bucket-oriented CR structure.

Example 4.2
<u>Example 4.2</u>

Consider the following bucket-query incidence matrix for these nine buckets and four queries is given by

$$
\begin{array}{c}
\quad\quad Q_1 \; Q_2 \; Q_3 \; Q_4 \\
\begin{array}{c}
B_1 \\ B_2 \\ B_3 \\ B_4 \\ B_5 \\ B_6 \\ B_7 \\ B_8 \\ B_9
\end{array}
\left(
\begin{array}{cccc}
1 & 0 & 0 & 0 \\
1 & 1 & 1 & 1 \\
1 & 1 & 1 & 0 \\
0 & 0 & 1 & 0 \\
0 & 0 & 0 & 1 \\
0 & 0 & 0 & 1 \\
0 & 0 & 0 & 1 \\
0 & 0 & 1 & 0 \\
0 & 0 & 0 & 1
\end{array}
\right)
\end{array}
$$

By trying all possible permutations of the rows of the above matrix, it is easy to see that there exists no permutation of the rows for which every column of the matrix has 1's in consecutive rows. Thus the query set has no bucket-oriented CR organization.

So it is important to investigate whether the bucket-oriented CR organization can be used as building buckets for the organization which preserves the consecutive storage of the buckets and has considerably small redundancy. These types of organizations are called the bucket-oriented consecutive retrieval with redundancy (CRWR) organization.

In Example 4.2, the organization $B_1,B_2,B_3,B_4,B_8,B_2,B_5,B_6,B_7,B_9$ contains the buckets pertinent to all four queries $\{Q_1,Q_2,Q_3,Q_4\}$ stored in consecutive storage. The redundancy of this bucket-oriented CRWR organization is $\frac{(10-9)}{9} = \frac{1}{9}$.

Applying the following algorithm to this new organization can also achieve the optimal response time for all four queries.

Algorithm B: The algorithm which arranges all buckets in disks
<u>Algorithm B</u>: The algorithm which arranges all buckets in disks

<u>Input</u> : A bucket-query incidence matrix $M_{n \times m}$ and the total number of disks (ND).

<u>Output</u>: The distribution of all buckets onto the ND disks.

<u>Step 1</u>: Rearrange $M_{n \times m}$ into $M'_{\ell \times m}$, $n < \ell$, (Let B'_i associated to the bucket

121

B_{j_i} be the i-th row of matrix $M'_{\ell \times m}$) a bucket-oriented CRWR organization.

Step 2: Assign bucket B_{j_i} to disk unit i mod ND, for all i=1,2,...,ℓ.

Example 4.3

Consider Example 4.2 again. Suppose these nine buckets are assigned into a 3-disk system. Since

$$M' = \begin{array}{c} \\ B_1 \\ B_2 \\ B_3 \\ B_4 \\ B_8 \\ B_2 \\ B_5 \\ B_6 \\ B_7 \\ B_8 \end{array} \begin{array}{cccc} Q_1 & Q_2 & Q_3 & Q_4 \\ \begin{pmatrix} 1 & 0 & 0 & 0 \\ 1 & 1 & 1 & 0 \\ 1 & 1 & 1 & 0 \\ 0 & 0 & 1 & 0 \\ 0 & 0 & 1 & 0 \\ 0 & 0 & 0 & 1 \\ 0 & 0 & 0 & 1 \\ 0 & 0 & 0 & 1 \\ 0 & 0 & 0 & 1 \\ 0 & 0 & 0 & 1 \end{pmatrix} \end{array}$$

is a bucket-oriented CRWR organization of M, we can use Algorithm B to arrange all buckets as follows

B_1 is assigned to disk unit 1 mod 3 = 1

B_2 is assigned to disk unit 2 mod 3 = 2

B_3 is assigned to disk unit 3 mod 3 = 0

B_4 is assigned to disk unit 4 mod 3 = 1

B_8 is assigned to disk unit 5 mod 3 = 2

B_2 is assigned to disk unit 6 mod 3 = 0

B_5 is assigned to disk unit 7 mod 3 = 1

B_6 is assigned to disk unit 8 mod 3 = 2

B_7 is assigned to disk unit 9 mod 3 = 0

B_9 is assigned to disk unit 10 mod 3 = 1.

The distribution of all buckets onto the 3 disks is listed below.

Assigned disk	Buckets
0	B_3, B_2, B_7
1	B_1, B_4, B_5, B_9
2	B_2, B_6, B_8

The time needed to respond to each query is shown in Table 4.1.

Table 4.1. Buckets to be examined and time required
to answer each query.

Query	Buckets to be examined	Response time
Q_1	B_1,B_2,B_3	1
Q_2	B_2,B_3	1
Q_3	B_2,B_3,B_4,B_8	2
Q_4	B_2,B_5,B_6,B_7,B_9	2

Similarly, from the proving procedures of Theorem 4.1, we can easily
conclude the following theorem.

Theorem 4.2

Given a well constructed multi-attribute file F and an ND-disk system.
Let $\{B_{j_1},B_{j_2},\ldots,B_{j_\ell}\}$ be a bucket-oriented CRWR organization of $\{B_1,B_2,\ldots,$
$B_n\}$, then Algorithm B can achieve the optimal response time for all possi-
ble partial match queries.

Perhaps, the most interesting thing of bucket-oriented CRWR organization is the reduction of the redundancy.

Section 5. Conclusions

In this paper, we presented the concept of bucket-oriented CR organization. We also proposed a very important algorithm based upon the bucket-oriented CR organization to distrubite all buckets into disks. We
showed that our algorithm indeed can achieve the optimal response time for
all possible partial match queries. The concept of using redundancy to
construct a bucket-oriented CRWR organization was introduced.

Our next job is to investigate how to minimize the redundancy of the
bucket-oriented CRWR organization when a bucket-query incidence matrix does
not have a bucket-oriented CR organization.

REFERENCES

1. Aho, A. V. and Ullman, J. D., (1979): Optimal Partial-match Retrieval
    when Fields are Independently Specified, ACM Transactions on Database
    Systems, Vol. 4, No. 2, June 1979, pp. 168-179.
2. Al-Fedaghi and Chin, Y. H., (1979): Algorithmic Approach to the Conse-
    cutive Retrieval Property, International Journal of Computer and
    Information Science, Vol. 8, No. 4, 1979, pp. 279-301.
3. Bentley, J. L. and Friedman, J. H., (1979): Data Structures for Range
    Searching, Computing Surveys, Vol. 11, No. 4, Dec. 1979, pp. 397-
    409.
4. Bolour, A., (1979): Optimality Properties of Multiple Key Hashing Func-
    tions, Journal of the Association for Computing Machinery, Vol. 26,
    No. 2, April 1979, pp. 196-210.
5. Chang, C. C., Du, M. W. and Lee, R. C. T., (1984): Performance Analysis
    of Cartesian Product Files and Random Files, IEEE Trans. Software
    Eng., Vol. SE-10, No. 1, Jan. 1984, pp. 88-99.
6. Chang, C. C., Lee, R. C. T. and Du, H. C., (1980): Some Properties of
    Cartesian Product Files, Proc. ACM-SIGMOD 1980 Conference, Santa
    Monica, Calif., May 1980, pp. 157-168.
7. Chang, C. C., Lee, R. C. T. and Du, M. W., (1982): Symbolic Gray Code
    as a Perfect Multi-attribute Hashing Scheme for Partial Match queries,

IEEE Transactions on Software Engineering, Vol. SE-8, No. 3, May 1982, pp. 235-249.

8. Chang, J. M. and Fu, K. S., (1981): Extended K-d Tree Database Organization: A Dynamic Multi-attribute Clustering Method, _IEEE Trans. Software Eng._, Vol. SE-7, No. 3, May 1981, pp. 284-290.

9. Chang, C. C. and Su, D. H., (1983a): Some Properties of Multiattribute File System Based upon Multiple Key Hashing Functions, _Proceedings of the 21-st Annual Allerton Conference on Communication, Control and Computing_, Urbana, Illinois, Oct. 1983, pp. 675-682.

10. Chang, C. C. and Su, D. H., (1983b): Application of a Heuristic Algorithm to Design Multiple Key Hashing Functions, _Proceedings of NCS 1983 Conference_, Hsinchu, Taiwan, Dec. 1983, pp. 218-231.

11. Chang, C. C. and Su, D. H., (1984): Performance Analysis of Multiattribute Files Based upon Multiple Key Hashing Functions and Haphazard Files, Institute of Applied Mathematics, National Chung Hsing University, Taichung, Taiwan, Tech. Report.

12. Du, H. C. and Sobolewski, J. S., (1982): Disk Allocation for Cartesian Product Files on Multiple Disk Systems, _ACM Trans. Database Systems_, Vol. 7, March, pp. 82-101.

13. Ghosh, S. P., (1972): File Organization: The Consecutive Retrieval Property, _Comm. ACM_, Vol. 15, pp. 802-808.

14. Ghosh, S. P., (1973): On the Theory of Consecutive Storage Relevant Records, _Information Sci._, Vol. 6, pp. 109.

15. Ghosh, S. P., (1977): _Data Base Organization for Data Management_, Academic Press, N. Y., 1977.

16. Jakobsson, M., (1980): Reducing Block Accesses in Inverted Files by Partial Clustering, _Inform. Systems_, Vol. 5, No. 1, 1980, pp. 1-5.

17. Lee, R. C. T., (1981): Clustering Analysis and Its Applications, _Advances in Information Systems Science_, Plenum Press, N. Y., (edited by J. T. Tou) 1981, pp. 169-287.

18. Lee, R. C. T. and Tseng, S. H., (1979): Multi-key Sorting, _Policy Analysis and Information Systems_, Vol. 3, No. 2, Dec. 1979, pp. 1-20.

19. Lin, W. C., Lee, R. C. T. and Du, H. C., (1979): Common Properties of Some Multi-attribute File Systems, _IEEE Transactions on Software Engineering_, Vol. SE-5, No. 2, March 1979, pp. 160-174.

20. Liou, J. H. and Yao, S. B., (1977): Multi-dimensional Clustering for Data Base Organizations, _Information Systems_, Vol. 2, 1977, pp. 187-198.

21. Nakano, T., (1973a): A Characterization of Intervals; The Consecutive (one's or retrieval) Property, _Comment Math. Univ. St. Paul._, 22, pp. 49-59.

22. Nakano, T., (1973b): A Remark on The Consecutivity of Incidence Matrices, _Comment Math. Univ. St. Paul._, Vol. 22, pp. 61-62.

23. Rivest, R. L., (1976): Partial-match Retrieval Algorithms, _SIAM Journal on Computing_, Vol. 14, No. 1, March 1976, pp. 19-50.

24. Rothnie, J. B. and Lozano, T., (1974): Attribute Based File Organization in a Paged Memory Environment, _Communications of the Association for Computing Machinery_, Vol. 17, No. 2, Feb. 1974, pp. 63-69.

TRACER: TRANSPOSED FILE ORGANIZATION SCHEME WITH CONSECUTIVE RETRIEVAL PROPERTY AND ITS APPLICATION TO STATISTICAL DATABASE SYSTEM

Hideto Ikeda, Yoshihumi Ohzawa and Kenji Onaga

Hiroshima University
1-1-89 Higashi-senda, Hiroshima 730 Japan

INTRODUCTION

According to the evaluation of database technology, various applications have been developed on existing database management system (DBMS), however, it is not sufficient for a conventional DMBS to organize statistical data. The problems involve as follows (Batory ,1984; Denning et al, 1984; Shoshani, 1982):
(1) The data model which is supported by a conventional DBMS is inadequate to the description of statistical data.
(2) The performance for access is unsuitable for statistical applications, which mainly depends on the physical storage structure.·
(3) Special security problems exist in statistical databases.
(4) Special interfaces to the existing statistical program packages are required.

Among the problems, the performance problem in (2) is essentially critical because it cannot be solved by additional development of a conventional DBMS. Some physical file organization schemes for statistical database management have been proposed, that is, transposed files (Burnett, 1981; Hoffer, 1975; March, 1982; Turner et al, 1979) and compression algorithms (Alsberg, 1975; Batory, 1983; Gey et al, 1983). These storage schemes are specialized to be efficient for accesses which are frequently required in statistical database systems.

A file organization scheme having consecutive retrieval property (CRFS) is physical storage structure of data which was a proposed and discussed by S. P. Ghosh (1972, 1973, 1975). This scheme enables us to access data efficiently by allocating relevant records to each query on consecutive locations of the atorage device. Ghosh discussed his "consecutive" concept on simplified storage devices, that is, one/two-dimensional storage devices. Ikeda (1983) expanded the concept to general storage devices by introducing a cencept of the "topology of a storage device".

There are many other works on physical file organization. Among then the works by Fuller (1974), Stone et al (1973) and Wilhelm (1976) on disk scheduling policies can be used actually, because they were done from the practical point of view and disks are the most important and popular storage devices as present database storages.

This paper proposes a new file organization scheme TRACER for statistical database systems, which is based on both the generalized CRFS and the transposed file structure. This scheme can fairly improve the access performance of a statistical database system.

In Section 2, we introduce a concept of "topology of a storage device" and show topology of actual devices. Section 3 defines a file organization scheme and discuss a familiar file organization scheme, i.e., a transposed file. Section 4 proposes a new file organization scheme TRACER based on both a transposed file and a CRFS. Section 5 demonstrates an experimental evaluation of TRACER on a floppy disk of a micro-computer.

TOPOLOGY OF STORAGE DEVICES

Ikeda (1983) proposed a concept of "topology of a storage device" which specified intentionally access time between two storage cells of the device. Formally we have the following definition:

<u>Definition</u> 1 (Topology of a storage device): A topology of a storage device $S = \{ s_i \mid i = 1, 2, ..., \ell \}$ is defined as a function $t: S \times S \longrightarrow R$ satisfying the following conditions:

$$(1)\ t(s,s') \geq 0$$

$$(2)\ t(s,s') \leq t(s,s'') + t(s'',s')$$

(2.1)

for any cells s, s' and s" in S, where R is the set of all real numbers.

The paper by Ikeda (1983) analyzed the topologies of three actual storage devices, that is a tape, a drum and a disk. Among them, a disk is the most important to organize statistical data, because disk storages are the most popular on the present computer systems. So we shall restrict storage devices to disks in the following discussions.

The topology of a disk was expressed in the paper as a function t as follows:

$$
t(s^k_{ij}, s^{k'}_{i'j'}) = \begin{cases} c_1(i'-i) + |k'-k|c_2 & \text{if } i' > i \\ c_2(\ell_1+i'-i) + |k'-k|c_2 & \text{otherwise} \end{cases}
$$

(2.2)

for i, i' = 1, 2, ..., ℓ_1; j, j' = 1, 2, ..., ℓ_2 and k, k' = 1, 2, ..., ℓ_3 where c_1 is the access time needed to move the heads to an adjacent sector and c_2 is the time needed to seek an adjacent cylinder. The notation $|p|$ means the absolute value for an integer p.

Although the formula (2.2) is easy to understand as topology of a disk, it is insufficient to describe topology of an actual disk. So we shall revise it from more practical sense by the following parameters in Table 1.

With the parameters in Table 1, we can describe topology of a disk as follows:

$$
\begin{aligned}
t(s^k_{ij}, s^{k'}_{i'j'}) &= \text{Tlate}/\ell_1(d_{ii'} + w) \\
&+ \text{Tseek}|j'-j| \\
&+ \delta_c(\text{Tsetl} + \text{Tload}) \\
&+ \delta_T \text{Tctrn} + \text{Tdtrn} + \text{Tproc}
\end{aligned}
$$

(2.3)

where i" = mod($\lceil a \rceil$ + 1 + i, ℓ_1),

$$
d_{ii'} = \begin{cases} i'-i" & \text{if } i' > i" \\ \ell_1 + i' - i" & \text{otherwise} \end{cases}
$$

a = ℓ_1/Tlate(Tseek$|j-j'|$+ δ_cTsetl +δ_cTload +δ_TTctrn + Tdtrn + Tproc),

Table 1 Characteristic Parameters of a Disk

Notation	Name	Explanation
Tseek	Seek time	Time for moving the heads to an adjacent track
Tsetl	Settling time	Time for settling the heads on the relevant track
Tlate	Latency time	Time for waiting until the head will be on a relevent sector
Tload	Loading time	Time for loading the head on the surface
Tdtrn	Data Transfer time	Time for transfering the data in a sector to the buffer memory
Tproc	Processing time	Expecting time for processing the data accessed
Tctrn	Control transfer time	Time for transfering the control of one head to another

$$w = \lceil a \rceil - a,$$

$$\delta_c = \begin{cases} 0 & \text{if } j = j' \text{ and } k = k' \\ 1 & \text{otherwise} \end{cases}$$

$$\delta_T = \begin{cases} 0 & \text{if } k = k' \\ 1 & \text{otherwise} \end{cases}$$

$\lceil p \rceil$: the minimun integer exceeding p.

Table 2 Values of characteristic parameters
for a floppy disk used in the experiment

Parameter	value	Parameter	value
Tseek	3 ms	Tsetl	15 ms
Tlate	166.6 ms	Tload	50 ms
Tdtrn	62.5 ms	Tproc	0 ms*
Tctrn	0 ms		

* This value is dependent on an individual I/O request.

In order to evaluate the theoretical model in (2.3) of the topology of a disk, we carried out an experiment to observe the access times between two sectors on a floppy disk. Table 2 shows the value of characteristic parameters for the floppy disk used in the experiment.

Figure 1 illustrates access times of data after an access of other data on a different surface. Since the read-write heads move through the adjacent sector while transfering the accessed data, the nearest sector is not the physical adjacent sector. In Figure 1, the connected straight lines are observed times in the experiment and the dotted lines are expected (theoretical) times based on the formula (2.3). The errors are caused by the linear

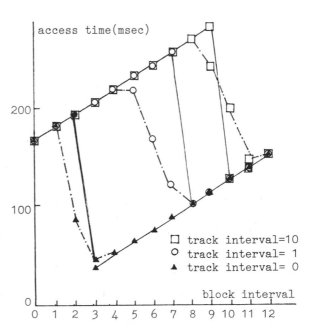

Figure 1. Access time between two sectors on different surface

127

contribution of the seek time Tseek to the access time. The actual
contribution were discussed by Teorey et al (1972). In spite of the
theoretical errors, the theoretical model in (2.3) is enough effective to
practical file organizations.

In the following sections, we use a notation adj(s) as the set of
storage cells;

$$\{ \ s' \ : \ \text{there exists no cell s'' such that } \ t(s,s'') < t(s,s') \}$$

FILE ORGANIZATION SCHEMES

Let A_1, A_2, ..., A_n be n attributes. Associated with each attribute
A_i is a subset of values, called a <u>domain</u> D_i. A <u>relation</u> R is a subset of
a Cartesian product $\underset{i \in I(R)}{\Pi} D_i$ where $I(R)$ is the index set of attributes
contributed in R. Each element of a relation is referred to a <u>record</u>. The
<u>i-th value of a record</u> r, denoted by $v_i(r)$, is defined as an i-th
coordinate of r when it is considered as a vector. The notation $v_{I'}(r)$
means a vector produced by projection of r on a subset I' of I(R). <u>Length
of a value</u> v is the size of storage which is required for the
representation of v and is denoted by $L(v)$.

A <u>database</u> B is a set of relations $\{R_1, R_2, ..., R_m\}$. In order to
implement a database system, it is necessary to organize all relations on
the storage devices. A database consisting of a relation is referred a
<u>single relation database</u>.

A <u>file organization scheme</u> is how to allocate all relations on the
target storage devices. Theoretically a file organization scheme can be
defined as follows:

<u>Definition</u> 2 (File Organization Scheme): A file organization scheme on
a storage device S is defined as a function f: v(B) ---> S satisfying the
following condition:

$$(\overset{V}{s} \ \varepsilon \ S)(L_0 \geq \underset{v_i(r) \ \varepsilon \ f^{-1}(s)}{\Sigma} L(v_i(r)) \ ) \tag{3.1}$$

where L_0 is the length of every storage cell s in S. According to the
definition above of a file organization scheme, we shall express a familiar
file organization schemes, that is, a transposed file organization scheme.
To simplify the following discussions, we assume every database be a single
relation database. The generality of the following discussions is not
however lossed by the assumption.

A <u>transposed file organization scheme</u> is a single relation database on
a disk is a file organization f_{TRN} having the following properties:
(1) If $f_{TRN}(v_p(r_q)) = s^k_{ij}$ for some $q \neq m$, then

$$f_{TRN}(v_p(r_{q+1})) = \begin{cases} s^k_{ij} & \text{if } \underset{v \ \varepsilon \ f^{-1}(s^k_{ij})}{\Sigma} L(v) \leq L_0 - L(v_p(r_{q+1})) \\ \\ next(s^k_{ij}) & \text{otherwise} \end{cases}$$

(2) If $f_{TRN}(v_p(r_m)) = s^k_{ij}$ for some $q \neq m$, then

$$f_{TRN}(v_{p+1}(r_1)) = \begin{cases} s^k_{ij} & \text{if } \underset{v \ \varepsilon \ f^{-1}(s^k_{ij})}{\Sigma} L(v) \leq L_0 - L(v_{p+1}(r_1)) \\ \\ next(s^k_{ij}) & \text{otherwise} \end{cases}$$

where

$$next(s^k_{ij}) = \begin{cases} s^k_{i+1,j} & \text{if } i < \ell_1 \\ s^k_{1,j+1} & \text{if } i = \ell_1 \text{ and } j < \ell_2 \\ s^{k+1}_{1,1} & \text{if } i = \ell_1 \text{ and } j = \ell_2. \end{cases}$$

NEW FILE ORGANIZATION SCHEME TRACER

We shall propose a new file organization scheme, which can improve the access performance of queries contributed substential attributes of all records. These queries appears frequently in statistical databases. A new file organization scheme, called an exact TRACER, is defined as a file organization scheme f_{ETC} having the following properties:
(1) If $f_{ETC}(v_p(r_q)) = s^k_{ij}$ for some $q \neq m$, then

$$f_{ETC}(v_p(r_{q+1})) = \begin{cases} s^k_{ij} & \text{if } \sum\limits_{v \in f^{-1}(s^k_{ij})} L(v) \leq L_0 - L(v_p(r_{q+1})) \\ adj(s^k_{ij}) & \text{otherwise} \end{cases}$$

(2) If $f_{ETC}(v_p(r_m)) = s^k_{ij}$ for some $q \neq m$, then

$$f_{ETC}(v_{p+1}(r_1)) = \begin{cases} s^k_{ij} & \text{if } \sum\limits_{v \in f^{-1}(s^k_{ij})} L(v) \leq L_0 - L(v_{p+1}(r_1)) \\ adj(s^k_{ij}) & \text{otherwise} \end{cases}$$

Sequential files and transposed files can always organized unless the size of file overflows to the storage capacity, however, exact TRACER files do not always exist. In order to avoid the disadvantage, we shall weaken the proprerties required in an exact TRACER.

If $f(v_p(r_q)) = s^k_{ij}$ for some $q \neq m$ and $f(v_p(r_{q+1})) \notin \{s^k_{ij}\} \cup adj(s^k_{ij})$, then allocateion fault of $v_p(r_{q+1})$ in f is defined by the formula:
$$t(f(v_p(r_q)), f(v_p(r_{q+1}))) - t(f(v_p(r_q)), s)$$

where s is an element of $adj(f(v_p(r_q)))$.

If $f(v_p(r_m)) = s^k_{ij}$ and $f(v_{p+1}(r_1)) \notin \{s^k_{ij}\} \cup adj(s^k_{ij})$, then allocation fault of $v_p(r_{q+1})$ in f is defined be the formula:

$$t(f(v_p(r_q)), f(v_{p+1}(r_1))) - t(f(v_1(r_m)), s)$$

where $s \in adj(f(v_1(r_m)))$.

A pseudo TRACER file is a file organization which minimizes the sum of allocation faults. A pseudo TRACER file is an optimal one that can be organized on any storage device and has desireable property as an exact TRACER file. So we shall call simply a TRACER file for a pseudo TRACER file.

A pseudo TRACER file can be constructed by the following alrorithm A. In the algorithm, $A(z, s^k_{ij})$ is the nearest sector that is unoccupied yet.

In order to show the difference of TRACER from the two organization schemes discussed in advance, we shall give examples in Figure 2, for a sample data in Table 3. In the examples, we assume that

$$adj(s^k_{ij}) = \{ s^k_{mod(i+2, 1),j}\} .$$

Table 3 A sample data used in Figure 2

Record No.	Attribute	A	B	C	D
1		a_1	b_1	c_1	d_1
2		a_2	b_2	c_2	d_2
3		a_3	b_3	c_3	d_3
4		a_4	b_4	c_4	d_4
5		a_5	b_5	c_5	d_5
6		a_6	b_6	c_6	d_6
7		a_7	b_7	c_7	d_7
Length of value		10	10	5	5

Algorithm A
begin
    q : = 1;
    z : = 1;
    $f(r_{DIC,\ 1})$: = $A(z,\ s^1{}_{1,1})$;
    while q < m do
        begin
            q : = q + 1;
            if $\sum_{v \in f^{-1}(A(z,\ s^1{}_{1,1}))} L(v) + L(r_{DIC,\ q}) \geq c(\ell_1)$
            then z : = z + 1;
            $f(r_{DIC,\ q})$: = $A(z,\ s^1{}_{1,1})$);
        end;
    while A = ∅ do
        begin
            Select attribute $A_{q'}$, as the following properties;
            $F(A_{q'}) > F(A^0)$ or $(F(A_{q'}) = F(A^0)$ and $L(A_{q'}) \leq L(A^0))$
                for all A^0 in A;
            Remove $A_{q'}$ from A;
            for $A_{q'}$ do
            z : = z + 1;
            for p = 1 to m do
                begin
                    if $\sum_{v \in f^{-1}(A(z,\ s^1{}_{1,1}))} L(v) + L(r_{p,q}) \geq c(\ell_1)$;
                    then z : = z + 1;
                    $f(r_{p,q})$: = $A(z,\ s^1{}_{1,1})$);
                end;
        end;
end.

Surface No.		1				2			
Sector No.		1	2	3	4	1	2	3	4
Track No.									
1		b_2	b_4	b_1	b_3	b_5	b_7	b_6	–
2		DIC_1	$c_1 c_2$	DIC_2	$c_3 c_4$	$c_7 d_1$	$d_4 d_5$	$c_5 c_6$	$d_2 d_3$
3		a_2	a_4	a_1	a_3	a_5	$d_6 d_7$	a_6	a_7

Figure 2 Example of TRACER

EXPERIMENTAL EVALUATION OF TRACER

    The new file organization scheme TRACER proposed in the last Section can be expected to the improvement of the access time for statistical queries. In order to show the improvement practically, we performed an experimentation as follows:

(1) The data used in the experimentation is shown in Table 4, which includes 4,000 records.
(2) On the data, three queries below were prepared,

Q_1 : Print avarage number of employees.

Q_2 : Print name of companies whose capital is greater then $5,000,000.

Q_3 : Print number of companies where the proceed is greater than 3,000,000 dollars and the regular profit is greater than 500,000 dollars and which do not attend the first market.

Table 4 Data used in the experiment

Attributes	Length of value	Probability of access
Name of company	40	0.5
Code of company	10	0.1
Location	80	0.1
Location code	3	0.3
Zip code	5	0.1
No. of employees	6	0.5
Average age of employees	4	0.3
Type of industry	20	0.3
Type code	3	0.1
Capital	8	0.5
Proceed	8	0.5
Profit	8	0.3
Market	1	0.3

Access time each query obtained in the experiment is shown in Table 5

Table 5 Access time obtained in the experiment

Query	Sequential file	Transposed file	TRACER file
Q_1	534	21	4
Q_2	534	128	23
Q_3	534	59	9

In the experiment, the improvement of TRACER for access performance for TRACER file is fairly good in the practical sense.

CONCLUSION AND DISCUSSION

This paper proposed a new file organization scheme TRACER, which enables us to improve the access time for statistical queries. The basic idea of the scheme depends on both transposed file and consecutive retrieval file. A concept of the "topology of a storage device" is also essential for the consideration of practical file organizations on the existing storage devices.

In order to improve the performance, further studies on TRACER are required, e.g., combination with buffering, pre-scheduling of accesses, update algorithms and quick loading techniques. At present, we have implemented TRACER as a file organization scheme for a personal statistical database system on a floppy disk of a micro-computer. Almost existing statistical packages on a micro-computer use a sequential file for the data management. Among these packages, it may be possible to realize 1000% innovation of the access performance by changing the access method.

On the other hand, TRACER is effective for the hard disk on a large-scale computer. When a disk is shared by various users, TRACER has a partial efficiency.

REFERENCES

Alsberg, P. A., 1975, Space and time saving through large database compression and dynamic restructuring, Proceedings of IEEE 63, 1114-1122.

Batory, D. S., 1983, A compression technique for large statistical databases, Proceedings of the 2nd international workshop on statistical database management, 306-314.

Batory, D. S., 1984, Physical storage and implementation issues, A quarterly bulletin of the IEEE computer society technical committee on database engineering 7, 49-52.

Burnett, R. A., 1981, A self-describing data file structure for large data sets, Computer science and statistics: Proceedings of the 13th symposium on the interface, 359-362.

Denning, D., Nicholson, W., Sande, G. and Shoshani, A., 1984, Research topics in statistical database management. A quarterly bulletin of the IEEE computer society technical committee on database engineering 7, 4-9.

Eggers, S.,Olken, F., and Shoshani, A., 1981, A compression for technique for large statistical databases, Proceedings of the VLDB'81, 424-434.

Fuller, S. H., 1974, Minimal-total--processing-time drum and disk scheduling disciplines, Comm. ACM 17, 376-381.

Gey, F., McCarthy, J. L., Merrill, D. and Holmes, H., 1983, Computer-independent data compression for large statistical databases, Proceedings of the 2nd international workshop on statistical databases management, 296-305.

Ghosh, S. P., 1972, File organization: Consecutive retrieval propery, Comm. ACM 15, 802-808.

Ghosh, S. P., 1973, On the theory of consecutive storage of relevant recods, Information Science 6, 1-9.

Ghosh, S. P.,1975, File organization: Consecutive storage of relevant records on drum-type storage, Information and Control 25, 145-165.

Ghosh, S. P., 1975, The consecutive storage of relevant records with redundancy, Comm. ACM 18, 464-471.

Hoffer, J. A., 1975, A clustering approach to the generation of subfiles for the design of a computer data base, Dept. of operations research, Cornell Univ., Ithaca, New York.

Ikeda, H., 1983, Topology of storage devices and generalization of consecutive retieval file organization, Data Base File Organization, Academic press,Inc. 1983, 179-198.

March, S. T., 1982, Techniques for structuring database records, ACM Computing Surveys 15,1, 45-80.

Shoshani, A., 1982, Statistical databases: Characteristics, problems and some solutions, Proceedings of 8-th VLDB, 208-227.

Stone, H. S. and Fuller, S. H., 1973, On the near optimality of the shortest-latency-time-first drum scheduling discipline. Comm. ACM 16, 352-353.

Teorey, T. J. and Pinkerton, 1972, A comparative analysis of disk scheduling policies. Comm. ACM 15, 177-184.

Turner, M. J., Hammond, R. and Cotton, P., 1979, A DBMS for large statistical data bases. Proceedings of the 5th VLDB, 319-327.

Wilhelm, N. C., 1976, An anomaly in disk scheduling: A comparison of FCFS and SSTF seek scheduling using an empirical model for disk accesses, Comm. ACM 19, 13-17.

MULTIPLE QUERY PROCESSING IN LOCAL AREA DATABASE SYSTEM

Shu Gang Shi

Department of Computer Science
Wuhan University
Wuhan, China

If a local area network is realized by a bus line, we have the follow_
ing two problems to be considered. (1)How to use its broadcasting capability.
(2)At the bus line transmission of files must be done serially, which may
cause a bottleneck of the system. thus the file transmission scheduling is
a very important problem for this case. In this paper, we will give procedures for file transmission schedule when global database queries are given.
Each query is specified by a set of files which are used. We also discuss
the case when a partial order on the files is also defined.

INTRODUCTION

Distributed database systems are expected to be used widely. Especially,
a local area network using a bus line seems to be very important for office
use. In this paper, query processing procedures for such a system is given.

Query processing for distributed database systems is handled by various
authors. Generally the following cases are treated.

(1)Optimization of a single given query.
(2)The cost is mainly determined by the data transmission.

For a local area distributed system, the results cannot be directly
applied by the following reasons.

(1)The bus line has a broadcasting capability.
(2)The cost of transmission is mainly determined by the bus occupation
   time independent of the distance between sites. If the bus line is used
   by one site, other sites cannot use the bus line.
(3)By the above property the problem to optimize one query is very much
   different from the problem to optimize multiple queries.

In the paper we will discuss the problem to handle multiple queries
for a local area network using a bus line. The problem is divided into the
two subproblems.

(1)Assignment of a host site for each global query.
(2)Selection of a file transmission order for the bus line.

We assume that each query is specified by a set of files and a partial order on them. If partial orders are not considered, the problem becomes simple, so we first discuss such a case. The file transmission schedule is determined under the condition, (1) to maximize the parallelism or (2) to minimize the total processing time. For the second case a consecutive retrieval file organization procedure can be used. When partial orders are considered, the file transmission schedule problem is shown to be a generalization of the well-known supersequence problem, which is known as NP-complete. Thus we showed a heuristic procedure for this problem.

BASIC DEFINTIONS

Let $F=\{f_1,f_2,\ldots,f_p\}$ be a set of files each of which contains a set of data (or records). We will consider a local area network consisting of sites $s_1, s_2, \ldots, s_n$. Each site s_i stores files expressed by F_i. For different sites s_i and s_j, F_i and F_j need not be disjoint (there may be a file stored at the both sites). F_i is dynamically modified due to the data transmission. We assume that the disk size at each site is not enough to store F. Every F_i is a proper subset of F.

In a distributed system, each query is processed as follows.

(1) Local processing of files at each site.
(2) Combining the result.
(3) Final processing on the result obtained at step (2).

As the cost for step(2) is very high, a set of files required to process a query can be used to represent a query. This kind of query modle is used in the study of the consecutive retrieval property. If we consider the processing cost of step(3), we need to consider ordering of files. A formal definition of a query is as follows.

Each query q_i is defined by a pair (G_i,P_i), Where G_i is the set of files used by q_i and p_i is a partial order on G_i. When p_i is $\emptyset$, query q_i is characterized by G_i only. In such a case q_i can be expressed by $\{f \mid f \in G_i$. We assume the number of queries is m.

Fig.1 shows an example of query $q_1=(G_1,P_1)$ where $G_1=\{f_1,f_2,f_3\}$ and $P_1=\{f_1\leqslant f_3, f_2\leqslant f_3\}$. q_1 requires three files f_1, f_2 and f_3. Since some data of f_1 and f_2 are used to process f_3, partial order p_1 gives a schedule of files to be retrieved.

f_1 f_2

f_3

Fig.1. A partial order on files

A local area network using Ether-type bus connection (see Fig. 2) is considered in this paper. The network has the following two characteristics.

(1) In the network there is always at most one site which can transmit data.
(2) The data sent from one site can be received by more than one site (broadcast).

Although data processing at different sites can be done parallelly, data thransmission on the bus is processed sequentially. The processing time is mainly determined by the bus occupation time (not by the distance of sites for which data transmission is required). Thus we have to reduce the total amount of data transmitted on the bus.

Consider a query which requires three files f_1, f_2 and f_3 which are located at s_1, s_2 and s_3, respectively. One possible mothed of query processing is as follows.

(1) Send f_1 from s_1.
(2) Process the query using f_1 and f_2 at site s_2.
(3) Send the result from s_2.
(4) Process the query at s_3.

HOST SITES FOR QUERIES

If there is only one query, to determine its host site is very easy.

Example 1: Let $q_1 = \{ f_1, f_2, f_3, f_4 \}$,

$\qquad F_1 = \{ f_1, f_2, f_3 \}$,

$\qquad F_2 = \{ f_3, f_4, f_5 \}$,

$\qquad F_3 = \{ f_1, f_5 \}$.

If s_1 is selected as the host site f_4 must be sent from s_2 to s_1. If s_2 is selected as the host site, f_1 and f_2 must be received from other sites. f_2, f_3 and f_4 must be transmitted if s_3 is the host site. If the sizes of files are same, the processing cost is minimum when s_1 is selected as the host site.

Procedure 1: Selection of the host site for a single query.

(1) Calculate the amount of data transmission for each site when it is selected as the host site of the given query.
(2) Select the site which will given the minimum value. The site s_i which does not have any files used by a query q_i cannot be a host site of q_i. This condition can be used to reduce the candidate at step(1).

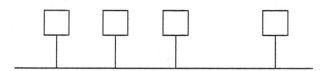

Fig. 2. A local area network using a bus line

When sizes of files are same, the site which has the maximam number of files used by the query is selected.

If there are more than two queries, the following problems must be considered.

(1) By procedure 1 more than one query is assigned to the same site(or the site which will give the minimum cost may be used by another query). In such a case we have two choices; process the queries at the same site or for each query find a different host site which may not give the minimum cost.

(2) If the same set of data is required from more than one site, broadcasting of the data is possible. The data transmission cost is reduced equivalently.

(3) When write operations are permitted, each query cannot use arbitrary copies of files, since "serializability" is required.

Example 2: Let us consider the following case.

$$q_1 = \{ f_1, f_2, f_3 \} \qquad q_2 = \{ f_3, f_4, f_5 \}$$
$$q_3 = \{ f_4, f_6 \} \qquad q_4 = \{ f_3, f_5, f_7 \}$$

for simplicity we assume that each f_i is stored at site s_i (i=1,...,7) and the size of each f_i is same. The hypergraph corresponding to these queries is shown in Fig.3(a). As f_3, f_4 and f_5 shared by more than one query, the cost to transmit these data can be equivalently reduced. The host site for q_1 is s_1 or s_2. The host sites of q_3 and q_4 are s_6 and s_7, respectively, since f_6 and f_7 are not shared. All the files required by q_2 are shared. Independent of the selection of the host site among s_3, s_4, s_5, files f_3, f_4, f_5 must be transmitted. Thus the host site for q_2 can be s_3, s_4, or s_5. Let the host sites of q_1, q_2, q_3 and q_4 be s_1, s_4, s_6 and s_7,respectively. The file at the host site need not be transmitted are shown in Fig.3(b), which will be used for file transmission scheduling.

In the above example, copies of files are not considered. In general, the host site assignment problem is as follows.

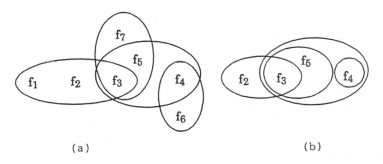

(a)   (b)

Fig. 3. Hypergraphs corresponds to queries

Procedure 2: Assignment of host sites for queries.

(1) Let h_{ij} be a boolean variable such that $h_{ij}{=}^{=1}_{=0}$ if s_i is the host site of q_i. Otherwise h_{ij} satisfies the following equation, since each query has just one site.

$$\text{for} \quad i=1,\ldots,m \quad \sum_{i=1}^{m} h_{ij}=1 \qquad\qquad (i)$$

(2) If there is a constraint such that each site can be a host of just one query, the following equation is required.

$$\text{for} \quad j=1,\ldots,n \quad \sum_{i=1}^{m} h_{ij}=1 \qquad\qquad (ii)$$

(3) When we permit that a site can be host sites of more than one query, it may be reasonable that the total size of files required by queries must not exceed some limit Z. Let c be the size of f.

$$\text{for} \quad j=1,\ldots,n \quad \sum_{i=1}^{m} h_{ij} (\sum_{f_i \in q_i} c_k) \leqslant Z \qquad\qquad (iii)$$

here $\sum c_k$ is the total size of files used by q_i.

(4) Under the constraints above we want to minimize the amount of data transmission on the bus line, which is shown as follows.

$$\sum_{f_i \in T} c_i \qquad\qquad (iv)$$

$$\text{where} \quad T= \bigcup_{i=1}^{m} (q_i-(q_i \cap F_j (h_{ij}=1)))$$

(5) We must solve the following Integer Programming problem.
    minimize : (iv)
    under the constraints : (i)
                            or (i) and (ii)
                            or (i) and (iii)
    If the constraints are (i) or (i) and (ii), the problem is (0,1)-Integer Programming problem.

(iv) is proved as follows. $q_i \cap F_j (h_{ij}=1)$ is the files which are in q_i and at s_j which is the host site of q_i. $q_i-(q_i \cap F_j(h_{ij}=1))$ is the set of files $F_j(h_{ij}=1)$ is the set of files to be transmitted for q_i. Union T of all such sets shows the files to be transmitted through the bus line. c_i shows the total cost of files in T.

Since the complexity of Integer Programming is known to NP, we have to find a method which can be solved easily. Procedure can be used to determine candidates of host sites. Use of broadcasting makes the problem difficult. The procedures can be applied even when there are partial orders.

FILE TRANSMISSION SCHEDULE FOR QUERIES WITHOUT PARTIAL ORDERS

Another problem is to determine the sequence of file transmission, since the bus line requires sequential processing. In this section we will discuss the case when the queries are defined without partial orders. Queries with partial orders will be discussed in next section.

There are the following two approaches for this problem.

(1) Maximize the parallelism among sites.
(2) Minimize the total processing time.

In a local area network usually there are many local processing besides global queries discussed so far. If global queries are dominant, (1) is important and in another case (2) is important. There is a tarde-off problem between (1) and (2). The following Example 3 and Procedure 3 are based on (1).

Example 3: Consider the same example as Example 2. Files to be transmitted are shown in Fig.3(b). We must determine the order of file transmission so that parallelism of query processing is maximized. One possible schedule is as follows.

(1)Transmit f_4 so that processing q_3 can be started.

(2)Transmit f_3 and f_5 so that q_2 and q_4 can be processed.

(3)Transmit f_2.

In order to increase parallelism, files shared by many queries should be transmitted first.

A heuristic method to determine the order is given as follows.

Procedure 3: Procedure to determine the order of file transmission.

(1)r_i of level k is defined to be the maximum number of queries which can be processed by transmitting f_i and k other files. Level 0.

(2)For the given level if there is nonzero r_i , select f_j which maximize r . If the level is not zero other k files must be also selected.

(3)Otherwise increase level by 1 and repeat (2). Repeat until all files are ordered.

If there are many local processing, parallel processing is realized by such local processing. In such a case reduction of the total processing time of global queries is important. This problem is reduced to the organization of the consecutive retrieval file considering buffers.[3,4]

Example 4: The sequence of files determined by Example 3 is shown in Fig.4(a) and (b). Each line corresponds to one query. The length of each line determines the processing time. The total processing is minimized if the order shown in Fig.4(b) is used. This sequence correspond to a consecutive retrieval organization for files required by global queries.

Procedure 4: Procedure to determine the order of file transmission to minimize the total processing time.

(1)For the set of files corresponding to queries, organize a consecutive retrieval file if possible. The result gives the sequence of files to be transmitted at the bus line which minimizes the total processing time.

| f_4 | f_2 | f_5 | f_3 |

(a)

| f_4 | f_5 | f_3 | f_2 |

(b)

Fig. 4. Schedules of file transmission

(2)If organization of a consecutive retrieval files is not possible, organize a consecutive retrieval file considering a buffer. In this case files required by a query need not be arranged consecutively. The span of a query is defined to be the length of a sebsequence which contains all the files required by the query. We need to determine the sequence which minimize the sum of spans of all queries. We can use redundant file transmission (a consecutive file with redundancy) to minimize the sum of spans.

The spans for the file sequence in Fig.4(a) is as follows.
q_1: 3, q_2: 3, q_3: 1, q_4: 2.
The span of q_1 for Fig.4(b) is 2.

FILE TRANSMISSION SCHEDULE FOR QUERIES WITH PARTIAL ORDERS

When each query is specified by a set of files G_i and a partial order p_i, we must construct a file transmission schedule satisfying the partial order. This problem is a general version of the supersequence problem.

Example 5: Consider the two queries $q_1=(G_1, P_1)$ and $q_2=(G_2, P_2)$. The partial orders are shown in Fig.5(a). We assume that q_1 and q_2 are processed at the sites having f_1 and f_2, respectively. By eliminating f_1 and f_2 we get the partial orders shown in Fig.5(b). The total order satisfying the partial order is the file transmission schedule, which is given by f_4, f_3, f_5.

Procedure 5: Procedure to determine a file transmission schedule when queries with partial orders are given.

(1)Determine the host site for eavh query.
(2)From each partial order erase the files which are stored at the host site of the query.
(3)Obtain a minimum seqrence which satisfies the partial orders.

There may be cases when partial orders cannot coexists. For example q_1 specifies $f_1 < f_2$ and q_2 specifies $f_2 < f_1$. There are the following approaches.

(1)Send the same file more than once.
(2)Use the buffer of the sites.

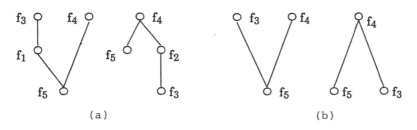

(a)   (b)

Fig. 5. Partial orders defined for queries

In the above example, by the first approach the order is f_1 f_2 f_1. If the host site for q_2 has enough buffer to store f_1 the order can be f_1 f_2. The supersequence problem is as follows.

Supersequence problem: For a set of sequences find a shortest sequence contains all sequence as aubsequences. Here a subsequence is defined as a sequence obtained from the given sequence by eliminating symbols. For example ab is subsequence of acb.

Our problem is a general version of the above problem. As the supersequence problem is known to be NP-complete, our problem is NP-hard. Thus we need to develop a procedure.

Procedure 6: Efficient procedure for a file transmission schedule.

(1),(2) same as procedure 5.
(3) In each partial order determine the minimum values, which correspond to vertices without any outgoing edges in the graph showing the partial order. These vertices are called level 0. Let the vertices which are connected to only vertices of level 0 be level 1.
(4) Consider only vertices of levels 0 and 1 and from a partial order (duplication of vertices is allowed). These values are eliminated from the partial orders for queries. These new partial orders are used by the next application of procedure.
(5) Any total order which obays the partial order at step(4) can be used.

After the transmission of files we may have new set of queries, and considering these queries with the original queries procedure 6 must be applied repeatedly.

CONCLUDING REMARKS

The procedure developed here can be generalized when each query has write operations. In such a case "serializability" must be considered.

Acknowledgement: The author expresses his sincere thanks to Prof. Y. Kambayashi of Kyusyu University, Japan for his discussion and help to prepare the final version of this paper.

REFERENCES

1. S.P.Ghosh et al., "Data Base File Organization-Theory and Applications of the Consecutive Retrieval Property," Academic press,1983.
2. S.P.Ghosh, "File Organization: The Consecutive Retrieval Property," CACM 15,802-808, 1972.
3. K.Tanaka and Y.Kambayashi, "Organization of Quasi-consecutive Retrieval Files," Information Systems 4, 1979.
4. Y.Kambayashi, "The Buffer Limited Quasi-consecutive Retrieval File Organization and its Application to Logic Circuit Layout," In 2.
5. A.R.Hevner and S.B.Yao, "Query Processing in Distributed Database Systems," IEEE Trans. Software Eng. SE-5,177-187,1979.

File Allocation
and
Distributed Databases

MANAGEMENT OF TABLE PARTITIONING AND REPLICATION

IN A DISTRIBUTED RELATIONAL DATABASE SYSTEM

Yoshifumi Masunaga

University of Library and
Information Science
Tsukuba Science-city, Ibaraki 305, Japan

Abstract. This paper investigates managerial issues which may appear when table partitioning and replication is introduced in a distributed relational database system, particularly, in the following four areas: (i) how to define table partitioning and replication, (ii) how to represent it internally, (iii) what is an appropriate method of system cataloging , and (iv) how to ensure database consistency. By introducing dummy replicas, both table partitioning and replication can be treated internally in a unified manner. A system catalog structure which is efficient to maintain the table partitioning and replication information is proposed. Lastly, investigation on database consistency problem is made and nested consistency control mechanism is proposed.

1. INTRODUCTION

In the last several years, great efforts were made in the research and development of distributed database systems. For example, the first distributed database system SDD-1 was implemented by Computer Corporation of America (CCA) /1/. INGRES was modified to a distributed version /2/ and its benchmark test was reported in /3/. IBM is developing a distributed version of System R named System R* /4/. Besides, many other activities also exist: SIRIUS-DELTA by INRIA /5/, ENCOMPASS by Tandem Computers Inc. /6/, DDM by CCA /7/, and others.

We can list up the following reasons why the research and development on distributed database systems were so prosperous in the latter half of the 1970's:

1) The architecture of single site relational database systems has been sucessfully established by that time and the relational systems are suitable for distributed purposes as it was pointed out in /8/.
2) The architecture of computer networks has been sucessfully established by that time. For example, ARPANET /9/ and ETHERNET /10/ were available.
3) The similar request to develop a distributed database system arose from the computer architectural side in order to develop highly available systems /11/.

4) The need of distributed database systems was requested from enterprises in order to raise their productivity.

It is very interesting to point out that most of the distributed database systems are designed and implemented using relational model of data /12/ except a few systems such as the Prime Computer's approach /13/.

In order to use distributed table resources more effectively, it is desirable to introduce table partitioning and replication method. Of course, table partitioning and replication concept had already been introduced in SDD-1 and the distributed version of INGRES. However, these systems do not allow the users to partition and replicate tables freely.

In this paper, we introduce a very relaxed way of table partitioning and replication mechanism. In genegal, the following is the list of topics where discussion should be made before going to table partitioning and replication activities:

a) How to define table partitioning and replication.
b) How to represent a partitioned and replicated tables internally.
c) How to design a system catalog which can accommodate the information about partitioned and replicated table.
d) How to guarantee the distributed database consistency.
e) How to process queries in such circumstances.
f) How to manage transactions in such circumstances.
g) How to authorize table partitioners and replicators.
h) How to define and maintain indexes of fragments and replicas.
i) What is an optimal No. of tables that should be created in partitioning and replication process?

We discuss the problems from a) to d) in detail in section 2 to 5 respectively. Section 6 summaries the theme of this paper. We choose System R* which is a typical homogeneous distributed relational database system /4/ as a distributed environment.

2. TABLE PARTITIONING AND REPLICATION

We introduce two types of table partitionings called horizontal and vertical partitioning, table replication, and table rebuilding concept in this section.

2.1 Horizontal Table Partitioning

Definition 1. A table T is said to be partitioned into n horizontal fragments T1, T2, ..., Tn if and only if (i) T1 APPEND T2 APPEND ... APPEND Tn = T, and (ii) Ti and Tj do not have any tuple in common for all i and j (i $\neq$ j).

Notice that APPEND is introduced as a table union operation which preserves duplicate tuples. Traditional UNION operation is a set theoretic operation and therefore it do not allow the existence of duplicate tuples: Suppose R={a, b} and S={b, c}. Then R APPEND S ={a,b,b,c}, while R UNION S = {a, b, c}. This distinction was paid since SQL tables allow duplicate tuples, while INGRES doesn't.

The next shows an SQL like statement to define horizontal table partitioning. We adopt a syntactic structure proposed in /4/.

Format 1. (Horizontal table partitioning)

```
DISTRIBUTE TABLE table-SWN HORIZONTALLY INTO
    fragment-name-1 WHERE search-condition-1
        (IN database-space-name-1@store-site-name-1)
    .....
    fragment-name-n WHERE search-condition-n
        (IN database-space-name-n@ store-site-name-n);
```

Table-SWN (System Wide Name) is the name of a table which is capable of identifying it uniquely in a distributed environment. In System R*, it is defined as a quadruple:
```
  (table-creator-name@table-creator-site-name.
                        table-name@table-birth-site-name) /4/.
```
For example, if John logged on at Kyoto created EMP(NAME, LOCATION, SALARY) table at Tokyo (and stored it in dbspace1@Osaka), then the table's SWN is John@Kyoto.EMP@Tokyo. The IN clauses are used to specify both a database space name, i.e. the name of the segment in which the fragment is contained and its site. But this is optionl and the default is same as of the parent table. For example, if we want to partition EMP into two fragments named EMP-TOKYO and EMP-OSAKA stored in dbspaceA@Tokyo and dbspace1@Osaka, respectively, then we may issue the following query:

Example 1

```
DISTRIBUTE TABLE John@Kyoto.EMP@Tokyo HORIZONTALLY INTO
    EMP-TOKYO WHERE LOCATION="Tokyo"
        IN dbspaceA@Tokyo
    EMP-OSAKA WHERE LOCATION="Osaka"
        IN dbspace1@Osaka;
```

Although search conditions given in the above example are very simple, unlike SDD-1 and the distributed version of INGRES, we do not restrict in writing arbitrary complex conditions as provided in the traditional SQL language. Therefore, we can use other tables than EMP to partition if it is necessary. The table which is partitioned is called the parent table and its descendants are called fragments.

Since a fragment is also a table, it must have a unique SWN. For this purpose we define the creator, the creator site, and the birth site of a fragment as follows:
```
    The creator of a horizontal fragment
        = the creator of the parent table.
    The creator site of a horizontal fragment
        = the creator site of the parent table.
    The birth site of a horizontal fragment
        = the birth site of the parent table.
```

In the above case, John@Kyoto.EMP-TOKYO@Tokyo is the SWN of fragment EMP-TOKYO. The reason why we define in this way for the creator and the creator site is that even though a table partitioner differs from the table creator, he must be granted to partition it by the creator directly or indirectly, and therefore we can regard virtually that the creator did it. The reason of inherient of parent's birth site information into the children which is a fragment birth site definition, is that we can manage both the parent table and its horizontal fragments information in a single system catalog table in order to process the table partitioning and replication information very efficiently (see section 4).

2.2 Vertical Table Partitioning

Definition 2. A table T is said to be partitioned into n vertical fragments

145

T1, T2, ..., Tn if and only if (i) Ti is a projection on a set of attributes of T for every i, and (ii) T1, T2, ..., Tn is an information lossless decomposition of T.

By information lossless decomposition, we mean the decomposition of a table having a way to rebuild the original table from its fragments. For example, a decomposition using functional dependency and multi-valued dependency is an information lossless decomposition. Moreover, decomposition using join dependency is also information lossless /12/. However, it is not our purpose to discuss which alternative should be choosen. By the term projection, we mean the traditional relational projection operation. In order to ensure that the information losslessness is maintained in the decomposition of tables with duplicate tuples (like SQL tables), we perform natural join operation ** which preserves tuple duplication. Tuple identifiers can be used to implement this operation. (detailed discussion is omitted here.)

An SQL like syntax to define vertical table partitioning is given next:

Format 2. (Vertical table partitioning)

```
DISTRIBUTE TABLE table-SWN VERTICALLY INTO
    fragment-name-1 WHERE search-condition-1
        (IN database-space-name-1@store-site-name-1)
    .....
    fragment-name-n WHERE search-condition-n
        (IN database-space-name-n@store-site-name-n);
```

Terms used in this definition have the same meaning as given in Format 1. Like horizontal partitioning, we can use arbitrary complex search conditions to partition a table vertically. A simple example is given next in which EMP(NAME, LOCATION, SALARY) is vertically partitioned into EMP-LOC(NAME, LOCATION) and EMP-SAL(NAME, SALARY) fragment.

Example 2

```
DISTRIBUTE TABLE EMPLOYEE VERTICALLY INTO
    EMP-LOC WHERE SELECT NAME, LOCATION
                FROM EMPLOYEE
    EMP-SAL WHERE SELECT NAME, SALARY
                FROM EMPLOYEE;
```

Here, we write EMPLOYEE instead of John@Kyoto.EMP@Tokyo assuming that it has already been registered as its synonym.

The system wide names of vertical fragments are defined in the same manner as was done for the horizontal partitioning case.

2.3 Table Replication

Table replication means to make replicas (copies) of a given table. The next shows an SQL like statement to define replicas.

Format 3 (Table replication)

```
DISTRIBUTED TABLE table-SWN REPLICATED INTO
    replica-name-1 (IN database-space-name-1@store-site-name-1)
    .....
    replica-name-n (IN database-space-name-n@store-site-name-n);
```

Of course each replica should have a unique system wide name. Like partitioning, we define the creator (and his site) of replicas and the replica birth site as that of the parent table. The following statement defines two replicas EMP–REP1 and EMP–REP2. (In this case EMP–REP1 and EMP–REP2 are stored in EMP store site by default since no IN clauses are specified.):

Example 3

DISTRIBUTE TABLE EMPLOYEE REPLICATED INTO
    EMP–REP1
    EMP–REP2;

2.4   REBUILD Statement

In order to cancel all partitioning and replication processes carried on a table we introduce a REBUILD statement with SQL like syntax and is given below:

Format 4

REBUILD TABLE table-SWN;

Notice that the effect of the cancellation should be propagated down into all its fragments and replicas recursively so that the designated table can be rebuild. For example if we issue
    REBUILD TABLE EMPLOYEE;
then all partitionings and replications issued to EMPLOYEE and its descendants will be cancelled and EMPLOYEE will be reconstructed.

3.   LINEAR REPRESENTATION OF TABLE PARTITIONING AND REPLICATION

3.1   Table Partitioning and Replication Rule

We can set up a simple and clear rule to dominate the table partitioning and replication considered in this paper. Thus it is : <u>Any actual tables are subject to partitioning or replication</u>. That is, when a table is partitioned, then it becomes virtual, while its fragments become actual tables. Therefore, they are subject to further partitioning and replication. In the case of table replication, both the parent table and its replicas are actual and therefore, both of them are subject to further partitioning and replication.

3.2   Linear Representation

In this section, we propose a <u>linear</u> tree method to represent table partitioning and replication structure. By linear, we mean a creation of tree of partitioning and replication in which a further partitioning and replication can be added recursively as a new tree to one of its leaves. In order to do so, we shall introduce <u>dummy</u> replicas when we duplicate tables:

First, it is clear that every table partitioning structure can be represented as a linear tree. For example, suppose table EMPLOYEE (=John@Kyoto.EMP@Tokyo) has partitioned into EMP–TOKYO and EMP–OSAKA horizontally. Then it is represented as follows:

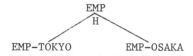

Fig. 1.   A horizontal partitioning tree.

In the tree, H denotes the "horizontal" table partitioning. We say that this tree representation is linear because only leaves, i.e. EMP-TOKYO and EMP-OSAKA are actual (and are subject to further partitioning and replication). (Notice that EMP is virtual.) Similarly, we use character V to represent "vertical" partitioning.

Now we discuss table replication. Suppose table EMPLOYEE is replicated into EMP-REP1 and EMP-REP2. One might try to represent it as follows:

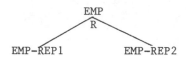

Fig. 2. A replication tree (tentative).

However, clearly, this representation is not linear because not only leaf tables but also the parent table EMP is actual.

In order to represent both table partitioning and replication in a unified mammar, we try to make replication trees also linear by introducing _dummy_ replicas. That is, whenever a table is replicated, we ask the database management system to create one more replica which is called a dummy replica, automatically, to make the parent table virtual. In the above example, a dummy replica named EMP-DUMMY will be created, and therefore EMP becomes virtual. Thus, a new replication representation tree of the above example is as follows:

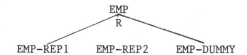

Fig. 3. A replication tree (linearlity assured).

The automatically generated replica, EMP-DUMMY will be stored in the same database space and the same site in which EMPLOYEE is stored. Management of dummy replicas is discussed in the next section.

Example 4

The following tree represents a sequence of actions on partitioning and replication : First T is replicated into TR1 (then T-DUMMY is created by the system), secondly, TR1 is again replicated into TR2 (then TR1-DUMMY is created by the system), thirdly, T is horizontally partitioned into TH1 and TH2 (this is interpreted by the system as if T-DUMMY is partitioned), and lastly, TH2 is replicated into TR3 (then TH2-DUMMY is created by the system).

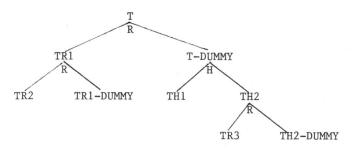

Fig. 4. A sample table partitioning and replication tree.

4. SYSTEM CATALOG STRUCTURE

4.1 An Existing System Catalog Structure

A system catalog is a set of system owned tables in which meta data about database is stored and is used for management purpose. We refer to the System R* system catalog structure defined in /4/ as an existing system catalog structure (in which no table partitioning and replication is supported). The system catalog consists of a bunch of relational tables in which meta data about tables, attributes, indexes, synonyms, authorization, etc. are registered. Among which one table called SYSCATALOG is maintained to store meta data about data objects such as tables, snapshots, and views. In this paper, we call it SYSTABLE because we are only concerned about the table (including fragment and replica) information. The next shows a part of the SYSTABLE structure:

Table 1. SYSTABLE schema (in which no table partitioning and replication information can be accommodated).

<u>SYSTABLE:</u>

> TABID (table identifier)
> TABNAME (table name)
> TABTYPE (table type)
> TBSITE (table birth site name)
> CREATOR (table creator name)
> CRSITE (table creator site name)
> TSSITE (table store site name)
> . . . . .

A table will be registered in the SYSTABLE of a site if and only if (a) it is born in that site, or (b) it is stored in that site, or (c) it is cached in that site /4/. The table information will be kept in its birth site until table is deleted.

4.2 NEW-SYSTABLE

To accommodate table partitioning and replication information, SYSTABLE is then modified to NEW-SYSTABLE and thus defined as follows:

Table 2. NEW-SYSTABLE Schema.

<u>NEW-SYSTABLE:</u>

> TABID
> TABNAME
> TABTYPE
> TBSITE
> CREATOR
> CRSITE
> TSSITE
> PRTYPE (partitioning or replication type)
> PARENT (parent table name)
> DUMMY (dummy table name)
> . . . . .

The three modifications that were made in SYSTABLE are:

1) New attributes called PRTYPE, PARENT, and DUMMY were added.
2) The TABTYPE domain was expanded so that it can take value AF.

149

3) The semantics of TSSITE (table store site) is modified in order to accommodate an afterimage store site.

Attribute PRTYPE takes value either H or V or R corresponding to the horizontal partitioning or vertical partitioning or replication, respectively. When a fragment or a replica is registered, the PARENT attribute in the NEW-SYSTABLE takes the value from the third component of its parent table. In our representation scheme, a dummy replica is introduced when a table is replicated. If a table is a dummy replica, then DUMMY takes Y (yes). If a table is an afterimage (defined in the next section), then its TABTYPE takes AF (AFterimage). In this case, TSSITE designates the afterimage store site.

4.3 Afterimages and Their Store Sites

In order to guarantee database consistency, we shall introduce a new consistency control mechanism in chapter 5. However, in order to execute this mechanism efficiently, we introduce afterimages and their store sites.

Definition 3.   A replica (including a dummy replica) which is further partitioned either horizontally or vertically is an afterimage of that replica. The site in which the replica is stored is called the afterimage store site. Similarly, a fragment which is further replicated is also an afterimage of that fragment. Its store site is also defined in a similar manner.

The creator, the creator site, and the birth site of an afterimage are set equivalent to those of the original replica or fragment. An afterimage is not an actual table, it will not be registered in the NEW-SYSTABLE in the afterimage store site. It is only registered in its birth site. Therefore, a table is registered in the NEW-SYSTABLE of a site if and only if (a) it is born in that site, or (b) it is actually stored in that site, or (c) it is cached in that site.   The table information in its birth site shall be kept until table is deleted.

4.4 Dummy Replicas

As we discussed in section 3.2, the creation of dummy replicas to make the replication tree linear we should know how much extra effort is necessary to create them. The following is a simple and easiest method where dummy replicas can be created with very little effort.

Suppose we have EMPLOYEE (=John@Kyoto.EMP@Tokyo) which is stored in Osaka, and want to make its replica EMP-REP at Tokyo. Then the dummy replica EMP-DUMMY is automatically created by the system at Osaka, and at this time EMP becomes virtual. This means that EMP entry must be deleted from Osaka NEW-SYSTABLE, and the new entry for EMP-DUMMY must be made at the same time. These two requirements can be accomplished within a single action: i.e. Take the EMP tuple, and modify TABNAME from EMP to EMP-DUMMY, and assign EMP and Y value to PARENT and DUMMY arrtibute of a NEW-SYSTABLE respectively. This means that we need not to do any actual table generation for EMP-DUMMY. Therefore the introduction of dummies do not cause any serious performance problem.

4.5 NEW-SYSTABLE Entries -An Example-

In order to explain how entries of NEW-SYSTABLEs changes corresponding to table partitioning and replication, we can illustrate in the example below. Suppose table EMPLOYEE whose SWN is John@Kyoto.EMP@Tokyo is created,

150

and stored at Osaka, then the Table 3 shows NEW-SYSTABLE entries at Tokyo and Osaka at that time.

Table 3.   Entries of NEW-SYSTABLEs when EMP was created.

NEW-SYSTABLE at Tokyo:

```
TABID       0
TABNAME     EMP
TABTYPE     R
TBSITE      Tokyo
CREATOR     John
CRSITE      Kyoto        ..........
TSSITE      Osaka
PRTYPE
PARENT
DUMMY
.....
```

NEW-SYSTABLE at Osaka:

```
TABID       1686
TABNAME     EMP
TABTYPE     R
TBSITE      Tokyo
CREATOR     John
CRSITE      Kyoto        ..........
TSSITE      Osaka
PRTYPE
PARENT
DUMMY
.....
```

Notice that the tuple in NEW-SYSTABLE at Tokyo states that table EMP was created by John from Kyoto at Tokyo site and is stored in Osaka. Therefore, the TABID value is zero. TABTYPE takes value R since the table is actual in the SYSTABLE site. The tuple in NEW-SYSTABLE at Osaka states that John@Kyoto.EMP@Tokyo is stored in this site, and therefore its TABID value takes a non-zero positive number, say 1686. (As the TABID value, it might be better to use system wide clock rather than a locally defined identification number since tables are subject to migrate under table partitioning and replication. We propose to use this approach, while SDD-1 use the local identifier /1/.)

Next, suppose a replica of EMP, named EMP-REP, is created and stored at Tokyo. Then the entries of NEW-SYSTABLEs look as follows:

Table 4.   Entries of NEW-SYSTABLEs after EMP was replicated.

NEW-SYSTABLE at Tokyo:

TABID	0	2011	0	
TABNAME	EMP	EMP-REP	EMP-DUMMY	
TABTYPE		R	R	
TBSITE	Tokyo	Tokyo	Tokyo	
CREATOR	John	John	John	
CRSITE	Kyoto	Kyoto	Kyoto	....
TSSITE		Tokyo	Osaka	
PRTYPE	R			
PARENT		EMP	EMP	
DUMMY			Y	
.....				

151

NEW-SYSTABLE at Osaka:

```
TABID      1686
TABNAME    EMP-DUMMY
TABTYPE    R
TBSITE     Tokyo
CREATOR    John
CRSITE     Kyoto          ..........
TSSITE     Osaka
PRTYPE
PARENT     EMP
DUMMY      Y
.....
```

   Notice that the first column of NEW-SYSTABLE at Tokyo states that EMP was replicated. The second column states that EMP-REP was created as a replica of EMP. Non zero TABID is given since it is stored in this site, Tokyo. The third column states that EMP-DUMMY was created as the system generated dummy replica of EMP and is stored at Osaka. If we look at Osaka site, then we can recognize that TABID value of EMP-DUMMY was set equal to that of EMP i.e. 1686.

   Next, suppose EMP-REP is horizontally partitioned into EMP-TOKYO and EMP-SAPPORO and stored at Tokyo and Sapporo respectively. Then NEW-SYSTABLE entries become as follows:

Table 5.   Entries of NEW-SYSTABLEs after EMP-REP was partitioned.

NEW-SYSTABLE at Tokyo:

TABID	0	0	0	6890	0
TABNAME	EMP	EMP-REP	EMP-DUMMY	EMP-TOKYO	EMP-SAPPORO
TABTYPE		AF	R	R	R
TBSITE	Tokyo	Tokyo	Tokyo	Tokyo	Tokyo
CREATOR	John	John	John	John	John
CRSITE	Kyoto	Kyoto	Kyoto	Kyoto	Kyoto
TSSITE		Tokyo	Osaka	Tokyo	Sapporo
PRTYPE	R	H			
PARENT		EMP	EMP	EMP-REP	EMP-REP
DUMMY			Y		
.....					

NEW-SYSTABLE at Osaka:

```
TABID      1686
TABNAME    EMP-DUMMY
TABTYPÉ    R
TBSITE     Tokyo
CREATOR    John
CRSITE     Kyoto          ..........
TSSITE     Osaka
PRTYPE
PARENT     EMP
DUMMY      Y
.....
```

NEW-SYSTABLE at Sapporo:

```
        TABID      8020
        TABNAME    EMP-SAPPORO
        TABTYPE    R
        TBSITE     Tokyo
        CREATOR    John
        CRSITE     Kyoto        ..........
        TSSITE     Sapporo
        PRTYPE
        PARENT     EMP-REP
        DUMMY
        .....
```

Notice that as the second column of Tokyo in NEW-SYSTABLE states that EMP-REP is now an afterimage (therefore TABTYPE takes value AF), and the EMP-REP store site remains as the afteriamge store site.

5. DATABASE CONSISTENCY CONTROL UNDER TABLE PARTITIONING AND REPLICATION

5.1 Database Consistency Problem

When tables are replicated in a distributed relational database system, the same data may appear in different sites. Therefore the problem is how to avoid inconsistent database state. In other words, the problem is how to avoid the situation in which a part of the duplicate data is new and the rest remains old, when both are offered to users. This is called the database consistency control problem.

The database consistency control problem has been discussed extensively by several authors as far as table replication is concerned. These strategies can be classified mainly into the following three classes /14/:

1) The unanimous agreement update strategy.
2) The primary-secondary update strategy.
3) The majority consensus update strategy.

The single primary update strategy which is a primary-secondary update strategy was expanded to the moving primary update strategy by Alsberg and Day /15/. The majority consensus approach was proposed by Thomas /16/. Each of them has merits and demerits. It is not our purpose to discuss it further.

In the following sections, we expand the single primary update strategy to work in our table replication and partitioning circumstance.

5.2 P-groups and C-groups

In any table partitioning and replication tree we can classify every nodes into two groups. They are p-groups and c-groups nodes. We define them as:

Definition 4 (p-groups). Suppose a table partitioning and replication tree is given. Then p-groups of node of the tree T is denoted by P-G(T) and is defined as follows: (i) If T is not partitioned, then P-G(T) is empty. (ii) If not, (a) set P-G(T) ={Ti | Ti is a fragment of T}. (b) For all i,

if P-G(Ti) is not empty, then set P-G(T) = (P-G(T) - { Ti }) UNION P-G(Ti).
(c) Do step (b) recursively until all P-G(Ti)s become empty.

<u>Definition 5 (c-groups).</u> Suppose a table partitioning and replication tree
is given. Then c-groups of node of the tree T is denoted by C-G(T) and is
defined as follows: (i) If T is not replicated, then C-G(T) is empty. (ii)
If not, (a) set C-G(T) ={Ti | Ti is a replica of T }. (b) For all i, if
C-G(Ti) is not empty, then set C-G(T) = (C-G(T) - { Ti }) UNION C-G(Ti). (c)
Do step (b) recursively until all C-G(Ti)s become empty.

   P-groups and c-groups stand for partitioning group and consistency
group, respectively. An example of it is shown below using a tree introduced
in Figure 4.

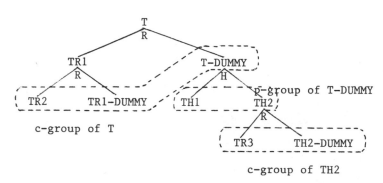

Fig. 5. P-groups and c-groups   - an example-.

5.3 Partition and Replication Transparency

   Tables are partitioned and replicated for various reasons such as to
enhance performance, availability, security, etc.. However, we provide
partitioning and replication transparency of tables to the users where user
can use tables as if they were not partitioned or replicated.   The
transparency can not be realized without providing a good consistency control
mechanism.

   In order to discuss this problem, we introduce a mechanism of
translating updates which are issued against a table and then propagate down
into its leaf tables, i.e. its fragments and replicas.   The basic update
translation scheme is:

1) If an update is issued against a table which is partitioned horizontally
   or vertically, then the update issued against it is propagated down into
   each member of its p-group so that the intended table update can be
   realized synchronously by composing results of the members.
2) If an update is issued against a table which is replicated, then ask its
   c-group representative to synchronize updates among members.
3) Do recursively 1) and 2) until no p-group and c-group is found.

   By the c-group representative we mean the primary replica of the single
primary update strategy. Notice that this scheme becomes a premise of the
consistency control mechanism proposed in the next section.

5.4 Nested Consistency Control Mechanism

   We propose a consistency control mechanism which is also called a nested consistency control mechanism for our purpose. First of all we will build up a p-c-skeleton of a table partitioning and replication tree using the following definition.

Definition 6. (p-c-skeleton). Given a table partitioning and replication tree of T.
    (i)   If T is also a leaf, then it is the p-c-skeleton.
   (ii)   If T is partitioned, then erase all paths from T to every P-G(T) members and span a new arc from T to each member of P-G(T).
  (iii)   If T is replicated, then erase all paths from T to every C-G(T) members and span a new arc from T to each member of C-G(T).
  (iv)   Perform step (ii) and (iii) recursively until every p-group and c-group becomes empty.

    Figure 6 shows the p-c-skeleton of Figure 5.

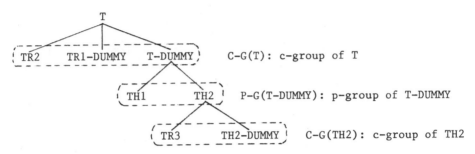

Fig. 6. P-c-skeleton of T - an example-.

    Based on the p-c-skeleton, we can derive a desired database consistency control algorithm. We call a site in which a query or an update request is issued the master site. If a user (logged on) at a certain site issues any query or update, then that site is the master site. The site which cooperates with the master site to perform the update is called the subsidiary sites.

Nested database consistency control algorithm

    Suppose a user at site A issues an update U to table T. Then the following series of actions will be performed:

   1)   Site A constructs the p-c-skeleton of T.
   2)   If it is a single node tree, i.e. T is neither partitioned nor replicated, then site A executes the update query U. Otherwise, it performs step 3).
   3)   If the next level of T constitutes a p-group, then site A asks every site of P-G(T) to be a subsidiary. If a subsidiary site is an afterimage site, then it behaves as the master site to perform update U to that afterimage. (Therefore, this control algorithm can be applied recursively, i.e. in the nested manner. The update result will be reported to the parent master site.) Otherwise, it performs step 4).
   4)   If the next level of T constitutes a c-group, then site A asks either the T-DUMMY stored site, where table is actual or T-DUMMY's afterimage stored site where table is virtual to be the primary site of the single primary update strategy. All other sites of C-G(T) become secondaries. If the primary or a secondary is an afterimage

155

store site, then it behaves as a new master site to perform the update request.

5) Step 3) and 4) can be performed recursively, until all p- and c-groups are exhausted.

For example, suppose update U is issued to table T whose p-c-skeleton is shown in Figure 6. Since T is replicated and its c-group consists of TR2, TR1-DUMMY and T-DUMMY, the master, i.e. the site which accepts update request U for T, asks T-DUMMY (afterimage store) site to be the primary site to carry on the single primary update strategy with association of its secondaries TR2 and TR1-DUMMY. In this case, since T-DUMMY is an afterimage, it becomes a master and asks TH1 store site and TH2 site to perform update U as subsidiaries. Again since TH2 is replicated, it asks TH2-DUMMY site to be the primary to synchronize update U for TR3 and TH2-DUMMY. T can be updated successfully if the success of update synchronization in C-G(TH2) is reported to TH2 site from TH2-DUMMY site, T-DUMMY site receives the success messages from both TH1 and TH2 sites, (notice that TH2 site can not report the success until it receives the success report from TH2-DUMMY site), and at last T site receives the success of update synchronization from T-DUMMY site. This process completes the desired update task. It is clear that the consistency of the first c-group, C-G(T), can not be ensured before the consistency of the second c-group, C-G(TH2), is ensured. Therefore we have a nested c-group structure. (By table T site, we mean either its store site or its afterimage site.)

The special features of the nested consistency control algorithm is presented here:

1) We introduce the p-groups and c-groups table which constitutes a tree hierarchy. The tree structure describes the nested control flow of update propagation down to fragments and replicas.
2) We introduce system generated dummy replicas. They play an important role in settleing down the update synchronization among replicas and update propagation to the fragments in lower level.
3) We introduce afterimages and their store sites (although no actual tables are stored!). They also play an important role in settleing down the update synchronization among replicas and update propagation to fragments in lower level.
4) We adopte the single primary update strategy to synchronize updates to a c-group. One of the reasons of doing so is that it sounds good to ask a dummy replica or its afterimage store site to be the primary site since the dummy is stored or was stored in that site before partitioning. (Notice that our primaries may change corresponding to the different c-groups.)
5) Basically, there is no reason to ask such a dummy replica site to be a primary. However, one reasonable way to select primary may depend on the overall cost analysis and it seems that this approach is quite complex. Although our choice is not optimal, we believe that our approach can be easily implemented.
6) By asking a dummy replica or its afterimage store site to be the primary, the system work load on update synchronization can be distributed widely. Many sites may cooperate with to perform it.
7) The algorithm has freedom to modify towards the utilization of other consistency control strategies than the single primary update strategy for local control.

We believe that our approach works well because first, it looks easy to be implemented, second, the work load balance among sites in order to achieve consistency control. There is a need of further discussion on the entire update propagation and committment control mechanism.

6. SUMMARY

We introduced table partitioning and replication in a distributed relational database system. We defined partitioning and replication process, its internal representation, system catalog structure, and nested database consistency control algorithm. Although the problems from a) to d) are discussed in this paper, the problems from e) to i) listed in chapter 1 are left for open discussion.

Acknowledgements

This work was initiated during the author's stay at the IBM Research Laboratory at San Jose, California. The author expresses his sincere thanks to Dr. Pat Selinger and all members of System R* project who encouraged him to investigate this research topic. The author also expresses his sincere thanks to Prof. S. Noguchi, Tohoku University, Dr. S. Uchinami, Osaka University, Prof. R. Hotaka, Tsukuba University, and Dr. S. Uemura, the Electro Technical Laboratory of Japan for their helpful discussions. This work is partly supported by the research grant of the Ministry of Education of Japan, under the grant number 58580017.

References

/1/ Rothnie, Jr. J.B., Bernstein, P.A., Fox, S., Goodman, M, Hammer, T.A., Landers, T.A., Reeve, C., Shipman, D.W. and Wong, E.: Introduction to a System for Distributed Databases (SDD-1), ACM TODS, Vol.5, No. 1, pp. 1-17 (1980).
/2/ Stonebraker, M. and Neuhold, E.: A Distributed Data Base Version of INGRES, Proc. of the 2nd Berkeley Workshop on Distributed Data Management and Computer Networks, pp. 19-36 (1977).
/3/ Stonebraker, M., Woodfill, J., Ranstrom, J., Murphy, M., Kalash, J., Carey, M. and Arnold, K.: Performance Analysis of Distributed Data Base Systems, Database Engineering, Vol. 5, No. 4, pp. 58-65 (1982).
/4/ Williams, R., Daniels, D., Hass, L., Lapis, G., Lindsay, B., Ng, P., Obermarck, R., Selinger, P., Walker, P., Wilms, P. and Yost, R.: R*: An Overview of the Architecture, Proc. International Conference on Database Systems, Jerusalem, Israel, pp. 1-27 (1982).
/5/ Le Bihan, J., Esculier, C., Le Lann, G., Litwin, W., Gardarin, G., Sedillot, S. and Treille, L.: SIRIUS: A freach Nationwide Project on Distributed Data Bases, Proc. VLDB 80, pp. 75-85 (1980).
/6/ Nauman, J.: ENCOMPASS: Evolution of a Distributed Database/Transaction System, Database Engineering, Vol. 5, No. 4, pp. 37-41 (1982).
/7/ Chan, A. and Ries, D. R.: Distributed Database Research at Computer Corporation of America, Database Engineering, Vol. 5, No. 4, pp. 14-19 (1982).
/8/ Codd, E. F. and Date, C. J.: The Relational and Network Approaches: Comparison of the Application Programming Interfaces, Proc. of the Workshop on Data Description, Access and Control, pp. 83-113 (1974).
/9/ Roberts, L. and Wessler, B.: Computer Network Development to Achieve Resource Sharing, Proc of the AFIPS SJCC, Vol. 36, pp. 543-549 (1970).
/10/ Metcalfe, R.M. and Boggs, D.R.: Ethernet: Distributed Packet Switching for Local Computer Networks, Comm. ACM, Vol. 19, No. 7, pp. 395-404 (1976).
/11/ Andler, S., Ding, I., Eswaran, E., Hauser, C., Kim, W., Mehl, J and Williams, R.: System D: A Distributed System for Availability, Proc. VLDB 82, pp. 33-44 (1982).
/12/ Date, C.J.: An Introduction to Database Systems, Third Edition, Addison-Wesley Pub. Co., Reading, Massachusetts, 1981.
/13/ DuBourdieu, D.: Survey of Current Research at Prime Computer, Inc. in

Distributed Database Management Systems, Database Engineering, Vol. 5, No. 4, pp. 20-22 (1982).

/14/ Selinger, P.G.: Replicated Data, in Distributed Data Bases, Draffan, I.W. and Poole, F. ed., Cambridge Univ. Press, pp. 223-231 (1980).

/15/ Alsberg, P.A. and Day, J.D.: A Principle for Resilient Sharing of Distributed Resources, Proc. 2nd International Conf. on Software Engineering, pp.562-570 (1976).

/16/ Thomas, R.H.: A Majority Consensus Approach to Concurrency Control for Multiple Copy Databases, ACM TODS, Vol. 4, No. 2, pp. 180-209 (1979).

/17/ Selinger, P.G. and Adiba, M.: Access Path Selection in Distributed Database Management Systems, Research Report, RJ2883, IBM Research Laboratory, San Jose, California (1980).

ON STRICT OPTIMALITY PROPERTY OF ALLOCATING BINARY

CARTESIAN PRODUCT FILES ON MULTIPLE DISK SYSTEMS

C. C. Chang and L. S. Lian

Institute of Applied Mathematics
National Chung Hsing University
Taichung, Taiwan

ABSTRACT

In this paper, we shall first review the progress made in the area of allocating multi-attribute file systems on multiple disks. We shall explore the strictly optimal allocation methods for binary Cartesian product files. A very important theorem concerning the sufficient and necessary conditions of whether a binary Cartesian product file have a strictly optimal allocation method on multiple disks or not shall be shown.

Section 1. Introduction

In a large database, a file cannot reside in memory, therefore, all of the records are divided into buckets and stored in disks. Since a record consists of several attributes, a retrieval query is often based on more than one attribute.

Up to now, we have many literatures concerning the design of multi-attribute file systems for partial match queries. [Aho and Ullman 1979, Al-Fedaghi and Chin 1979, Bolour 1979, Chang 1982, Chang, Du and Lee 1983, Chang, Du and Lee 1984, Chang, Lee and Du 1980, Chang, Lee and Du 1982, Chang and Su 1983a, Chang and Su 1983b, Chang and Su 1984, Du and Lee 1980, Du and Sobolewski 1982, Ghosh 1977, Jakobssen 1980, Lee and Tseng 1979, Lin, Lee and Du 1979, Liou and Yao 1977, Rivest 1976, Rothnie and Lozano 1974]. By a multi-attribute file system, we mean a file system whose records are characterized by more than one attribute. By partial match queries, we mean queries of the form $(A_1=a_1, A_2=a_2, \ldots, A_N=a_N)$, where for $1 \leq i \leq N$, a_i is either a key belonging to D_i, the i-th attribute domain, or is unspecified (i.e., a don't care condition) in which case it is denoted by * and where the number of unspecified attributes is j where $1 \leq j \leq N-1$.

We shall assume that all of the records are divided into buckets and stored in disks. Each time a partial match query is processed, one or more disk accesses is performed. Since the disk access time dominates the response time of a given query, we shall measure the performance of our file structure by the average number of buckets necessary to be examined over all possible partial match queries.

Let us consider the file system shown in Table 1.1.

Table 1.1(a). Simple 2-attribute file system.

Bucket 1	Bucket 2	Bucket 3	Bucket 4
(a,a)	(a,c)	(c,a)	(c,c)
(a,b)	(a,d)	(c,b)	(c,d)
(b,a)	(b,c)	(d,a)	(d,c)
(b,b)	(b,d)	(d,b)	(d,d)

Table 1.1(b). The corresponding inverted list of the file shown in Table 1.1(a).

Queries	Buckets to be examined
(a,*)	1,2
(b,*)	1,2
(c,*)	3,4
(d,*)	3,4
(*,a)	1,3
(*,b)	1,3
(*,c)	2,4
(*,d)	2,4

Let us consider another file system shown in Table 1.2.

Table 1.2(a). Simple 2-attribute file system.

Bucket 1	Bucket 2	Bucket 3	Bucket 4
(a,a)	(a,b)	(a,c)	(a,d)
(b,b)	(b,c)	(b,d)	(b,a)
(c,c)	(c,d)	(c,a)	(c,b)
(d,d)	(d,a)	(d,b)	(d,c)

Table 1.2(b). The corresponding inverted list of the file shown in Table 1.2(a).

Queries	Buckets to be examined
(a,*)	1,2,3,4
(b,*)	1,2,3,4
(c,*)	1,2,3,4
(d,*)	1,2,3,4
(*,a)	1,2,3,4
(*,b)	1,2,3,4
(*,c)	1,2,3,4
(*,d)	1,2,3,4

It is easy to see that the average number of buckets to be examined, over all possible partial match queries, is 2 for the file system in Table 1.1 and 4 for that in Table 1.2.

Hence the file design problem can be viewed as: Given a set of records, arrange these records into buckets in such a way that the average number of buckets to be examined is minimized.

Suppose that we have several disk systems which can be used to store buckets of a file system. Given a file system which is designed for partial match queries, it is desirable to achieve some degree of parallelism in examining required buckets in order to reduce the response time. Assuming that one bucket can be accessed in one unit of time, several buckets can be accessed in one unit of time if they are on distinct and independently accessible disk units. The response time to a query in this case is no longer proportional to the total number of buckets which need to be examined but becomes proportional to the number of buckets which need to be examined on a particular disk unit. Consider Table 1.1 again. Imagine that we have two disks and the buckets are assigned to these two disks as follows:

Table 1.3.   Disks and buckets assigned.

Disk	Buckets
1	1,4
2	2,3

In this case, for query (a,*), buckets 1 and 2 are to be examined. Since they reside on two disks and they can be accessed in parallel, only one unit of time is needed. Similarly, for query (*,c), buckets 2 and 4 are to be examined. Again, only one unit of time is needed. In fact, the reader should be able to verify by himself that the average response time is one unit of time for the above allocation.

If we allocate these buckets as follows:

Table 1.4.   Disks and buckets assigned.

Disk	Buckets
1	1,2
2	3,4

then it takes one unit of time to respond to four queries and two units of time to respond to the other four queries. Thus the performance under this allocation method is inferior to that under the previous allocation.

We now know that in a multi-disk system, the response time will be minimized if we maximize concurrency of disk accesses. Thus, the file allocation problem can be stated as: Given a set of buckets, allocate these buckets into disks in such a way that the maximal disk access concurrency can be achieved.

In this paper, we discuss the optimality property of allocating the buckets in binary Cartesian product files on multiple disk systems. In Section 2, we introduce the concept of binary Cartesian product files. In Section 3, we review some heuristic allocation methods for binary Cartesian product files. In Section 4, the strict optimality property of file allocation for binary Cartesian product files is discussed. Conclusions and further researches are presented in Section 5.

Section 2.   Binary Cartesian Product Files

A very important concept in designing multi-attribute files is the Cartesian product file which was first introduced in [Lin, Lee and Du 1979]. Lin, Lee and Du [1979] pointed out that under certain condition, the files designed by Rivest [1976], Rothine and Lozano [1974], Liou and Yao [1977], and Lee and Tseng [1979] were all Cartesian product files. We shall assume that all possible records are present throughout this paper. Let each record is characterized by $A_1, A_2, \ldots, A_N$. Each A_i is associated with a domain D_i.

Definition 2.1

A file system is called a Cartesian product file if each domain is divided into m_i subdivisions $D_{i1}, D_{i2}, \ldots, D_{im_i}$, $|D_{i1}| = |D_{i2}| = \ldots = |D_{im_i}| = z_i$ and all records in every bucket are in the form of $D_{1s_1} \times D_{2s_2} \times \ldots \times D_{Ns_N}$ where D_{js_j} is a subset of D_j. This bucket is denoted as $[s_1, s_2, \ldots, s_N]$.

Example 2.1

Let $D_1 = \{a,b,c,d,e,f\}$ and $D_2 = \{a,b,c,d\}$. Let $D_{11} = \{a,b,c\}$, $D_{12} = \{d,e,f\}$, $D_{21} = \{a,b\}$ and $D_{22} = \{c,d\}$. Then the following file is a Cartesian product file:

Bucket 1 = Bucket [1,1]: $D_{11} \times D_{21} = \{(a,a),(a,b),(b,a),(b,b),(c,a),(c,b)\}$

Bucket 2 = Bucket [1,2]: $D_{11} \times D_{22} = \{(a,c),(a,d),(b,c),(b,d),(c,c),(c,d)\}$

Bucket 3 = Bucket [2,1]: $D_{12} \times D_{21} = \{(d,a),(d,b),(e,a),(e,b),(f,a),(f,b)\}$

Bucket 4 = Bucket [2,2]: $D_{12} \times D_{22} = \{(d,c),(d,d),(e,c),(e,d),(f,c),(f,d)\}$.

Table 2.1 shows this Cartesian product file.

Table 2.1. A 2-attribute Cartesian product file.

Bucket 1	Bucket 2	Bucket 3	Bucket 4
(a,a)	(a,c)	(d,a)	(d,c)
(a,b)	(a,d)	(d,b)	(d,d)
(b,a)	(b,c)	(e,a)	(e,c)
(b,b)	(b,d)	(e,b)	(e,d)
(c,a)	(c,c)	(f,a)	(f,c)
(c,b)	(c,d)	(f,b)	(f,d)

In [Lin, Lee and Du 1979], it was pointed out that good multi-attribute file systems all exhibit some kind of clustering property. That is, within each bucket, records should be similar to one another. It just happens that Cartesian product files do cluster similar records together.

In a file F, if each attribute domain D_i contains only two elements, then F is a binary file. Since any record type can be encoded as a binary string, the binary file seems to be popular. Several papers concerning the file design problem for partial match queries concentrated on binary files [Rivest 1976, Burkhard 1976, Burkhard 1979].

Since in a binary file, there are only two elements in each attribute domain, to design a Cartesian product file for a multi-attribute binary file, each domain can be partitioned into either 1 or 2 subsets (i.e., m_i =1 or 2). Therefore the binary Cartesian product file can be defined as following:

Definition 2.2 [Du 1982]

In a Cartesian product file F if each attribute domain D_i contains only two elements, then file F is a binary Cartesian product file.

Note that a binary Cartesian product file is potentially a Cartesian product file.

Section 3. Allocation of Binary Cartesian Product Files in Multi-disk Systems

The problem of disk allocation for Cartesian product files on multiple disk systems was first considered by Du and Sobolewski [1982]. They proposed an allocation method named Disk Modulo (DM) allocation method which assigns all buckets of a Cartesian product file to an ND-disk system (disks labelled as units $0,1,\ldots,$ND-1) and showed that under certain conditions this allocation method is strictly optimal. They defined a query to be strictly optimal, if a maximum of $\left\lceil \dfrac{BA}{ND} \right\rceil$ buckets need to be accessed on any one of ND independently accessible disks to examine the necessary BA buckets in response to this query, where $\left\lceil \dfrac{BA}{ND} \right\rceil$ denotes the smallest integer equal to or greater than $\dfrac{BA}{ND}$. An allocation is called strictly optimal if and only if for all partial match queries under this allocation are all strictly optimal. Evidently a strictly optimal method is superior to all other possible allocation methods.

The DM allocation method can be stated as follows: Each bucket $[s_1, s_2, \ldots, s_N]$ in a Cartesian product file F is assigned to disk unit $(s_1 + s_2 + \ldots + s_N)$ mod ND.

For example, let us first consider the file in Table 1.1 again. The four buckets are defined as follows:

$$\text{Bucket } [1,1]: D_{11} \times D_{21}$$
$$\text{Bucket } [1,2]: D_{11} \times D_{22}$$
$$\text{Bucket } [2,1]: D_{12} \times D_{21}$$
$$\text{Bucket } [2,2]: D_{12} \times D_{22}$$

Assume that these four buckets are assigned into a 2-disk (disks labelled as units 0,1) system. Each bucket $[s_1, s_2]$ in file F is assigned to disk unit $(s_1 + s_2)$ mod 2. The distribution of assigning all buckets to two disks is the same as that shown in Table 1.3. Thus, the average response time to a query is 1 unit, which is the minimum time needed to respond to a query.

Du and Sobolewski [1982] showed the DM allocation method is strictly optimal under some cases.

Theorem 3.1 [Du and Sobolewski 1982]

The DM allocation method is strictly optimal in the following cases:

(1) all partial match queries with only one unspecified attribute,
(2) all partial match queries with at least one unspecified attribute j for which m_j mod ND=0,
(3) all possible partial match queries when m_i mod ND=0 or m_i=1 for all $1 \leq i \leq N$,
(4) all partial match queries when ND=2 or 3.

In the following example, we use the DM allocation method to allocate buckets of a binary Cartesian product file to some disks.

Example 3.1

Consider the case where F is a binary Cartesian product file, $m_1 = m_2 = m_3 = 2$ and ND=7.

Table 3.1 shows the distribution of all buckets onto the 7 disks by applying the DM allocation method. As can be seen, disk 4, 5 and 6 are never used, the distribution is not uniform. Hence the DM method has a relatively poor performance in this case. In other words, DM allocation method is not suitable for binary Cartesian product files.

A better allocation method for binary Cartesian product file was proposed by Du [1982].

Table 3.1. The distribution of all buckets onto the 4 disks by applying the DM allocation method.

Bucket	Assigned Disk
[0,0,0]	0
[0,0,1]	1
[0,1,0]	1
[0,1,1]	2
[1,0,0]	1
[1,0,1]	2
[1,1,0]	2
[1,1,1]	3

Disk #	# of Buckets Assigned to Disks
0	1
1	3
2	3
3	1
4	0
5	0
6	0

Let F be an N-attribute binary Cartesian product file (therefore, m_i =1 or 2 for $1 \leq i \leq N$). Let $ND=2^h$ be the number of disks to be used. Let $T=\{j_1,j_2,\ldots,j_k\}$ be the set of all attribute i with $m_i=2$. Assume that j_i =i for $1 \leq i \leq k$. A heuristic allocation method (HEU allocation method in short) proposed by Du [1982] was defined as follows: Bucket [$s_1,s_2,\ldots,$ s_N] is assigned to disk unit ($\sum_{j=1}^{N} s_j \cdot P_j$) mod ND, where $P_j=2^{(j \bmod \log_2 ND)}$ for $1 \leq j \leq k$ and $P_j=1$ for $k+1 \leq j \leq N$.

Example 3.2

Let F be a 5-attribute binary Cartesian product file with $m_1=m_2=m_3$ $=m_4=2$ and $m_5=1$. Let $ND=2^2=4$ be the number of disks available. Then $\log_2 ND$ $=2$, $P_1=P_3=2^{(1 \bmod 2)}=2^{(3 \bmod 2)}=2^1=2$, $P_2=P_4=2^{(2 \bmod 2)}=2^{(4 \bmod 2)}=2^0=1$, and $P_5=1$ (Since $m_5=1$). Assume that bucket [s_1,s_2,s_3,s_4,s_5] is stored by using the HEU method. The distribution of assigning all buckets to 5 disks is shown in Table 3.2. Note that all 16 buckets are uniformly distributed among four disks. Moreover, Du [1982] pointed that the number of buckets needed to be searched to respond to a query will be the same no matter the j-th attribute with $m_j=1$ is specified or not. For the reason of simplicity, in the rest of this section, we assume that each binary Cartesian product file is with $m_i=2$ for all i.

Let T_i, for $0 \leq i < \log_2 ND$, be the set of all attribute j with i=j mod $\log_2 ND$ and $m_j=2$. In Example 3.2, we have $T_0=\{2,4\}$ and $T_1=\{1,3\}$. Du [1982] showed the following theorem:

164

Table 3.2. The distribution of all buckets onto the
4 disks by applying the HEU method.

Bucket	Assigned disk	Bucket	Assigned disk
[0,0,0,0,0]	0	[1,0,0,0,0]	2
[0,0,0,1,0]	1	[1,0,0,1,0]	3
[0,0,1,0,0]	2	[1,0,1,0,0]	0
[0,0,1,1,0]	3	[1,0,1,1,0]	1
[0,1,0,0,0]	1	[1,1,0,0,0]	3
[0,1,0,1,0]	2	[1,1,0,1,0]	0
[0,1,1,0,0]	3	[1,1,1,0,0]	1
[0,1,1,1,0]	0	[1,1,1,1,0]	2

Disk #	# of Buckets Assgined to That Disk
0	4
1	4
2	4
3	4

Theorem 3.2 [Du 1982]

Let $ND=2^h$ and q be a partial match query containing at least one un-specified attribute from each T_i for $0 \leq i < \log_2 ND=h$. Then the heuristic allocation method is strictly optimal for query q.

Consider Example 3.2 again, queries $q=(A_1=*, A_2=*, A_3=a_3, A_4=a_4, A_5=a_5)$, where $a_i \in D_i$ for $3 \leq i \leq 5$, and queries $q'=(A_1=*, A_2=a_2, A_3=*, A_4=*, A_5=a_5)$, where $a_2 \in D_2$ and $a_5 \in D_5$ are strictly optimal if allocation method is HEU, since these queries contain at least two unspecified attributes, one belongs to $T_0=\{2,4\}$ and the other belongs to $T_1=\{1,3\}$.

Du [1982] also proposed the following corollary:

Corollary 3.1 [Du 1982]

Let $m_i=2$ for $1 \leq i \leq k$ and $ND=2^h$, and k is a multiple of h. The heuristic allocation method is strictly optimal for all partial match queries which contain more than $(h-1) \cdot k/h$ unspecified attributes.

In a binary Cartesian product file system, it has usually very large number of attributes and a query being asked usually has small number of attributes being specified. And it has usually some number of attributes will never or have a very little chance being specified in a query. If there are $\log_2 ND=h$ attributes never being specified, Du [1982] suggested that we can assign one of such attributes to each T_i for $0 \leq i \leq h$ and hence the HEU method is becoming strictly optimal for all possible partial match queries (with those attributes unspecified). Therefore, the above theorem and corollary seem to be useful in practice.

When the number of available disks (ND) is not a power of 2, Du [1982] proposed two heuristic methods HEU1 allocation method and HEU2 allocation method for distributing all buckets onto the ND disks. The two proposed allocation methods was very similar to HEU allocation method by replacing $p_j=2^{(j \bmod \log_2 ND)}$ with either $p_j=2^{(j \bmod \lfloor \log_2 ND \rfloor)}$ or $p_j=2^{(j \bmod \lceil \log_2 ND \rceil)}$. Let HEU1 and HEU2 denote the former and the latter modified HEU allocation methods respectively. If $\log_2 ND - \lfloor \log_2 ND \rfloor < \lceil \log_2 ND \rceil - \log_2 ND$, Du [1982] suggested to use HEU1 allocation method otherwise use HEU2 allocation method.

However, the HEU, HEU1, and HEU2 allocation methods do not guarantee strict optimality for all possible partial match queries.

Section 4.   The Existence of the Strictly Optimal Performance

Let F be an N-attribute binary Cartesian product file, and ND be the number of available disk units.   Does there exist a strictly optimal allocation method of F for all possible partial match queries or not?

For convenience, we assume that each domain is divided into two subdivisions (therefore, $m_1=m_2=\ldots=m_N=2$).

Definition 4.1

Let F be a binary Cartesian product file.   Bucket $[s_1,s_2,\ldots,s_N]$ and Bucket $[s_1',s_2',\ldots,s_N']$ are mutually complemented, if and only if $s_i \neq s_i'$ for $1 \leq i \leq N$.   We shall call the Bucket $[s_1',s_2',\ldots,s_N']$ is the complement bucket of Bucket $[s_1,s_2,\ldots,s_N]$.

Example 4.1

Let F be a 3-attribute binary Cartesian product file.   The following are all buckets to be used in F.

                Bucket 1 : Bucket [0,0,0]
                Bucket 2 : Bucket [0,0,1]
                Bucket 3 : Bucket [0,1,0]
                Bucket 4 : Bucket [0,1,1]
                Bucket 5 : Bucket [1,0,0]
                Bucket 6 : Bucket [1,0,1]
                Bucket 7 : Bucket [1,1,0]
                Bucket 8 : Bucket [1,1,1]

Among these buckets, Bucket 1 and Bucket 8, Bucket 2 and Bucket 7, Bucket 3 and Bucket 6, and Bucket 4 and Bucket 5 are mutually complemented respectively.

Table 4.1(a).   The distribution of all buckets onto the 4 disks.

Bucket	Assigned Disk
[0,0,0]	0
[0,0,1]	1
[0,1,0]	2
[0,1,1]	3
[1,0,0]	3
[1,0,1]	2
[1,1,0]	1
[1,1,1]	0

Table 4.1(b).   Buckets assigned to each disk.

Disk Number	Buckets Assigned to This Disk
0	[0,0,0],[1,1,1]
1	[0,0,1],[1,1,0]
2	[0,1,0],[1,0,1]
3	[0,1,1],[1,0,0]

The readers can verify that the allocation method in Table 4.1(b) satisfies the condition described in Theorem 4.1 and hence is strictly optimal.

166

Let $q = (A_{i_1} = a_{i_1}, A_{i_2} = a_{i_2}, \ldots, A_{i_k} = a_{i_k})$, where $1 \leq k \leq N$ and a_{i_j} belongs to the i_j-th attribute domain D_{i_j}, be a partial match query. Moreover, $a_{i_j} \in D_{i_j, s_{i_j}}$, s_{i_j} is either 0 or 1. Let $BR = \{Bucket\ [c_1, c_2, \ldots, c_N] \mid c_{i_j} = s_{i_j}$ for $j=1,2,\ldots,k$ and $c_{i_j} = 0$ or 1 for $j=k+1,\ldots,N\}$ be the response to q, we mean the set of buckets to be examined by q. If Bucket $[c_1, c_2, \ldots, c_N]$, which belongs to BR, then Bucket $[c_1', c_2', \ldots, c_N'] \notin BR$. That is, any two mutually complemented buckets can not be examined by the same partial match query. Hence, the following theorem can be easily concluded.

Theorem 4.1

Let F be an N-attribute binary Cartesian product file, and let ND be the number of available disk units. If M is an allocation method satisfying the following condition: each disk unit which contains one bucket or two mutually complemented buckets, then M is a strictly optimal allocation method.

Example 4.2

Consider the binary Cartesian product file system in Example 4.1 again, let the total number of available disk units ND=4. The distribution of assigning all buckets to ND disks is shown in Table 4.1.

Definition 4.2

Let F be an N-attribute binary Cartesian product file system, and let ND be the total number of available disk units (disks labelled as units $0, 1, \ldots, ND-1$). The Folding allocation method is defined as follows:

Bucket $[s_1, s_2, \ldots, s_N]$ (note $s_i = 0$ or 1 for $1 \leq i \leq N$) is assigned to disk unit $(2^{N-1}-1) s_1 + (-1)^{s_1} (\sum\limits_{k=2}^{N} 2^{N-k} s_k)$ mod ND.

Example 4.3

Consider the same 3-attribute file system of Example 4.1, let ND=8. In this case, our Folding allocation method can be performed as $(2^{3-1}-1) s_1 + (-1)^{s_1} (2^{3-2} s_2 + s_3)$ mod $8 = 3 s_1 + (-1)^{s_1} (2 s_2 + s_3)$ mod 8.

Table 4.2 shows the distribution of all buckets in the above example. According to Theorem 4.1, the allocation method presented in this example is strictly optimal for all possible partial match queries.

Table 4.2(a). The distribution of all buckets onto the 4 disks.

Bucket	Assigned Disk
[0,0,0]	0
[0,0,1]	1
[0,1,0]	2
[0,1,1]	3
[1,0,0]	3
[1,0,1]	2
[1,1,0]	1
[1,1,1]	0

Table 4.2(b). Buckets assigned to each disk.

Disk Number	Buckets Assigned to This Disk
0	[0,0,0],[1,1,1]
1	[0,0,1],[1,1,0]
2	[0,1,0],[1,0,1]
3	[0,1,1],[1,0,0]
4	
5	
6	
7	

Example 4.4

Consider the 3-attribute binary Cartesian product file system in Example 4.1 again, let ND=6. Assume that bucket $[s_1,s_2,s_3]$ is stored on disk $3s_1+(-1)^{s_1}(2s_2+s_3)$ mod 6. The distribution of assigning all buckets to ND disks is shown in Table 4.3.

Table 4.3(a). The distribution of all buckets onto the 4 disks.

Bucket	Assigned Disk
[0,0,0]	0
[0,0,1]	1
[0,1,0]	2
[0,1,1]	3
[1,0,0]	3
[1,0,1]	2
[1,1,0]	1
[1,1,1]	0

Table 4.3(b). Buckets assigned to each disk.

Disk Number	Buckets Assigned to This Disk
0	[0,0,0],[1,1,1]
1	[0,0,1],[1,1,0]
2	[0,1,0],[1,0,1]
3	[0,1,1],[1,0,0]
4	
5	

In this case, the performance under the Folding allocation method is still strictly optimal.

Example 4.5

For the case in Example 4.2 in which ND=4, we may also apply the Folding allocation method, which assigns bucket $[s_1,s_2,\ldots,s_N]$ into disk unit $3s_1+(-1)^{s_1}(2s_2+s_3)$ mod 4, for all buckets to ND disks. The distribution of all buckets among all disk units are the same as that was shown in Table 4.1. As can be seen, the performance of the Folding allocation method in this case is strictly optimal.

Lemma 4.1

Let NB $\leq$ 2ND. The Folding allocation method in which bucket $[s_1,s_2,$

168

$\ldots,s_N]$ is stored on disk unit $(2^{N-1}-1)s_1+(-1)^{s_1}(\sum\limits_{k=2}^{N}2^{N-k}s_k)$ mod ND can be reduced to that each bucket $[s_1,s_2,\ldots,s_N]$ is stored on disk unit $(2^{N-1}-1)s_1+(-1)^{s_1}(\sum\limits_{k=2}^{N}2^{N-k}s_k)$.

Proof:

For $s_1=0$, we have $\max\{(2^{N-1}-1)s_1+(-1)^{s_1}(\sum\limits_{k=2}^{N}2^{N-k}s_k)\}=\max\{\sum\limits_{k=2}^{N}2^{N-k}s_k\}$
$=2^{N-1}-1$ and $\min\{\sum\limits_{k=2}^{N}2^{N-k}s_k\}=0$. $\hspace{3cm}$ (1)

For $s_1=1$, we have $\max\{(2^{N-1}-1)s_1+(-1)^{s_1}(\sum\limits_{k=2}^{N}2^{N-k}s_k)\}=\max\{(2^{N-1}-1)$
$-(\sum\limits_{k=2}^{N}2^{N-k}s_k)\}=2^{N-1}-1$ and $\min\{(2^{N-1}-1)-\sum\limits_{k=2}^{N}2^{N-k}s_k\}=\min\{(2^{N-1}-1)-(2^{N-1}-1)\}$
$=0$. $\hspace{8cm}$ (2)

From (1) and (2), the following inequality holds: $0\leq(2^{N-1}-1)s_1$
$+(-1)^{s_1}(\sum\limits_{k=2}^{N}2^{N-k}s_k)\leq2^{N-1}-1$. $\hspace{3cm}$ (3)

Since $NB=2^N\leq2ND$, it is obvious that $2^{N-1}\leq ND$. Therefore, (3) can be rewritten as $0\leq(2^{N-1}-1)s_1+(-1)^{s_1}(\sum\limits_{k=2}^{N}2^{N-k}s_k)\leq ND-1$.

That is, $(2^{N-1}-1)s_1+(-1)^{s_1}(\sum\limits_{k=2}^{N}2^{N-k}s_k)$ mod $ND=(2^{N-1}-1)s_1(\sum\limits_{k=2}^{N}2^{N-k}s_k)$.

$\hspace{10cm}$ Q.E.D.

We now prove the following lemma.

Lemma 4.2

Let s_k, $s_k'=0$ or 1 for $0\leq k\leq N$. If $\sum\limits_{k=0}^{N}2^k(s_k-s_k')=0$, then $s_k=s_k'$, for $0\leq k\leq N$.

Proof:

(1) Let $N=0$. $\sum\limits_{k=0}^{N}2^k(s_k-s_k')=s_0-s_0'=0$, then $s_0=s_0'$.

(2) Let $N=1$. $\sum\limits_{k=0}^{N}2^k(s_k-s_k')=2(s_1-s_1')+(s_0-s_0')=0$, then $s_1=s_1'$ and $s_0=s_0'$.

(3) Assume that $N=m-1$, our statement holds. That is, $\sum\limits_{k=0}^{m-1}2^k(s_k-s_k')=0$, then $s_k=s_k'$ for all $k=0,1,\ldots,m-1$.

(4) Let $N=m$. Suppose that $\sum\limits_{k=0}^{m}2^k(s_k-s_k')=0$ or $\sum\limits_{k=0}^{m-1}2^k(s_k-s_k')+2^m(s_m-s_m')$
$=0$. That is $2^m(s_m-s_m')=-\sum\limits_{k=0}^{m-1}2^k(s_k-s_k')$. $\hspace{2cm}$ (i)

From (i), we have $|2^m(s_m-s_m')|=|\sum\limits_{k=0}^{m-1}2^k(s_k-s_k')|\leq\sum\limits_{k=0}^{m-1}2^k|s_k-s_k'|\leq\sum\limits_{k=0}^{m-1}2^k$
$=2^m-1$.

$|s_m - s'_m| = 1$, if $s_m \neq s'_m$. Therefore, if $s_m \neq s'_m$, we have $|2^m(s_m - s'_m)| = 2^m \leq$

$2^m - 1$, which is impossible. So $s_m = s'_m$.

From (i), we have $\sum\limits_{k=0}^{m-1} 2^k(s_k - s'_k) = 0$. Furthermore, from (3), we know that

$\sum\limits_{k=0}^{m-1} 2^k(s_k - s'_k) = 0$, then $s_k = s'_k$, for all $k = 0, 1, \ldots, m-1$.

Thus, if s_k, $s'_k = 0$ or 1 and $\sum\limits_{k=0}^{m} 2^k(s_k - s'_k) = 0$, then $s_k = s'_k$, $0 \leq k \leq m$.

Q.E.D.

In the following, we shall formally show that the Folding allocation method is a strictly optimal allocation method in the case that NB $\leq$ 2ND.

<u>Theorem 4.2</u>

Let F be an N-attribute binary Cartesian product file, NB be the total number of buckets and ND be the number of available disk units. If NB $\leq$ 2ND, then the Folding allocation method is a strictly optimal allocation method of F for all possible partial match queries.

<u>Proof</u>

We shall show that each disk unit contains only one or two mutually complemented buckets.

Assume that bucket $[s_1, s_2, \ldots, s_N]$ and bucket $[s'_1, s'_2, \ldots, s'_N]$ be stored

on the same disk under the Folding allocation method. Therefore, $(2^{N-1}-1)s_1$

$+(-1)^{s_1}(\sum\limits_{k=2}^{N} 2^{N-k} s_k) \bmod ND = (2^{N-1}-1)s'_1 + (-1)^{s'_1}(\sum\limits_{k=2}^{N} 2^{N-k} s'_k) \bmod ND$. Because NB

$\leq$ 2ND, by Lemma 4.1, we know that $(2^{N-1}-1)s_1 + (-1)^{s_1}(\sum\limits_{k=2}^{N} 2^{N-k} s_k) = (2^{N-1}-1)s'_1$

$+(-1)^{s'_1}(\sum\limits_{k=2}^{N} 2^{N-k} s'_k)$.

(i)

(1) Case 1: Let $s_1 = s'_1$.

From (i), we have $\sum\limits_{k=2}^{N} 2^{N-k} s_k = \sum\limits_{k=2}^{N} 2^{N-k} s'_k$. That is, $\sum\limits_{k=2}^{N} 2^{N-k}(s_k - s'_k) = 0$.

By Lemma 4.2, we have $s_k = s'_k$ for all $k = 2, 3, \ldots, N$.

(2) Case 2: Let $s_1 \neq s'_1$.

For convenience, let $s_1 = 1$ and $s'_1 = 0$. Equation (i) can be rewritten

as $(2^{N-1}-1) - (\sum\limits_{k=2}^{N} 2^{N-k} s_k) = \sum\limits_{k=2}^{N} 2^{N-k} s'_k$. Since $2^{N-1} - 1 = \sum\limits_{k=2}^{N} 2^{N-k}$, we have $\sum\limits_{k=2}^{N} 2^{N-k}$

$- \sum\limits_{k=2}^{N} 2^{N-k} s_k = \sum\limits_{k=2}^{N} 2^{N-k} s'_k$. That is $\sum\limits_{k=2}^{N} 2^{N-k}(1-s_k) = \sum\limits_{k=2}^{N} 2^{N-k} s'_k$.

Let $t_k = 1 - s_k$. Since $s_k = 0$ or 1, we have $t_k = 0$ or 1 for all $k = 2, 3, \ldots, N$.

So $\sum\limits_{k=2}^{N} 2^{N-k}(t_k - s'_k) = 0$. By Lemma 4.2, we have $t_k = s'_k$. That is, $1 - s_k = s'_k$. So $s_k \neq s'_k$ for all $k = 2, 3, \ldots, N$.

Therefore, for bucket $[s_1, s_2, \ldots, s_N]$ and bucket $[s'_1, s'_2, \ldots, s'_N]$ stored on the same disk unit, we can conclude that either they are the same or they are mutually complemented. So the Folding allocation method is a strictly optimal allocation method of F for all possible partial match queries.

Q.E.D.

We now know that the Folding allocation method is a strictly optimal allocation method in the case that $NB \leq 2ND$. For the case $NB > 2ND$ and $ND \geq 4$, does there exist a strictly optimal allocation method or not?

Lemma 4.3

Let $F'=D_1 \times D_2 \times \ldots \times D_{N'}$ be an N'-attribute Cartesian product file with each domain D_i is divided into m_i subdivisions. Let $F=D_1 \times D_2 \times \ldots \times D_{N'} \times D_{N'+1} \times \ldots \times D_N$ be an N-attribute Cartesian product file with each domain D_i is divided into n_i subdivisions, where $n_i=m_i$ for $i=1,2,\ldots,N'$. Let ND be the number of available disk units. If there is no strictly optimal allocation method of file F' for all possible partial match queries then there is no strictly optimal allocation method of file F for all possible partial match queries.

Proof

Assume that there exists an allocation method M which is strictly optimal of F over all possible partial match queries.

Let $q=(A_{i_1}=a_{i_1}, A_{i_2}=a_{i_2}, \ldots, A_{i_j}=a_{i_j}, A_{N'+1}=a_{N'+1}, A_{N'+2}=a_{N'+2}, \ldots, A_N=a_N)$, where $a_i \in D_i$, $1 \leq i_j \leq N'$ and $1 \leq j \leq N'-1$, be a partial match query in F. The expected response time of this query q is strictly optimal under the allocation method M. Therefore, the expected response time of a corresponding partial match query $q'=(A_{i_1}=a_{i_1}, A_{i_2}=a_{i_2}, \ldots, A_{i_j}=a_{i_j})$ in F', where $a_{i_j} \in D_{i_j}$, $1 \leq i_j \leq N'$ and $1 \leq j \leq N'-1$, can also achieve strictly optimal. That is, there must exist a strictly optimal allocation method of F' over all possible partial match queries. That is contradicted to our condition. Thus it is impossible to have a strictly optimal allocation method of F.

Q.E.D.

Theorem 4.3

Let F be an N-attribute binary Cartesian product file and NB be the total number of buckets. Let ND be the total number of available disks and $ND \geq 4$. If $NB > 2ND$, then there is no strictly optimal allocation method of F over all possible partial match queries.

Proof

Because $ND \geq 4$, we have $2^h \leq ND < 2^{h+1}$ with $h \geq 2$. For $NB > 2ND$, we have $NB \geq 2^{h+2}$. We shall first consider the case where $NB = 2^{h+2}$. In this case, $\dfrac{NB}{ND} > \dfrac{2^{h+2}}{2^{h+1}} = 2$. So there must exist a disk unit which contains at least three buckets: bucket $[a_1,a_2,\ldots,a_{h+2}]$, bucket $[b_1,b_2,\ldots,b_{h+2}]$ and bucket $[c_1,c_2,\ldots,c_{h+2}]$.

(1) Case 1: Let $(a_1,a_2)=(b_1,b_2)$

In this case, bucket $[a_1,a_2,\ldots,a_{h+2}]$ and bucket $[b_1,b_2,\ldots,b_{h+2}]$ will be examined by the partial match query $q=(a_1,a_2,*,*,\ldots,*)$. Since these two buckets reside on the same disk unit, the response time of q is greater than or equal to 2. In fact, the strictly optimal response time of q should be $\left\lceil \dfrac{2^h}{ND} \right\rceil = 1$. Consequently, this case would not be a strictly optimal arrangement.

(2) Case 2: Let $(a_1,a_2)\neq(b_1,b_2)\neq(c_1,c_2)$

Because a_i, b_i and c_i are either 0's or 1's, this case has only four possibilities.

	(i)	$(0,0)\neq(0,1)\neq(1,0)$
	(ii)	$(0,0)\neq(0,1)\neq(1,1)$
	(iii)	$(0,0)\neq(1,0)\neq(1,1)$
and	(iv)	$(0,1)\neq(1,0)\neq(1,1)$.

For convenience, this case can be considered as $a_1=b_1\neq c_1$ and $a_2=c_2\neq b_2$.

For bucket $[a_1,a_2,\ldots,a_{h+2}]$ and bucket $[b_1,b_2,\ldots,b_{h+2}]$, if $a_i=b_i$ for $1<i\leq h+2$, then bucket $[a_1,a_2,\ldots,a_{h+2}]$ and bucket $[b_1,b_2,\ldots,b_{h+2}]$ have to be examined by the partial match query $q=(a_1,*,\ldots,*,a_i,*,\ldots,*)$. Since these two buckets reside on the same disk unit, the response time of q is greater than or equal to 2. As a matter of fact, the strictly optimal response time of q should be $\left\lceil \dfrac{2^h}{ND} \right\rceil = 1$. So this case would not be a strictly optimal arrangement. That is $a_i\neq b_i$ for all $i=3,4,\ldots,h+2$. (1)

Similarly, let us consider bucket $[a_1,a_2,\ldots,a_{h+2}]$ and bucket $[c_1,c_2,\ldots,c_{h+2}]$. If there exists a j, $3\leq j\leq h+2$, such that $a_j=c_j$, then bucket $[a_1,a_2,\ldots,a_{h+2}]$ and bucket $[c_1,c_2,\ldots,c_{h+2}]$ need to be examined by the partial match query $q=(*,a_2,*,\ldots,*,a_j,*,\ldots,*)$. Since these two buckets reside on the same disk unit, the response time of q is again greater than or equal to 2. In fact, the strictly optimal response time of q should be $\left\lceil \dfrac{2^h}{ND} \right\rceil = 1$. Hence, this case would not be a strictly optimal arrangement. That is $a_i\neq c_i$ for $i=3,4,\ldots,h+2$. (2)

From (1) and (2), we know that $b_i=c_i$ for $i=3,4,\ldots,h+2$. Both bucket $[b_1,b_2,\ldots,b_{h+2}]$ and bucket $[c_1,c_2,\ldots,c_{h+2}]$ need to be examined by the partial match query $q=(*,*,b_3,b_4,\ldots,b_{h+2})$. Since these two buckets reside on the same disk unit, the response time of q is greater than or equal to 2. However, the strictly optimal response time of q should be $\left\lceil \dfrac{2^2}{ND} \right\rceil = 1$. Thus, we conclude that there is no strictly optimal allocation method of F over all possible partial match queries in the cases $2^h \leq ND < 2^{h+1}$ and $NB = 2^{h+2}$.

For the case of $NB > 2^{h+2}$, by Lemma 4.3, there is impossible to have a strictly optimal allocation method.

Q.E.D.

From Theorem 4.2 and Theorem 4.3, we can easily conclude the following theorem.

Theorem 4.4

Let F be an N-attribute binary Cartesian product file and NB be the total number of buckets. Assume that we have ND available disk units, where $ND \geq 4$. Then

(1) for $NB \leq 2ND$, there is a strictly optimal allocation method of F among ND disk systems.

(2) for $NB > 2ND$, there is no strictly optimal allocation method of F among ND disk systems.

Let $ND \geq 4$, we have $2^h \leq ND < 2^{h+1}$, where $h \geq 2$. In other words, $h = \lfloor \log_2 ND \rfloor \geq 2$. Since $NB > 2ND$, we have $NB > 2^{h+1} > 2^{\lfloor \log_2 ND \rfloor + 1}$. From Lemma 4.3, the following theorem can also be concluded.

Theorem 4.5

Let F be an N-attribute Cartesian product file. Assume that we have ND available disk units and $ND \geq 4$. If there are more than $\lfloor \log_2 ND \rfloor + 1$ attributes satisfying the condition that the number of subdivisions of each attribute is equal to 2, then there is no strictly optimal allocation method of F among ND disk systems.

Section 5. Conclusions and Further Researches

In this paper, we explore the strict optimality property of the disk allocation problem for partial match retrieval. Particularly, we concentrate on the strictly optimal allocation problem for binary Cartesian product files for the following two reasons: Any file can be converted into its corresponding binary file and Cartesian product files have been shown effective for partial match queries. The existing DM allocation method was shown to be not suitable for binary Cartesian product files. The proposed HEU method, HEU1 method and HEU2 method by Du [1982] were shown to be "near" strictly optimal under some cases. We have proved a very important theorem concerning the sufficient and necessary conditions of whether a binary Cartesian product file have a strictly optimal allocation method on multiple disks or not? We also provide some strictly optimal allocation methods for some binary Cartesian product file systems.

Up to now, the following problems still remain to be open problems:

(1) Problem 1 (The File Allocation Problem for Multi-disk Systems)

Given an arbitrary file structure and a multiple disk system. How are we going to allocate buckets into different disks so that the response time is minimized?

(2) Problem 2 (The Complexity of Problem 1)

We would like to know how difficult of Problem 1.

(3) Problem 3 (The Expected Optimal Performance of File Allocation)

What is the performance that can be expected for an optimal allocation method?

(4) Problem 4 (The Combination of File Design Problem
    and File Allocation Problem

The results obtained for the file allocation problem can be affected
by the file design problem. In this case, how can we first allocate records into buckets and then assign buckets into disks in an optimal way?

(5) Problem 5 (The Complexity of Problem 4)

We like to know whether Problem 4 is an NP-hard problem.

(6) Problem 6 (The Strict Optimality Criteria of Cartesian Product Files)

In Section 4, we proved a theorem concerning the sufficient and necessary conditions of whether a binary Cartesian product file have a
strictly optimal allocation method on multiple disks or not? Again, we
would like to know the strict optimality criteria of general Cartesian
product file systems.

REFERENCES

1. Aho, A. V. and Ullman, J. D., (1979): Optimal Partial-match Retrieval
     when Fields are Independently Specified, ACM Transactions on Data-
     base Systems, Vol. 4, No. 2, June 1979, pp. 168-179.
2. Al-Fedaghi and Chin, Y. H., (1979): Algorithmic Approach to the Con-
     secutive Retrieval Property, International Journal of Information
     Sciences, Vol. 8, No. 4, 1979, pp. 279-301.
3. Bolour, A., (1979): Optimality Properties of Multiple Key Hashing Func-
     tions, Journal of the Association for Computing Machinery, Vol. 26,
     No. 2, April 1979, pp. 196-210.
4. Burkhard, W. A., (1976): Hashing and Trie Algorithms for Partial Match
     Retrieval, ACM Transactions on Database Systems, Vol. 1, No. 2, June
     1976, pp. 175-187.
5. Burkhard, W. A., (1979): Partial Match Hashing Coding: Benefits of
     Redundancy, ACM Transactions on Database Systems, Vol. 4, No. 2,
     June 1979, pp. 228-239.
6. Chang, C. C., (1982): A Survey of Multi-Key File Design, Proceedings
     of the First Conference on Computer Algorithms, Hsinchu, Taiwan,
     July 1982, pp. 3-1 - 3-62.
7. Chang, C. C., Du, M. W., and Lee, R. C. T., (1983): The Hierarchical
     Structure in Multi-attribute Files, to appear in Information Science.
8. Chang, C. C., Du, M. W. and Lee, R. C. T., (1984): Performance Analysis
     of Cartesian Product Files and Random Files, IEEE Transactions on
     Software Engineering, Vol. SE-10, No. 1, Jan. 1984, pp. 88-99.
9. Chang, C. C., Lee, R. C. T. and Du, H. C., (1980): Proc. ACM-SIGMOD
     1980 Conference, Santa Monica, Calif., May 1980, pp. 157-168.
10. Chang, C. C., Lee, R. C. T. and Du, M. W., (1982): Symbolic Gray Code
     as a Perfect Multi-attribute Hashing Scheme for Partial Match Queries,
     IEEE Transactions on Software Engineering, Vol. SE-8, No. 3, May
     1982, pp. 235-249.
11. Chang, C. C. and Su, D. H., (1983a): Some Properties of Multi-attribute
     Files System Based upon Multiple Key Hashing Functions, Proceedings
     of the 21-st Annual Allerton Conference on Communication, Control
     and Computing, Urbana, Illinois, Oct. 1983, pp. 675-682.
12. Chang, C. C. and Su, D. H., (1983b): Application of a Heuristic Algo-
     rithm to Design Multiple Key Hashing Functions, Proceedings of NCS
     1983 Conference, Hsinchu, Taiwan, Dec. 1983, pp. 218-231.
13. Chang, C. C. and Su, D. H., (1984): Performance Analysis of Multi-
     attribute Files Based upon Multiple Key Hashing Functions and
     Haphazard Files, Institute of Applied Mathematics, National Chung
     Hsing University, Taichung, Taiwan, Tech. Report.

14. Du, H. C., (1982): Concurrent Disk Accessing for Partial Match Retrieval, Computer Science Department, University of Minnesota, Minneapolis, 1982.
15. Du, H. C. and Lee, R. C. T., (1980): Symbolic Gray Code as a Multi-Key Hashing Functions, IEEE Transactions on Pattern Analysis and Machine Intelligence, Vol. PAMI-2, No. 1, Jan. 1980, pp. 83-90.
16. Du, H. C. and Sobolewski, J. S., (1982): Disk Allocation for Cartesian Product Files on Multiple Disk Systems, ACM Trans. Database Systems, Vol. 7, March 1982, pp. 82-101.
17. Ghosh, S. P., (1977): Data Base Organization for Data Management, Academic Press, N. Y., 1977.
18. Jakobsson, M., (1980): Reducing Block Accesses in Inverted Files by Partial Clustering, Inform. Systems, Vol. 5, No. 1, 1980, pp. 1-5.
19. Lee, R. C. T. and Tseng, S. H., (1979): Multi-Key Sorting, Policy Analysis and Information Systems, Vol. 3, No. 2, Dec. 1979, pp. 1-20.
20. Lin, W. C., Lee, R. C. T. and Du, H. C., (1979): Common Properties of some Multi-attribute File Systems, IEEE Transactions on Software Engineering, Vol. SE-5, No. 2, March 1979, pp. 160-174.
21. Liou, J. H. and Yao, S. B., (1977): Multi-dimensional Clustering for Data Base Organizations, Information Systems, Vol. 2, 1977, pp. 187-198.
22. Rivest, R. L., (1976): Partial-match Retrieval Algorithms, SIAM Journal on Computing, Vol. 14, No. 1, March 1976, pp. 19-50.
23. Rothine, J. B. and Lozano, T., (1974): Attribute Based File Organization in a Paged Memory Environment, Communications of the Association for Computing Machinery, Vol. 17, No. 2, Feb. 1974, pp. 63-69.

ON THE COMPLEXITY OF FILE ALLOCATION PROBLEM

C. C. Chang and J. C. Shieh

Institute of Applied Mathematics
National Chung Hsing University
Taichung, Taiwan

ABSTRACT

In this paper, we shall analyze the problem of finding an optimal way to distribute buckets among several disks to facilitate parallel searching for all possible queries. It is quite unlikely that there exists a polynomial algorithm to solve the above mentioned problem will be shown.

Section 1 Introduction

The design of multi-attribute file systems for partial match queries has been widely explored [Aho and Ullman 1979, Bentley and Friedman 1979, Bolour 1979, Chang, Du and Lee 1984, Chang and Fu 1978, Chang, Lee and Du 1980, Chang, Lee and Du 1981, Ghosh 1977, Lee and Tseng 1979, Lin, Lee and Du 1979, Liou and Yao 1977, Rivest 1976, Rothnie and Lozano 1974].

By a multi-attribute file system, we mean a file system whose records are characterized by more than one attribute. By partial match queries, we mean queries of the following form: Retrieve all records where $A_{i_1} = a_{i_1}$, $A_{i_2} = a_{i_2}$, $\ldots$, $A_{i_j} = a_{i_j}$, $i_1 \neq i_2 \neq \ldots \neq i_j$ and a_{i_k} is an attribute value belonging to the domain of the i_k-th attribute.

We shall assume that every file is divided into buckets and stored in disks. Each time the disk is accessed, an entire bucket is brought into primary memory. The problem of multi-attribute file design can be explained by considering two file systems shown in Table 1.1 and Table 1.2.

Table 1.1(a). Simple 2-attribute file system.

Bucket 1	Bucket 2	Bucket 3	Bucket 4
(a,a)	(a,b)	(a,c)	(a,d)
(b,b)	(b,c)	(b,d)	(b,a)
(c,c)	(c,d)	(c,a)	(c,b)
(d,d)	(d,a)	(d,b)	(d,c)

177

Table 1.1(b). The corresponding inverted list of the
file shown in Table 1.1(a).

Queries	Buckets to be Examined
(a,*)	1,2,3,4
(b,*)	1,2,3,4
(c,*)	1,2,3,4
(d,*)	1,2,3,4
(*,a)	1,2,3,4
(*,b)	1,2,3,4
(*,c)	1,2,3,4
(*,d)	1,2,3,4

Table 1.2(a). Simple 2-attribute file system.

Bucket 1	Bucket 2	Bucket 3	Bucket 4
(a,a)	(a,c)	(c,a)	(c,c)
(a,b)	(a,d)	(c,b)	(c,d)
(b,a)	(b,c)	(d,a)	(d,c)
(b,b)	(b,d)	(d,b)	(d,d)

Table 1.2(b). The corresponding inverted list of
the file shown in Table 1.2(a).

Queries	Buckets to be Examined
(a,*)	1,2
(b,*)	1,2
(c,*)	3,4
(d,*)	3,4
(*,a)	1,3
(*,b)	1,3
(*,c)	2,4
(*,d)	2,4

It can be seen that the average number of buckets to be examined, over
all possible partial match queries, is 2 for the file system in Table 1.2
and 4 for that in Table 1.1.

Thus the problem of multi-attribute file system design for partial
match queries is as follows: We are given a set of multi-attribute records,
arrange the records into NB buckets in such a way that the average number
of buckets to be examined, over all possible partial match queries, is minimized. Tang, Buehrer and Lee [1983] pointed out that the general problem
stated above is an NP-hard problem.

Given a file system which is designed for partial match retrieval,
it is desirable to achieve some degree of parallelism in examining required
buckets in order to reduce the response time. Assume that one bucket can
be accessed in one unit of time, several buckets can be accessed in one
unit of time if they are on distinct and independently accessible disk
units. Consider Table 1.2 again, Imagine that we have two disks and the
buckets are assigned to these two disks as follows:

Disk	Buckets
1	1,4
2	2,3

In this case, for query (a,*), buckets 1 and 2 are to be examined. Since
they reside on two disks and they can be accessed in parallel, only one
unit of time is needed. Similarly, for query (*,c), buckets 2 and 4 are to

178

be examined.Again, only one unit of time is needed. In fact, the reader should be able to verify by himself that the average response time is one unit of time for the above allocation.

If we allocate these buckets as follows:

Disk	Buckets
1	1,2
2	3,4

then it takes one unit of time to responed to four queries and two units of time to respond to the other four queries. Thus the performance under this allocation method is inferior to that under the previous allocation.

We now understand that in a multi-disk system, assigning different buckets to disks influences the performance.

Let us describe the File Allocation Problem formally.

Definition 1.1

The File Allocation Problem is defined as follows:

Given a multi-attribute file system, designed primarily for partial match queries and m disk system, find a way to allocate buckets among the m disk units such that for all partial match queries, the response time is minimized.

Du and Sobolewski [1980] discussed the above mentioned problem. The interested reader is suggested to read [Du and Sobolewski 1980].

In the following section, we shall explore the complexity of File Allocation Problem.

Section 2 The Complexity of Allocating Files Among Multiple Disk Systems

Let us consider the following colorability problem of a graph.

Definition 2.1 [Karp. 1972]

The Colorability Problem: Let $G=(V,E)$ be an undirected graph. Find a way to colour the vertex of the graph G with k distinct colours such that the number of adjacent vertices which have the same colour is minimum.

The above problem is a very difficult one as shown by the following theorem:

Theorem 2.1 [Karp. 1972]

The Colorability Problem is NP-hard.

Let us restate our File Allocation Problem. Given a set of queries $Q=\{q_1,q_2,\ldots,q_k\}$, each of which need examine different buckets from a file system F which is designed for partial match retrieval, find an optimal distribution of buckets into m disks, such that the average number of disk accesses in respond to all partial match queries is minimized.

Let F consist of a set of buckets and $F_A=\{(F_{A1}),(F_{A2}),\ldots,(F_{Am})\}$, where $F_{Aj} \subseteq F$, be one of the possible allocation of buckets in F into m disks. Therefore, the time to respond to the partial match query q_i in the above allocation is $\max_{1\leq j\leq m} |(F)_i \cap F_{Aj}|$, where $(F)_i$ denotes the set of

179

buckets required to respond to q_i and $|\cdot|$ indicates the cardinality. So it is obvious to see the average response time is

$$\frac{1}{k} \sum_{i=1}^{k} \max_{1 \leq j \leq m} \left| (F)_i \cap F_{Aj} \right|.$$

In the following theorem, we shall show that our File Allocation Problem can be easily reduced into the Colorability Problem. In view of the fact that the Colorability Problem is NP-hard, thus the File Allocation Problem is NP-hard too.

Theorem 2.2

The Colorability Problem α the File Allocation Problem.

Proof

For each Colorability Problem, we can construct a corresponding File Allocation Problem as follows:

Associate each vertex V_i in G with a bucket B_i in F and the corresponding buckets B_i and B_j are required to respond to any query if an edge exists between two vertices V_i and V_j. Then the File Allocation Problem with m disks is equivalent to find a way to colour the vertices of the corresponding graph with m distinct colours, such that the number of adjacent vertices which has the same colour is minimized. Clearly the Colorability Problem has a polynomial solution if and only if the corresponding File Allocation Problem has a polynomial solution.

Q.E.D.

Theorem 2.3

The File Allocation Problem is NP-hard.

Proof

(1) The Colorability Problem is NP-hard.
(2) According to Theorem 2.2, the Colorability Problem α the File Allocation Problem.
(3) We can conclude from (1) and (2) that the File Allocation Problem is NP-hard.

Section 3   Concluding Remarks

In this paper, we have shown that the File Allocation Problem is indeed hard to solve.

A problem cannot be solved in polynomial time does not mean that we should not try to solve them. In fact, the simplex method for linear programming is a well known exponential time algorithm, it still works very well. Our next job is to investigate a method suitable for allocating file among disks in the optimal or near optimal way.

REFERENCES

1. Aho, A. V. and Ullman, J. D., (1979): Optimal Partial-match Retrieval when Fields are Independently Specified, ACM Transactions of Database Systems, Vol. 4, No. 2, June 1979, pp. 168-179.
2. Bentley, J. L. and Friedman, J. H., (1979): Data Structures for Range Searching, Computing Surveys, Vol. 11, No. 4, Dec. 1979, pp. 397-409.
3. Bolour, A., (1979): Optimality Properties of Multiple Key Hashing Functions, Journal of the Association for Computing Machinery, Vol. 26, No. 2, April 1979, pp. 196-210.

4. Chang, C. C., Du, M. W. and Lee, R. C. T., (1982): The Hierarchical Ordering in Multi-attribute Files, to appear in Information Sciences.
5. Chang, C. C., Du, M. W. and Lee, R. C. T., (1984): Performance Analysis of Cartesian Product Files and Random Files, IEEE Transactions on Software Engineering, Vol. SE-10, No. 1, Jan. 1984, pp. 88-99.
6. Chang, C. C., Lee, R. C. T. and Du, H. C., (1980): Some Properties of Cartesian Product Files, Proc. ACM-SIGMOD 1980 Conference, Santa Monica, Calif., May 1980, pp. 157-168.
7. Chang, C. C., Lee, R. C. T. and Du, M. W., (1981): Symbolic Gray Code as a Perfect Multi-attribute Hashing for Partial Match Queries, IEEE Transactions on Software Engineering, Vol. SE-8, No. 3, May 1982, pp. 235-249.
8. Chang, J. M. and Fu, K. S., (1978): On the Retrieval Time and the Storage Space of Doubly-chained Multiple-attribute Tree Data Base Organization, Policy Analysis and Information Systems, Vol. 1, No. 2, Jan. 1978, pp. 22-48.
9. Du, H. C. and Sobolewski, J. S., (1982): Disk Allocation for Cartesian Product Files on Multiple Disk Systems, ACM Trans. Database Systems, Vol. 7, March 1982, pp. 82-101.
10. Ghosh, S. P., (1977): Data Base Organization for Data Management, Academic Press, N. Y., 1977.
11. Karp, R. M., (1972): Reducibility among Combinatorial Problems, Complexity of Computer Computations, Plenum Press, New York, pp. 85-103.
12. Lee, R. C. T. and Tseng, S. H., (1979): Multi-key Sorting, Policy Analysis and Information Systems, Vol. 3, No. 2, Dec. 1979, pp. 1-20.
13. Lin, W. C., Lee, R. C. T. and Du, H. C., (1979): Common Properties of Some Multi-attribute File Systems, IEEE Transactions on Software Engineering, Vol. SE-5, No. 2, March 1979, pp. 160-174.
14. Liou, J. H. and Yao, S. B., (1977): Multi-dimensional Clustering for Data Base Organizations, Information Systems, Vol. 2, 1977, pp. 187-198.
15. Rivest, R. L., (1976): Partial-match Retrieval Algorithms, SIAM Journal on Computing, Vol. 14, No. 1, March 1976, pp. 19-50.
16. Rothnie, J. B. and Lozano, T., (1974): Attribute Based File Organization in a Paged Memory Environment, Communications of the Association for Computing Machinery, Vol. 17, No. 2, Feb. 1974, pp. 63-69.
17. Tang, T. Y., Buehrer, D. J. and Lee, R. C. T., (1982): On the Complexity of Some Multi-attribute File Design Problems, Proceeding of ICS 1982 Conference, Taichung, Taiwan, Dec. 1982, pp. 649-655.

A QUANTITATIVE EVALUATION OF SCHEDULING SYSTEMS FOR THE PHYSICAL LOCKING SCHEME IN A DATABASE SYSTEM

Yutaka Matushita, Toshifumi Tatsumi and Katsuhiko Kawamura

OKI Electric Industry Co., Ltd.
Computer and Communication Systems Division
Warabi City, Saitama Pref., Japan

Abstract

Two scheduling systems for the physical locking scheme are discussed in this paper,in which one column (attribute) of a flat-table is assumed to be the granule size.

A transaction generally refers to several flat-tables. So,it can be divided into several units,called actions, each of which refers to only one flat-tables. Whenever a transaction arrives, the transaction locks all the attributes needed. This is called "transaction-oriented scheduling". On the other hand, the scheduling system in which each action locks the required attributes independently of other related actions is called "action-oriented scheduling".

These two scheduling systems are analyzed in this paper. In particular, this paper discusses how the system performance is affected by both the deadlock frequency and the ratio of update request transactions to xall transactions.

1. Introduction

In an environment where simultaneous access by multiple users to a database system is permitted, it is necessary to have a certain concurrency control for preserving data consistency.

A number of methods using various locking schemes have been proposed for concurrency control.

They can be classified into two categories; physical locks(1),(2) and predicate locks(3). Physical locking means the operation of locking granules, including the entities referred to by a transaction. A granule of the database may be a record (tuple)(4), a page(5), a segment(6), a subset of the entire database(7), etc.

The system performance usually depends on the granule size. In general, fine granularity allows a higher degree of parallelism at a greater cost of managing locks. Coarse granularity, on the other hand, lowers the degree of parallelism but minimizes the cost of managing locks.

According to D.R. Rise, et al(1), a small number of granules is sufficient to allow enough parallelism for efficient machine utilization, and a large number of granules is extremely costly. Furthermore, they point out in their revised paper(2) that fine granualarity is preferable only in the case of small transactions (i.e. the number of entities referred to by a transaction is small).

Predicate locking(3),(8) means the operation of locking only the entities to be handled in each transaction, in other words, the entities that meet the predicates specifying the range to be referred to by the transaction. In other words, the predicate locking scheme makes its granules size change dynamically on each transaction in contrast to the fixed granule size of physical locking scheme. The predicate locking scheme requires more complicated processing and overhead in managing locks than physical locking scheme does. Therefore, if the predicate locking scheme is used, the number of locked predicates should be kept as small as possible for minimizing the overhead although it deteriorates the degree of parallelism.

A quantitative comparison of these two schemes is discussed in another paper(9). In this paper, the physical locking scheme is used. Particularly, two scheduling systems of the physical locking scheme will be discussed, in which one column (attribute) of a flat-table is assumed to be the granule size.

A transaction generally refers to several flat-tables. So,it can be divide into several units, called actions each of which refers only to one flat-table. Where a transaction arrives, any conflict which arises between the attributes referred to by it and those locked by all processing (active) transactions is checked. If there is no conflict, the transaction will be accepted for processing. The accepted transaction locks all the attributes needed. This is called "transaction-oriented scheduling". On the other hand, the scheduling in which each action locks the required attributes independently of other related actions which belong to the same transZaction is called "action-oriented scheduling".

Transaction-oriented scheduling can avoid conflicts among all active transactions and can process each transaction smoothly at a lower degree of parallelism. On the other hand, action-oriented scheduling has a higher degree of parallelism at a greater cost in managing "deadlocks" (that is, the detection and process of making void the effect caused by one of the transactions being in deadlock).

Assume that a transaction (TR1) is decomposed into two actions A1,A2, and another transaction (TR2) is decomposed into actions B1, B2. Furthermore, assume that conflicts between A2 and B1, and between A1, B2 occur after actions A1 and B1 are accepted. IN such a case, both TR1 and TR2 can never finish because A1 or B1 cannot release the attributes locked by itself, even if either one finishes its processing. Such o situation is called deadlock. The violation of the 2-phase principle(3) may cause an inconsistent situation.

In this paper, two scheduling systems, namely, the transaction-oriented and the action-oriented, are analyzed. The analysis forcuses on how the system performance is affected by both the deadlock frequency and the ratio of update request transaction to all transactions (called UR below).

184

2. Simulation Model

1) Transaction-oriented scheduling

This is a system in which all the attributes of the flat-tables referred to by a transaction are locked before it is decomposed into several actions. Each decomposed action can release the attributes locked by itself independently of its related actions, whenever its processing is finished.

The model of this scheduling is shown in Fig.1. Inter-arrival time of transactions is assumed to be an exponential distribution with a mean value of ta (1/). An arrived transaction enters the queue (Qa). If window-A is not in a busy state, a transaction is selected from Qa for conflict checking. a transaction accepted in window-A is decomposed into several actions each of which enters the queue (Qb) for database processing. An action is selected from Qb on the first-in first-out basis and is processed if window-B is not in a busy state. The processing time (Tb) is given by the random number of an exponential distribution. The attributes locked by the action are released after it has finished database processing.

2) Action-oriented scheduling

The model of this scheduling is shown in Fig.2. This is a system in which each action can lock the required attributes independently of other related actions. However,the attributes locked by action can be released only when all the actions which belong to the same transaction are accepted in window-A. An action which is accepted in window-A is called active action. Otherwise, the action which finishes its processing must wait until all other related actions become active.

Furthermore, a database system equipped with this scheduling must have a deadlock detection and recovery mechanism.

An arrived transaction enters Qa after it is decomposed into several actions. Any conflict that may arise between the attributes rewuired by an action and those needed by all active actions in Qb is checked. An accepted action enters Qb. The granules locked by an action cannot be released until all other related actions enter Qb. As data consistency requires that transactions be 2-phase(3), it is necessary to have a third queue (Qc) in this model.

Deadlock possibility is checked whenever an action finishes its database processing (either update or retrieval). There exists a possibility of deadlock if there is in Qa a related action (A1) which belongs to the same transaction as an action (A2) which has just finished processing in window-B. If an action (A1) in Qa refers to the same attributes as those locked by any action in Qb, a deadlock situation is said to have arised. When a deadlock is detected, all other related actions in Qa which belong to the same transaction as action A2 are removed from Qa, and the effect creates by action (A2) is ignored in this model.

The time which is required for checking the deadlock possibility, TD1, is proportional to the number of actions in Qa. This is because it is necessary to search through all the actions in Qa since there exists a possibility of deadlock if there is in Qa a related action which belongs to the same transaction as the action which has just finished in window-B. Similarly, the time required for deadlock detection, TD2, is proportional to the number of actions in Qb. Furthermore, it is assumed that the time required for deadlock recovery is constant.

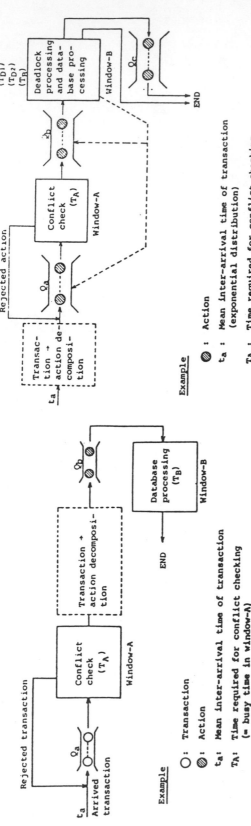

Example (Fig. 1)

○ : Transaction
◉ : Action
t_a: Mean inter-arrival time of transaction
T_A: Time required for conflict checking (= busy time in window-A)
T_B: Time required for database processing (= busy time in window-B)

Note 1: T_A is given by a random number depending on the number of actions in Q_b.

2: T_B is given by an exponential distribution.

3: The number of actions into which a transaction is decomposed is given as an input data. The time to decompose a transaction into actions is ignored.

Fig. 1 Model of the transaction-oriented scheduling

Example (Fig. 2)

◉ : Action
t_a : Mean inter-arrival time of transaction (exponential distribution)
T_A : Time required for conflict checking (= busy time in window-A)
T_{D1}: Processing time required for checking deadlock possibility
T_{D2}: Processing time required for deadlock detection
T_B : Time required for database processing (= busy time in window-B)
⟶ : Flow of Action
····▸ : Reference by the deadlock processing

Note: The time required by an action in window-B is one of three cases, T_B, T_B+T_{D1} or $T_B+T_{D1}+T_{D2}$.

Fig. 2 Model of the action-oriented scheduling

3) Assumptions used in this simulation model

    A database handled in this model is composed of two flat-tables, R1
and R2. R1 and R2 have 10 and 4 columns respectively. R1 and R2 have a same
attribute.

    The size of granule for locking is assumed to be equivalent to a
column of each flat-table. Therefore, the number of attributes is 14.

3. Simulation Results and Evaluations

1) Comparison of the two scheduling systems

    The average processing times (from the moment when a transaction is
input to the model to the moment when it is output from the model) for both
the transaction-oriented and action-oriented scheduling are estimated (AT
and AT are respectively for mean inter-arrival time i). Relative indices,
S   and S for indicating superiorty regarding the two schedulings are as
follows;

$$S_A = \sum_i \left(\frac{AT_i^T}{AT_i^A} - 1 \right) \text{ if } \frac{AT_i^T}{AT_i^A} \geq 1 \qquad\qquad S_T = \sum_i \left| \frac{AT_i^T}{AT_i^A} - 1 \right| \text{ if } \frac{AT_i^T}{AT_i^A} < 1$$

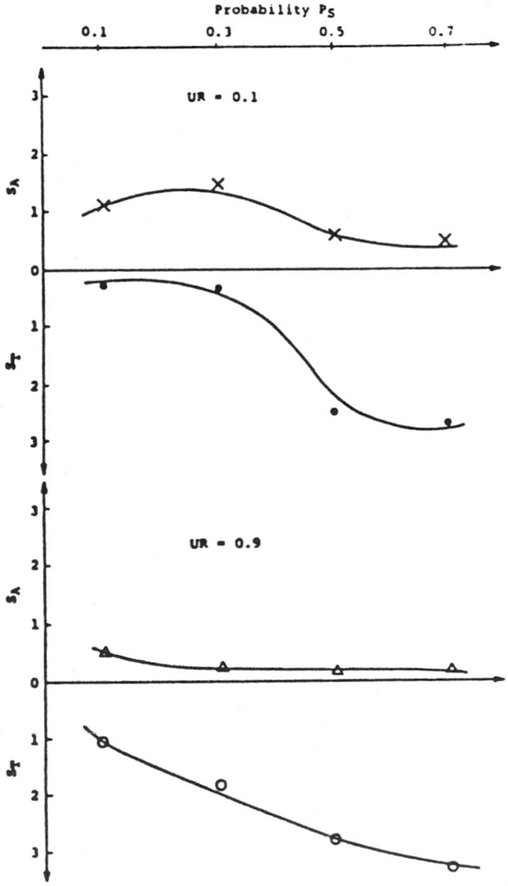

Fig. 3 Relative indices (S_A, S_T) for superiority regarding two
        scheduling systems

SA and ST versus Ps are shown Fig.3, in which the probability that both flat-tables R1 and R2 are referred to simultaneously by a transaction is called Ps.

If SA   ST, the action-oriented scheduling is said to be superior to the transaction-oriented one. Otherwise, the transaction-oriented schedu ling is preferable.

The following points can be deduced from Fig.3

If Ps is large (larger than 0.4), the transaction-oriented scheduling is preferable regardless of the ratio of update request transactions to all transactions (called UR).

If Ps is small (smaller than 0.4), the superiority of these two schedulings depends on UR. Namely, if UR is larger, the transaction-oriented scheduling is preferable. Otherwise, the action-oriented scheduling is preferable.

Fig.4 (a) and (b) show the average processing time versus 1/ta with UR=0.1 and UR=0.9, respectively, where Ps is fixed to 0.3 . It can deduced from Fig.4 that if UR is large enough, the transaction-oriented scheduling is always preferable, or else, the action-oriented scheduling is more favourable.

Fig.5 shows the average processing times (AT and AT) versus Ps, where the mean inter-arrival time (ta) is fixed. THe same fact which is deduced from Fig.3 , can also be deduced from Fig.5. That is, in case of small UR, if Ps is small (smaller than 0.4), the action-oriented scheduling is preferable, otherwise,. the transaction-oriented scheduling is preferable.

If UR is large, the transaction-oriented scheduling is preferable regardless of value of Ps.

The smaller the mean inter-arrival time of transactions is, the clearer the tendency shown in Fig.5 becomes. When the mean inter-arrival time becomes larger, the Ps coordinate of the point of intersection between the curves AT and AT also becomes larger. If ta is greater than a certain value, the curves AT and AT will never cross each other.

2) Deadlock frequency on the action-oriented scheduling

Fig.6 shows the graph of deadlock frequency versus 1/ta, where Ps is fixed to 0.7 and in addition UR is fixed to 0.1 and 0.9 . In the case of small UR (0.1), the deadlock frequency is inversely proportional to ta (dotted line in Fig.6). If UR is large (0.9), then the curve of the deadlock frequency has a peak value at a certain value of ta. The reason for having o peak is described below.

In general, the smaller the mean inter-arrival time is , the larger the length of Qb becomes, and furthermore the number of actions which will finish database processing also becomes larger. However, if the mean inter-arrival time is below a certain value, much longer CPU time will be allocated for conflict checking, and the CPU time allocated for database processing will be much shorter. Therefore, the number of actions which will finish database processing within a unit of time becomes smaller although the number of actions which will arrive at the model becomes larger, and the deadlock frequency goes down correspondingly.

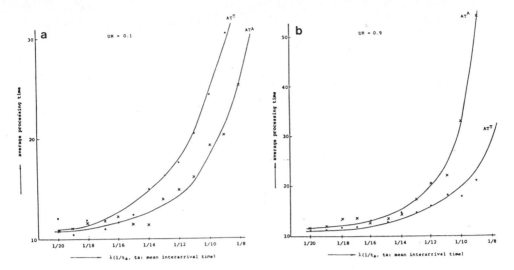

Fig. 4(a) Average processing time versus $\lambda(1/t_a)$

Fig. 4(b) Average processing time versus $\lambda(1/t_a)$

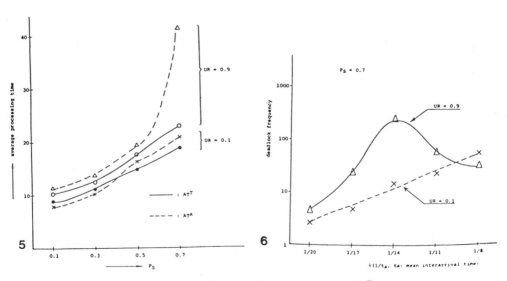

Fig. 5 Average processing time versus P_S
with the fixed interarrival time (t_a)

Fig. 6 Deadlock frequency versus $\lambda(1/t_a)$

4. Conclusion

The result of this simulation gives an insight into the performance of the two scheduling systems for the physical locking scheme. If the query structure of a transaction is simple (that is, the number of flat-tables referred to by it is small) and the ratio of update request transactions to all transactions is small, the action-oriented scheduling is preferable. Otherwise, the transaction-oriented scheduling is to be preferred.

References

(1) D.R. Ries, M. Stonebraker. "Effects of Locking Granularity in a Database Managemeny System," ACM Transactions on Database System., Vol.2 , No.3, Sept., 1977.

(2) D.R. Ries, M.R. Stonebraker. "Locking Granularity Revised," ACM Transaction on Database System., Vol.4, No.2, June., 1979.

(3) K.P. Eswaren, J.N. Gray., et. al. "The notion of consistency and Predicate Locks in a Database System," Comm. of the ACM., Vol.19,No.11,Nov., 1976.

(4) M.M. Astrahan, et. al. "System R: Relational Approach to database Management," ACM Transaction on Database System., June., 1976.

(5) J.N. Gray, et.al. "Granularity of Locks in a shared database," Proc. Int. Conf. on Very Large Database., Sept.,1975

(6) M. Stonebraker, et. al. "The Design and Implementation of INGRES," ACM Trns. on Database System.,Sept., 1976.

(7) J.F. Spitzer. "performance prototyping of data management applications," Proc. of ACM., 1976.

(8) J.J. Florentin. "Consistency auditing of database," Computer Journal., Feb., 1974.

(9) Y. Matsushita, et. al. "An Evaluation of the Physical and Predicate Locking Schemes in a Database System" to be published.

ALLOCATION OF DOCUMENTS IN TWO LEVEL MEMORY

FOR INFORMATION RETRIEVAL SYSTEMS

Wojciech Ziarko and S.K.M. Wong

Dept. of Computer Science
University of Regina
Regina, Saskatchewan S4S 0A2 Canada

ABSTRACT

We show in this paper that the problem of storage allocation in an information retrieval system can be formulated in terms of the network flow model. The optimal solution to this problem is obtained by identifying the cut with minimum capacity, which can be easily found by the standard maximum flow algorithm.

1.   INTRODUCTION

In the physical storage design for an information retrieval system, we often have to select suitable storage media characterized by different memory costs and access times. For example, magnetic tapes are slow and cheap while magnetic disks are much faster but more costly. In general, the cost per bit increases when the average memory access time decreases. The criterion of storage cost versus performance leads to the classification of storage media into distinct memory levels with the fastest memory being assigned to the top level (corresponding to the highest cost also) and followed by devices with decreasing storage costs and performance [1]. Due to large volumes of data involved in information retrieval systems, an optimal selection of memory levels for different groups of data is crucial to the overall system performance particularly from a cost and benefit point of view.

The basic objective of every information retrieval system is to retrieve and present to the user in a readable form all documents which are relevant to the query submitted to the system. In a well designed system the documents should be stored in a way to satisfy the user in terms of the retrieval time and to minimize the storage and retrieval

This research was supported in part by a grant from The National Sciences and Engineering Research Council of Canada

costs. Intuitively, to meet these requirements in a
multilevel storage environment the groups of documents which
constitute answers to some common (or some important) queries
should be stored in the higher levels of the memory
hierarchy. In fact, the strategy for storage allocation
indeed would be rather straightforward if the families of
subsets of documents, each of which corresponds to a
different query, were disjoint. Under such circumstances, we
can adopt, for instance, the allocation strategy based on the
access frequency characteristics of the queries as proposed
by Chen [1] (who dealt with the problem of file allocation in
multilevel memories). However, in practice, the subsets of
documents corresponding to different queries may have nonempty intersections. The statistical method in [1] can no
longer be applied to these situations.

    In this paper we propose an optimal storage allocation
strategy (applicable to a two-level memory hierarchy, for
example, consisting of disk storage and a "cache" memory) for
the information retrieval model suggested by Marek and Pawlak
[2]. It should be noted that our method is applicable to
other Boolean information retrieval systems as well. In
essence, our method is based on the network flow model [3]
which enables us to transform an apparent 2^n combinatorial
problem into a polynomial time algorithm.

2.   THE INFORMATION SYSTEM MODEL

    In order to illustrate our ideas, we choose to apply our
method to the information retrieval (IR) model proposed by
Marek and Pawlak [2], which is outlined below for future
reference. We can specify the IR system by a quadruple:

$$S = <X, D, R_I, f> \quad , \tag{1}$$

where X is a set of documents, D is a set of descriptors, R_I
is an equivalence relation in D, (its equivalence classes are
indexed by a set I of attributes), and

$$f : D \longrightarrow 2^X \quad , \tag{2}$$

where 2^X denotes the power set formed from X. That is, f is
a function which associates with every descriptor a D a
subset of documents f(a) relevent to this descriptor. By
definition, the function f satisfies the following
conditions:

$$aR_I b \quad \text{and} \quad a \neq b \quad => \quad f(a) \cap f(b) = \emptyset \quad , \tag{3}$$

$$\bigcup \{f(c) : aR_I c\} = X \quad , \quad \text{for all } a,c \in D . \tag{4}$$

    Condition (3) specifies the fact that each document is
characterized by no more than one value of an attribute
(where a value of an attribute refers to a descriptor
belonging to the corresponding equivalence class of R_I).
Condition (4) implies that every document is characterized by
a certain value of each attribute in I.

192

The query language of the above information retrieval system consists of a collection of terms Tr which are boolean expressions constructed from descriptors D, boolean operators $\neg, +, ., ->$, and constants 0 and 1. The semantic of each query submitted to such a system is determined by an extension $||.||$ of the function f defined inductively in the following way:

$$||a|| = f(a) \text{ for } a \in D, \quad ||0|| = \emptyset , \quad ||1|| = X , \tag{5}$$

$$||\neg t|| = X - ||t|| , \quad t \in Tr, \tag{6}$$

$$||t_1 . t_2|| = ||t_1|| \cap ||t_2|| , \tag{7}$$

$$||t+t_2|| = ||t_1|| \cup ||t_2|| , \tag{8}$$

$$||t_1 -> t_2|| = (X-||t_1||) \cap ||t_2|| , \quad t_1, t_2 \in Tr. \tag{9}$$

Note that two terms t,t' are equivalent (denoted by $t \sim t'$) if $||t||=||t'||$. It can be shown [2] that $t \sim t'$ iff the equality $t=t'$ can be deduced from the axioms of boolean algebra enriched by the following axioms:

$$a = \neg \sum_{bR_I a, \ b \neq a} b \qquad \text{for all } a \in D . \tag{10}$$

Let $\{D_i\}_{i \in I}$ be the partition induced on D by R_I. It can be proved, by using the above axioms, that each term t can be transformed into the standard form

$$t = \sum_{j \in J} \hat{t}_j , \tag{11}$$

where each $\hat{t}_j$ is a simple term, i.e.

$$\hat{t}_j = \prod_{i \in I} a_i , \qquad \text{for every } i \in I , \ a_i \in D_i \tag{12}$$

3.   METHOD OF ATOMIC COMPONENTS AND TWO LEVEL MEMORY

For any simple term t defined by equation (12), there exists a subset of documents $||t||$, which is referred to as the component corresponding to t.   Note that all components (derived from all the simple terms) are distinct and they are in fact the atoms of the boolean algebra of subsets generated by the family of sets $\{||a||\}_{a \in D}$.   Since any query can be transformed into the standard form (sum of simple terms), each query can, therefore, be mapped into a group of components.   Based on this observation, Marek and Pawlak suggested that all the documents in a component should be stored closely together (e.g. in consecutive memory locations) to facilitate the retrieval of all the components belonging to a particular query.   By this method (referred to as the method of atomic components), it is therefore sufficient to know the storage address of every component in order to process any query efficiently.   Since query preprocessing is much faster than performing set theoretical operations on groups of documents, the above method is much more efficient than other methods like, for example, inverted

lists.   Furthermore,   we   can   enhance   the   retrieval
performances by employing a consistent lexicographic scheme
[2] for ordering the components in a linear store which can
subsequently be coupled with an index or hashing method to
determine the address of each component.   This method was
later improved by Lipski [4] where he exploited the
lexicographic ordering of components to minimize the number
of memory accesses required in query processing.

        In this paper, we look at the original method of atomic
components from a different point of view.   We propose an
extension of this method in a two-level memory hierarchy
environment.   Our   method   is   based   on   the   following
assumptions:

> (i)   All the documents belonging to the same component
>   are stored in the same level of memory.

> (ii) For certain queries, "split" components may occur.
>   That is, some components of a certain query may
>   reside in the higher speed medium whereas the other
>   components are allocated to a slower device.   In
>   this case, we assume that the processing times for
>   this category of queries would not increase (in
>   fact, they could decrease instead).

4.   THE PROBLEM MODEL

        In   our   model   the   optimal   storage   allocation   of
components will be obtained with respect to a predefined
class of common terms, namely,

$$T = \{t_1, t_2, \ldots, t_k\} \subset Tr \tag{13}$$

Without loss of generality, we may assume that all terms in
T are already in standard form.   The storage consists of two
levels characterized by cost per bit c_i (i=1,2).   Assume that
$c_2 > c_1$.   For any given term t $\in$ T, denote by t* the set of
all simple terms participating in the expansion of t, and let
$s(\hat{t})$ denote the size (in terms of bits) of the component
$||\hat{t}||$.   The additional cost d(t) incurred by storing all
documents relevent to a query t in the faster memory level is
given by

$$d(t) = \sum_{\hat{t} \in t*} d(\hat{t}) = (c_2 - c_1) \sum_{\hat{t} \in t*} s(\hat{t}) \ . \tag{14}$$

        Now we have to find a quantitative measure of the
benefit as a result of storing all the components of a
particular term in the more expensive memory of level two.
Obviously, reducing the processing time for t, say, would be
beneficial to all those users who use such a query.   However,
some of these users may not be willing to accept the extra
cost because faster response time may not be necessary in
their particular kinds of applications (e.g. in a
bibliographic systems, a slower response time is acceptable).
Therefore, a compromise solution must be found in order to
satisfy all users with respect to their willingness to accept

higher cost. We propose the following criterion as a measure
of the value w(t) of a query t, i.e. ,

$$w(t) = k(c_2 - c_1)s(t)m(t,k) ,$$ (15)

where k is a constant and m(k,t) is the number of users who
would accept the extra charge $k(c_2 - c_1)$ s(t) for the query t.
Note that the choice of the value for k depends on many
factors such as profit margin and other environmental issues.
We only assume here that a reasonable value for k can be
chosen based on prudent management decisions. Another
possible way of assigning importance factors to queries is by
gathering their frequency characteristics. In this approach
(proposed by Eisner and Severance [6] in somewhat differnet
context), a user's perceived value of a query t is
proportional to its frequency $\mathcal{V}$(t), i.e.

$$w(t) = k \, \mathcal{V}(t)$$ (16)

where k is a constant and w(t) is in the same units as
storage cost d(t). Again, as in the previous case, the
problem is to choose an appropriate value for k. This
question is discussed in detail in [6]. It should be
emphasized, however, that our document allocation method does
not depend on the choice of the function w(t). The
assumption we make is that such a function exists and is
known.

At this point, the problem of document storage
allocation can be formulated in terms of the benefits and
costs as follows. The optimal selection of a subset T' of
terms in T (i.e. T' $\subseteq$ T) to be stored in the faster memory
is determined by maximizing the total "profit" P defined by

$$P = \sum_{t \in T'} w(t) - (c_2 - c_1) \sum_{t \in T'} s(\hat{t}) ,$$ (17)

where $U_{T'} = \bigcup_{t \in T'} t^*.$

Note that the second term in equation (17) represents the
total increase in memory cost due to transferring the
components of all terms belonging to T' to the higher memory
level. We will describe a general method to solve such an
optimal selection problem in section 7.

5.   FLOWS IN NETWORK

We briefly review the concepts of the network model in
this section. A flow network is a triple N=<V,A,c>, where V
is a finite set of vertices (nodes), A$\subseteq$VxV is a set of
arcs, and the function c from A to non-negative reals is the
capacity function. For each arc $(v_i,v_j) \in$ A, the value
$c_{ij} = c(v_i,v_j)$ is called the capacity of the arc (v_i,v_j).
(Sometimes we write (v_i,v_j) as (i,j)). A flow in a network
from a source s to a sink e (s,e $\in$ V) is a non-negative
function f on the arcs, with $f_{ij} = f(i,j) \leq c_{ij}$ for all arcs

(i,j), satisfying

$$\sum_{(i,j)\,\in\,A} f_{ij} - \sum_{(k,i)\,\in\,A} f_{ki} = 0 \qquad \text{for all } i \neq s,e, \qquad (18)$$

and $\quad r = \displaystyle\sum_{(s,j)\,\in\,A} f_{sj} = \sum_{(k,e)\,\in\,A} f_{ke}$. $\qquad\qquad$ (19)

The number r is called the value of the flow and is the net flow out of s. Equation (18) implies that for any node i≠s,e, the flow into i equals the flow out of i.

If X and Y are two subsets of nodes in V, define (X,Y)={(i,j) ∈ A: i ∈ X, j ∈ Y}. Consider a partition of the set of nodes V into sets, say, Z and $\bar{Z}$, with s∈ Z and e ∈ $\bar{Z}$. The set (Z,$\bar{Z}$) is refered to as cut of the network N (disconnects all paths leading from s to e). The value of a cut is the sum of capacities on arcs (i,j) with i∈ Z and j∈ $\bar{Z}$, namely,

$$c(Z,\bar{Z}) = \sum_{(i,j)\,\in\,(Z,\bar{Z})} c_{ij}. \qquad\qquad (20)$$

The fundamental theorem in the network flow model states that the value of the maximum flow is equal to the value of the minimum cut [3]. Moreover, the same computational procedure (labelling algorithm) simultaneously determines maximum flow and minimum cut. We will show how the storage allocation problem can be reduced to minimum a cut problem.

6.   THE SOLUTION OF THE DOCUMENT ALLOCATION
      PROBLEM IN TWO LEVEL MEMORY

Based on the network flow model, the optimal storage allocation in the information retrieval system introduced in section 2 can be stated as follows. For any finite set of terms T consider the selection network N = <V ,A ,c >.

The set of nodes V consists of a source node s, a sink e, a node for every term t ∈ T, and a node for each distinct simple term $\hat{t}$ in ∪T, i.e.

V = {s,e} ∪ T ∪ (∪T) , $\qquad\qquad$ (21)

where ∪T denotes the set of simple terms {$\hat{t}$} which span all the terms in T. There are three different types of arcs that may exist in A:

   (i)   (s,t) , t ∈ T,

   (ii)   (t,$\hat{t}$) , if $\hat{t}$ ∈ t* , $\qquad\qquad$ (22)

   (iii) ($\hat{t}$,e) , $\hat{t}$ ∈ ∪T.

The capacity function c on A is defined by

   (i)   c(s,t) = w(t) ,

196

(ii) $c(t,\hat{t}) = \infty$,   (23)

(iii) $c(\hat{t},e) = d(\hat{t})$.

By definition, a selection is a subset of terms, T' of T together with all the simple terms which belong to this collection. The value of the selection (T', $\cup$ T') is the sum of the profits minus the sum of the costs of the distinct simple terms in $\cup$ T', namely,

$$val(T, \cup T') = \sum_{t \in T'} w(t) - \sum_{\hat{t} \in \cup T'} d(\hat{t}) . \qquad (24)$$

Our objective here is to find a selection which gives the maximum value defined by equation (24) i.e. to identify terms whose components ought to be stored in higher memory level to achieve the maximal benefits from the system as a whole. The following two Lemmas [5] explain the relationship between selections and cuts in such a selection network.

Lemma 1. There is a one-to-one correspondence between cuts containing no arcs of type $(t,\hat{t})$ in network N and selections.

Proof: Consider such a cut. Let the subset of nodes still connected to s be G, and the rest $\bar{G}$; and the set of nodes of $\cup$ T still connected to G be H and the rest $\bar{H}$. Obviously, (G,H) is a selection.

Suppose now that (G,H) is a selection. This means that no arcs $(t,\hat{t})$, with $t \in G$ and $\hat{t} \in \bar{H}$ or $t \in \bar{G}$ and $\hat{t} \in H$, may exist in N . Therefore, there exists a cut $(Z,\bar{Z})$ corresponding to the selection (G,H) where $Z=\{s\}\cup H \cup G$. Clearly, there is no arc of type $(t,\hat{t})$ in this cut.

Lemma 2. The minimum cut corresponds to the selection of maximum value.

Proof: Let (G,H) corresponds to a minimum cut containing no arc of type $(t,\hat{t})$. Let (C',H') correspond to any other cut. It follows that

$$\sum_{t \in \bar{G}} w(t) + \sum_{\hat{t} \in H} d(\hat{t}) \leq \sum_{t \in \bar{G'}} w(t) + \sum_{\hat{t} \in H'} d(\hat{t}) . \qquad (25)$$

Let $P = \sum_{t \in T} w(t)$. Note that

$$P = \sum_{t \in G} w(t) + \sum_{t \in \bar{G}} w(t) = \sum_{t \in G'} w(t) + \sum_{t \in \bar{G'}} w(t) \qquad (26)$$

Equation (25) can be rewritten as

$$P - \sum_{t \in G} w(t) + \sum_{\hat{t} \in H} d(\hat{t}) \leq P - \sum_{t \in G'} w(t) + \sum_{\hat{t} \in H'} d(\hat{t}) , \qquad (27)$$

or

$$\sum_{t \in G'} w(t) - \sum_{t \in H'} d(\hat{t}) \leq \sum_{t \in G} w(t) - \sum_{t \in H} d(\hat{t}) . \quad (28)$$

which completes the proof.

Therefore, we have demonstrated that to solve the optimal storage allocation problem is equivalent to maximize the flow in the network N which simultaneously identifies a minimum cut and hence a selection of maximum value. The labelling algorithm, for determining the maximum flow in a network, is well known and has many implementations [3,8]. The algorithm produces the solution in $O(|V|^3)$ elementary steps [8]. Better results can be achieved by using a specialized selection algorithm which we propose in [7]. In that algorithm a linear time process is used to perform a preliminary selection which leads to a sub-network of reduced dimension.

7. CONCLUSIONS

In this paper we have demonstrated, within the framework of the information retrieval system proposed by Marek and Pawlak [2], how the well known techniques in the network flow model can be applied to the document allocation problem in two-level memory. We have explicitly shown that the storage allocation problem can be reduced to identifying a minimum cut in a suitably constructed network. Therefore, an optimal solution can be obtained directly from the standard maximum flow algorithm which simultaneously identifies a minimum cut. One of the open questions is how to generalize this approach to multilevel memory systems.

REFERENCES

1. P.P.S., Chen Optimal File Allocation in Multilevel Storage Systems. IEEE National Conference (1973) pp. 277-281.
2. W. Marek, Pawlak, Z. Information Storage and Retrieval Systems: Mathematical Foundations. Theoretical Computer Science (1976).
3. L.R. Ford, Fulkerson, D.R. Flows in Networks. Princeton University Press (1962).
4. W. Lipski, On an Efficient Method of Information Retrieval. Fundamental Informaticae, Vol. 2, No. 3 (1979) pp. 227-243.
5. M.L. Balinski, On Selection Problem. Management Science, Vol. 17, No. 3, (1970) pp. 230-231.
6. M.J. Eisner, Severance, D.D. Mathematical Techniques for Efficient Record Segmentation in Large Shared Databases. Journal of the ACM, Vol. 23, No. 4 (1976) pp. 613-635.
7. S.K.M. Wong, W. Ziarko, On Specialized Optimal Selection Algorithm. Computer Science Dept. University of Regina Technical Report No. CS-84-15.
8. J. Picard, M. Queyranne, Selected Applications of Minimum Cuts in Networks. Canadian Journal of Operational Research and Information Processing, Vol. 20, No. 4, (1982) pp. 394-422.

Mathematical File Organization and Computational Geometry

FURTHER RESULTS ON HYPERCLAW DECOMPOSITION AND

BALANCED FILING SCHEMES

Sumiyasu Yamamoto

Department of Applied Mathematics
Okayama University of Science
Ridai-cho, Okayama 700, Japan

INTRODUCTION

A balanced file organization scheme (BFS) introduced first as a generalization of inverted filing scheme by Abraham, Ghosh and Ray-Chaudhuri [1] can be defined formally in terms of a many-to-one addressing function from a set of canonical queries to be inverted to a set of bucket addresses. It can, therefore, be organized by defining a multiple queries to an address transformation (MQAT, see Yamamoto [7,10]).

Since there are so many possibilities of defining a many to one addressing function from a set of queries to a set of bucket addresses, the redundancy of organized scheme which is defined by the average number of buckets, or average times of storing a record, can be used for selecting the best scheme in a given situation (see Yamamoto, Ikeda, Shigeeda, Ushio and Hamada [4], Yamamoto, Tazawa, Ushio and Ikeda [6] and Yamamoto and Tazawa [8]).

In this paper, we shall restrict our attention to the balanced file organization schemes of order k (BFS_k) where every record is characterized by m binary-valued attributes and the set of canonical queries, $Q^{(k)}$, is composed of all queries of order k, i.e., each query requires to retrieve all records pertinent to specified k attributes simultaneously. A BFS_k is said to be of degree c if its MQAT defined over $Q^{(k)}$ is c-to-one. It is easily seen that a BFS_k of degree c can be constructed if and only if $\binom{m}{k}$ is an integral multiple of c.

Among those BFS_k of degree c, the redundancy of a scheme is the least if every set of c queries of order k corresponding to a bucket is composed of such c queries specifying the same k-1 except one in common, provided the probability distribution of records is invariant under every permutation of attributes (see Yamamoto and Tazawa [8]). The possibility of the existence and the construction of such scheme having the least redundancy, called $HUBFS_k$, have been investigated in [8] in some detail, especially for the case k=3.

The purpose of this paper is (i) to present further results on the possibility of constructing $HUBFS_k$ stated in the terminology of combinatorial theory, and, in passing, (ii) to revise a misleading statement which leads to erroneous results for the case $k \geq 4$ in the necessary condition of construction given in Yamamoto and Tazawa [8,9].

Let $\mathbf{E} = \{E_i | i \varepsilon I\}$ be a family of nonempty subsets of a finite set $\mathbf{V} = \{v_j | j=1,2,\dots,m\}$. The couple $\mathbf{H} = (\mathbf{V},\mathbf{E})$ is called a hypergraph on $\mathbf{V}$ if $\cup_{i \varepsilon I} E_i = \mathbf{V}$. The cardinality $|\mathbf{V}| = m$ is called the order of $\mathbf{H}$. The elements of $\mathbf{V}$ and $\mathbf{E}$ are called the vertices and the edges of $\mathbf{H}$, respectively. A hypergraph $\mathbf{H}$ is said to be simple if its edges are all distinct. A simple hypergraph on $\mathbf{V}$ is called a k–graph if every edge has the same cardinality k. A k–graph on $\mathbf{V}$ is said to be complete and denoted by $H_{m,k}$ if $\mathbf{E}$ is composed of all $\binom{m}{k}$ edges. A collection of c edges $E_1, E_2, \dots, E_c$ of $H_{m,k}$ ($k \geq 2, c \geq 2$) is called a k–hyperclaw, (abbreviated as k–HC) or simply, hyperclaw, of degree c if $|\cap_{j=1}^{c} E_j| = k-1$. This intersection set is called the root and the vertices $E_i - \cap_{j=1}^{c} E_j$, $i=1,2,\dots,c$, are called the leaves of the k–HC, respectively.

If a one-to-one correspondence between the set $\mathbf{V} = \{v_j | j=1,2,\dots,m\}$ of m vertices of a hypergraph and the set of m attributes characterizing the records is established, then a query of order k can be translated as a k–edge or a subset of k vertices. A BFS_k of degree c can, therefore, be identified with a c-to-one transformation defined over the set of all k–edges of $H_{m,k}$. This is equivalent to a partition of the set of all k–edges of $H_{m,k}$ into mutually disjoint subsets of cardinality c each. An $H_{m,k}$ is said to be k–hyperclaw decomposable of degree c if its set of edges can be partitioned into pairwise edge-disjoint k–hyperclaws of the same degree c. This is equivalent to the existence of $HUBFS_k$ of degree c since every bucket corresponds to a k–HC and has the least redundancy with respect to the permutation invariant distribution of records. Thus we have:

THEOREM 1. The existence of $HUBFS_k$ of degree c for m binary-valued records is equivalent to the decomposability of $H_{m,k}$ into edge-disjoint k–HC of degree c.

A NECESSARY CONDITION

The problem of the hyperclaw decomposability of complete hypergraph $H_{m,k}$ treated here turns out to that of claw decomposability of complete graph K_m in the case k=2. The latter has been raised and solved completely by Yamamoto, Ikeda, Shige-eda, Ushio, and Hamada [5]. For those cases $k \geq 3$, the problem has been treated in Yamamoto and Tazawa [9]. A necessary condition and several sufficient conditions for the decomposability of complete hypergraph $H_{m,k}$ into edge-disjoint k–HC of degree c are given there. Especially, the decomposition of $H_{m,3}$ into edge-disjoint 3–HC has been discussed in some detail. It has been shown that the lower bound on m for the existence of 3–HC decomposition given in the necessary condition is really tight. The necessary condition and its proof given in [9] and cited also in [8] are, however, valid inasmuch as k=3. They are erroneous for those cases $k \geq 4$.

Revising the necessary condition and summarizing the results obtained for k=2 and k=3, we have the following theorem.

THEOREM 2. The following two conditions (i) and (ii) are necessary for the decomposability of complete hypergraph $H_{m,k}$ into edge-disjoint k–hyperclaws of degree c i.e.,

(i) $\binom{m}{k}$ is an integral multiple of c,

(ii) $m \geq$

$2c$,	if $k=2$;
$\frac{3}{2}c + 1$,	if $k=3$;
$\frac{1}{2}\{5 - 8c + \sqrt{(112c^2+1)}\}$,	if $k=4$;
$c + k - 1$,	if $k\geq5$.

Proof. The condition (i) is obviously necessary.

It has been shown in [5] that $m\geq2c$ is not only necessary but also sufficient if $k=2$. With respect to the case $k=3$, it has been shown that $m \geq \frac{3}{2}c + 1$ is necessary and this lower bound is really tight for every evenly odd number c, i.e., $c=4n+2$ for $n=0,1,2,...$ (see [9]).

Consider the case $k\geq4$ and suppose $H_{m,k}$ is decomposed into pairwise edge-disjoint k-HC of degree c. Let $P=\{x_1,x_2,...,x_{k-2}\}$ be an arbitrarily selected set of k-2 vertices. Then, the remaining m-k+2 vertices can be classified into two sets $Q=\{y_1,y_2,...,y_d\}$ and $R=\{z_1,z_2,...,z_{m-k-d+2}\}$ in such a way that every set of k-1 vertices $P\cup\{y_i\}$ is the root of some k-HC of degree c, whereas every $P\cup\{z_j\}$ is not.

If $|R|=m-k-d+2$ is either 0,1 or 2, we have $d=m-k+2$, $m-k+1$ or $m-k$. Suppose $|R|\geq3$ and consider a k-edge $P\cup\{z_i,z_j\}$ for every pair of vertices in R. One of the k-1 subsets containing z_i, z_j and a certain set of k-3 vertices of P must be the root of a k-HC including the k-edge composed of P and $\{z_i,z_j\}$ since both $P\cup\{z_i\}$ and $P\cup\{z_j\}$ are not the root of any k-HC of degree c. Thus for every pair of $\{z_i,z_j\}$, there corresponds just one k-HC of degree c. Enumerating the number of k-edges which may possibly be members of $\binom{m-k-d+2}{2}$ k-HC's of degree c by classifying them into those having exactly two or three vertices of R, we have,

(1) $$\binom{m-k-d+2}{2}c \leq \binom{m-k-d+2}{2}(d + 1) + \min(3,k-2)\binom{m-k-d+2}{3}.$$

Thus we have,

(2) $\qquad d \geq 3c - 2m + 5 \qquad$ for the case $k=4$, and

(3) $\qquad m \geq c + k - 1 \qquad$ for the case $k\geq5$.

Note that the coefficient $\min(3,k-2)$ in (1) which is the upper bound of the number of different k edges among three for the case $k\geq4$ has been dropped in the formula (2.1)) of [9]. The arguments developed in [9], however, can be applied if $k=3$, and we have $d \geq \frac{3}{2}c - \frac{m}{2}$. This leads us to

(4) $\qquad m \geq \frac{3}{2}c + 1.$

If $k=4$, since $m \geq c+3$ is necessary for the existence of 4-HC of degree c, the inequality (2) holds whenever $|R|\geq3$ or not. Thus for every selection of the set P, we have

(5) $\qquad \binom{m}{2}(3c-2m+5) \leq \sum d \leq 3\binom{m}{4}/c.$

This leads us to

(6) $\qquad m^2 + (8c-5)m - 2(6c^2+10c-3) \geq 0,$

and we have

(7) $\qquad m \geq \frac{1}{2}\{5 - 8c + \sqrt{(112c^2+1)}\}.$

Summarizing those, we have (ii). This completes the proof.

Let V be a finite set of points and let B be a family of subsets B_i (called blocks -- not necessary distinct) of V. The pair (V,B) is called a block design. Let v^*, k^*, t, and λ_t be four positive integers satisfying $v^* \geq k^* \geq t$. A block design (V,B) is called a $t-(v^*,k^*,\lambda_t)$ design if and only if

(a) $|V| = v^*$;

(b) $|B_i| = k^*$ for every $B_i \varepsilon B$; and,

(c) every t subset of V is contained in exactly λ_t blocks of B.

THEOREM 3. A complete hypergraph $H_{m,k}$ is k-hyperclaw decomposable of degree $c=m-k+1$, if and only if there exists a $t-(v^*,k^*,\lambda_t)$ design with $\lambda_t=1$ and $k^*=t+1$. Where $v^*=m$, and $k^*=c=m-k+1$. Moreover, such decomposition satisfies the lower bound on m given in Theorem 2 for the case $k \geq 5$.

Proof. Suppose a $t-(v^*,t+1,1)$ design exists. Construct a k-hyperclaw of degree c for every block B_i of $k^*=t+1$ points in such a way that $v^*-k^*(=k-1)$ points of $V-B_i$ and $k^*(=t+1=c)$ points in B_i be the root of cardinality $k-1$ and c leaves of the hyperclaw, respectively. Since $\lambda_t=1$ and $k^*-1=t$, a k-HC decomposition of degree c is obtained for a complete hypergraph $H_{m,k}$ with $m=v^*$. Conversely, if an $H_{m,k}$ is k-HC decomposable of degree c, then a $t-(v^*,t+1,1)$ design is obtained by identifying the set of c leaves of every k-HC with a block of the design. Clearly, this decomposition attains the lower bound given in Theorem 1 for the case $k \geq 5$.

Table 1 shows a hyperclaw decomposition of $H_{7,5}$ into 3-HC of degree 3 using well known 2-(7,3,1) design. Similarly, Table 2 shows a hyperclaw decomposition of $H_{8,5}$ into 5-HC decomposition of degree 4 using 3-(8,4,1) design listed in Kageyama and Hedayat [2].

TABLE 1. Decomposition of $H_{7,5}$ into 5-HC of degree 3

Root	Leaves
(2, 4, 5, 6)	0, 1, 3
(3, 5, 6, 0)	1, 2, 4
(4, 6, 0, 1)	2, 3, 5
(5, 0, 1, 2)	3, 4, 6
(6, 1, 2, 3)	4, 5, 0
(0, 2, 3, 4)	5, 6, 1
(1, 3, 4, 5)	6, 0, 2

TABLE 2. Decomposition of $H_{8,5}$ into 5-HC of degree 4

Root	Leaves
(4, 6, 7, 8)	1, 2, 3, 5
(5, 7, 1, 8)	2, 3, 4, 6
(6, 1, 2, 8)	3, 4, 5, 7
(7, 2, 3, 8)	4, 5, 6, 1
(1, 3, 4, 8)	5, 6, 7, 2
(2, 4, 5, 8)	6, 7, 1, 3
(3, 5, 6, 8)	7, 1, 2, 4
(1, 2, 4, 7)	3, 5, 6, 8
(2, 3, 5, 1)	4, 6, 7, 8
(3, 4, 6, 2)	5, 7, 1, 8
(4, 5, 7, 3)	6, 1, 2, 8
(5, 6, 1, 4)	7, 2, 3, 8
(6, 7, 2, 5)	1, 3, 4, 8
(7, 1, 3, 6)	2, 4, 5, 8

A list of parameters of small size t-designs satisfying the condition of Theorem 2 is given in Table 3. The construction of those designs can be seen in the literatures (see [3]).

TABLE 3. Parameters of known small size t-designs and corresponding
hypergraphs satisfying the condition of Theorem 2.

$2-(v^*,3,1)$ designs

v^*	k^*	b^*	r^*	λ		m	k	c
7	3	7	3	1		7	5	3
9	3	12	4	1		9	7	3
13	3	26	6	1		13	11	3
15	3	35	7	1		15	13	3
19	3	57	9	1		19	17	3
21	3	70	10	1		21	19	3
.	.	.	.	.		.	.	.

$3-(v^*,4,1)$ designs

8	4	14	7	1		8	5	4
14	4	91	26	1		14	11	4
20	4	285	57	1		20	17	4
26	4	285	100	1		26	23	4
50	4	4900	392	1		50	47	4
.	.	.	.	.		.	.	.

$b^* =$ No. of k-HC of degree c.

REFERENCES

[1] Abraham, C.T., Ghosh, S.P. and Ray-Chaudhuri, D.K. (1968). File organization schemes based on finite geometries. Information and control
12, 143-163.
[2] Hedayat, A. and Kageyama, S. (1980). The family of t-designs--Part I.
J. Statist. Plann. Inference 4, 173-212.
[3] Kageyama, S. and Hedayat, A. (1983). The family of t-designs--Part II.
J. Statist. Plann. Inference 7, 257-287.
[4] Yamamoto, S., Ikeda, H., Shige-eda, S., Ushio, K. and Hamada, N.(1975).
Design of a new balanced file organization scheme with the least redundancy. Information and Control 28, 156-175.
[5] Yamamoto, S., Ikeda, H., Shige-eda, S., Ushio, K. and Hamada, N.(1975).
On claw decomposition of complete graphs and complete bigraphs.
Hiroshima Math. J. 5, 33-42.
[6] Yamamoto, S.,Tazawa, S., Ushio, K. and Ikeda, H.(1978). Design of a
generalized balanced multiple-valued file organization schemes of order
two. Pro. 8th ACM-SIGMOD, 47-51.
[7] Yamamoto, S. (1979). Design of optimal balanced filing schemes. Proc.
IBM Working Conf. of Database Engineering, II-23-36.
[8] Yamamoto, S. and Tazawa, S. (1979). Combinatorial aspects of balanced
file organization schemes. J. Information Processing 2, 127-133.
[9] Yamamoto, S. and Tazawa, S. (1980). Hyperclaw decomposition of complete
hypergraphs. Annals of Discrete Mathematics 6, 385-391.
[10] Yamamoto, S. (1982). Design of optimal balanced filing schemes. Lecture
Notes in Computer Science No.113, 253-265.

PLACING TILES IN THE PLANE

Andrei Broder[1]
DEC Systems Research Center
130 Lytton Avenue, Palo Alto, California 94301

Barbara Simons
IBM Research Laboratory K55-281
5600 Cottle Road, San Jose, California 95193

ABSTRACT

The VLSI placement problem consists of finding an optimum placement of the VLSI components in the plane of the chip. A standard optimization goal is to minimize the total amount of space occupied by the wires on the chip. We model the VLSI placement problem by considering the problem of placing tiles on the plane when each tile has a preassigned area into which it must be placed. We show that for most cases this problem appears computationally infeasible, and we present some fast algorithms for some special cases.

1. INTRODUCTION

One of the major problems in the design of VLSI chips is the placement problem, which consists of finding an optimal placement of the VLSI components in the plane. An important optimization goal for the placement problem is to minimize the total amount of space occupied by the wires on the chip. Such a goal could be achieved by placing components with many wires between them relatively close together. This could be accomplished by determining an area (at least as large as the component) in which each component can be placed on the chip, with components having many wires between them being assigned areas which are close together or intersecting. The question is then reduced to determining a placement of the components of the chip so that each component is placed in the area to which it has been assigned, and no two components are placed so that they overlap.

We model the VLSI placement problem by considering the problem of placing tiles on the plane when each tile has a preassigned area into which it must be placed. To simplify the problem, we assume that all tiles are identical rectangles and that the tiles have an orientation. In addition, the areas in which the tiles must be placed are assumed to be rectangular. It is also assumed that both the tiles and the regions have

---

[1]This research was done while the author was a student research assistant at IBM Research, San Jose.

integer lengths and widths. Finally, the boundaries of the regions are assumed to be straight lines that are parallel to either the x-axis or to the y-axis and which intersect the opposite axis at some integer point.

For some limited versions of the problem there are "fast" algorithms. Unfortunately, however, slight generalizations appear to make the problem computationally infeasible (NP-complete). An implication of these results is that if the heuristic approach mentioned above it to be used for VLSI layout, then the chip designer is limited to using a simplified version. Nonetheless, for some problems an algorithm for the simplified version used interactively might be a convenient tool.

Since our model can be viewed as a generalization of classical multiple machine scheduling theory, we can apply known results from scheduling theory to obtain fast algorithms. Conversely, some of our negative results imply negative scheduling results.

2. DEFINITIONS

The class of problems studied in this paper all have the following form. We are given a set of n oriented *tiles,* T_1, T_2, ..., T_n. Each tile T_i has dimension $p_i \times q_i$, where both p_i and q_i are integers. For each tile T_i we are given an oriented rectangular *region* R_i, where R_i has dimension $u_i \times v_i$. The location of R_i in the plane is denoted by (x_i, y_i), the bottom left corner of R_i. As in the case of the tiles, u_i, v_i, x_i, and y_i are integers. The *feasibility* problem consists of determining if there is a non-overlapping placement of the tiles in the plane such that each tile is placed entirely within its region without violating the orientation of either the region or the tile.

A problem is in the class NP if it is possible to verify in time which is polynomial in the length of the input that a proposed solution to the problem is feasible. An algorithm is *polynomial* if it runs in time which is polynomial in the length of the input. For the tile problem mentioned above, if we were given a proposed layout of the tiles, we could determine in time which is polynomial in n, the number of tiles, if the proposed layout is feasible. For example, for each tile we could first verify that it is placed entirely within its region and then compare its location with that of the other n-1 tiles to see if there is any overlapping. This procedure can be accomplished with $O(n^2)$ comparisons, which is clearly polynomial in n, the number of tiles in the input. (Alternatively, the tiles can be sorted lexicographically according to their coordinates, in which case the verification can be accomplished in $O(n \log n)$ time by a "sweeping line" technique. However, the $O(n^2)$ algorithm is sufficient to demonstrate that the problem is in the class NP).

A problem which is in the class NP is called *NP-complete* if it is at least as hard as any problem in the class NP. Unfortunately, there is no polynomial time algorithm known for any NP-complete problem. Furthermore, if a polynomial time algorithm were discovered for one NP-complete problem, then because of the technical definition of NP-complete problems (cf [GJ]), there would be polynomial time algorithms for all NP-complete problems. Since a lot of effort has been devoted to finding polynomial time algorithms for a variety of NP-complete problems, the general consensus is that none of these problems has a polynomial time algorithm.

To show that a problem is NP-complete, it is sufficient to first show that it is in the class NP and to then produce a *reduction* of a problem that is already known to be NP-complete to the given problem. Such a reduction involves constructing a problem instance of the test problem and then demonstrating that its solution

208

provides a solution to the known NP-complete problem. An additional constraint is that the size of the test problem should grow by at most a polynomial factor in the size of the problem we are trying to prove NP-complete.

The first problem which was shown to be NP-complete is the *satisfiability* problem. The satisfiability problem is defined as follows. We are given a set of *variables,* $a_1, a_2, ..., a_n$, and a set of *clauses,* $C_1, C_2, ..., C_m$. Each clause consists of the conjunction of some of the variables, with each occurrence of a variable being in either its positive or negative form. The feasibility question is: given a set of m clauses, each of which consists of a subset of the n variables with each occurrence of a variable being in either the positive or the negative form, does there exists a truth assignment of the variables so that every clause evaluates to "true"? The 3-satisfiability problem, abbreviated as 3-SAT, is the same as the satisfiability problem except that each clause can have at most three occurrences of variables. Nonetheless, 3-SAT remains NP-complete. For a more thorough and rigorous discussion of NP-complete problems and satisfiability, see [GJ].

3. NP-COMPLETENESS RESULTS FOR THE TILE PLACEMENT PROBLEM

We have already shown above that the tile placement problem is in the class NP. The difficulty lies in showing that the tile placement problem is at least as hard as any NP-complete problem. To prove this, we reduce 3-SAT to the tile placement problem. To demonstrate the generality of the result, we prove that two restrictions of the problem, one for the case in which the tiles are identical squares and the other in which they are identical rectangles, are NP-complete. It follows from the NP-completeness of even one of the restrictions that the general case, namely that in which the size of the tiles is part of the problem input, is NP-complete.

Problem SQUARE.
Input: a set of n oriented rectangular regions R_1, R_2, ..., R_n. Each region R_i is defined by its bottom left corner (x_i, y_i) and its dimensions $u_i \times v_i$, with x_i, y_i, u_i, and v_i integers.
Question: Does there exist a placement of n square tiles T_1, T_2, ..., T_n of size 2×2 so that the following conditions hold:
  1. each tile T_i is entirely contained within its region R_i,
  2. no two tiles overlap.

We prove that SQUARE is NP-complete by encoding an arbitrary instance of 3-SAT as a version of SQUARE in which all the tiles can be placed if and only if the instance of 3-SAT is has a satisfying truth assignment. Given an instance of 3-SAT with n variables and m clauses, we construct an instance of SQUARE which has a total of $15mn - 2n + m$ tiles and regions. Note that this is polynomial in the size of the input. The tiles, which are each 2×2, are divided into the following groups:

  m *c-tiles* (clause tiles)
  12mn *v-tiles* (variable tiles)
  2(m-1)n *b-tiles* (bonding tiles)
  mn *f-tiles* (forcing tiles).

To encode the truth assignment of an instance of a variable, we use a construction called a *VAR*. A VAR consists of 12 overlapping v-regions, 10 of which have dimension 3×2 and 2 of which have dimension 2×3. Figure 1 is an illustration of a VAR. For each variable, there is a chain of m VARs, one for each clause. Two VARs in the chain are linked by a pair of b-regions, each of which has a

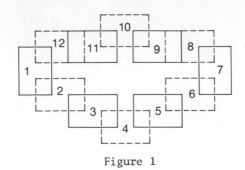

Figure 1

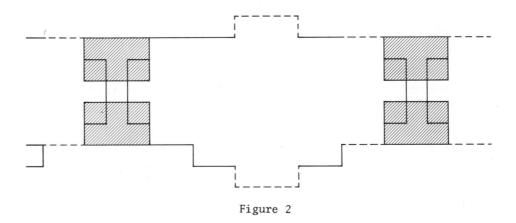

Figure 2

Figure 3

corresponding b-tile. The b-regions linking VAR(i,j) and VAR(i,j+1) are referred to as $B(i,j)_1$ and $B(i,j)_2$. Figure 2 shows a VAR linked to its neighbors by b-regions. Finally, the VAR chains are placed one on top of another in a grid-like formation.

The VARs are labelled VAR(i,j), where $1 \leq i \leq n$ and $1 \leq j \leq m$. The regions in a VAR are referred to as $V(i,j)_k$, $1 \leq k \leq 12$, and the tiles corresponding to them are referred to as $v(i,j)_k$. Figure 3 is an illustration of a 3×3 grid of VARs.

Note that if a v-tile is placed in its corresponding v-region, it must overlap at least one other v-region. Furthermore, a v-tile may overlap at most one other v-region if all the v-tiles in the same VAR are to be placed. Therefore, the following lemma holds.

Lemma 1. The placement of any v-tile in its region uniquely determines the placement of the remaining 11 v-tiles in the same VAR.

We call a placement of a v-tile a *feasible placement* if it overlaps exactly one other v-region and intersects with no other v-tile.

Lemma 2. Let i be some integer such that $1 \leq i \leq n$. If $v(i,j)_k$ is placed in a feasible placement, then there is only one possible feasible placement for all other v-tiles in the set $v(i,j)_k$, $1 \leq j \leq m$, $1 \leq k \leq 12$.

Proof. Suppose j=1. We show that there is only one feasible placement for the v-tiles of VAR(i,2). It follows from lemma 1 that there is a unique placement of $v(i,1)_7$. Because $v(i,1)_7$ intersects both $B(i,1)_1$ and $B(i,1)_2$, there is a unique placement of both $b(i,1)_1$ and $b(i,1)_2$. This in turn implies a unique placement of $v(i,2)_1$. The proof follows from repeated applications of lemma 1 together with the above observations. □

We make use of lemma 2 to encode the truth assignment of variables. If $a_i :=$ "true," we place $v(i,1)_1$ in the upper two thirds of $V(i,1)_1$. This is illustrated in Figure 4a. Conversely, if $a_i :=$ "false," then $v(i,1)_1$ is place in the bottom two thirds of $V(i,1)_1$, as is illustrated in Figure 4b.

Lemma 3. Let i be some integer such that $1 \leq i \leq n$. If v_i is placed in the upper (resp. lower) two thirds of $V(i,j)_1$, then $v(i,j)_1$ for $1 \leq j \leq m$ is also placed in the upper (resp. lower) two thirds of its regions.

The proof follows from the construction and from lemma 2. □

The construction is designed so that each row of the grid encodes the truth assignment of one of the variables and each column represents one of the clauses. Each c-region intersects a unique column of n VARs, and each column of VARs has a unique c-region. The c-regions are labelled C(j) and the corresponding c-tiles are labelled c(j). Figure 5 illustrates the c-regions for Figure 3.

The f-regions which overlap C(j) are used to force the placement of c-tile c(j) in space which is available if and only if the truth assignment causes at least one of the variables in C_j to evaluate to "true." Figure 6a illustrates the case in which the corresponding clause will evaluate to "true" if the variable has "true" as its truth assignment; Figure 6b is the case in which an assignment of "false" to the variable results in an assignment of "true" to the clause; Figure 6c is the case in which the variable does not occur in the clause.

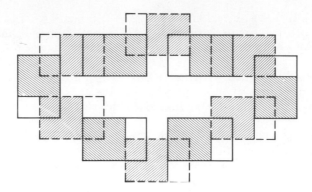

Figure 4a) "true"

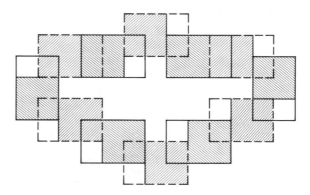

Figure 4b) "false"

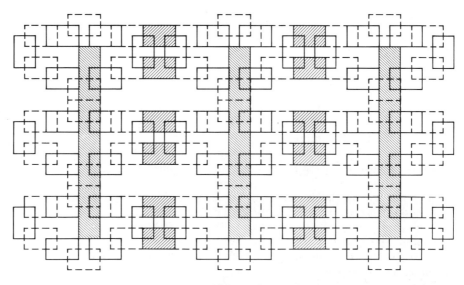

Figure 5

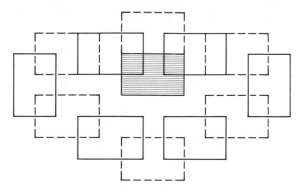

Figure 6a) f region for "true"

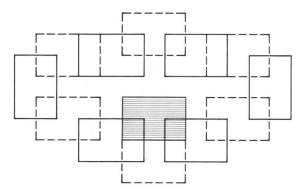

Figure 6b) f region for "false"

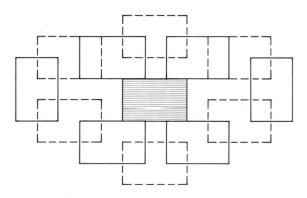

Figure 6c) f region for "not in clause"

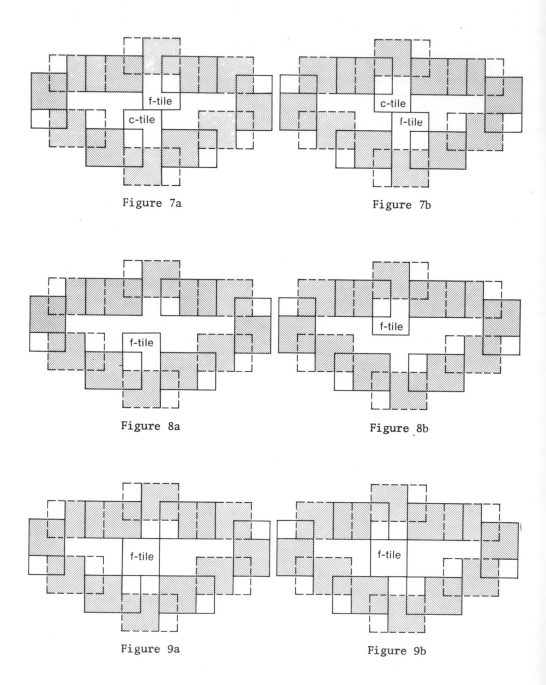

Figure 7a

Figure 7b

Figure 8a

Figure 8b

Figure 9a

Figure 9b

Lemma 4. Let $C_h = a_i \cup a_j \cup a_k$, where a_i, a_j, and a_k are variables which occur in either their positive or negative form. Then $c(h)$ can be placed if and only if at least one of a_i, a_j, and a_k evaluates to "true."

Proof. Without loss of generality, let $C_h = a_1 \cup \bar{a}_2 \cup a_3$. First assume that at least one of $a_1, \bar{a}_2, a_3$ evaluates to "true." Figure 7a demonstrates the placement of the f-tile if either a_1 or $a_3 =$ true; Figure 7b shows the placement of the f-tile if $\bar{a}_2 = $ false. In either case, both the f-tile and c_h can be placed.

Conversely, suppose that none of $a_1, \bar{a}_2, a_3$ evaluates to "true." The case a_1 or $a_3 = $ "false" is illustrated in figure 8a; $\bar{a}_2 = $ "true" is illustrated in Figures 8b. Note that in both cases the forced placement of the f-tile prevents the c-tile from being placed. Figures 9a and b illustrate the two cases in which the variable does not appear in the clause. In 9a the variable represented by the VAR has an assignment of "true," in 9b it is assigned "false." Once again, in both cases the f-tile prevents the c-tile from being placed, but the f-tile itself can be placed. □

Theorem 1. SQUARE is NP-complete.

Proof. The proof follows directly from lemmas 3 and 4. □

As was stated earlier, the tiling problem remains NP-complete if all the tiles are identical oriented rectangles. Because the proof is similar to the proof for the squares, we present a sketch of the case in which the tiles are 2×1 rectangles.

Problem **RECTANGLE**.
Input: a set of n oriented rectangular regions R_1, R_2, ..., R_n. Each region R_i is defined by its bottom left corner (x_i, y_i) and its dimensions $u_i \times v_i$, with x_i, y_i, u_i, and v_i integers.
Question: Does there exist a placement of n oriented rectangular tiles T_1, T_2, ..., T_n of size 2×1 so that, the following conditions hold:
    1. each tile T_i is entirely contained within its region R_i,
    2. no two tiles overlap.

As in the case of SQUARE, we reduce 3-SAT to RECTANGLE. Once again, there is an $n \times m$ grid of regions called VARs, with each row of VARs being used to encode the truth assignment of one of the variables and each column representing one of the clauses. In this case, however, the VAR construction requires 18 v-regions which are 2×2 squares. Figure 10 is an illustration of a VAR for RECTANGLE.

The VAR construction is such that we can encode the truth assignment of a variable by overlapping adjacent VARs; therefore, we do not require the b-tiles to connect the VARs. Figure 11 illustrates a 3×3 grid of VARs; Figure 12 illustrates the c-regions on the 3×3 grid. Figure 13a illustrates the encoding of "true" for a VAR; Fgiure 13b shows the "false" encoding. Each VAR contains three f-regions, one for each of three f-tiles. Figures 14a shows the three f-regions for the case in which the variable occurs in the "true" form; Figure 14b shows the f-regions for the case in which the variable occurs in the "false" form. Figure 14c shows a single f-region which is shared by three f-tiles for the case in which the variable does not appear in the clause being represented by the column. Figures 15a and b show the placement of the c-tiles when the truth assignment to the variable causes the clause to evaluate to "true." Figures 16a and b show the placement of the f-tiles when the truth assignment to the variable does not cause the clause to evaluate to "true," and Figures 17 a and b show the placement of the f-tiles when the variable does not occur in the clause.

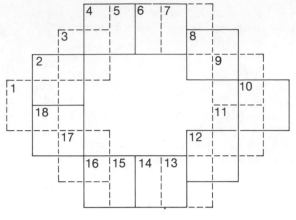

Figure 10

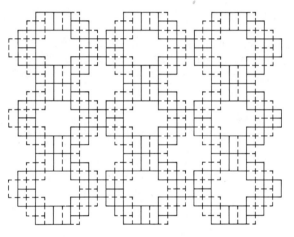

Figure 11

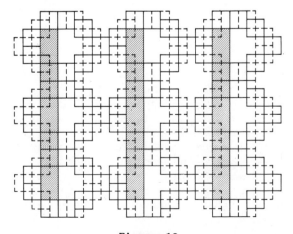

Figure 12

216

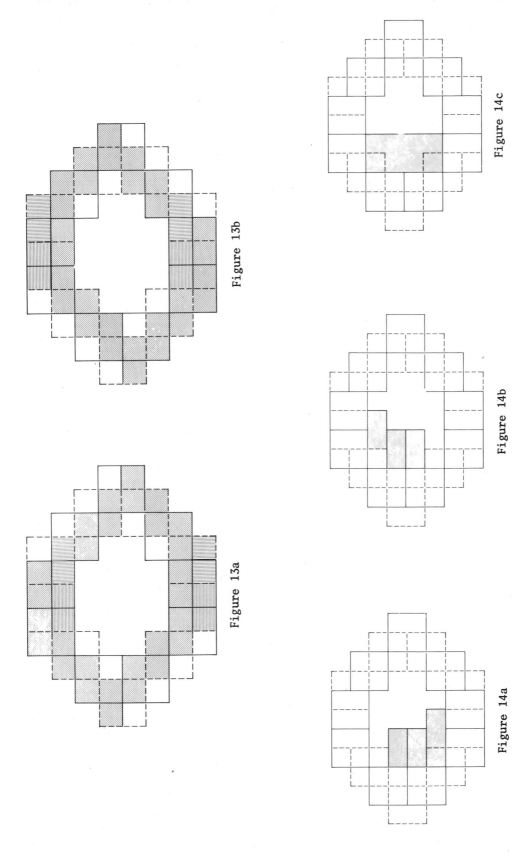

Figure 13a

Figure 13b

Figure 14a

Figure 14b

Figure 14c

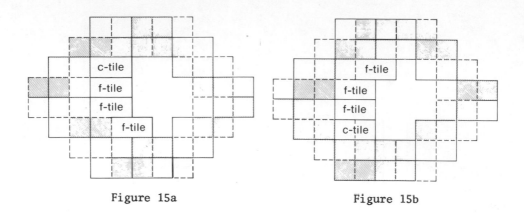

Figure 15a

Figure 15b

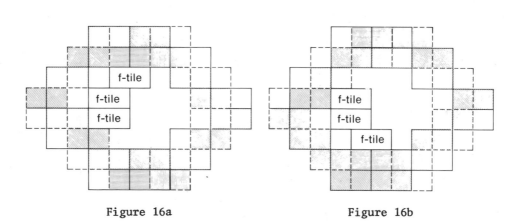

Figure 16a

Figure 16b

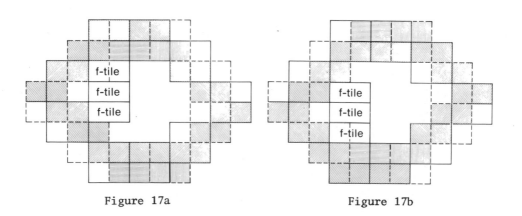

Figure 17a

Figure 17b

Theorem 2. RECTANGLE is NP-complete.

We note that theorems 1 and 2 show that even some simple versions of the problem are NP-complete. It follows that the problem is NP-complete in its most general cases, including for example non-rectangular regions and/or tiles, non-oriented regions and/or tiles, regions that are not connected, and generalizations to more than two dimensions.

As was mentioned earlier, the tiling problem has parallels in scheduling theory. Define the *release time* of a job to be the earliest time at which that job can be run, and the *deadline* of the job as the time by which it must be completed. Assume that a job may not be interrupted once processing has begun on it. The following scheduling problem models the situation in which jobs have release times and deadlines and the machines have different characteristics, thereby limiting the choice of machines on which a job can run.

Problem CONTIGUOUS ASSIGNED MACHINE SCHEDULING.
Input: the integers p and M and two sets of n ordered pairs of integers (r_i, d_i) and (m_{i1}, m_{i2}), with $0 \leq m_{i1} \leq m_{i2} \leq M$.
Question: Does there exist a schedule of n jobs $J_1, J_2, ..., J_n$ so that the following conditions hold:
1. once a job begins running, it runs for p consecutive units of time on the same machine,
2. J_i is run on some machine m_k with $m_{i1} \leq m_k \leq m_{i2}$,
3. J_i is started no earlier than r_i completed no later than d_i,
4. no two jobs overlap on the same machine.

Note that ASSIGNED MACHINE SCHEDULING, which is obtained by removing the constraint that the machines to which a job is assigned be contiguous, is more general than CONTIGUOUS ASSIGNED MACHINE SCHEDULING. Therefore, if CONTIGUOUS ASSIGNED MACHINE SCHEDULING is hard, so is ASSIGNED MACHINE SCHEDULING. It is easy to show that an algorithm for CONTIGUOUS ASSIGNED MACHINE SCHEDULING could be used to solve RECTANGLE. Therefore, the following theorem holds.

Theorem 3. CONTIGUOUS ASSIGNED MACHINE SCHEDULING is NP-complete.

4. SPECIAL CASES ALLOWING POLYNOMIAL TIME ALGORITHMS

Although the most general version of the tiling problem is NP-complete, there are some special cases for which fast algorithms are known to exist. In this section we discuss those cases.

A bipartite graph is a graph $G = (S, T, E)$, where $S \cup T$ is the set of nodes, $S \cap T = \emptyset$, E is the set of edges, and each edge is adjacent to a node in S and a node in T. We shall consider only bipartite graphs in which the edges are undirected. A subset $E' \subseteq E$ is called a *matching* if no two elements of E' are adjacent to the same node. A *maximum cardinality* matching is a matching with the maximum possible number of edges. If each edge has a weight, then a *maximum weighted* matching is a matching such that the sum of the weights of the edges in the matching is maximum over all possible matchings.

Problem **UNIT SQUARES.**

Input: a set of n oriented rectangular regions R_1, R_2, ..., R_n. Each region R_i is defined by its bottom left corner (x_i, y_i) and its dimensions $u_i \times v_i$, with u_i, v_i, u_i, v_i, x_i, and y_i integers.

Question: Does there exist a placement of n square tiles T_1, T_2, ..., T_n of size 1×1 so that the following conditions hold:

1. each tile T_i is entirely contained within its region R_i,
2. no two tiles overlap.

Theorem 4. UNIT SQUARE can be solved in $O(n^4)$ time.

Proof. We transform the tiling problem into a matching problem by constructing a bipartite graph $G = (S, T, E)$. T consists of the tiles, S consists of 1×1 regions which are obtained by partitioning the original possibly larger regions, and an edge (s,t) is in E if and only if tile t can be feasibly placed in the region represented by s. Note that if a tile is adjacent to at least n non-intersecting regions, then it can always be placed since the other tiles can occupy at most n-1 of those regions. Therefore, to avoid a blow-up of the running time, we make a tile node adjacent to at most n region nodes, thereby eliminating "surplus" regions.

Let $|S| = p$ and $|T| = q$. The maximum cardinality matching for G can be computed in $O(p^2 q)$ time. In the transformation p=n and $q \le n^2$. Therefore, the running time of the algorithm is $O(n^4)$. For a more detailed description of this technique see [SS]. $\square$

We note that the approach used in Theorem 4 can be generalized to obtain a tile placement in which the sum of the distances of the tiles from the origin (or any other point) is minimized. Choose some metric, for example the sum of the x-coordinate and the y-coordinate. Let each $e_i \in E$ have weight w_i, where w_i is the distance of the region from the origin according to the chosen metric. As in theorem 4 we keep only the n regions of smallest weight which are adjacent to a node.

Corollary 1. A feasible solution to UNIT SQUARE which minimizes the maximum distance of any of the tiles from the origin is computable in $O(n^4)$ time.

We use the max-min matching algorithm, which is presented in detail in [La]. This algorithm produces a maximum cardinality matching for which the largest edge weight is minimized. The running time is $O(n^4)$ in our model. $\square$

Corollary 2. A solution to UNIT SQUARE which both maximizes the number of tiles placed and minimizes the sum of the distances of the tiles from the origin is computable in $O(n^4)$.

Proof. This result is obtained by constructing the same bipartite graph as that used in corollary 1 and finding for this graph the maximum cardinality matching of minimum total weight. This type of problem has been discussed in the literature (see [La, SS]). $\square$

Theorem 4 depends only on the fact that the regions can be partitioned into smaller non-overlapping regions in a unique manner. Therefore, the results of theorem 4 apply to the problems in which the tiles are $C \times C$ and u_i, v_i, x_{i1}, and y_{i1} are all multiples of C.

Techniques similar to those in theorem 4 can also be used to solve the AXIS BOUNDED REGIONS problem which is defined below.

Problem **AXIS BOUNDED REGIONS.**

Input: a set of n oriented rectangular regions, R_1, R_2, ..., R_n. Each region R_i has dimensions $u_i \times v_i$, with u_i and v_i integers. The bottom left corner of every region is the point $(0,0)$.

Question: Does there exist a placement of n identical oriented rectangular tiles, T_1, T_2, ..., T_n so that the following conditions hold:

1. each tile T_i is entirely contained within its region R_i,
2. no two tiles overlap.

Theorem 5. AXIS BOUNDED REGIONS can be solved in $O(n^4)$ time.

Proof. Beginning at the origin, the regions are partitioned into non-overlapping subregions, each of which has the same dimensions and orientation as the tiles. It is easy to verify that any portion of a region which is not included in the partition can be discarded without affecting feasibility. Once this observation has been made, it is straightforward to apply the techniques of theorem 3 and corollaries 1 and 2. ☐

COLUMN REGIONS, defined below, easily translates to a non-trivial scheduling problem by a partition of the plane into "machines."

Problem **COLUMN REGIONS.**

Input: a set of n oriented rectangular regions, R_1, R_2, ..., R_n. Each region R_i has dimensions $u_i \times M$, where M is a fixed integer. The bottom left corner of region R_i is the point $(x_i,0)$, with x_i an integer.

Question: Does there exist a placement of a set of oriented rectangular tiles of size of size $p \times 1$, T_1, T_2, ..., T_n so that the following conditions hold:

1. each tile T_i is entirely contained within its region R_i,
2. no two tiles overlap.

Theorem 6. COLUMN REGIONS can be solved in $O(Mn^2)$ time.

Proof. This version of the problem can be modeled as a multiprocessor scheduling problem as follows. We partition the region into M machines, m_0, m_2, ..., m_{M-1}, by assigning the interval between the x-axis and the line $y=1$ to m_0, the interval between the lines $y=1$ and $y=2$ to m_1, ..., the interval between the lines $y=M-1$ and $y=M$ to m_{M-1}.

Each tile T_i is now considered to be a job of length p which must be scheduled, with each job having release time x_i and deadline x_i+u_i. Furthermore, each job can be run on any of the M machines. There is a very simple $O(n^3 \log n)$ algorithm [Si] for computing a feasible schedule for this problem. In addition, there is a more complicated $O(Mn^2)$ algorithm for the same problem [SW]. If $M=1$, then the problem can be solved in $O(n \log n)$ time [GJST]. ☐

Consider the following modification of COLUMN REGIONS, which we call MODIFIED COLUMN REGIONS. Instead of insisting that all regions R_i have the same value M for v_i, we only require that v_i have an integer value less than or equal to M. Note that this is equivalent to a special case of CONTIGUOUS ASSIGNED MACHINE SCHEDULING in which the set of machines on which a job can run is a contiguous set of machines which always contain machine m_0 but which contain machine m_{M-1} only if all of the machines are in the set.

We define below the JOB REMOVAL SCHEDULING problem, which is an open problem in scheduling theory.

Problem JOB REMOVAL SCHEDULING.

Input: an integer p and a set of n ordered pairs of integers (r_i, d_i).
Question: Assuming the infeasibility of the problem of scheduling n jobs $J_1, J_2, ..., J_n$ so that conditions 1 - 3 below are satisfied, what is the minimum number of the jobs which must be removed from $J_1, J_2, ..., J_n$ so that the following conditions hold:

1. once a job begins running, it runs for p consecutive units of time on the same machine,
2. each job J_i which remains in the set starts no earlier than r_i and is completed no later than d_i,
3. no two jobs overlap.

Theorem 7. MODIFIED COLUMN REGIONS is at least as hard as JOB REMOVAL SCHEDULING.

Proof. We present a reduction which determines whether or not the infeasible scheduling problem instance can be made feasible by the removal of only C jobs. If such an algorithm exists, then by doing binary search on the value of C, one can find the minimum C which allows feasibility.

Let $T_1, T_2, ..., T_n$ be n tiles of dimension $p \times 1$. The region for T_i has $x_i = r_i$ and $y_i = 0$. Also set $u_i = d_i - r_i$ and $v_i = C+1$, $1 \le i \le n$. Let d_{max} be the maximum deadline of any of the jobs in the original problem instance of problem Q, and let $D = \lfloor d_{max}/p \rfloor - 1$. We add to the tiling problem a set of $D \times C$ "filler" tiles $F_{i,j}$, $1 \le i \le D$, $1 \le j \le C$, all of which are $p \times 1$ and have regions with bottom left corner $(0,0)$ and dimensions $u_{i,j} = p \lfloor d_{max}/p \rfloor + p - 1$ and $v_{i,j} = j$.

All filler tiles of the form $F_{i,1}$ must be placed in the row between $y=0$ and $y=1$. Since the placement of D tiles in that row leaves $2p-1$ units with no tiles, one additional tile can be placed in the row without violating the constraints of the $F_{i,1}$ tiles. In general, all tiles of the form $F_{i,j}$ must be placed in the first j rows, and there is space for exactly j additional tiles in the first j rows.

If no more than C tiles from the set $T_{i,j}$ are placed in the first C rows, then it is easy to see that the scheduling problem has a feasible solution which involves the removal of no more than C jobs. Conversely, if it is not possible to obtain a feasible tiling for the modified tiling problem, then there is no feasible schedule which involves the removal of only C jobs. □

5. OPEN PROBLEMS

1. Find an algorithm for the MODIFIED COLUMN REGIONS problem or prove it is computationally infeasible. Are the MODIFIED COLUMN REGIONS problem and the JOB REMOVAL problem computationally equivalent?

2. In the NP-completeness reductions of theorems 1 and 2 we needed much larger regions for the c-tiles than for the others. Is the problem still NP-complete if all the regions have the same dimensions but possibly different orientations? What about the same dimensions and the same orientations?

3. Find additional special cases of the tiling problem for which polynomial time algorithms can be constructed.

6. REFERENCES

[GJ] Garey, M. R. and Johnson, D. S., 1978, *Computers and Intractability: A Guide to the Theory of NP-completeness.* W. H. Freeman, San Francisco.

[GJST] Garey, M. R., Johnson, D. S., Simons, B. B. and Tarjan, R. E., 1981, Scheduling Unit-Time Tasks with Arbitrary Release Times and Deadlines. *SIAM J. Comput.* **10**, pp. 256-269.

[La] Lawler, E. L., 1976,. *Combinatorial Optimization: Networks and Matroids.* Holt, Rinehart & Winston, New York.

[Si] Simons, B. B., 1983, Multiprocessor Scheduling of Unit-time Jobs with Arbitrary Release Times and Deadlines, *Siam J. Comput.* **12**, pp. 294-299.

[SS] Simons, B. B. and Sipser, M. F., 1984, On Scheduling Unit-length jobs with Multiple Release Time/Deadline Intervals, *Op. Res.,* **32**, pp. 80-88.

[SW] Simons, B. B. and Warmuth, W., A Fast Algorithm for Multiprocessor Scheduling of Unit-time Jobs with Arbitrary Release Times and Deadlines, to appear.

MIXED-TYPE MULTIPLE-VALUED FILING SCHEME OF

ORDER ONE AND TWO

Shinsei Tazawa

Department of Mathematics
Kinki University
Osaka, Japan

ABSTRACT

This paper will present a filing scheme for multiple-valued records. It is assumed that records are characterized by m attributes having n values each, and the query set consists of two kinds of queries, where one (called a first-order query) specifies a value of one attribute and another one (called a second-order query) specifies values of two attributes. The number of buckets to be organized and the redundancy of the file are less than those of an inverted file, provided the distribution of records is uniform. The number of pointer fields to be attached to the accession number stored in a bucket is considerably reduced in this scheme. Every first-order query can be answered by the access to one bucket. Every second-order query can be answered by the access to one bucket and the traversal of its contents without any set operations.

INTRODUCTION

In an information storage and retrieval system, it is required to retrieve quickly the corresponding set of records pertinent to each of the considerable number of quires specifying values of given attributes. For such purpose, it is popular to provide buckets and organize secondary index file or a directory file for values of qualified attributes. As for designed file organization schemes, the amount of required storage space and the retrieval time are important parameters used to measure their performances. Unfortunately, these two parameters cannot be reduced simultaneously. In other words, reduction in one is usually achieved at the expense of the other. Consider, for example, the performance of a generalized inverted file (or query inverted file) in which for every query, a list of the accession numbers of pertinent records is organized. It might be an efficient scheme in so far as the retrieval time is concerned, since the list corresponding to each query can be organized in consecutive locations. However, the redundancy of the filing scheme (the average number of times the accession number of a record is stored repeatedly) cannot be disregarded, in the case where the size of the file and the total number of attribute values are larger.
Combinatorial filing schemes have been developed in an attempt to solve this problem. The balanced filing schemes of order two (BFS_2) by Abraham, Ghosh and Ray-Chaudhuri[1]; Ray-Chaudhuri[11]; Chow[6]; Yamamoto et al.[13]; Tazawa[12]

and the balanced multiple-valued filing schemes of order two (BMFS$_2$) by Ghosh and Abraham[9]; Bose, Abraham and Ghosh[4]; Bose and Koch[5]; Ghosh[8]; Yamamoto et al.[15]; Yamamoto, Teramoto and Futagami[16]; Berman[2,3] and, recently, Donaldson and Hawkes[7] vividly reflect a tradeoff between retrieval efficiency and storage efficiency. In order to reduce both the number of buckets to be prepared and the redundancy of a filing scheme, they are designed combinatorically so that every bucket is associated with a part (subset) in a partition of the query-set and contains the accession numbers of records pertinent to at least one of the queries in the subset. The Hiroshima University balanced filing scheme of order two (HUBFS$_2$)[13] and the Hiroshima University balanced multiple-valued filing scheme of order two (HUBMFS$_2$)[15] are designed by modeling on the decompositions of graphs. These two schemes were proved optimal in that they have the least redundancy among all BFS$_2$'s and BMFS$_2$'s, respectively, having the same parameters, under certain theoretical assumptions on the distribution of records.

Those filing schemes of order two primarilly designed for quick retrieval of second-order queries, each specifying values of two attributes. They have, therefore, much slower response time for first-order queries, each specifying one attribute value. A primitive way of overcoming such deficiency may be to combine schemes of order one and order two, such as the combined organization of an inverted filing scheme of order one (IFS$_1$) which is designed for first-order queries and an inverted filing scheme of order two (IFS$_2$) which is designed for second-order queries. Such several attempts have been made. Among others, a generalized HUBFS$_2$ (GHUBFS$_2$)[13] is an example of an embedded organization scheme designed for queries of order two as well as those of order one. A GHUBFS$_2$ which has the same number of buckets and the same redundancy as those of IFS$_1$ has been implemented and evaluated experimentally by Ikeda[10]. In the case that the file is multiple-valued, a means of combining schemes of order one and order two was considered by Yamamoto et al.[14]. The scheme constructed there is called a generalized HUBMFS$_2$ (GHUBMFS$_2$). This scheme has the following specific properties: (1) The number of buckets is equal to the number of attribute values and is the same as that of the IFS$_1$. (2) The redundancy of the filing scheme is the same as that of the IFS$_1$. (3) Every first-order query can be retrieved almost as quickly as IFS$_1$. (4) Every second-order query can be answered by the access to one bucket and the traversal of its contents without any set operations.

In this paper, we shall present a filing scheme for multiple-valued records. This scheme is more generalized than GHUBMFS$_2$ in that the number of buckets and the redundancy of the file are less than those of the corresponding IFS$_1$ (the above (1) and (2)), provided that all records are equally likely, and that the properties (3) and (4) hold.

MIXED-TYPE CLAW BUCKET AND REDUNDANCY

In most information storage and retrieval system, each record stored in a file, Ω, consists of two parts. One part is provided to store the accession number of the record. In another part some attribute values which characterize the record may be stored or some corresponding pointers instead of their values may be stored. This part may involve, if necessary, some other information. On a mathematical model for filing schemes, a record ω can be characterized formally by m-vector, called the characteristic vector of ω, $x(\omega)=(x_1, x_2, \ldots, x_m)$, belonging to the Cartesian product space $R^{(m)}$ of m finite sets $R_1, R_2, \ldots, R_m$, i.e., $R^{(m)} = R_1 \times R_2 \times \cdots \times R_m$, where R_i is the domain of the i-th attribute A_i and x_i is the i-th attribute value of the record ω. We assume that all m attributes vary over the same number of values, i.e., $|R_i|=n$, $i=1,2,\ldots,m$, where n is fixed. If n=2, then a record is binary-valued. We here consider the multiple-valued case where n is greater than 2.

A retrieval request or query is a request to retrieve from the file the subset of all records having certain specific attribute values. It can be specified by a subset Q of $R^{(m)}$. A procedure of retrieving a query Q can be described by that of finding a subset Ω_Q of Ω characterized by

$$\Omega_Q = \{\omega \in \Omega \mid x(\omega) \in Q\}.$$

A query which specifies all records whose i-th attribute value is p, i.e.,

$$Q_p^i \equiv \{x \mid x_i = p\}$$

is called a first-order query. A query which specifies values of different attributes A_i and A_j, simultaneously, i.e.,

$$Q_{p;q}^{i;j} \equiv \{x \mid x_i = p, \ x_j = q\} = Q_p^i \cap Q_q^j$$

is called a second-order query.

Consider a set which is composed of c first-order queries $Q_{p_\alpha}^{i_\alpha}$, $\alpha = 1,2,$ ...,c and d second-order queries $Q_{q_\beta, r_\beta}^{j_\beta, k_\beta}$, $\beta = 1,2,\ldots,d$. The present paper extends the results in Yamamoto et al.[14], where the case of c=1 is treated. A storage area prepared for the storage of the information for access to a record pertinent to at least one of those queries will be called a <u>mixed-type bucket</u> associated with the query-set. It is called a <u>mixed-type claw bucket</u> if i_1, i_2,..., and i_c are mutually distinct and if either

$Q_{q_\beta}^{j_\beta}$ or $Q_{r_\beta}^{k_\beta}$ coincides with at least one of $Q_{p_1}^{i_1}, Q_{p_2}^{i_2}, \ldots, Q_{p_c}^{i_c}$ for all β.

The probability of a record being stored in such a bucket B will be called the redundancy of the bucket B and it is denoted by $R(B)$. The empirical redundancy of a filing scheme, which is defined by the sum of the redundancies of the buckets to be prepared, is dependent on the distribution of the records which are actually stored. Therefore, a generalized calculation of redundancy is difficult. Thus, assume that the distribution of records is uniform, i.e.,

$$\Pr\{x(\omega) = (x_1, x_2, \ldots, x_m)\} = 1/n^m$$

for all $(x_1, x_2, \ldots, x_m) \in R^{(m)}$. Then we have

Theorem. Let c and d be positive integers satisfying $d \leq c(2mn-c-2n+1)/2$. Let B be a mixed-type bucket associated with c first-order queries and d second-order queries. Then

$$R(B) > 1 - \left(1 - \frac{1}{n}\right)^c, \tag{2.1}$$

whenever B is not a mixed-type claw bucket.

Proof. Suppose that B is a mixed-type bucket associated with c+d queries $Q_{p_1}^{i_1}, \ldots, Q_{p_c}^{i_c}, Q_{q_1, r_1}^{j_1, k_1}, \ldots, Q_{q_d, r_d}^{j_d, k_d}$. Then we have

$$R(B) = \Pr\left\{ \left(\bigcup_{\alpha=1}^{c} Q_{p_\alpha}^{i_\alpha} \right) \cup \left(\bigcup_{\beta=1}^{d} Q_{q_\beta, r_\beta}^{j_\beta, k_\beta} \right) \right\} = \Pr\left\{ \bigcup_{\alpha=1}^{c} Q_{p_\alpha}^{i_\alpha} \right\} + \Pr\left\{ \overline{\left(\bigcup_{\alpha=1}^{c} Q_{p_\alpha}^{i_\alpha} \right)} \left(\bigcup_{\beta=1}^{d} Q_{q_\beta, r_\beta}^{j_\beta, k_\beta} \right) \right\},$$

where $\overline{Q}$ denotes the complement or negation of Q. Since we have

$$\Pr\left\{ \overline{\left(\bigcup_{\alpha=1}^{c} Q_{p_\alpha}^{i_\alpha} \right)} \left(\bigcup_{\beta=1}^{d} Q_{q_\beta, r_\beta}^{j_\beta, k_\beta} \right) \right\} = \Pr\left\{ \bigcup_{\beta=1}^{d} \left(\bigcap_{\alpha=1}^{c} \overline{Q_{p_\alpha}^{i_\alpha}} Q_{q_\beta, r_\beta}^{j_\beta, k_\beta} \right) \right\} = 0$$

if and only if $\Pr\left\{ \bigcap_{\alpha=1}^{c} \overline{Q_{p_\alpha}^{i_\alpha}} Q_{q_\beta, r_\beta}^{j_\beta, k_\beta} \right\} = 0$ for all β, the following inequality holds:

$$R(B) \geq \Pr\{ \bigcup_{\alpha=1}^{c} Q_{P_{\alpha}}^{i_{\alpha}} \},\tag{2.2}$$

where the equality holds if and only if either $Q_{q_{\beta}}^{j_{\beta}}$ or $Q_{r_{\beta}}^{k_{\beta}}$ coincides with at least one of $Q_{P_1}^{i_1}, \ldots, Q_{P_c}^{i_c}$ for all β. Furthermore, by following the proof of Theorem 4.1 in Yamamoto et al.[15], it can be easily shown that

$$\Pr\{ \bigcup_{\alpha=1}^{c} Q_{P_{\alpha}}^{i_{\alpha}} \} \geq 1 - (1-\tfrac{1}{n})^{c}\tag{2.3}$$

holds and that the equality holds if and only if $i_1, i_2, \ldots, i_c$ are mutually distinct. Hence it follows that $R(B) \geq 1-(1-1/n)^{c}$ and that the equality holds if and only if B is a mixed-type claw bucket associated with c first-order queries and d second-order queries.

As seen in (2.2), note that the right-hand side of (2.1) is independent on the number of second-order queries, d.

Let Q be the set of c first-order queries $Q_{P_1}^{i_1}, Q_{P_2}^{i_2}, \ldots, Q_{P_c}^{i_c}$ and d second-order queries $Q_{q_1,r_1}^{j_1,k_1}, Q_{q_2,r_2}^{j_2,k_2}, \ldots, Q_{q_d,r_d}^{j_d,k_d}$, where $i_1, i_2, \ldots,$ and i_c are mutually distinct and either $Q_{q_{\beta}}^{j_{\beta}}$ or $Q_{r_{\beta}}^{k_{\beta}}$ coincides with at least one of $Q_{P_1}^{i_1}, \ldots, Q_{P_c}^{i_c}$ for all $\beta=1,2,\ldots,d$. Then a mixed-type claw bucket associated with the query-set Q may be organized as a multilist having access paths in order to tell which first-order query is pertinent to which record and which second-order query is pertinent to which record. The number of access paths in each bucket is nearly equal to $(c+m/2)n+c$, as seen presently. To complete the retrieval of second-order query, we add one of the attribute numbers $i_1, i_2, \ldots, i_c$ (corresponding to c first-order queries) to each accession number, as seen in Fig. 1. It is desirable to let d second-order queries correspond to a mixed-type claw bucket in such a way that the number of related attributes becomes as small as possible, in order to reduce the information to be stored in the bucket.

Fig. 1 shows an example of a multilist structure in a mixed-type claw bucket which is the bucket B_1 given in Table 2, where those associated queries are as follows:

$$Q_{1,1}^{1,2} \quad Q_{1,1}^{1,3} \quad Q_{1,1}^{1,4} \quad Q_{1,1}^{2,3} \quad Q_{1,1}^{2,4} \quad Q_{1,1}^{2,5}$$

$$Q_1^1, \quad Q_1^2 \quad Q_{1,2}^{1,2} \quad Q_{1,2}^{1,3} \quad Q_{1,2}^{1,4} \quad Q_{1,2}^{2,3} \quad Q_{1,2}^{2,4} \quad Q_{1,2}^{2,5}$$

$$Q_{1,3}^{1,2} \quad Q_{1,3}^{1,3} \quad Q_{1,3}^{1,4} \quad Q_{1,3}^{2,3} \quad Q_{1,3}^{2,4} \quad Q_{1,3}^{2,5}$$

Note that this bucket has two kinds of pointer fields, namely, one is provided for first-order query and another one is done for second-order query. Since the first-order queries associated with this bucket are Q_1^1 and Q_1^2, furthermore, note that one of attribute numbers 2 and 1 is added to each accession number. If we want the records pertinent to the first-order query Q_1^1, those records may be obtained by traversing A_1 pointer for the first-order query such as #46 → #42 → #30 → #21 → #16 → #11 in this bucket. A_i pointer is written as A_i-ptr for short in Fig. 1. In the case of the second-order queries, for example, $Q_{1,1}^{2,4}$, A_4 pointer indicates 13 in the directory part. We first go to the accession number #42. Since the attribute number in this number #42 is 1, it is seen that the record having #42 is not pertinent to the query $Q_{1,1}^{2,4}$. As A_4 pointer in #42 indicates 10, we next go to #35 and the attribute number in #35 is 2. So the record hav-

ing #35 is pertinent to $Q_{1,1}^{2,4}$. By continuing this procedure, the records
pertinent to this query can be obtained such as #35 →#8 →#4 in this bucket.

ORGANIZATION OF THE MIXED-TYPE SCHEMES

A multiple-valued filing scheme which has the least redundancy under the assumption of the uniform distribution of records can be organized if we can construct mn/c mixed-type claw buckets in such a way that (1) every bucket is associated with c first-order queries, (2) every first-order query is associated uniquely with a bucket and (3) every second-order query is associated uniquely with a bucket. The least redundancy of the filing scheme is given by

$$R = \frac{mn}{c}\{1-(1-\frac{1}{n})^c\}.$$

Such a scheme can be constructed in a number of ways. It is, therefore, desirable to construct a scheme which satisfies the following requirements: (a) The number of pointer fields to be attached to the accession number of a record is as small as possible. (b) The second-order queries are associated with mixed-type claw buckets as evenly as possible. We call a scheme satisfying these requirements a mixed-type multiple-valued filing scheme of order one and two which is denoted by $F(1,2)$.
    In constructing an $F(1,2)$, we consider in this paper only when the number, c, of first-order queries associated with a bucket is a factor of m. The construction of the associated queries and the access method to a mixed-type claw bucket of $F(1,2)$ are given in the following:

    (1) Case m is odd.
    It can be easily verified that if the (in+p)-th bucket B_{ip} associated with the set of c first-order queries Q_p^{ic+1}, Q_p^{ic+2},..., $Q_p^{(i+1)c}$ and d (= (m-1)nc/2) second-order queries

$$\bigcup_{j=1}^{c}\{Q_{p,q}^{ic+j,k} \mid \begin{array}{l} k-1\equiv ic+j,ic+j+1,...,ic+j+\frac{m-3}{2}(\bmod\ m)\\ q=1,2,...,n \end{array}\}$$

is constructed for every i=0,1,...,(m/c)-1 and p=1,2,...,n, then an $F(1,2)$ with parameters m, n, c, and d (=(m-1)nc/2) is obtained. Put $K_i=\bigcup_{j=1}^{c}\{k\mid k-1$ $\equiv ic+j,ic+j+1,...,ic+j+\frac{m-3}{2}(\bmod\ m)\}$, i=0,1,...,(m/c)-1. Then the cardinality $|K_i|$ is the number of pointer fields for second-order queries to be attached to the accession number of a record and it is given by (m-3)/2+c. Thus the number of pointer fields in a record is N=(m-3)/2+c+1, considering a pointer for first-order queries. An example of this series is in Table 1.
    For a first-order query Q_p^h, the access to the ([(h-1)/c]n+p)-th bucket $B_{[(h-1)/c],p}$ is sufficient, where [x] is the greatest integer not exceeding x. For a second-order query $Q_{p,q}^{h,k}$, put $\Delta\equiv k-h$ (mod m). If $\Delta\leq(m-1)/2$, then the access to the ([(h-1)/c]n+p)-th bucket $B_{[(h-1)/c],p}$ is sufficient, and if $\Delta>(m-1)/2$, then the access to the ([(k-1)/c]n+q)-th bucket $B_{[(k-1)/c],q}$ is sufficient.

    (2) Case m is even.
    Put m/2=ct+s (0≤s<c). Let the (in+p)-th bucket B_{ip}, i=0,1,...,(m/c)-1; p=1,2,...,n, associate with the query-set $Q_{ip}^{(1)}\cup Q_{ip}^{(2)}$ given in the following, where $Q_{ip}^{(1)}$ and $Q_{ip}^{(2)}$ are the set of c first-order queries and the set of second-order queries, respectively:
    For 0≤i≤t-1 and 1≤p≤n,

$$Q_{ip}^{(1)} = \{Q_p^{ic+1}, \; Q_p^{ic+2}, \ldots, \; Q_p^{(i+1)c}\},$$

$$Q_{ip}^{(2)} = \overset{c}{\underset{j=1}{\cup}} \{Q_{p,\;q}^{ic+j,k} \mid \begin{array}{l} k=ic+j+1,ic+j+2,\ldots,ic+j+\frac{m}{2} \\ q=1,2,\ldots,n \end{array}\},$$

where $|Q_{ip}^{(2)}| = mnc/2 \; (=d_1)$ and $N=c+(m/2)$.

For i=t and $1 \leq p \leq n$,

$$Q_{tp}^{(1)} = \{Q_p^{tc+1}, \; Q_p^{tc+2}, \ldots, \; Q_p^{(t+1)c}\},$$

$$Q_{tp}^{(2)} = (\overset{s}{\underset{j=1}{\cup}} \{Q_{p,\;q}^{tc+j,k} \mid \begin{array}{l} k=tc+j+1,tc+j+2,\ldots,tc+j+\frac{m}{2} \\ q=1,2,\ldots,n \end{array}\})$$

$$\cup (\overset{c-s}{\underset{j=1}{\cup}} \{Q_{p,\;q}^{\frac{m}{2}+j,k} \mid \begin{array}{l} k=1,2,\ldots,j-1,\frac{m}{2}+j+1,\frac{m}{2}+j+2,\ldots,m \\ q=1,2,\ldots,n \end{array}\}),$$

where $|Q_{tp}^{(2)}| = (\frac{m-2}{2}c+s)n \; (=d_2)$ and $N=c+(m/2)-1$.

For $t+1 \leq i \leq (m/c)-1$ and $1 \leq p \leq n$,

$$Q_{ip}^{(1)} = \{Q_p^{ic+1}, \; Q_p^{ic+2}, \ldots, \; Q_p^{(i+1)c}\},$$

$$Q_{ip}^{(2)} = \overset{c}{\underset{j=1}{\cup}} \{Q_{p,\;q}^{ic+j,k} \mid \begin{array}{l} k=1,2,\ldots,ic+j-(m/2)-1,ic+j+1,ic+j+2,\ldots,m \\ q=1,2,\ldots,n \end{array}\},$$

where $|Q_{ip}^{(2)}| = \frac{m-2}{2}nc \; (=d_3)$ and $N=c+(m/2)-1$. Then an $F(1,2)$ with parameters m,n,c,d_1,d_2 and d_3 is obtained. When s=0, of course, this is a scheme with parameters m,n,c,d_1 and d_3. An example of this series is in Table 2.

For a first-order query Q_p^h, the access to the $([(h-1)/c]n+p)$-th bucket $B_{[(h-1)/c],p}$ is sufficient. For a second-order query $Q_{p,q}^{h,k}$, if either h<k and $k-h \leq m/2$ or h>k and $h-k > m/2$ holds, then the access to the $([(h-1)/c]n+p)$-th bucket $B_{[(h-1)/c],p}$ is sufficient. If either h<k and $k-h > m/2$ or h>k and $h-k \leq m/2$ holds, then the access to the $([(k-1)/c]n+q)$-th bucket $B_{[(k-1)/c],q}$ is sufficient.

CONCLUSION

A new information storage and retrieval scheme is presented in this paper. It is assumed that records are characterized by m attributes having n values each, and the query-set consists of two kinds of queries, where one (called a first-order query) specifies one attribute value and another one (called a second-order query) specifies values of two attributes.

Usually, a secondary file or a directory file, which is called a bucket, is organized for each of the attribute values in order to retrieve every first-order query quickly. Although such inverted filing scheme of order one (IFS_1) is an efficient scheme for the retrieval of every first-order query, it is not always efficient for the retrieval of the second-order query, because after the access to two buckets the set operations of them have to be executed. To organize an inverted bucket for every second-order query in order to reduce the retrieval time, however, is impractical in many situations in that both the number of buckets to be prepared and the redundancy of the file, which is defined by the average number of buckets in which the

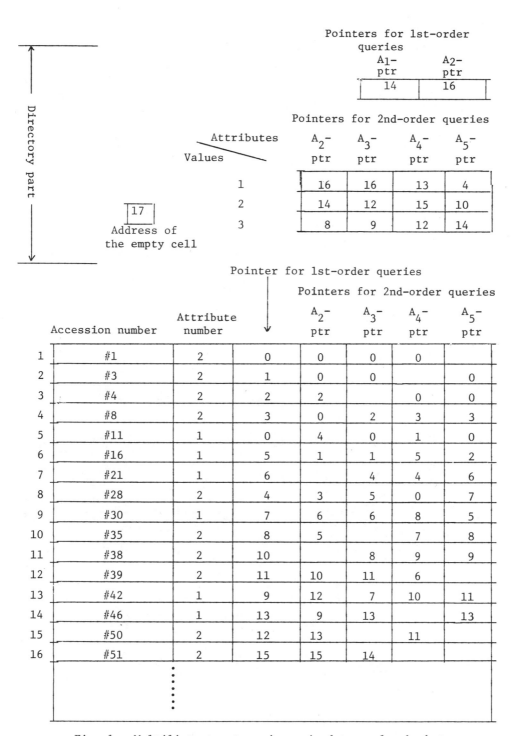

Fig. 1.   Multilist structure in a mixed-type claw bucket

Table 1. Mixed-type multiple-valued filing scheme of order one and two
(m=9, n=3, c=3, d=36)

bucket   first-order and second-order queries

B_1

Q_1^1, $Q_{1,1}^{1,2}$, $Q_{1,2}^{1,2}$, $Q_{1,3}^{1,2}$, $Q_{1,1}^{1,3}$, $Q_{1,2}^{1,3}$, $Q_{1,3}^{1,3}$, $Q_{1,1}^{1,4}$, $Q_{1,2}^{1,4}$, $Q_{1,3}^{1,4}$, $Q_{1,1}^{1,5}$, $Q_{1,2}^{1,5}$, $Q_{1,3}^{1,5}$

Q_1^2, $Q_{1,1}^{2,3}$, $Q_{1,2}^{2,3}$, $Q_{1,3}^{2,3}$, $Q_{1,1}^{2,4}$, $Q_{1,2}^{2,4}$, $Q_{1,3}^{2,4}$, $Q_{1,1}^{2,5}$, $Q_{1,2}^{2,5}$, $Q_{1,3}^{2,5}$, $Q_{1,1}^{2,6}$, $Q_{1,2}^{2,6}$, $Q_{1,3}^{2,6}$

Q_1^3, $Q_{1,1}^{3,4}$, $Q_{1,2}^{3,4}$, $Q_{1,3}^{3,4}$, $Q_{1,1}^{3,5}$, $Q_{1,2}^{3,5}$, $Q_{1,3}^{3,5}$, $Q_{1,1}^{3,6}$, $Q_{1,2}^{3,6}$, $Q_{1,3}^{3,6}$, $Q_{1,1}^{3,7}$, $Q_{1,2}^{3,7}$, $Q_{1,3}^{3,7}$

B_2

Q_2^1, $Q_{2,1}^{1,2}$, $Q_{2,2}^{1,2}$, $Q_{2,3}^{1,2}$, $Q_{2,1}^{1,3}$, $Q_{2,2}^{1,3}$, $Q_{2,3}^{1,3}$, $Q_{2,1}^{1,4}$, $Q_{2,2}^{1,4}$, $Q_{2,3}^{1,4}$, $Q_{2,1}^{1,5}$, $Q_{2,2}^{1,5}$, $Q_{2,3}^{1,5}$

Q_2^2, $Q_{2,1}^{2,3}$, $Q_{2,2}^{2,3}$, $Q_{2,3}^{2,3}$, $Q_{2,1}^{2,4}$, $Q_{2,2}^{2,4}$, $Q_{2,3}^{2,4}$, $Q_{2,1}^{2,5}$, $Q_{2,2}^{2,5}$, $Q_{2,3}^{2,5}$, $Q_{2,1}^{2,6}$, $Q_{2,2}^{2,6}$, $Q_{2,3}^{2,6}$

Q_2^3, $Q_{2,1}^{3,4}$, $Q_{2,2}^{3,4}$, $Q_{2,3}^{3,4}$, $Q_{2,1}^{3,5}$, $Q_{2,2}^{3,5}$, $Q_{2,3}^{3,5}$, $Q_{2,1}^{3,6}$, $Q_{2,2}^{3,6}$, $Q_{2,3}^{3,6}$, $Q_{2,1}^{3,7}$, $Q_{2,2}^{3,7}$, $Q_{2,3}^{3,7}$

B_3

Q_3^1, $Q_{3,1}^{1,2}$, $Q_{3,2}^{1,2}$, $Q_{3,3}^{1,2}$, $Q_{3,1}^{1,3}$, $Q_{3,2}^{1,3}$, $Q_{3,3}^{1,3}$, $Q_{3,1}^{1,4}$, $Q_{3,2}^{1,4}$, $Q_{3,3}^{1,4}$, $Q_{3,1}^{1,5}$, $Q_{3,2}^{1,5}$, $Q_{3,3}^{1,5}$

Q_3^2, $Q_{3,1}^{2,3}$, $Q_{3,2}^{2,3}$, $Q_{3,3}^{2,3}$, $Q_{3,1}^{2,4}$, $Q_{3,2}^{2,4}$, $Q_{3,3}^{2,4}$, $Q_{3,1}^{2,5}$, $Q_{3,2}^{2,5}$, $Q_{3,3}^{2,5}$, $Q_{3,1}^{2,6}$, $Q_{3,2}^{2,6}$, $Q_{3,3}^{2,6}$

Q_3^3, $Q_{3,1}^{3,4}$, $Q_{3,2}^{3,4}$, $Q_{3,3}^{3,4}$, $Q_{3,1}^{3,5}$, $Q_{3,2}^{3,5}$, $Q_{3,3}^{3,5}$, $Q_{3,1}^{3,6}$, $Q_{3,2}^{3,6}$, $Q_{3,3}^{3,6}$, $Q_{3,1}^{3,7}$, $Q_{3,2}^{3,7}$, $Q_{3,3}^{3,7}$

B_4

Q_1^4, $Q_{1,1}^{4,5}$, $Q_{1,2}^{4,5}$, $Q_{1,3}^{4,5}$, $Q_{1,1}^{4,6}$, $Q_{1,2}^{4,6}$, $Q_{1,3}^{4,6}$, $Q_{1,1}^{4,7}$, $Q_{1,2}^{4,7}$, $Q_{1,3}^{4,7}$, $Q_{1,1}^{4,8}$, $Q_{1,2}^{4,8}$, $Q_{1,3}^{4,8}$

Q_1^5, $Q_{1,1}^{5,6}$, $Q_{1,2}^{5,6}$, $Q_{1,3}^{5,6}$, $Q_{1,1}^{5,7}$, $Q_{1,2}^{5,7}$, $Q_{1,3}^{5,7}$, $Q_{1,1}^{5,8}$, $Q_{1,2}^{5,8}$, $Q_{1,3}^{5,8}$, $Q_{1,1}^{5,9}$, $Q_{1,2}^{5,9}$, $Q_{1,3}^{5,9}$

Q_1^6, $Q_{1,1}^{6,1}$, $Q_{1,2}^{6,1}$, $Q_{1,3}^{6,1}$, $Q_{1,1}^{6,7}$, $Q_{1,2}^{6,7}$, $Q_{1,3}^{6,7}$, $Q_{1,1}^{6,8}$, $Q_{1,2}^{6,8}$, $Q_{1,3}^{6,8}$, $Q_{1,1}^{6,9}$, $Q_{1,2}^{6,9}$, $Q_{1,3}^{6,9}$

B_5

Q_2^4, $Q_{2,1}^{4,5}$, $Q_{2,2}^{4,5}$, $Q_{2,3}^{4,5}$, $Q_{2,1}^{4,6}$, $Q_{2,2}^{4,6}$, $Q_{2,3}^{4,6}$, $Q_{2,1}^{4,7}$, $Q_{2,2}^{4,7}$, $Q_{2,3}^{4,7}$, $Q_{2,1}^{4,8}$, $Q_{2,2}^{4,8}$, $Q_{2,3}^{4,8}$

Q_2^5, $Q_{2,1}^{5,6}$, $Q_{2,2}^{5,6}$, $Q_{2,3}^{5,6}$, $Q_{2,1}^{5,7}$, $Q_{2,2}^{5,7}$, $Q_{2,3}^{5,7}$, $Q_{2,1}^{5,8}$, $Q_{2,2}^{5,8}$, $Q_{2,3}^{5,8}$, $Q_{2,1}^{5,9}$, $Q_{2,2}^{5,9}$, $Q_{2,3}^{5,9}$

Q_2^6, $Q_{2,1}^{6,1}$, $Q_{2,2}^{6,1}$, $Q_{2,3}^{6,1}$, $Q_{2,1}^{6,7}$, $Q_{2,2}^{6,7}$, $Q_{2,3}^{6,7}$, $Q_{2,1}^{6,8}$, $Q_{2,2}^{6,8}$, $Q_{2,3}^{6,8}$, $Q_{2,1}^{6,9}$, $Q_{2,2}^{6,9}$, $Q_{2,3}^{6,9}$

B_6

Q_3^4, $Q_{3,1}^{4,5}$, $Q_{3,2}^{4,5}$, $Q_{3,3}^{4,5}$, $Q_{3,1}^{4,6}$, $Q_{3,2}^{4,6}$, $Q_{3,3}^{4,6}$, $Q_{3,1}^{4,7}$, $Q_{3,2}^{4,7}$, $Q_{3,3}^{4,7}$, $Q_{3,1}^{4,8}$, $Q_{3,2}^{4,8}$, $Q_{3,3}^{4,8}$

Q_3^5, $Q_{3,1}^{5,6}$, $Q_{3,2}^{5,6}$, $Q_{3,3}^{5,6}$, $Q_{3,1}^{5,7}$, $Q_{3,2}^{5,7}$, $Q_{3,3}^{5,7}$, $Q_{3,1}^{5,8}$, $Q_{3,2}^{5,8}$, $Q_{3,3}^{5,8}$, $Q_{3,1}^{5,9}$, $Q_{3,2}^{5,9}$, $Q_{3,3}^{5,9}$

Q_3^6, $Q_{3,1}^{6,1}$, $Q_{3,2}^{6,1}$, $Q_{3,3}^{6,1}$, $Q_{3,1}^{6,7}$, $Q_{3,2}^{6,7}$, $Q_{3,3}^{6,7}$, $Q_{3,1}^{6,8}$, $Q_{3,2}^{6,8}$, $Q_{3,3}^{6,8}$, $Q_{3,1}^{6,9}$, $Q_{3,2}^{6,9}$, $Q_{3,3}^{6,9}$

first-order and second-order queries

B_7

$Q_1^7,\ Q_{1,1}^{7,1}, Q_{1,2}^{7,1}, Q_{1,3}^{7,1}, Q_{1,1}^{7,2}, Q_{1,2}^{7,2}, Q_{1,3}^{7,2}, Q_{1,1}^{7,8}, Q_{1,2}^{7,8}, Q_{1,3}^{7,8}, Q_{1,1}^{7,9}, Q_{1,2}^{7,9}, Q_{1,3}^{7,9}$

$Q_1^8,\ Q_{1,1}^{8,1}, Q_{1,2}^{8,1}, Q_{1,3}^{8,1}, Q_{1,1}^{8,2}, Q_{1,2}^{8,2}, Q_{1,3}^{8,2}, Q_{1,1}^{8,3}, Q_{1,2}^{8,3}, Q_{1,3}^{8,3}, Q_{1,1}^{8,9}, Q_{1,2}^{8,9}, Q_{1,3}^{8,9}$

$Q_1^9,\ Q_{1,1}^{9,1}, Q_{1,2}^{9,1}, Q_{1,3}^{9,1}, Q_{1,1}^{9,2}, Q_{1,2}^{9,2}, Q_{1,3}^{9,2}, Q_{1,1}^{9,3}, Q_{1,2}^{9,3}, Q_{1,3}^{9,3}, Q_{1,1}^{9,4}, Q_{1,2}^{9,4}, Q_{1,3}^{9,4}$

B_8

$Q_2^7,\ Q_{2,1}^{7,1}, Q_{2,2}^{7,1}, Q_{2,3}^{7,1}, Q_{2,1}^{7,2}, Q_{2,2}^{7,2}, Q_{2,3}^{7,2}, Q_{2,1}^{7,8}, Q_{2,2}^{7,8}, Q_{2,3}^{7,8}, Q_{2,1}^{7,9}, Q_{2,2}^{7,9}, Q_{2,3}^{7,9}$

$Q_2^8,\ Q_{2,1}^{8,1}, Q_{2,2}^{8,1}, Q_{2,3}^{8,1}, Q_{2,1}^{8,2}, Q_{2,2}^{8,2}, Q_{2,3}^{8,2}, Q_{2,1}^{8,3}, Q_{2,2}^{8,3}, Q_{2,3}^{8,3}, Q_{2,1}^{8,9}, Q_{2,2}^{8,9}, Q_{2,3}^{8,9}$

$Q_2^9,\ Q_{2,1}^{9,1}, Q_{2,2}^{9,1}, Q_{2,3}^{9,1}, Q_{2,1}^{9,2}, Q_{2,2}^{9,2}, Q_{2,3}^{9,2}, Q_{2,1}^{9,3}, Q_{2,2}^{9,3}, Q_{2,3}^{9,3}, Q_{2,1}^{9,4}, Q_{2,2}^{9,4}, Q_{2,3}^{9,4}$

B_9

$Q_3^7,\ Q_{3,1}^{7,1}, Q_{3,2}^{7,1}, Q_{3,3}^{7,1}, Q_{3,1}^{7,2}, Q_{3,2}^{7,2}, Q_{3,3}^{7,2}, Q_{3,1}^{7,8}, Q_{3,2}^{7,8}, Q_{3,3}^{7,8}, Q_{3,1}^{7,9}, Q_{3,2}^{7,9}, Q_{3,3}^{7,9}$

$Q_3^8,\ Q_{3,1}^{8,1}, Q_{3,2}^{8,1}, Q_{3,3}^{8,1}, Q_{3,1}^{8,2}, Q_{3,2}^{8,2}, Q_{3,3}^{8,2}, Q_{3,1}^{8,3}, Q_{3,2}^{8,3}, Q_{3,3}^{8,3}, Q_{3,1}^{8,9}, Q_{3,2}^{8,9}, Q_{3,3}^{8,9}$

$Q_3^9,\ Q_{3,1}^{9,1}, Q_{3,2}^{9,1}, Q_{3,3}^{9,1}, Q_{3,1}^{9,2}, Q_{3,2}^{9,2}, Q_{3,3}^{9,2}, Q_{3,1}^{9,3}, Q_{3,2}^{9,3}, Q_{3,3}^{9,3}, Q_{3,1}^{9,4}, Q_{3,2}^{9,4}, Q_{3,3}^{9,4}$

Table 2. Mixed-type multiple-valued filing scheme of order one and two
(m=6, n=3, c=2, d_1=18, d_2=15, d_3=12)

bucket   first-order and second-order queries

B_1

$Q_1^1,\ Q_{1,1}^{1,2}, Q_{1,2}^{1,2}, Q_{1,3}^{1,2}, Q_{1,1}^{1,3}, Q_{1,2}^{1,3}, Q_{1,3}^{1,3}, Q_{1,1}^{1,4}, Q_{1,2}^{1,4}, Q_{1,3}^{1,4}$

$Q_1^2,\ Q_{1,1}^{2,3}, Q_{1,2}^{2,3}, Q_{1,3}^{2,3}, Q_{1,1}^{2,4}, Q_{1,2}^{2,4}, Q_{1,3}^{2,4}, Q_{1,1}^{2,5}, Q_{1,2}^{2,5}, Q_{1,3}^{2,5}$

B_2

$Q_2^1,\ Q_{2,1}^{1,2}, Q_{2,2}^{1,2}, Q_{2,3}^{1,2}, Q_{2,1}^{1,3}, Q_{2,2}^{1,3}, Q_{2,3}^{1,3}, Q_{2,1}^{1,4}, Q_{2,2}^{1,4}, Q_{2,3}^{1,4}$

$Q_2^2,\ Q_{2,1}^{2,3}, Q_{2,2}^{2,3}, Q_{2,3}^{2,3}, Q_{2,1}^{2,4}, Q_{2,2}^{2,4}, Q_{2,3}^{2,4}, Q_{2,1}^{2,5}, Q_{2,2}^{2,5}, Q_{2,3}^{2,5}$

B_3

$Q_3^1,\ Q_{3,1}^{1,2}, Q_{3,2}^{1,2}, Q_{3,3}^{1,2}, Q_{3,1}^{1,3}, Q_{3,2}^{1,3}, Q_{3,3}^{1,3}, Q_{3,1}^{1,4}, Q_{3,2}^{1,4}, Q_{3,3}^{1,4}$

$Q_3^2,\ Q_{3,1}^{2,3}, Q_{3,2}^{2,3}, Q_{3,3}^{2,3}, Q_{3,1}^{2,4}, Q_{3,2}^{2,4}, Q_{3,3}^{2,4}, Q_{3,1}^{2,5}, Q_{3,2}^{2,5}, Q_{3,3}^{2,5}$

B_4

$Q_1^3,\ Q_{1,1}^{3,4}, Q_{1,2}^{3,4}, Q_{1,3}^{3,4}, Q_{1,1}^{3,5}, Q_{1,2}^{3,5}, Q_{1,3}^{3,5}, Q_{1,1}^{3,6}, Q_{1,2}^{3,6}, Q_{1,3}^{3,6}$

$Q_1^4,\ Q_{1,1}^{4,5}, Q_{1,2}^{4,5}, Q_{1,3}^{4,5}, Q_{1,1}^{4,6}, Q_{1,2}^{4,6}, Q_{1,3}^{4,6}$

(cont'd)

Table 2. (Continued)

bucket	first-order and second-order queries
B_5	Q_2^3, $Q_{2,1}^{3,4}$, $Q_{2,2}^{3,4}$, $Q_{2,3}^{3,4}$, $Q_{2,1}^{3,5}$, $Q_{2,2}^{3,5}$, $Q_{2,3}^{3,5}$, $Q_{2,1}^{3,6}$, $Q_{2,2}^{3,6}$, $Q_{2,3}^{3,6}$ Q_2^4, $Q_{2,1}^{4,5}$, $Q_{2,2}^{4,5}$, $Q_{2,3}^{4,5}$, $Q_{2,1}^{4,6}$, $Q_{2,2}^{4,6}$, $Q_{2,3}^{4,6}$
B_6	Q_3^3, $Q_{3,1}^{3,4}$, $Q_{3,2}^{3,4}$, $Q_{3,3}^{3,4}$, $Q_{3,1}^{3,5}$, $Q_{3,2}^{3,5}$, $Q_{3,3}^{3,5}$, $Q_{3,1}^{3,6}$, $Q_{3,2}^{3,6}$, $Q_{3,3}^{3,6}$ Q_3^4, $Q_{3,1}^{4,5}$, $Q_{3,2}^{4,5}$, $Q_{3,3}^{4,5}$, $Q_{3,1}^{4,6}$, $Q_{3,2}^{4,6}$, $Q_{3,3}^{4,6}$
B_7	Q_1^5, $Q_{1,1}^{5,1}$, $Q_{1,2}^{5,1}$, $Q_{1,3}^{5,1}$, $Q_{1,1}^{5,6}$, $Q_{1,2}^{5,6}$, $Q_{1,3}^{5,6}$, $Q_{1,1}^{6,1}$, $Q_{1,2}^{6,1}$, $Q_{1,3}^{6,1}$ Q_1^6, $Q_{1,1}^{6,2}$, $Q_{1,2}^{6,2}$, $Q_{1,3}^{6,2}$
B_8	Q_2^5, $Q_{2,1}^{5,1}$, $Q_{2,2}^{5,1}$, $Q_{2,3}^{5,1}$, $Q_{2,1}^{5,6}$, $Q_{2,2}^{5,6}$, $Q_{2,3}^{5,6}$, $Q_{2,1}^{6,1}$, $Q_{2,2}^{6,1}$, $Q_{2,3}^{6,1}$ Q_2^6, $Q_{2,1}^{6,2}$, $Q_{2,2}^{6,2}$, $Q_{2,3}^{6,2}$
B_9	Q_3^5, $Q_{3,1}^{5,1}$, $Q_{3,2}^{5,1}$, $Q_{3,3}^{5,1}$, $Q_{3,1}^{5,6}$, $Q_{3,2}^{5,6}$, $Q_{3,3}^{5,6}$, $Q_{3,1}^{6,1}$, $Q_{3,2}^{6,1}$, $Q_{3,3}^{6,1}$ Q_3^6, $Q_{3,1}^{6,2}$, $Q_{3,2}^{6,2}$, $Q_{3,3}^{6,2}$

accession number of a record has to be stored, are very large and cannot be disregarded.

Combinatorial filing schemes have been developed in an attempt to solve this problem. As a solution in such an attempt, this paper presents a filing scheme. In this scheme, organization of a multilist structure among the stored accession numbers in the buckets associated with the sub-sets of the query-set which consists of the first-order and the second-order queries is proposed. The structure within a bucket constructed by the added pointers makes it possible to retrieve second-order queries as well as first-order queries without any set operations. The introduction of such structure can be expected to improve the retrieval performance of the second-order query considerably, without any lowering of efficiency of first-order query retrieval in the IFS_1. The scheme has the following specific properties:

(1) The number of buckets to be organized is less than that of the IFS_1.

(2) The redundancy of the file is less than that of the IFS_1, provided all records are equally likely.

(3) Every first-order query can be retrieved almost as quickly as IFS_1.

(4) Every second-order query can be answered by the access to one bucket and the traversal of its contents without any set operations.

ACKNOWLEDGMENTS

The author wishes to express their thanks to the referees for very valuable criticism and helpful comments.

REFERENCES

1. C. T. Abraham, S. P. Ghosh, and D. K. Ray-Chaudhuri, File organization schemes based on finite geometries, Information and Control 12, 143–163 (1968).
2. G. Berman, The application of difference sets to the design of a balanced multiple-valued filing scheme, Information and Control 32, 128–138 (1976).
3. G. Berman, Combinatorial multiple-valued filing systems for multiattribute queries, Information and Control 36, 119–132 (1978).
4. R. C. Bose, C. T. Abraham, and S. P. Ghosh, File organization of records with multiple-valued attributes for multi-attributes queries, in: "Combinatorial Mathematics and Its Applications", University of North Carolina Press, Chapel Hill, N.C., pp.277–297 (1969).
5. R. C. Bose and G. G. Koch, The design of combinatorial information retrieval systems for files with multiple-valued attributes, SIAM J. Appl. Math. 17, 1203–1214 (1969).
6. D. K. Chow, New balanced-file organization schemes, Information and Control 15, 377–396 (1969).
7. V. Donaldson and L. W. Hawkes, Cyclic multiple-valued filing schemes for higher-order queries, Information Sciences 32, 47–74 (1984).
8. S. P. Ghosh, Organization of records with unequal multiple-valued attributes and combinatorial queries of order 2, Information Sciences 1, 363–380 (1969).
9. S. P. Ghosh and C. T. Abraham, Application of finite geometry in file organization for records with multiple-valued attributes, IBM J. Res. Develop. 12, 180–187 (1968).
10. H. Ikeda, Combinatorial file organization schemes and their experimental evaluation, Hiroshima Math. J. 8, 515–544 (1978).
11. D. K. Ray-Chaudhuri, Combinatorial information retrieval systems for files, SIAM J. Appl. Math. 16, 973–992 (1968).
12. S. Tazawa, Design of file organization schemes for selected query set on binary-valued attributes and their redundancies, Information and Control 50, 23–33 (1981).
13. S. Yamamoto, H. Ikeda, S. Shige-eda, K. Ushio, and N. Hamada, Design of a new balanced file organization scheme with the least redundancy, Information and Control 28, 156–175 (1975).
14. S. Yamamoto, S. Tazawa, K. Ushio, and H. Ikeda, Design of a generalized balanced multiple-valued file organization scheme of order two, Proc. of 8-th ACM-SIGMOD Int. Conf. on Management of Data (1978).
15. S. Yamamoto, S. Tazawa, K. Ushio, and H. Ikeda, Design of a balanced multiple-valued file-organization scheme with the least redundancy, ACM Trans. Database Systems 4, 518–530 (1979).
16. S. Yamamoto, T. Teramoto, and K. Futagami, Design of a balanced multiple-valued filing scheme of order two based on cyclically generated spread in finite projective geometry, Information and Control 21, 72–91 (1972).

SPACE PARTITIONING AND ITS APPLICATION
TO GENERALIZED RETRIEVAL PROBLEMS

David Avis

School of Computer Science
McGill University
805 Sherbrooke Street West
Montreal, Quebec, Canada   H3A 2K6

1. Introduction

A fundamental problem in the area of database systems is the retrieval of data satisfying certain criteria. The simplest such problem is the retrieval of data based on a single key, and a comprehensive treatment of this problem is contained in Knuth[1]. A more difficult problem is the retrieval of data based on criteria for several keys, the so-called multidimensional search problem. Suppose that each of n records of a *file* contains a fixed number, d, of *keys*. A *query* asks for all records for which the d keys satisfy certain criteria. For example, upper and lower bounds maybe specified for some or all keys, yielding the so-called *orthogonal range query problem.*

In this paper we study the multidimensional search problem from a geometric standpoint. We consider the $d-tuple$ of keys for a given record as a point in d-dimensional space, R^d. The previously mentioned orthogonal range query problem can now be interpreted as finding all points in a given $d-dimensional$ hyperrectangle. Suppose we are given upper bounds, u_i, and lower bounds l_i for each of the $i=1,2,\cdots,d$ keys. Then we are asked to find every record whose key $(x_1,\cdots,x_d)$ satisfies:

$$l_i \leq x_i \leq u_i \qquad i=1,\ldots,d.$$

A more general query could consist of a linear function: $a_1x_1+a_2x_2+\cdots+a_dx_d+a_{d+1}$. The *half-space query problem* is to find all records whose keys satisfy:

$$a_1x_1+a_2x_2+\cdots+a_dx_d+a_{d+1}\leq 0.$$

Even more generally, we may be given a set of k such linear functions, which we represent in matrix form:

$$A = \begin{bmatrix} a_{11} & \cdots & a_{1d} \\ \vdots & & \vdots \\ a_{k1} & \cdots & a_{kd} \end{bmatrix}, \; b = \begin{bmatrix} a_{1,d+1} \\ \vdots \\ a_{k,d+1} \end{bmatrix}$$

Letting $x = (x_1, x_2, \ldots, x_d)$ be the $d-vector$ of keys for a record, we search for all such vectors satisfying

$$Ax + b \leq 0.$$

Such a query is called a *polyhedral range query.* Note that the orthogonal range query problem is just a special case of this problem obtained by letting $k = 2d$, and by setting

$$
\left.\begin{array}{rcl}
a_{2i-1,i} & = & 1 \\[4pt]
a_{2i,i} & = & -1 \\[4pt]
a_{2i-1,d+1} & = & -u_i \\[4pt]
a_{2i,d+1} & = & l_i
\end{array}\right\} \quad i = 1, 2, \ldots, d
$$

$$
a_{ij} = 0 \quad otherwise.
$$

An excellent survey of orthogonal range searching is contained in the paper by Bentley and Friedman[2].

As an example of when a polyhedral range query may be necessary, consider the following student database. Each record contains 6 keys representing the student ID number, and the number of A, B, C, D and F grades obtained. The ID number is a 7 digit integer of the form, $yyzzzzz$, where yy indicates the student first registered in 19yy. Assume a grade point scheme where A=4, B=3, C=2, D=1 and F=0. The grade point average (GPA) is defined to be the average of the grades obtained, using the above numerical conversion. An orthogonal range query may be used to list, for example, all students with more than 4 F's who started in or before 1979:

$$
x_1 \leq 7999999
$$

$$
x_6 \geq 4.
$$

But suppose we want to find all students with GPA at least 3.2 and who started in 1980 or 1981. Then we need a general half plane query:

$$
.8x_2 - .2x_3 - 1.2x_4 - 2.2x_5 - 3.2x_6 \geq 0
$$

$$
x_1 \geq 8000000
$$

$$
x_2 \leq 8199999
$$

Similar examples can be found in any application involving several numeric fields: payroll, inventory control, accounting etc.

A completely different application for polyhedral query problems arises when the data is itself geometrical. In a geometrical database, the data corresponds to physical objects in space. For example, we may have a digital representation of a satellite photograph, where we have x, y coordinates for every building or road in the picture (see [3]). We may be required to retrieve all points lying inside a given region, for example, a municipality. If the region can be described by straight-line segments, then this query is a 2-dimensional polyhedral query (or perhaps several queries if the region is non-convex). This type of 2-dimensional query is called a *polygonal query* [4]. Sometimes even regions with curved boundaries can be transformed into polyhedral queries. Yao [5] gives the example of circular queries in 2-dimensions. We wish to find all records satisfying

$$
(x_1 - a_1)^2 + (x_2 - a_2)^2 - a_3^2 \leq 0. \tag{1}
$$

For example, we may be interested in all dwellings within 50km of a TV transmitter. This query problem can be mapped to a polyhedral query in 3 dimensions. For each key (x_1, x_2) create the 3-dimensional key $(x_1, x_2, x_3) = (x_1, x_2, x_1^2 + x_2^2)$. The inequality (1) can be rewritten:

$$
-2a_1 x_1 - 2a_2 x_2 + x_3 + (a_1^2 + a_2^2 - a_3^2) \leq 0
$$

which is a half-space query. Similarly, any d-dimensional spherical query can be reformulated as a $d+1$-dimensional half-space query.

In this paper we consider data structures for storing the keys for each record in a file. Using the geometrical approach, we consider d-dimensional space to be partitioned into regions. All keys falling into the same region of space are thus associated together in the data structure. In the next section, we review several successful space partitioning techniques, including quad trees, $k-d$ trees and multipaging. Whilst these methods work well for orthogonal range queries, they do not seem well suited for the more general half-space queries. For these types of queries, a new space partitioning approach introduced by Willard [4] for 2-dimensions, and generalized by Yao [5] for 3-dimensions is discussed. Section 3 contains theoretical results relating to the new type of data structure. The results of Willard and Yao are reviewed and new results for higher dimensions are given. The best known algorithms for the various basic data structure operations are also reviewed. Finally, in the fourth section, we mention some of the many open problems related to this data structure. Some conclusions as to its applicability are also given.

2. Space Partition

In this section, we discuss various techniques for partitioning a set of records given as points in d-dimensional space. Let $S = \{ p_1, ..., p_n \}$ be a set of points in R^d, each corresponding to a record of a file with d-keys. The first three partitioning techniques that we will consider all consist of partitions of S by hyperplanes that are parallel to the coordinate axes. These techniques are respectively: quad trees, $k-d$ trees and multipaging. In analysing these structures we will consider the following cost functions:

$P(n,d)$: the cost of preprocessing S into a data structure,

$S(n,k)$: the storage required by the data structure,

$I(n,k)$: the cost of inserting a new point into the data structure,

$D(n,k)$: The cost of deleting a point from the data structure.

One of the earliest data structures of a space partitioning nature is the quad tree [6]. A quad tree for S is built as follows. First take some point p_i as the root. Taking p_i as the origin, d-space is split into 2^d orthants by considering the d hyperplanes parallel to the coordinate axes. Each orthant is then split again using the same technique. This is continued until each region has exactly one point, corresponding to a leaf of the tree. An example is given in Figure 1. In a quad tree, the internal nodes each have degree at most four. In degenerate cases, as for example when all the points lie on a diagonal, the internal nodes have degree two and the depth of the tree is $\log_2 n$ rather than the best possible bound of $\log_{2^d} n$. In [6] an expected time insertion algorithm of $O(\log n)$ is given, however no efficient deletion algorithm is given. Using the subscript Q to denote quad tree we have

$$P_Q (n,d) = O (n \log n)$$

$$S_Q (n,d) = O (n d)$$

$$I_Q (n,d) = O (\log n).$$

Overmars and van Leeuwen [7] describe a variant of the quad tree called the pseudo-quad tree that allows both efficient insertions and deletions. Denoting this data structure by PQ, they prove

$$P_{PQ} (n,d) = O (n \log n)$$

$$S_{PQ} (n,d) = O (n d)$$

$$I_{PQ} (n,d) = D_{PQ} (n,d) = O(\frac{1}{\delta} \log^2 n)$$

for *every constant* $0 < \delta < 1$ *(average).*

Another space-partitioning data structure is the multidimensional binary search tree or $k-d$ tree due to Bentley [8]. A $k-d$ tree for S is built as follows. Again we choose some point p_i from S. Using the first coordinate only, we split S into two subsets. The point p_i is made the root of the $k-d$ tree and the two subsets are made its two sons. In each subset we again choose a point and split according to the second coordinate. This procedure is repeated for the first d

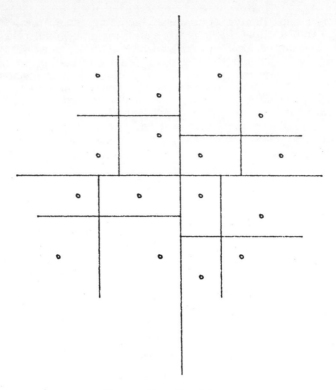

Figure 1.   Quad Tree

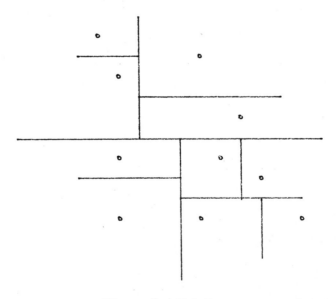

Figure 2.   K-D Tree

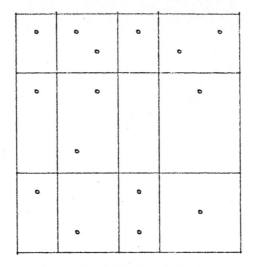

Figure 3. Cell Partitioning

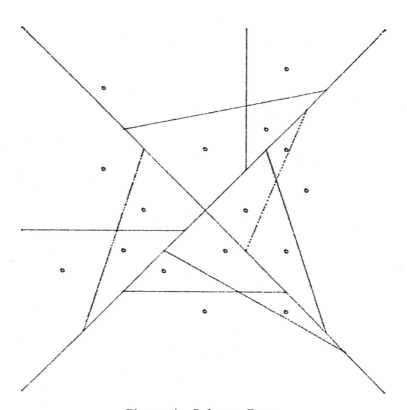

Figure 4. Polygon Tree

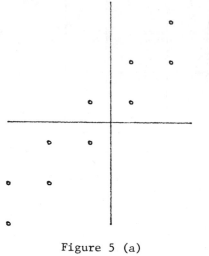

Figure 5 (a)

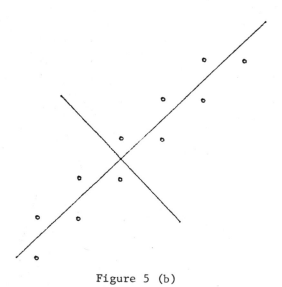

Figure 5 (b)

levels in the tree, and is then repeated for each set of d levels, as required until each region contains one point. An example for $d = 2$ is shown in Figure 2. Bentley gives algorithms for $k-d$ trees which yield

$$P_{kd}\ (n,d)\ =\ O\ (\ dn\ \log\ n\ )$$

$$S_{kd}\ (n,d)\ =\ O\ (\ n\ d\ )$$

$$I_{kd}\ (n,d)\ =\ O\ (\ \log\ n\ )\qquad average$$

$$D_{kd}\ (n,d)\ =\ O\ (\ n^{1-\frac{1}{d}}\ )\qquad worst\ case$$

$$=\ O\ (\ \log\ n\ )\qquad average$$

Again, Overmars and van Leeuwen [7] are able to improve on the insertion deletion times by introducing a modification to the data structure. They obtain results similar to those already quoted for quad trees.

In contrast to the two tree based methods described here are the so-called cell based methods for direct access to multidimensional data. Schemes are given by Rothnie and Lozano [9], Liou and Yao [10] and Merrett [11]. A convenient description of these techniques and detailed analysis of the latter technique, called multipaging, is contained in the thesis of Otoo[12]. The basic idea of the cell method is to partition the space by families of hyperplanes parallel to the given coordinate axes. For the ith axis, $i = 1, ..., d$, some number k_i of values are chosen. Suppose these values are $a_{i\,1}, a_{i\,2}, ..., a_{ik_i}$. Then d-space is partitioned by the $k_1 + \cdots + k_d$ hyperplanes of the form:

$$x_i\ =\ a_{ij}\qquad j = 1, ..., k_i\ ,\qquad i = 1, 2, ..., d\ .$$

This produces a partition of space into

$$(\ k_1 + 1\ )\ (\ k_2 + 1\ )\quad\cdots\quad (\ k_d + 1\ )$$

hyperrectangular regions. An example for $d = 2$ and $k_1 = 2, k_2 = 3$ is shown in Figure 3. The idea is to make the cells small enough so that all of the points in S lying in a single cell can fit in one page of memory. In cases where the points are highly clustered, some overflow mechanism is required to make use of underoccupied cells. If n pages or cells are generated the directory requires

$O(\ dn^{\frac{1}{d}}\ )$ space. For this type of data structure, worst case results are usually very bad, however they give good average time performance.

The common feature of all of the above schemes is that the partitioning hyperplanes are all parallel to the coordinate axes. Thus the regions created are always hyperrectangles. In the fourth data structure to be described this restriction is dropped. This data structure was first described by Willard[4] who called it a *polygon tree* in two dimensions. In general, we will call it a *polyhedron tree*. In d-space, d mutually intersecting hyperplanes are chosen. These decompose space into 2^d regions. The d hyperplanes are associated with the root of the tree. The subsets of S that lie in each of the 2^d regions form the 2^d sons of the root. Each of these regions is again subdivided by d mutually intersecting hyperplanes. The process is continued until each region contains at most one point of S. An example for $d = 2$ is shown in Figure 4. The similarity with quad trees is readily apparent. The power of polyhedron trees is the ability to partition points of S into subsets of essentially the same cardinality. Consider figure 5(a) and (b). In the quad tree, 5(a), two of the sons of the root contain almost no points. In a polygon tree, 5(b), however, each son has one quarter of the points. It is perhaps somewhat surprising that in two dimensions, S can always be partitioned by two lines so that each open region contains at most a quarter of the points[4]. Yao[5] gives a similar result for 3 dimensions. These theoretical results are described in the next section. We now outline how polyhedral trees can be used to solve the half-space query problem in sublinear time.

For simplicity, we consider the 2-dimensional problem. Assume that we have built a polygon tree as illustrated in figure 4. The algorithm for half-plane queries consists of a standard top down walk through the polygon tree. All nodes whose ranges intersect with the requested half-plane are visited. Each leaf visited will have the coordinates of the point it contains tested against the given query. The success of the algorithm depends on the following observation:

A half-plane can intersect at most three of the four regions formed by a partition of the plane by two lines.

This means that at most three of the four subtrees of an internal node need to be visited. Let h be the height of the polygon tree. Then a simple analysis shows that the total number of nodes visited is

$$\sum_{d=0}^{h} 3^d \quad - \quad O\left(n^{\log_4 3}\right), \qquad \log_4 3 = .792... ,$$

where n is the number of data items in the tree. If k data items are contained in the output of the query, then the algorithm has sublinear complexity $O\left(n^{.792} + k\right)$. By partitioning the data into 6 equal cardinality regions by three lines, this complexity can be reduced to $O\left(n^{.774} + k\right)$,[13]. Furthur optimizations were obtained by Edelsbrunner and Welzl [13]. By reusing some subdividing lines they reduce the exponent to 0.694.

Analagous results hold in three dimensions[5]. A half-space in three dimensions can intersect at most 7 of the 8 regions formed by a partition by three intersecting planes. This gives rise to a half-space query algorithm of complexity $O\left(n^{.936} + k\right)$, since $\log_8 7 = .936... $.

It is clear that the success of these methods depends on efficient space partition. Generalizations to higher dimensions will depend on generalizing the results on space partitioning. These topics are discussed in the next section.

3. Geometrical Results

In this section we discuss the geometrical results which are the basis for fast search algorithms that use polyhedron trees. Some of these results are rather technical, and the reader who is not mathematically inclined may skip most of this material. From an applications point of view, the important results are corollary 2.1, theorem 3, and theorem 6. The reader is referred to the corresponding references to see how these results are applied. For the interested reader, the development in this section will be done informally to give some intuition into the results. For formal proofs, the reader is referred to the original sources. We begin with some terminology.

Let S be a set of n points in R^d and let $0 < \alpha < 2^{-d}$. We say that S is α-partitionable if there exist a set of d intersecting hyperplanes that partition S so that no open region contains more than αn points of S. We say that S is non-partitionable if for every partition of S by d intersecting hyperplanes, some set of $2^d - 1$ open regions together contain all the points of S. A point x in space is called a *centre* of S if every hyperplane H through x contains at least $\dfrac{n}{(d+1)}$ points in each half-space bounded by H. Two classical results are the Centre Theorem and the Ham Sandwich Theorem.

Theorem 1. Centre Theorem[14]

Every set S in R^d has a centre.

Theorem 2. Ham-Sandwich Theorem[15]

Given d disjoint sets of points in R^d, there is a hyperplane which simultaneously divides each point set evenly.

From theorem 2, we get the immediate corollary:

Corollary 2.1[4]

Every set S in the plane can be $\frac{1}{4}$ - partitioned. Furthermore one of the partitioning lines may be chosen to be an arbitrary bisector of S.

Proof: Choose a line l that bisects S. This can easily be found by choosing the median point in one of the coordinate directions and taking a line parallel to the appropriate coordinate axis. This divides the set S into two disjoint sets, to which theorem 2 may be applied

Corollary 2.1 thus shows that a perfectly balanced polygon tree can always be constructed. This justifies the sublinear half-plane algorithm described in the previous section. Before we move to three dimensions, another 2-dimensional result is required.

Lemma 1[5]

Let R and B be two n point sets in the plane with a common centre. Then R and B can be simultaneously $\frac{1}{12}$ - partitioned by two lines.

It can now be shown that all sets in R^3 can be partitioned.

Theorem 3[5]

Every set S in 3-dimensions can be $\frac{1}{24}$ - partitioned. Furthermore, one of the partitioning planes may be chosen to be any bisector of S.

Proof: Let H be any bisecting plane through S, splitting S into sets S_1 and S_2. Let x_1 be a centre for S_1 and x_2 be a centre for S_2. Map every point of S onto H by projecting parallel to the line through x_1 and x_2. This gives rise to sets R and B of size $\frac{n}{2}$ that satisfy the conditions of lemma 1. Therefore R and B can be $\frac{1}{12}$ - partitioned by two lines l_1 and l_2. Let H_i be the plane containing l_i, x_1 and x_2, $i = 1, 2$. Then H, H_1, H_2 form a $\frac{1}{24}$ - partition of S.

This result is sufficient to get an $O(n^{.98} + k)$ algorithm for half-space queries in three dimensions. To get the faster algorithm described in section 2, we require the following stronger and more difficult result.

Theorem 4[16] [17]

Every set S in 3-dimensions can be $\frac{1}{8}$ - partitioned.

In [17] , a proof is given based on Theorem 2. The proof is rather technical and is not included here. Yao has apparently proved that theorem 4 can be strengthened to allow one hyperplane to be chosen as an arbitrary bisector of S [16].

In four dimensions, it is not known whether an analogue of corollary 2.1 and theorem 4 holds. In the strengthened form they definitely do not hold.

Theorem 5[18]

For any $\alpha > 0$ and any n, there exists a set S in R^4, of cardinality n, that cannot be α - partitioned by a prescribed bisecting hyperplane and any three other hyperplanes.

In dimensions of 5 and higher, the situation becomes worse. Let $\delta_d = d \bmod 2$.

Theorem 6[18]

For any integer n, and $d \geq 5$, there exist non-partitionable sets of cardinality n. Furthermore, sets exist for which every partition by d hyperplanes leaves at least $2^d - d^2 - \delta_d$ open regions with no points of S.

The "nasty" point sets of theorems 5 and 6 are easily described. For any integer t, let $s_t = (t, t^2, ..., t^d)$. For any integer n we set

$$S = \{ s_t : 1 \leq t \leq n \}. \tag{2}$$

These sets play a fundamental role in the theory of convex polyhedra, as they form the vertices of the *cyclic polytopes*. An accessible introduction to these fascinating polytopes is the paper by Gale[19]. Let

$$f(x_1, ..., x_d) = \sum_{i=1}^{d} a_i x_i + a_{d+1}.$$

We denote by f the hyperplane $f(x_1, x_2, ..., x_d) = 0$ in d-space. The set S is partitioned by f according to the sign of the function

$$g(t) = \sum_{i=1}^{d} a_i t^i + a_{d+1}.$$

Integers t for which $g(t)$ is positive correspond to points s_t in S that lie in one of the open half-spaces bounded by f, and integers t for which $g(t)$ is negative correspond to points in the other open half-space. Let H and L, respectively denote these point sets. Integers for which $g(t) = 0$ correspond to points of S on f. Such points are denoted by M. Since g is a polynomial of degree d, it has some number k, which is at most d, of real roots. Therefore f can contain at most d points of S. If k is zero, then all points of S lie on the same side of f. If k is positive, let $t_1 \leq t_2 \leq \cdots \leq t_k$ denote the real roots of g. Then

$$M = \{ t_1, ..., t_k \},$$

and H and L contain $k + 1 \leq d + 1$ strings of consecutive points of S. If we consider the d partitioning hyperplanes together, it can be shown using this fact[18] that at most $d^2 + 1$ of the 2^d open regions contain points of S. In fact, if d is even, only d^2 regions can contain points. This outlines the proof of theorem 6.

For theorem 5, we observe that $d = 4$ implies $d^2 = 2^d$, and so the previous argument does not yield anything of interest. However the situation is very "tight" in the following sense. The set S defined by (2) with $d = 4$ can be partitioned if and only if there exists a binary matrix of dimension 4×16 with the following properties:

P1: Every column of A is distinct.

P2: Each row of A consists of at most 5 strings of zeroes and ones.

P3: Two consecutive columns of A differ in exactly one row.

Properties P_1 and P_3 define a Gray code. Gilbert[20] has published a list of all Gray codes of length 16. Of the 9 such codes, one satisfies P_2. It is:

$$A = \begin{bmatrix} 0 & 1 & 1 & 0 & 0 & 0 & 1 & 1 & 1 & 1 & 1 & 1 & 0 & 0 & 0 & 0 \\ 0 & 0 & 1 & 1 & 1 & 1 & 1 & 1 & 0 & 0 & 0 & 1 & 1 & 0 & 0 & 0 \\ 0 & 0 & 0 & 0 & 1 & 1 & 1 & 1 & 1 & 1 & 0 & 0 & 0 & 0 & 1 & 1 \\ 0 & 0 & 0 & 0 & 0 & 1 & 1 & 0 & 0 & 1 & 1 & 1 & 1 & 1 & 1 & 0 \end{bmatrix}$$

This matrix determines how the four intersecting hyperplanes must partition S. Suppose S contains 16 points. Row 1 of A can be interpreted as saying that points s_1, s_4, s_5, s_6, s_{13}, s_{14}, s_{15}, and s_{16} must lie above hyperplane H_1. Row 2 can be similarly interpreted for H_2, etc. Following this prescription, a separating set of four hyperplanes can be contructed. But suppose we begin with a bisecting hyperplane H such that s_1, s_3, s_5, ..., s_{15} lie above H and s_2, s_4, ..., s_{16} lie below H. Such a pattern does not appear in A (or any cyclic permutation of the columns of A). Since A is essentially unique, this shows that H cannot be used in any partition of S by 4 hyperplanes. A similar argument can be shown for arbitrary n, demonstrating theorem 5.

As a final result of this type, we show that lemma 1 cannot be generalized to 3-dimensions.

Theorem 7

There exist two n point sets R and B in 3 dimensions, with a common centre, such that R and B are not simultaneously partitionable.

Proof: (Outline)

Consider the set S and hyperplane H described in the last paragraph. Let S_1 and S_2 be the sets separated by H. Let x_1 and x_2 be points in the centres of S_1 and S_2. Project S_1 and S_2 onto H parallel to the line through x_1 and x_2 giving R and B respectively. Now suppose that R and B could be simultaneously partitioned by three planes. Then as in lemma 1, this partition could be extended to a partition of S in 4-dimensions by 4 hyperplanes, one of which was H. But by the remarks preceding the theorem, this is impossible. Hence R and B may not be simultaneously partitioned.

The preceding results are negative and tell us that the algorithms for 2 and 3 dimensions may not be extended easily to high dimensions. Theorem 6 tells us that sets in 5-dimensional space may not be separable by 5 hyperplanes. This leads us to ask: Is there some suitably high dimensional space, such that every set of points can be partitioned by 5 hyperplanes? This leads to the following definition. Let k be any positive integer. Then $f(k)$ is the smallest dimension such that every point set in $f(k)$ - dimensional space can be 2^{-k} - partitioned by k mutually intersecting hypeplanes. Then we can prove:

Theorem 8

$$\frac{2^k - 1}{k} \leq f(k) \leq 2^{k-1} .$$

4. Conclusions

In this paper we have surveyed various techniques for partitioning multidimensional data for efficient searching. The standard methods involve partitioning data by hyperplanes parallel to the coordinate axes. These techniques prove very efficient for orthogonal range queries but are not useful for more general half-space queries. This type of query can be more efficiently answered by a new type of data structure called a polyhedron tree, when the dimension is 2 or 3. For higher dimensions, it has been demonstrated that these trees may not always be constructible.

Many open questions remain. In particular, the non-partitionable data set constructed in dimensions 5 and above is highly pathological. One wonders if in practice, most data sets in dimensions 4, or 5 may be partitionable. There also remains the problem of finding fast partitioning algorithms, and the problem of updating the data base. Perhaps there is a role for heuristics here. Finally, there is the theoretical question of obtaining better bounds on the function $f(k)$ described at the end of the last section.

5. Acknowledgements

Many of the results reported in this paper were obtained whilst the author was a visiting researcher at the University of Tokyo, supported by a grant from the Japanese Society for the Promotion of Science. The author expresses his thanks to Professor Masao Iri for providing such a stimulating research environment. The author would also like to thank an anonymous referee for a carefull reading of the paper.

References

1. D. E. Knuth, *The Art of Computer Programming, Vol 3: Sorting and Searching,* Addison-Wesley, Reading, MA, 1973.

2. J.L. Bentley and J. H. Friedman, "Data Structures for Range Searching," *Comp. Surveys,* vol. 11, pp. 397-409, 4, 1979.

3. M. Iri et al., "Fundamental Algorithms for Geographic Data Processing," *Operations Research Society of Japan,* vol. TR 83-1, (In Japanese), 1983.

4. D. E. Willard, "Polygon Retrieval," *SIAM Journal on Computing,* vol. 11, pp. 149-165, 1982.

5. F. Yao, "3-Space Partition and Its Applications," in *Proceedings of the 15th STOC,* pp. 258-263, A.C.M., Boston, Massachusets , 1983.

6. R.A. Finkel and J. L. Bentley, "Quad Trees: A Data structure for retreival on composite keys," *Acta Informatica,* vol. 4, pp. 1-9, 1974.

7. M. H. Overmars and J. van Leeuwen, "Dynamic Multi-Dimensinal Data Structures Based on Quad- and K-D Trees," *Acta Informatica,* vol. 17, pp. 267-285, 1982.

8. J. L. Bentley, "Multidimensional Binary Search Trees used for Associative Searching," *Comm. ACM,* vol. 18, pp. 509-517, 1975.

9. J. B. Rothnie and T. Lozano, "Attribute Based File Organization in Paged Memory Environment," *Comm. of ACM,* vol. 17, pp. 63-69, 2, 1974.

10. J. H. Liou and S. B. Yao, "Multidimensional Clustering for Data Base Organization," *Information Systems,* vol. 2, pp. 187-198, 1977.

11. T. Merrett, "Multidimensional Paging for Efficient Data Base Querying," in *Proc. Int. Conf. on Management of Data, ICMOD,* pp. 277-290, milano, 1978.

12. E. J. Otoo, "Low Level Structures in the Implementation of the Relational Algebra," SOCS 83.17, McGill University, School of Computer Science, August 1983.

13. H. Edelsbrunner and E. Welzl, "Halfplane Range Search in Linear Space and $O(n^{0.695})$ Query Time," F111, Technical University of Graz , 1982.

14. I.M. Yaglom and V.G. Boltyanskii, *Convex Figures,* Holt, Rinehart and Winston, 1961.

15. B. Mendelson, *Introduction to Topology,* Allyn & Bacon, Boston, 1962.

16. F. Yao, *Private Communication,* 1983.

17. D.P. Dobkin and H. Edelsbrunner, "Space Searching for Intersecting Objects," in *Proceedings of the 25th FOCS,* pp. 387-391, IEEE, Ocober, 1984.

18. D. Avis, "Non-Partitionable Point Sets," *Information Proc. Letters,* vol. 19, pp. 125-129, 1984.

19. D. Gale, "Neighborly and Cyclic Polytopes," *Proceedings of the Symp. in Pure Mathematics,* vol. 7, pp. 225-232, 1963.

20. E. N. Gilbert, "Gray Codes and Paths on the n-cube," *Bell System Technical Journal,* pp. 815-826, 1958.

ON COMPUTING AND UPDATING TRIANGULATIONS

Hossam A. ElGindy
Department of Computer Science,
University of Pennsylvania,
Philadelphia, PA 19104
U.S.A.

Godfried T. Toussaint
School of Computer Science
McGill University
Montreal, Quebec, H3A 2K6
Canada.

1. Introduction

Much attention has been given to triangulating sets of points and polygons (see [1] for a survey) but the problem of triangulating line segments has not been previously explored. It is well known that a polygon can always be triangulated and a simple proof of this can be found in [2]. Furthermore, efficient algorithms exist for carrying out this task [3,4]. Thus at first glance one may wonder why not just construct a simple polygon through the set of line segments and subsequently apply the algorithms of [3] or [4]. Unfortunately a set of line segments does not necessarily admit a simple circuit [5]. The reader can easily construct such an example with three parallel line segments. In the following section we provide optimal O(nlogn) algorithm for triangulating a set of n line segments. Optimality follows from the fact that Ω(nlogn) time is a lower bound for triangulating a set of points [6, p. 187] which is a set of line segments of zero length. Section 3 is devoted to presenting algorithms for inserting and deleting edges from triangulations.

2. An Algorithm for Triangulating a Set of Line Segments

Let $L=\{l_1,l_2, \ldots ,l_n\}$ be a set of non intersecting line segments specified by the cartesian coordinates of their end points, i.e., by the n pairs (p_i,q_i), i= 1, 2,..., n. A triangulation of L, denoted by T(L), is a triangulation of the point set $P \cup Q$, where $P=\{p_1,p_2, \ldots ,p_n\}$ and $Q=\{q_1,q_2, \ldots ,q_n\}$ with the constraint that L is a subset of $T(P \cup Q)$. Figure 1 illustrates a set of line segments (shown in bold lines) and a triangulation of their end points. That such a triangulation always exists follows from Fary's lemma [7]. The convex hull of L, denoted by CH(L), is the convex hull of the set of end points $P \cup Q$. The *next element subdivision* of the plane with respect to L is a partition of the plane into regions obtained by constructing lines from each end point of L in the -x direction until they "hit" a line segment of L (see Figure 2).

2.1. Preliminaries

First we highlight a result that will be used in several places in this paper. It concerns triangulating certain types of regions. Let $S = (s_1, s_2,..., s_n)$ be a simple polygon with n vertices. A polygon S is termed *edge visible* if there exists an edge of S, say $[s_i, s_{i+1}]$, such that for every point $x \in S$ there exists a point $y \in [s_i, s_{i+1}]$ such that the line segment [x, y] lies in S. Figure 3 illustrates an *edge visible* polygon. In [8] an algorithm is presented for internally triangulating *edge visible* polygons in linear time. In the following paragraph we show that this algorithm can be used to triangulate more general regions.

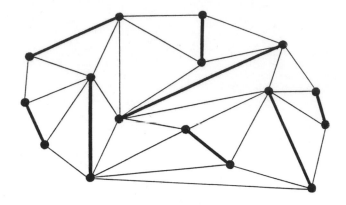

Figure 1:   A triangulation of a set of line segments.

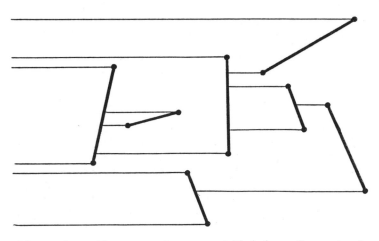

Figure 2:   The next-element-subdivision of a set of
             line segments.

We generalize the notion of a *simple polygon* to allow vertices and edges to be non-unique. In other words some vertices and edges of S can appear more than once on the boundary of the region. However, we retain the property that the interior of the generalized polygon (an open connected set) always lies to the right of the edges as they are traversed in clockwise manner. We call such a generalized polygon a *simple polygonal region* or *region* for short. An example of a simple polygonal region $R = (r_1, r_2,..., r_n)$ is illustrated in Figure 4.

Lemma 1 An *edge visible* region (an example of such region is shown in Figure 5) can be triangulated in linear time if the edge of visibility is known.

Proof The proof is identical to that for *edge visible* polygons given in [8].

Lemma 2 [9] The *next element subdivision* for the set of line segments L can be computed in O(nlogn) time.

2.2 Triangulating line segments

We now give a description of the algorithm for triangulating a set of line segments.

Algorithm TR-SEGMENTS

Input. A set $L = \{l_1, l_2,..., l_n\}$.
Output. A triangulation of L.
Begin
    Step 1: Compute the convex hull of L.
    Step 2: Augment the set L to obtain L' by adding the convex hull edges of L.
    Step 3: Compute the *next element subdivision* of L' inside CH(L).
    Step 4: Triangulate the *next element subdivision*.
    Step 5: Remove the *Steiner* points.
End

Theorem 1 Algorithm TR-SEGMENTS triangulates a set of n line segments in O(nlogn) time.

Proof To compute the convex hull of L in step 1 we just compute the convex hull of the set of end-points P ∪ Q, which can be performed in O(nlogn) time with a variety of algorithms [e.g, 10]. Step 2 can be done in linear time since the convex hull of L cannot have more than 2n edges. Therefore the cardinality of L' is O(n). From lemma 2, step 3 can be performed in O(nlogn) time. The results of these steps are illustrated in Figure 6. Note that the *next element subdivision* of L' in CH(L) is a decomposition of CH(L) into *monotone edge visible regions*. Each region is edge visible from a line segment of L or an edge of CH(L). It follows from lemma 1 that we can triangulate each region in time proportional to the number of its vertices. Note that new vertices, or Steiner points, have been added (2, 4, 5 and 7 for R_i in Figure 6), but the number of these new vertices is bounded by 2n. Therefore, step 4 can be completed in O(n) time.

We now must remove the *Steiner* points. The *Steiner* points occur on either edges of L or edges of CH(L). The algorithm in [9] gives these points in sorted order along each edge. Therefore for each such edge we traverse the list of *Steiner* points removing the edges emanating from them and creating an *edge visible simple polygonal region*. We now triangulate this edge visible region in linear time. The total time required for step 5 is thus O(n) and the entire algorithm runs in O(nlogn) time.

Q.E.D.

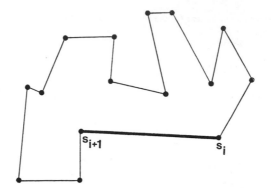

Figure 3:   An edge-visible polygon.

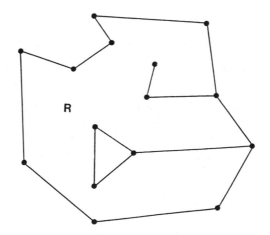

Figure 4:   A simple polygonal region.

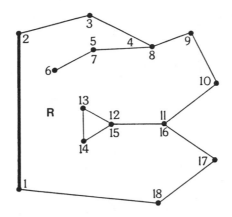

Figure 5:   An edge-visible simple polygonal region.

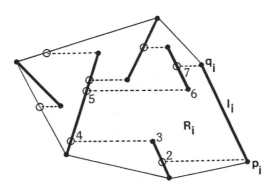

Figure 6:   The next-element subdivision of a set of line
             segments inside their convex hull.

3. Inserting and Deleting Edges from Triangulations

3.1. Introduction

In this section we address the problem of updating a triangulation of a point set efficiently. Given a finite set $P = \{p_1, p_2,..., p_n\}$ and a suitable representation of its triangulation T(P), we want to identify T'(P) so that an edge exists between a selected pair of points (edge insertion problem), or to identify T'(P) so that a selected pair of points are not connected (edge deletion problem).

3.2. Inserting an Open Edge in a Triangulation

In this problem we are given a pair of points p_i and p_j in P, and are interested in updating T(P) such that the new triangulation T'(P) contains an edge between p_i and p_j. (See Figure 7(a)). The following algorithm accomplishes the task.

Algorithm INSERT-OPEN-EDGE

Input A triangulation of a set of points $P=\{p_1, p_2,..., p_n\}$ and a pair of points p_i and p_j in P.
Output A triangulation T'(P) with an edge connecting p_i and p_j.

Begin

Step 1: Insert $[p_i, p_j]$ as an edge in T(P).

Step 2: Delete all edges in T(P) that intersect the line segment (p_i, p_j) (This operation results in a planar subdivision of the interior of the convex hull of the set P with bounded regions consisting of triangles and at most two *edge visible regions*).

Step 3: Triangulate the two edge-visible regions to obtain T'(P).

End

Theorem 2 Algorithm INSERT-OPEN-EDGE inserts an edge between two specified vertices of a triangulation of n points in O(n) time.

Proof Step 1 can be done in O(n) time. To determine all the edges of T(P) that intersect $[p_i, p_j]$ we traverse the triangles of T(P) that are intersected by $[p_i, p_j]$. Initializing this scan can be done in O(n) time by linearly scanning all the triangles connected to p_i. Thereafter, every time we encounter an edge of a triangle intersected by $[p_i, p_j]$ we need only two line-segment intersection tests to determine the next triangle. Therefore the traversal and deletion of all intersecting edges can be done in O(n) time. During this traversal we also create two lists of vertices of the edges that are deleted, those on the right of $[p_i, p_j]$ and those on the left. This operation yields a planar subdivision of the interior of the convex hull of the set P with bounded regions consisting of triangles and at most two regions R_1 and R_2 as illustrated in Figure 7(b). $R_1 \cup R_2$ consists of a union of triangles all of which intersect $[p_i, p_j]$. Therefore R_1 and R_2 are *edge visible regions* with respect to $[p_i, p_j]$. By lemma 1, they can be triangulated in O(n) time and thus the entire algorithm runs in O(n) time.

Q.E.D.

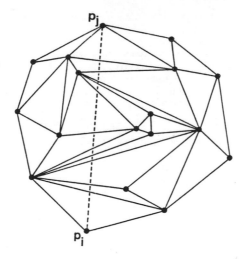

Figure 7(a)   Edges intersecting the added edge between p_i and p_j are deleted.

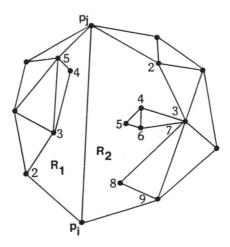

Figure 7(b)   The edge-visible simple polygonal regions R_1 and R_2 are to be triangulated.

3.3. Inserting a Closed Edge in a Triangulation

In this problem we are given a pair of points q_1 and q_2, and are interested in updating T(P) to obtain a new triangulation T'(P $\cup$ {q_1} $\cup$ {q_2}) with an edge connecting q_1 and q_2.

This problem can easily be reduced to the previous problem by merely determining in which triangles of T(P) the new points lie and joining each one to the three vertices of the containing triangle. This operation can be completed in O(n) time by performing two point-inclusion test for each triangle. (See Figure 8). It remains to insert an open edge between q_1 and q_2 which by theorem 2 can be done in O(n) time. We have therefore proved the following theorem:

Theorem 3 A closed edge can be inserted into a triangulation of n points in O(n) time.

3.4. Deleting a Closed Edge from a Triangulation

In this problem we want to delete an edge $[p_i, p_j]$ ε T(P) from the triangulation of P along with its end-points p_i and p_j, resulting in a triangulation T'(P') where P'= P - {p_i, p_j}.

Definition A polygon S is *weakly internally visible* if there exists a convex region in S, termed the kernel of S and denoted by K(S), such that for every point x ε S there exists a point y ε K(S) such that [x,y] lies in S.

Theorem 4 A closed edge can be deleted from a triangulation of n points in O(n) time.

Proof First delete the edge in question from T(P) by deleting p_i, p_j, the edge $[p_i, p_j]$, and all the edges emanating from p_i and p_j. This can be done in O(n) time by using a triangle traversal operation as in section 3.2 (See Figure 9(a)). Furthermore, this traversal yields a list of ordered vertices being pointed to by the deleted edges (See Figure 9(b)). This operation yields a subdivision of the interior of the convex hull of P whose bounded regions consist of triangles and a *weakly internally visible* region R = R_1 $\cup$ R_2. To see this, note that R is the union of a set of triangles of T(P). Each such triangle has either p_i or p_j as one of its vertices. Therefore $[p_i, p_j]$ is the kernel K(P). We must now triangulate R. First we decompose R into two *edge visible regions* R_1 and R_2 by adding an edge between r_i and r_j where r_i and r_j denote the first intersection points obtained by extending a line through p_i and p_j in both directions (see Figure 9(b)). Note that r_i and r_j may not be vertices of P and therefore may have to be deleted. Finding r_i and r_j can be done in O(n) time by intersecting the line segments of T(P) with the line passing through p_i and p_j. By lemma 1 we can now triangulate R_1 and R_2 in O(n) time. All that remains is to remove the *Steiner* points r_i and r_j from the triangulation obtained so far. Removing a vertex from a triangulation yields a *star shaped* polygon to be triangulated and this can be done in O(n) time [11]. Therefore the entire procedure runs in O(n) time.

Q.E.D.

3.5. Deleting an Open Edge from a Triangulation

Given a triangulation of a set of points P , denoted by T(P), and two points which are adjacent in the set P, the *open-edge deletion* problem requires updating T(P) such that the given two points are *not* adjacent in a new triangulation T'(P), if possible. There exists a triangulation T'(P) with the property that the two points p_i and p_j are not adjacent in T'(P) if and only if there exists a pair of points, p_k and p_l, such that the open line segment (p_k, p_l) intersects the open line segment (p_i, p_j). The pair of points p_k and p_l is called a *stabbing pair*.

In this section we present two solutions to the open-edge deletion problem. The first approach requires O(| P |^4) preprocessing time and O(| P |^2) storage. However it allows for each open-edge deletion to be performed in O(K) time, where K is the number of edge changes from T(P) to T'(P). The second method solves the open-edge deletion problem for a set of n points in O(nlogn) running time per update. However, it uses O(n) storage only.

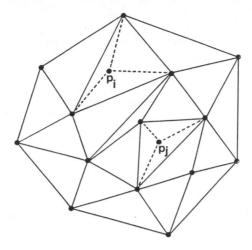

Figure 8:   If p_i or p_j lie outside the convex hull of P then
             such a point is joined to all visible vertices of
             the convex hull of P. Otherwise the point is joined
             to the vertices of the triangle in which it lies.

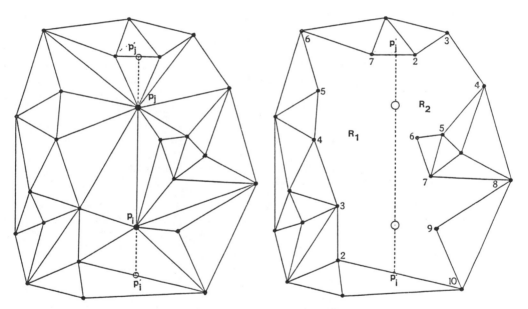

Figure 9(a)   Deleting the closed edge   Figure 9(b)   The region $R_1 \cup R_2$
               between p_i and p_j.   is a weakly-internally
                                                          visible region.

Algorithm I

As a preprocessing step, the algorithm locates a stabbing pair for each pair of points in the set, if one exists. A two-dimensional array is used as the data structure to store the stabbing pairs. A complete description of the preprocessing procedure is given in Figure 10.

Lemma 3 The preprocessing procedure, Preprocess(P), runs in $O(|P|^4)$ time and requires $O(|P|^2)$ storage.

Proof Trivial.

When a request for deleting an open-edge is received, the algorithm checks the array for the corresponding stabbing pair. If a stabbing pair exists, then the algorithm computes the new triangulation by inserting the open line segment joining the stabbing pair into T(P). Otherwise, no change of the current triangulation is needed.

We now describe a function Delete-Open-Edge(T(P), p_i, p_j) which given a triangulation of the set P and a pair of points of P returns a triangulation such that p_i and p_j are not adjacent, if possible.

Delete-Open-Edge (T(P), p_i , p_j)

> *global variable stabbing-pairs[n][n]*
> *local variable p_k , p_l;*
>
> (p_k , p_l) = stabbing pair stored in the entry stabbing-pairs[i][j];
>
> *if* (p_k , p_l) == (Nil,Nil)
>   *return* (Nil);
> *else*
>   *return* (Insert-Open-Edge(T(P), p_k , p_l));

Theorem 5 Let T(P) be a triangulation of a set P, and p_i and p_j be a pair of points of P. Given an array containing the stabbing pairs, the function Delete-Open-Edge returns a triangulation of P such that the two points p_i and p_j are not adjacent using O(K) time, where K is the number of changed edges.

Algorithm II

Given a set of points $P=\{p_1, p_2,..., p_n\}$ and a triangulation T(P), one would like to update the triangulation such that the two points p_i and p_j are not adjacent in a new triangulation T'(P).

There are two cases to consider. First, if p_i and p_j are not adjacent in T(P), then T'(P) equals T(P). On the other hand, if p_i and p_j are adjacent in T(P), then we must find a stabbing pair p_k and p_l and compute the triangulation T'(P) by inserting the open line segment (p_k,p_l) into T(P). It is easy to see that the insertion of the open line segment (p_k,p_l) will result in the deletion of the open line segment (p_i,p_j) (see Figure 11).

The algorithm for solving the open-edge insertion problem has been described in section 3.2. Now we concentrate on how to locate a stabbing pair, if one exists.

For convenience, we assume that the point p_i has a larger y-coordinate than the point p_j. Let P_r and P_l be the subsets of P such that points lie respectively to the right and left of the directed line connecting p_i to the

Preprocess(P)

global variable stabbing-pairs[n][n]
local variable p_i, p_j, p_k, p_l;

for each pair (p_i and p_j) of end points in P *do*

if a line segment (p_k and p_l) intersect the current pair
store p_k and p_l as the stabbing pair in the entry stabbing-pair[i][j];

Figure 10

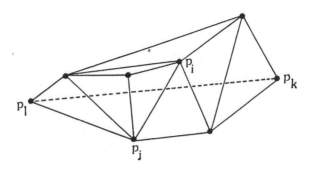

Figure 11

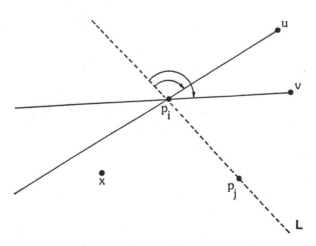

Figure 12

259

point p_j. Let "u" be a point in P_l, and let CH(P_r, u) denote the convex hull of the set of points of P_r to the left of the directed line connecting "u" to the point p_i.

Informally, the algorithm sorts the points of the set P in clockwise order around the point p_i and computes the two subsets P_r and P_l. Starting at the first point in P_l, P_{l_1}, we compute the intersection of the halfplane to the left of the directed line connecting p_j to the point P_{l_1} and the convex polygon CH(P_r, P_{l_1}). If the intersection is not empty, then we report a stabbing pair and stop. Otherwise, we proceed to examine the next point in P_l. If the intersection is found to be empty for all the points in P_l, then no stabbing pair exists.

The key properties for locating a stabbing pair efficiently are given in the following facts:

Fact 1 (see Figure 12). Let L be the directed line connecting p_i to the point p_j, and let "u" and "v" be two points in P_l such that the directed line connecting p_i to "u" makes a clockwise turn with the line L smaller than that of the directed line connecting p_i to the point "v". If a point "x" in P_r lies to the left of the directed line connecting "u" to p_i, then "x" also lies to the left of the directed line connecting "v" to p_i.

Fact 2 (see Figure 13). If the intersection of the halfplane to the left of the directed line connecting p_j to the point "u" and CH(P_r, u) is not empty, then the boundary of the intersection contains a point of P_r which forms a stabbing pair together with the point "u".

In Figure 14 we give a detailed description of the function for reporting a stabbing pair, if one exists. The functions Insert_Point(q, CH) and Insert_Halfplane(CH,u,v) efficiently solve the point insertion and the halfplane insertion problems respectively. (Refer to [12] for complete descriptions.)

Lemma 4 The function *Locate-Stabbing-Pair* runs in O(nlogn) time and uses O(n) storage.

Proof When the function examines a point of the subset P_r, it updates the convex hull of the points of P_r scanned so far. A process that can be completed in O(log| P_r|) time. It follows from **Fact 1** that no point need be deleted. Therefore, the total time required to process the points of P_r is O(| P_r| log| P_r|). For each point of the subset P_l the function checks the intersection of a halfplane with a convex polygon. A process which can be performed in O(log| P_r|). Therefore, the total time required to process the points of P_l is O(| P_l| log| P_r|). Finally, reporting a stabbing pair requires time proportional to the number of vertices in the convex hull at that time. Therefore, the running time of the function Locate-Stabbing-Pair is dominated by the time required for sorting points of the set P, i.e., O(nlogn). Thus the lemma follows.

Q.E.D.

Now we can present a function Delete-Open-Edge-2(T(P), p_i, p_j) which given a triangulation of the set P and a pair of points of P returns a triangulation of P such that p_i and p_j are not adjacent, if possible.

Delete-Open-Edge-2 (T(P), p_i, p_j)

    local variable p_k, p_l;

    (p_k, p_l) = Locate-Stabbing-Pair(P, p_i, p_l);
    if (p_k, p_l) == (Nil,Nil)
        return (Nil);
    else
        return (Insert-Open-Edge(T(P), p_k, p_l));

Theorem 6 Let T(P) be a triangulation of a set P, and p_i and p_j be a pair of points of P. The function Delete-Open-Edge-2 returns a triangulation of P such that the two points p_i and p_j are not adjacent using O(nlogn) time and O(n) storage.

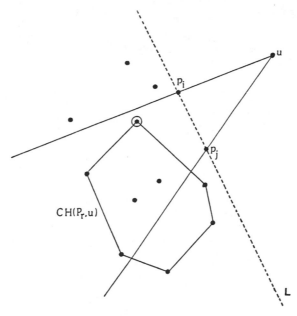

CH(P_r,u)

Figure 13

Locate-Stabbing-Pair (P, p_i, p_j)

    local variable CH;

    Sort the point of P in clockwise order around the point p_i.

    Divide P into the two subsets P_r and P_l. Points of P that lie on the line passing through p_i and p_j are ignored.

    b = 1; a = 0;
    CH = Nil;

    while b ≤ | P_l |

        while $P_{r_{a+1}}$ is left of the directed line passing through P_{l_b} and p_i
            CH = Insert_Point ($P_{r_{a+1}}$, CH);
            a = a + 1;

        if ((CH = Insert_Halfplane (CH, p_j, P_{l_b})) != nil)

            fetch a point of the set P_r from the current CH, say P_{r_k}.
            return(P_{l_b}, P_{r_k});
        else
            b = b + 1;
    return (Nil,Nil);

Figure 14

261

4. Concluding Remarks

We close with an interesting open problem. We have presented an $O(n\log n)$ time and $O(n)$ space algorithm for checking whether one specified open edge can be deleted from a triangulation. Given a set of specified open edges $E=\{e_1,e_2,\ldots,e_k\}$ it is an interesting open problem to design an efficient algorithm for checking whether E can be deleted from the triangulation.

One may wonder why not just check each open edge separately for deletion. However deleting an open edge from a triangulation of a n point set restricts another edge (a stabbing pair) to remain in the triangulation. It is easy to construct an example where each specified open edge can be deleted separately, but the set can not be deleted simultaneously.

REFERENCES

[1] G.T. Toussaint, "Pattern recognition and geometric complexity," *Proc. Fifth International Conference on Pattern Recognition*, Miami Beach,Dec. 1980, pp. 1324-1347.

[2] G. Meisters, "Polygons have ears," *AMerican Mathematical Monthly*, June-July 1975, pp. 648-651.

[3] M. Garey, D. Johnson, F. Preparata and R. Tarjan, "Triangulating a simple polygon," *Information Processing Letters*, Vol.7, 1978, pp.175-179.

[4] B. Chazelle, "A theorem on polygon cutting with applications," *23rd Annual IEEE Symposium on Foundations of Computer Science*, 1982, pp. 339-349.

[5] D. Avis, H. Imai, D. Rappaport and G.T. Toussaint, "Finding simple circuits in sets of line segments," manuscript in preparation.

[6] F. Preparata and M. Shamos, *Computational Geometry: An Introduction*. Springer-Verlag, 1985.

[7] I. Fary, "On straight-line representation of planar graphs," *Acta. Sci. Math. Szeged.*," Vol. 11, 1948, pp. 229-233.

[8] G.T. Toussaint and D. Avis, "On a convex hull algorithm for polygons and its application to triangulation problems," *Pattern Recognition*, Vol.15, 1982, pp.32-29.

[9] H. Edelsbrunner, M.H. Overmars, and R. Seidel, "Some methods of computational geometry applied to computer graphics," Technical Report F117, Technical University of Graz, June 1983.

[10] S.G. Akl and G.T. Toussaint, "A fast convex hull algorithm," *Information Processing Letters*, Vol. 7, No. 5, August 1978, pp. 219-222.

[11] A.A. Schoone and J. van-Leeuwan, "Triangulating a star-shaped polygon," Technical Report RUV-CS-80-3, University of Utrecht, April 1980.

[12] D. Avis, H.A. ElGindy and R. Seidel, "Simple on-line algorithms for planar convex polygons," to appear in *Computational Geometry*, ed. G.T. Toussaint, 1985.

Database Machines

VLSI TREES FOR FILE ORGANIZATION

Fabrizio Luccio

University of Pisa
Pisa, Italy

ABSTRACT

File organization is an important field for coming VLSI applications. VLSI tree architectures seem to be most suitable to store and process files, in particular for dictionary and relational operations.

Due to the vastity of this subject, our discussion is confined to a new VLSI structure called TOT (Tree of Trees). TOT has the form of a binary tree, whose nodes are in turn binary trees. It stores tables, treated as relations in the data base sense, and performs efficiently various input, output and dictionary operations, and a complete set of relational operations. General considerations on file processing in VLSI trees stem from this discussion.

INTRODUCTION

The current research stream in VLSI algorithms and architectures leads to the development of functional units, for the execution of operations which are frequently required and clearly consolidated.

The high cost of chip development causes lack of flexibility in the definition of tasks to be performed by a VLSI modulus, where late changes are very difficult, if not totally impracticable. Hence, the major impact of VLSI technology in the foreseeable future should be in the realization of blocks implementing basic functions, with a limited role for more flexible structures.

Several books and surveys have been published on this subject. Let us merely recall the original effort by Mead and Conway,[15] for chip organization; Muroga[17] for a discussion of hardware problems; Rice,[19] and an issue of IEEE Computer[8] for system implications; and the record of the latest workshop on VLSI,[3] for the most recent algorithmic developments. Reconfigurable structures have their major justification in a work of Valiant,[23] while different examples have been given in [13, 20].

File processing may be an important field for VLSI applications, since increasingly large chips will be adequate, in the near future, to support large sets of data, like the ones required in data base systems. VLSI tree structures have been studied for searching, sorting and various dictionary operations [2, 12, 18]. VLSI systolic arrays [9, 10]; and trees [4, 21], have been considered to process relations. Powerful meshes of trees [11] have proven to be suitable for sorting [5, 11] and various data base probelms [6]. On a more practical side, several proposals for data base applications have been presented in journals and conferences, as for example [1, 7, 16].

In this paper we will focus our attention on the use of VLSI tree architectures to store and process files, defined as sets of multi-valued records, or bidimensional tables. In addition to storage, input/output and update operations, tabular operations will also be treated in the framework of the relational theory.

Due to the vastity of this subject, our investigation will be confined to a specific stucture. In fact, we propose here a novel chip architecture, called TOT, and discuss file storage and file operations on TOT's. General considerations on file organization for VLSI trees will stem from this discussion.

THE TOT CHIP

The **Tree of Trees** (**TOT**) structure has the form of a binary tree, whose nodes are in turn binary trees (fig.1). The main frame is a **Broadcast Tree** (**BT**), whose nodes act as dispatchers of commands and data. Each node of the BT is also the root of a local **Terminal Tree** (**TT**), whose nodes hold and process data items. Therefore, BT nodes are connected to the left and right son in the BT, and to the left and right son in the TT; while TT nodes (except for the root) are simply connected to their sons.

As in all the VLSI tree structures, all TOT nodes are elementary processors connected via bidirectional busses. All the processors are

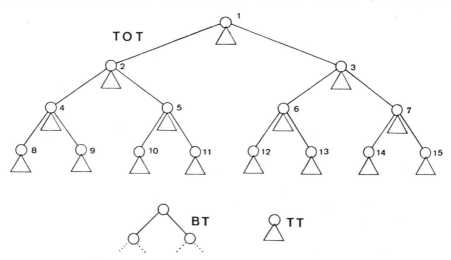

Fig. 1. The Tree of Trees (TOT) structure, consisting of a Broadcast Tree (BT) with a Terminal Tree (TT) at each node.

synchronously driven by the same clock, which is distributed through the structure.

In principle a TOT can be realized on a single rectangular chip, as shown in fig.2. The BT has the well known "H layout",[15] connecting square regions within a binary tree. Each region contains a TT, whose "double-Y layout" is feasible because of the fixed, and small, number of nodes in a TT (see below). This is a tightly packed structure with rectangular nodes of equal area, where, in the upper and lower sub-structure, each node has double height and half width of its father. Several other layouts are possible for TOT's, as described in a detailed report[14]

Direct access to subtrees can be added to the H layout,[13] thereby using a single chip to host different smaller TOT's. In fig.2 we show the two accesses to half-chip TOT's, whose use will be explained later. If instead a TOT does not fit in a single chip, it can be displayed in different chips as indicated in[12].

Let BT contain N nodes, and each TT contain K nodes. For increasingly large chips, we consider N to be the increasing parameter, while K is fixed. The area of a TT is clearly linear in K. The arcs of BT will transmit BT node numberings, hence, their whidth must be $O(\log N)$. As a consequence, the area of TT is of $O(M)$, where $M= \max (K, \log^2 N)$. Due to the properties of the H layout, the total TOT area is then of $O(N \cdot M)$.

As already indicated, the files to be stored in TOT's consist of n elements with k values each. A file will be treated as a **table** (or relation, in the data base sense) of n k **elements**, arranged in n **rows** and k **columns**. In each row, (or k-tuple) one of the elements is called the **label**.

Unless differently stated, we assume that a table is stored in a

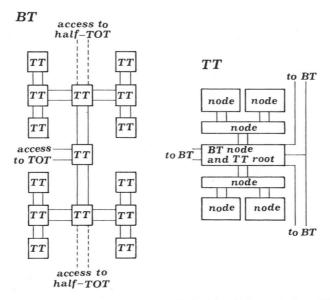

Fig. 2. H layout for BT, and double-Y layout for TT.

TOT, with n≤N and k≤K. Each row is stored in a TT, with each element in a
node and the label in the root. (A more flexible use of TOT's will be
indicated in section 5). Elements are coded by pairs of integers, namely:
element.type (name of the "domain" from which the element is taken, or a
prefixed constant for the label) and **element.value**. The specification of
the element.type relieves from the burden of storing elements in fixed
nodes of the TT (note, however, that the label will always be stored in
the root; and we espect elements of different rows and same types, to be
stored in the same positions in the TT's).

The stucture of TOT nodes [14] will not be detailedly discussed here.
We simply mention some operations that both BT nodes (TT roots) and TT
nodes must perform, with reference to the register structure of fig.3.
Namely:

1. store a permanent element (in a storage register SR);
2. store a transient element (in a buffer register BR);
3. compare permanent and transient elements;
4. activate microprograms to perform a set of elementary operations, and
   establish connections with father, left and right sons in the TT (via
   busses FB, LSB and RSB, respectively).

In addition, BT nodes must have the following abilities:

5. keep information on available TT's (in a counter register CR);
6. activate specific BT operations, and establish connections with left
   and right sons in the BT (via busses LSBB and RSBB, respectively).

Table elements are fed into the TOT sequentially, one per clock
time. They are dispatched through the BT, in parallel, to the TT's, which
are reached at consecutive times, depending on their distance from the
root of TOT. Then, the elements traverse the TT's in parallel, and
storage and comparisons are performed in the TT nodes.

In addition to **store** tables, TOT's are apt to perform a set of basic
operations to **input, output** and **update** tables; and a complete set of
relational operations on tables; as it will be explained in the next
sections. Each operation is activated by an instruction, whose code is
issued by an external Control Processor, and broadcast to all nodes via

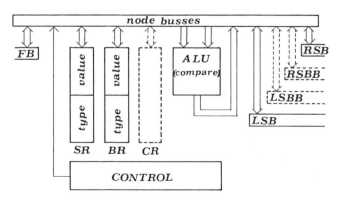

Fig. 3. Register structure of a TOT node (dashed parts belong to BT
nodes).

BT. For brevity all operations will be described in their main features, making use of specific BT and TT operations discussed elsewhere [14]. On the basis of the structures of fig.2 and 3, however, BT and TT operations will be easily understandable, and their stated execution times clearly attainable.

BASIC OPERATIONS OF TOT's

To describe the operations that can be performed in a TOT, let us consider two numberings of the TT's.

Breadth numbering is shown in fig.1. The binary representation b of each TT number can be obviously interpreted as the description of the path through TOT, from the root of TOT to the root of the TT. For example, for the TT 13 we have: b = 1101. Extract from b the substring s to the right of the most significant 1 of b, that is: s = 101. Substitute, in s, "right" for 1, and "left" for 0, thus obtaining the sequence of moves: right-left-right, leading from the root of TOT to the root of TT 13.

Depth numbering is instead obtained assigning to the TT's the consecutive integers representing the position of their roots, in a preorder binary visit of BT. Also from these numbers, the paths reaching the TT's from the root of TOT are easily identified.

The basic operations of TOT follow a breadth, or a depth strategy, if the n rows of T are loaded in the first n TT's according to the breadth, or to the depth, numbering. To indicate occupancy, the TT roots contain an availability flag **AF**, which is set to 1 (or to 0) to indicate that the TT is available (or occupied).

The first operation will be now described in some detail, to give a close idea of the TOT procedures.

B.LOAD a table T in TOT (Breadth strategy).

Input. A table T (n rows of k elements), presented at the TOT root row by row, and element by element, with the label as the last element of each row.
Result. The rows of T are stored in the TT's 1 to n (breadth numbering).
Presetting by instruction code. In each TT node, the type field of register SR (fig.3) is preset to the element.type of the element to be stored in that node. The TT roots have the AF flag preset to 1 (available), and the CR Register preset to 1 (i.e., to the breadth number of the first available TT).
Procedure. The rows r_1, ..., r_n of T are broadcast in parallel to all the TT's. Each row r_i is loaded into all the TT's with AF = 1, by allocating the element values in the TT nodes with matching element types. Each TT root stores in CR the number of the first available TT. Based on this information, a label flag **LF** is set to 1 in the label of the only copy of r_i that is sent to the first available TT. In such a TT, AF is turned to 0 upon arrival of the signal LF = 1, that is, r_i is actually retained in that TT only. Meanwhile, the contents of all CR registers are increased by one, to route the next row of T to the next available TT.

Execution time. $O(k \cdot n)$ to input T in the TOT, plus $O(\log n + \log k)$ to traverse the TOT. Overall order is then $k \cdot n$.

Other basic operations are the following.

LOAD a table T in TOT (Depth Strategy).
Input. A table T, as in B.LOAD.
Result. The rows of T are stored in the TT's 1 to n (depth numbering).
Presetting by instruction code. TT nodes are preset as in B.LOAD, except for the TT roots that have AF=1 and CR=2^i, where (recalling that BT has' N=2^h-1 nodes) i=h-1 at the root of BT; i=h-2 at the sons of the root; ...; i=0 at the leaves of BT. (I.e., each TT root specifies in CR the number of TT's in its left BT subtree, plus one).
Procedure. As in B.LOAD, except that each TT root stores, in CR, the number of available TT's in its left BT subtree, plus one if the TT itself is available. Based on this information, the first available TT (in preorder) is the one with CR=1 (and AF=1), that is found starting from the root of TOT, and moving left where CR 1, right where CR=0. For each row r_i, only the copy that is sent to the first available TT is actually loaded, and the CR contents in the path followed by such a copy are properly updated (decrease by one, if a left move is performed in that position).
Execution time. $O(k \cdot n)$ to input T, plus $O(\log N + \log k)$ to traverse the TOT. We may expect that $n \cdot k \gg \log N$, that is, overall order is $k \cdot n$ in most cases.

A comment is needed at this point, that will affect some of the next operations. CR contents may be used, as in LOAD, to indicate the first available TT in preorder, even if TOT is randomly filled. In fact, let CR contain the number of available TT's in the left BT subtree of the present TT, plus one if the TT itself is available. Such CR contents will be called **depth indicators** (fig.4). The first available TT, in preorder, is reached with the same rule given in LOAD. When such a TT is occupied, depth indicators are accordingly modified, as shown in fig 4 The use of depth indicators to reach the first available TT will still be called "depth strategy".

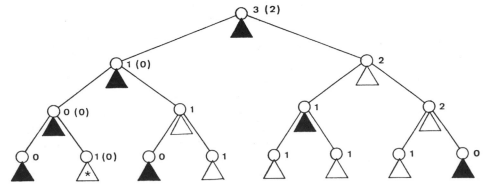

Fig. 4. Depth indicators to keep track of available TT's (shown in white). "*" marks the first available TT, in preorder: if such a TT is occupied, indicators are modified as shown in parentheses.

B.EXTRACT a table T from TOT (Breadth strategy).

Comment. The instruction may specify a subset S of h$\leq$k element types, to indicate, in each row of T, a subset R of elements whose types belong to S.

Input. T is already stored in TOT, whith breadth strategy.

Result. The rows of T, or the subsets R of the rows, are sequentially extracted from the root of TOT.

Procedure. According to the request conveyed by the instruction code, the rows of T, or the subsets R of the rows, are sent up the TOT through the TT roots, following their breadth numbering. Departure delays are computed by the TT roots, as to form a continuous stream of elements travelling in TOT.

Execution time. $O(\log k + \log n) + O(k \cdot n)$ to extract the elements of T in sequence. Overall order is $k \cdot n$. (If a subset S is specified, overall order is $h \cdot n$: see comment above).

EXTRACT a table T from TOT (Depth strategy).

Comment. As in B.EXTRACT.

Input. T is already stored in TOT, in **any** ordering. TT roots contain depth indicators.

Result. As in B.EXTRACT.

Procedure. As in B.EXTRACT, with departure delays computed by the TT roots on the basis of depth indicators.

Execution time. $O(\log k + \log N) + O(k \cdot n)$. Then, $O(k \cdot n)$ in most cases. ($O(h \cdot n)$, if a subset S is specified: see comment above).

SEARCH a (sub)row r (Any strategy).

Input. A table T of $n \times k$ elements already stored in TOT. A (sub)row r of h elements, h$\leq$ k, presented at the root of TOT.

Result. A signal is received at the root of TOT, if there exists at least one row of T whose elements are equal to the ones of r of corresponding types.

Procedure. r is broadcast in parallel through the TOT, and signals of matching (if any) are sent back to the root of TOT. These signals are OR-ed, if they meet at some BT node.

Execution time. $O(h) + O(\log N + \log k)$ to insert, broadcast and compare r; plus $O(\log N + \log k)$ to transmit the matching signal. (If no such a signal is received after a proper delay, SEARCH yields a negative result). Overall order is $h + \log N + \log k$.

Several variatious of SEARCH are obviously possible. In particular, the operation could specify the position(s) of the row(s) of T matching with r.

INSERT a row r (Depth strategy).

Input. A table T already stored in TOT in any ordering, with depth indicators. A row r, with element types matching the ones of T, presented at the root of TOT.

Result. r is inserted into T, in the first available TT (in preorder). If r is already present in T, the previous instance of r is deleted to avoid duplications.

Procedure. r is broadcast to all TT's. As in B.LOAD, the copy of r that reaches the first available TT is actually allocated. All other copies are compared with resident rows: if a matching is found, the TT is made available by turning AF to 1 in its root, and a signal is sent up the TOT to update depth indicators.

Execution time. $O(k) + O(\log N + \log k)$ to insert, broadcast and allocate r; plus $O(\log N + \log k)$ to update depth indicators (in case of matching). Overall order is $k + \log N$.

INSERT can be obviously modified in several ways. We could also define operations to **extract rows** from TOT; however, such operations will instead be regarded as subcases of the relational operation of SELECTION, to be treated in the next section.

INSERT can be iteratively used to load a **retundant table**, that is a table whose rows appear in multiple copies; and to **pack** the table, by retaining only one copy for each row. As we shall see, this new operation is an important module in relational operations.

PACKLOAD a redundant table T in TOT (Depth strategy).

Input. A redundant table T, presented at the root of TOT as in B.LOAD. TT roots have preset depth indicators.

Result. The rows of T are stored in TOT, without duplications. Final occupancy of TOT cannot be predicted.

Procedure. Iterative pipelined applications of INSERT, for each row of T.

Comment. If the TOT is initially nonempty, the rows of T are stored (without duplications) in the available TT's. Each row originally contained in TOT, which is equal to a row of T, is cancelled from TOT.

Execution time. Each of the n applications of INSERT requires time of $O(k + \log N)$, however, a global order $k \cdot n + \log N$ is attained by pipelining. Overall order is $k \cdot n$ (assuming $k \cdot n \gg \log N$).

The last basic operation presented here is aimed to the cancelation of rows from a table. This is also an important module in relational operations.

DELETE rows (Depth strategy).

Input. A table T already stored in TOT, with depth indicators. A sub(row) r of h elements, $h \leqslant k$ presented at the root of TOT.

Result. All rows of T whose elements are equal to the ones of r of corresponding types, are deleted from T.

Procedure. r is broadcast through TOT. Each TT where matching is found is made available by turning AF to 1 in its root, and a signal is sent up the TOT to update depth indicators.

Execution time. $O(h) + O(\log N + \log k)$ to insert, broadcast and compare r; plus $O(\log N + \log k)$ to update depth indicators (this operation is made in parallel, for all deletd rows). Overall order is $h + \log N + \log k$.

RELATIONAL OPERATIONS

A table T stored in TOT can be treated as a data base **relation** of n **k–tuples**. This interpretation is realistic if n k, hence N K as in our basic assumptions on TOT. We shall now examine how a TOT can handle a standard set of relational operations, namely UNION, PRODUCT, DIFFERENCE,

PROJECTION and SELECTION. As known [22], this set permits the implementation of all features of relational agebra, including the realization of JOIN. Due to the practical importance of this latter operation, however, an efficient dedicated algorithm can be devised, as discussed later.

The algorithms indicated below derive from the ones presented in [4], for a simple tree structure. Some interesting possibilities of TOT, however, are granted by the separate accesses to the two half-TOT trees (called **HALFTOT1** and **HALFTOT2**), indicated in fig.2. Two tables can be concurrently and independently loaded and extracted in HALFTOT1 and HALFTOT2. Operations involving two tables, or requiring to move a table from one TOT to another, can be handled in HALFTOT1, HALFTOT2 inside the same chip (i.e., off the external processors), where the original TOT root is used as a communication channel. This happens in the following relational operations, where two relations (tables) X, Y, respectively containing n_x and n_y tuples of k_x and k_y elements, are loaded in HALFTOT1, HALFTOT2. (We assume that $n_x, n_y \leq (N-1)/2$, that is, X and Y fit in the half-TOT trees. If not, each relation is loaded in a separate TOT, with obvious minor changes in the algorithms).

UNION of two relations X, Y (Denoted by X∪Y).

Comment. This operation is defined only if the tuples of X and Y have the same element types ($k_x = k_y$). X∪Y is a new relation consisting of the tuples belonging to X or Y or both.

Input. Relations X and Y, respectively stored in HALFTOT1 and HALFTOT2 with depth indicators.

Result. X∪Y is stored in HALFTOT1.

Procedure. EXTRACT Y from HALFTOT2 and PACKLOAD Y into HALFTOT1.

Execution Time. $O(k_y n_y)$, needed for the pipelined execution of EXTRACT and PACKLOAD.

PRODUCT of two relations X, Y (Denoted by X×Y).

Comment. X×Y is a new relation composed of $(k_x + k_y)$-tuples, which are obtained as all the conbinations of the tuples of X with the tuples of Y.

Input. Relations X and Y respectively stored in HALFTOT1 and HALFTOT2 with depth indicators. A new TOT, called TOT3, with $N_3 \geq n_x n_y$ TT's of $K_3 \geq k_x + k_y$ nodes.

Result. X×Y is stored in TOT3.

Presetting by instruction code. Each TT node of TOT3 stores the element types of the tuples of X and Y, to receive a tuple of X×Y.

Procedure.

1. EXTRACT X from HALFTOT1, and broadcast its n_x tuples through TOT3. For each tuple, the n_y copies sent to the first n_y available TT's have sequential copy numbers $1, \ldots, n_y$ inserted in their labels, and LF set to 1, by the BT nodes. Each one of these copies is allocated in a TT, with copy number stored in the root, and its elements stored in the nodes of corresponding element types.

2. EXTRACT Y from HALFTOT2, and broadcast its n_y tuples through TOT3. Allocate each tuple in all the TT's with copy number currently equal to 1, to merge that tuple with all the resident tuples of X; and decrease by one the copy number in **all** TT's when they are reached by a

new tuple. (With this operation exactly n_y TT's have copy number equal to 1, when they have to store the same tuple of Y).

Execution time. $O(\log N + \log k_x) + O(k_x n_x) + O(\log N_3 + \log K_3)$, to EXTRACT X from HALFTOT1 and store it in TOT3. Plus $O(\log N + \log k_y) + O(k_y n_y) + O(\log N_3 + \log K_3)$, to EXTRACT and store Y. (To diminish this time, the two phases could be pipelined). Overall time is of order $k_x n_x + k_y n_y$.

DIFFERENCE of two relations **X and Y** (Denoted by X–Y),

Comment. This operation is defined only if X and Y have the same element types ($k_x = k_y$). X–Y is a new relation consisting of the tuples of X not present in Y.

Input. Relations X and Y respectively stored in HALFTOT1 and HALFTOT2 with depth indicators.

Result. X–Y is stored in HALFTOT1.

Procedure. EXTRACT Y from HALFTOT2, and DELETE the rows of Y in HALFTOT1.

Execution time. Overall time is $O(k_y n_y)$, for the pipelined execution of EXTRACT, and DELETE n_y times.

PROJECTION of relation X onto domain set D (Denoted by $PROJ_D(X)$).

Comment. $D = (d_1,...,d_h)$ is a subset of the set of domains (i.e. element types) of X ($h \leqslant k_x$).
$PROJ_D(X)$ is a new non redundant relation of h-tuples, which are obtained as restrictions of the k_x-tuples of X to the domains of D (if the same h-tuples derives from different tuples of X, only one copy is retained in $PROJ_D(X)$).

Input. Relation X stored in HALFTOT1 with depth indicators.

Result. $PROJ_D(X)$ is stored in HALFTOT2.

Procedure. EXTRACT, from HALFTOT1, the subsets of the rows of X whose elements have types belonging D; PACKLOAD the output stream of HALFTOT1 into HALFTOT2.

Execution time. Overall time is $O(h \cdot n_x)$, for the pipelined execution of EXTRACT and PACKLOAD.

The operation of SELECTION, to be lastly examined, is aimed to extract from a relation all the tuples satisfying specific queries. Each query Q will be expressed in Boolean form, for example (and without loss of generality), as a disjunction of conjuntive clauses. That is:

$$Q = C_1 \vee C_2 \vee ... \vee C_q \quad , \quad q \geqslant 1,$$

where:

$$C_i = b_1 \wedge b_2 \wedge ... \wedge b_{p_i} \quad , \quad 1 \leqslant i \leqslant q, \; p_i \geqslant 1.$$

Each b_j is a Boolean variable of the form::

$$b_j: r_1 \; op \; r_2,$$

where r_1 is an element type; r_2 is an element type or a costant value; op is an arithmetic comparison operator.

The interpretation of Q is easily explained with an example. Consider the query:

$$Q1 = b_1 \vee (b_2 \wedge b_3)$$
$$= (d_1 > c_1) \vee ((d_1 \geqslant d_2) \wedge (d_3 = c_2)),$$

274

where d_1, d_2, d_3 are element types (domains), and c_1, c_2 are costants. With reference to any tuple, let v_1, v_2, v_3 be the values of the elements in the domains d_1, d_2, d_3, respectively. Q1 requires that either $v_1 > c_1$; or $v_1 \geqslant v_2$ and $v_3 = c_2$.

SELECTION of a relation X under a query Q (Denoted by $SEL_Q(X)$).

Comment. The Boolean variables of Q are built with element types (domains) of X. $SEL_Q(X)$ is defined as the set of tuples of X satisfying Q.

Input. Relation X stored in TOT, with any strategy.

Result. The tuples of $SEL_Q(X)$ are sequentially extracted from the root of TOT.

Procedure.

1. Broadcast Q through TOT. The consecutive Boolean variables are properly delayed, to allow each TT to decide if a variable is locally true, before the next variable is received. (Computing a variable $b:r_1$ op r_2 requires $O(\log k_x)$ time in a TT, to bring the elements of domains r_1, r_2 to the root of TT, for comparison). Q is therefore evaluated, variable by variable, and clause by clause, in the root of all TT's, where a Selection Flag **SF** is set to 1 if Q is verified.

2. EXTRACT from TOT all the tuples with SF = 1, using an obvious variation of the operation EXTRACT.

Execution time. Let v be the number of Boolean variables in Q, and n_{SEL} be the number of tuples in $SEL_Q(X)$ (usually, $n_{SEL} \ll n_x$). Phase 1 requires $O(v \cdot \log k_x) + O(\log N)$ time, for communicating and computing Q in all TT's; phase 2 requires $O(k_x n_{SEL}) + O(\log N)$ for EXTRACT. Overall time is of order $v \cdot \log k_x + \log N + k_x n_{SEL}$. (If X is stored with breadth strategy, $\log n_x$ must be considered instead of $\log N$).

CONCLUDING REMARKS

Most of the operations discussed in this paper have their bottleneck in the necessity of communicating data through the root of a tree. This implies sequential input and output of words of a costant number of bits. In fact, using external connections of fixed bandwidth seems to be a realistic assumption for present and future chip technology, where the number of external pins cannot be increased as a function of the number of internal components.

As noted in the introduction, the use of VLSI trees for file operations has been proposed frequently. The TOT structure introduced here has features in common with almost all other tree structure. It allows dictionary operations as in [2, 18], and relational operations as in [4, 21], yet attaining some better efficiency because of its peculiar organization. In particular, a unique feature of TOT is the realization of some relational operations inside the chip, without using external resources. A summary of the characteristics of TOT operations is reported in Table I. Note that the slowest operations are performed in time $O(n \cdot k)$, which is the size of data.

Alike most tree structures, TOT's are somehow inflexible in storage capabilities. As already noted, tables with $n \ll N$ rows can be stored in

Table I. Summary of operations for a TOT with N TT's, storing tables of
size n x k. h ≤ k is the cardinality of a subset of element
types. For the execution time of SELECTION, see such operation.
The "In-chip" column contains "*", if the corresponding
operation is totally executed inside the chip.

Type	Operation	In-chip	Execution time (order of magnitude)
Input/output	B.LOAD/LOAD		$k \cdot n$
	B.EXTRACT/EXTRACT		$k \cdot n$
	PACKLOAD		$k \cdot n$
Dictionary	SEARCH		$h + \log N + \log k$
	INSERT		$k + \log N$
	DELETE		$h + \log N + \log k$
Relational	UNION	*	$k \cdot n$
	PRODUCT		$k \cdot n$
	DIFFERENCE	*	$k \cdot n$
	PROJECTION	*	$k \cdot n$
	SELECTION		$v \cdot \log k + \log N + k \cdot n_{SEL}$

portions of TOT's independently accessible, and tables with $n > N$ can be
stored in groups of connected TOT's. However, a serious limitation
derives from the fixed size K of the TT's, that implies storage of tables
with $k \leq K$ columns. Moreover, k must be close to K for tight filling. A
possible method to overcome this point is to build TOT's with a small
value of K (e.g., K = 7 as in fig.2), and to use subtrees of TOT
containing $H \geqslant 1$ TT's to store table rows, where H can be treated as a
parameter. This would obviously complicate all operations, with great
benefit in storage efficiency.

An interesting application of TOT in relational algebra, is the
efficient realization of the operation of JOIN. As well known, two
relations X,Y can be "joined" through a sequence of PRODUCT, SELECTION
and PROJECTION [22]. However, this procedure generates very large
intermediate data, due to the execution of PRODUCT, while the result of
JOIN is likely to be much smaller. With some ingenuity, an efficient ad
hoc algorithm for JOIN can be added to the repertoire of standard
relational operations [14].

Some important points have been left open in this paper. In
particular, the realizability of significantly large TOT's with present
technology, and system implications in the use of TOT's, could be
discussed along the lines indicated for VLSI trees in the referenced
literature. A challanging problem would be the extension of the TOT
structure to mesh of trees arrangements.

REFERENCES

1. B.W. Arden and R. Ginosar, A single-relation module for a data base

machine, Proc. 8-th Annual Symp. on Computer Architecture, May 1981.

2. J.L. Bentley and H.T. Kung, A tree machine for searching problems, Proc. IEEE Int. Conference on Parallel Processing, Bellaire, MI, August 1979.

3. P. Bertolazzi and F. Luccio (Eds.), "VLSI Algorithms and Architectures", North Holland, Amsterdam 1985.

4. M. Bonuccelli, E. Lodi, F. Luccio, P. Maestrini and L. Pagli, A VLSI tree machine for relational data bases, Proc. 10-th Annual Symp. on Computer Architecture, Stokholm, June 1983, 67-73.

5. M.A. Bonuccelli, E. Lodi and L. Pagli. External Sorting in VLSI, IEEE Trans Comp. 33, 10 (1984) 931-934.

6. M.A. Bonuccelli, E. Lodi, F. Luccio, P. Maestrini and L. Pagli, VLSI Mesh of Trees for Data Base Processing, Proc. CAAP 83, Springer Verlag 1983.

7. Y. Dohi, A. Suzuki and N. Matcui, Hardware Sorter and its Application to Data Base Machine, Proc. 9-th Annual Symp. on Computer Architecture, April 1982.

8. IEEE Computer. Issue on "Higly Parallel Computing", January 1982.

9. H.T. Kung and P.L. Lehman, Systolic (VLSI) Arrays for Relational Data Base Operations, Proc. ACM-SIGMOD 1980 Int. Conf. on Management of Data, ACM, May 1980, 105-116.

10. P.L. Lehman, A Systolic (VLSI) Array for Processing Simple Relational Queries, in "VLSI Systems and Computations" H.T. Kung et al. eds: Computer Science Press, Rockville, MD, 1981, 285-295.

11. F.T. Leighton, New Lower Bound Techniques for VLSI, Proc. 22nd Ann. Symp. Foundations of Computer Science, 1981, 1-12.

12. C.E. Leiserson, Systolic Priority Queues, Proc. Caltech Conf. on VLSI, 1979, 199-214.

13. F. Luccio and L. Pagli, A Versatile Interconnection Pattern Laid on O(n) Area, Proc. 10-th IFIP Conf. on System Modeling and Optimization, Springer Verlag 1982, 596-604.

14. F. Luccio, Layout and Operations for TOT's, Tech. Rep. Dipartimento di Informatica, Pisa, Italy, 1984.

15. C. Mead and L. Conway, "Introduction to VLSI Systems", Addison Wesley, Reading 1980.

16. M.J. Menon and D.K. Hsiao, Design and Analysis of a Relational Join Operation for VLSI, Proc. Conf. on Very Large Data Bases, August 1981.

17. S. Muroga, "LSI Systems Design", Wiley, New York 1982.

18. T.A. Ottman, A.L. Rosenberg and L.J. Stockmeyer, A Dictionary Machine (for VLSI), IEEE Trans. Comp. 31, 9 (1982) 892-897.

19. R. Rice, "VLSI: The Coming Revolution in Applications and Design", IEEE, New York 1980.

20. L. Snyder, Introduction to the Configurable Highly Parallel Computer, IEEE Computer 15,1 (1982) 47-56.

21. S.W. Song, A Highly Concurrent Tree Machine for Data Base Applications, Proc. IEEE 1980 Int. Conf. on Parallel Processing, 1980, 259-260.

22. J.D. Ullman, "Principles of Data Base Systems", Pitman, London, 1980.

23. L.G. Valiant, Universality Considerations on VLSI Circuits, IEEE Trans. Comp., 30,2 (1981) 135-140.

THE INVERTED FILE TREE MACHINE:

EFFICIENT MULTI-KEY RETRIEVAL FOR VLSI

Hans-Peter Kriegel*, Rita Mannss†
and Mark Overmars‡

*Praktische Informatik
 Universität Bremen
 D-2800 Bremen 33, West Germany

†Institut für Informatik
 ETH Zürich
 CH-8092 Zürich, Switzerland

‡Vakgroep Informatica
 Universiteit Utrecht
 NL-3508 TA Utrecht, The Netherlands

ABSTRACT

    In this paper we review index structures suggested for implementation
in silicon. These structures do not sufficiently support multi-key retrieval.
We present two variants of the inverted file tree machine as efficient VLSI
solutions to this problem improving query time of previous machines at least
by the factor number of attributes specified in the query. Concerning non-
redundant and redundant insertions and deletions, we suggest two phase methods to handle these update operations while continuously pipelining following operations. Both variants are compared with respect to their performance and space requirement. Furthermore, we describe a large class of
multi-key queries all of which are handled on our machine with the speed
up factor number of specified attributes.

1. INTRODUCTION

    The problem of retrieving all the records satisfying a query involving
a multiplicity of attributes, known as multi-key retrieval or associative
retrieval, is a major concern in physical database organization. Physical
database organization deals with the assignment of physical data records
into pages of secondary storage and the methods of retrieving the associated pages and records in response to a given query. The most commonly
used method to support retrieval is an index or directory which relates
combinations of attribute values to the physical records which have these
value combinations. Various software methods for the construction of such
indexes for different types of queries are at the disposal of the database
designer; for a survey see [5, 9, 6].

    Search for hardware-oriented solutions to database problems, in the con-
text of design and implementation of the so-called database machines, has

been around for more than ten years. With recent advances in Very Large
Scale Integration (VLSI) technology, the focus has shifted to hardware solutions directly implementable in silicon.

In this paper we will review index structures suggested for implementation in silicon. None of these structures efficiently supports multi-key
retrieval. We present the inverted file tree machine as an efficient VLSI
solution to this problem.

For completeness sake, let us review some terminology. In the following, we consider a collection of N records which we call a <u>file</u> or a <u>data-
base</u>. Each <u>record</u>, more exactly <u>k-dimensional record</u>, consists of an ordered k-tuple $x = (x_1, \ldots, x_k)$ of values and some associated information.
x_i is called value of attribute A_i, $1 \leq i \leq k$. An exact match query specifies a value for each attribute.

2. SURVEY OF PREVIOUS WORK

VLSI technology offers the potential of implementing complex structures and algorithms directly in hardware. With VLSI circuitry increasing
in speed and density at an amazing rate, there is ample motivation in developing customized designs of algorithms implementable in silicon. However,
since computation is cheap in VLSI and communication determines the performance, structures and algorithms have to be reconsidered. A lot of attention has been paid to the problem of designing an index structure implementable in silicon. Bentley and Kung [2] were the first to suggest a tree
machine which can solve all of the decomposable searching problems. Let us
consider a file of one-dimensional records consisting of a key (primary
key) and a location where the record is stored on external storage. The
following dictionary operations are of basic interest: determine whether
the record with a given key is a MEMBER in the file, INSERT a new record
in the file and DELETE an existing record from the file. The tree machine
suggested in [2] consists of a collection of processors connected as a
binary tree (see Fig. 1). There are three kinds of processors in the machine: circles, squares, and triangles. Each record (i.e., the key and
location) resides in a square which is provided with some logic to carry
out a limited repertoire of instructions. The circles broadcast streams
of instructions and data to the squares where the instructions are executed
in parallel. The squares compute preliminary results which are then combined by the triangles to produce the final result being output through the
root triangle. The structure of the tree machine is that of two complete
binary trees, one being the mirror image of the other. Consider now the
problem of performing the operation "Is the record with key 17 a MEMBER
in the file?" We accomplish this by inserting 17 into the input root and
broadcasting it together with a code for the MEMBER operation down the
tree - $\log_2 N$ steps later the value and the operation code will arrive at
all of the N squares. At that point we compare the key stored in each
square to 17 and set a bit to one if the value is equal to 17 and zero
otherwise. We can now combine the bits together through the bottom portion
of the tree machine by letting each triangle compute the logic OR of its
two inputs. So after $2 \log_2 N$ time units have passed since the query was
posed, a single bit emerges from the output root telling whether or not the
record with key 17 is a MEMBER of the file. For a single MEMBER operation,
the performance of the tree machine is not exciting. However if a sequence
of MEMBER operations has to be performed, then we can take advantage of the
very regular data flow by pipelining the process of answering those queries.
As the key of the first operation is going down the tree, the next operation
can follow one step behind, and so on. Thus a sequence of m MEMBER opera-
-tions can be answered in $2 \log_2 N + m$ time units whwre N is the number of
squares in the tree machine. In [2] it is shown how the tree machine can
solve all of the decomposable searching problems.

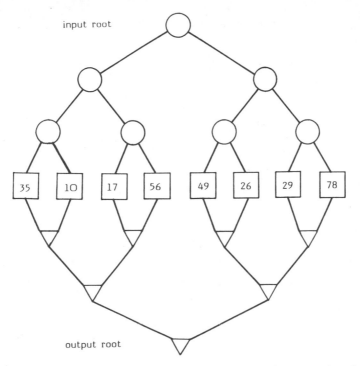

input root

output root

35 10 17 56 49 26 29 78

Fig. 1.   Basic tree machine for primary key retrieval.

Since 1979, many improvements have been suggested to the Bentley-Kung machine. A major drawback is that the Bentley-Kung machine does not allow deletions to be pipelined. Suggesting a new space allocation scheme, this problem was solved by Song [12] besides many practical implementation problems. However, Song's suggestion does not allow the pipelining of redundant insertions and deletions. We call an insertion redundant if it tries to add an already existing key, and we call a deletion redundant if it tries to remove a non-existing key. The problem of handling redundant insertions and deletions was independently solved by [1, 10, 11]. In addition to hand--ling redundant insertions and deletions, all machines exhibit logarithmic performance in the problem size and not in the hardware size thus mimicking the performance of software implementations more closely.

There is one problem, however, for which (with one exception, see remark at the end of this paper) no major improvements have been suggested since the Bentley-Kung machine: the problem of multi-key retrieval. Consider an exact match query with k attributes specified where we want to count the satisfying records. In [2] the following approach is suggested: we store the k-dimensional records in the (larger) square processors and sequentially broadcast all the k specified attribute values down the tree keeping track in each square processor whether it has satisfied all the attribute values shipped so far. We load a one if all conditions have been satisfied and a zero otherwise and combine the partial results by having the triangles sum their inputs. Thus m exact match queries with k attributes specified can be performed in $2 \log_2 N + m \cdot k$ time units. This is obviously not very efficient. The idea of serially feeding the k specified attributes into the input root clearly contradicts the paradigm of VLSI algorithms of exploiting parallelism as best as possible. On top of that, the square processors are at least k times as large as in the case of a single-key retrieval and need some additional logic to ensure comparing values of the same attribute. This contradicts the design guideline that the logic and storage needed at each processor be as small as possible [7].

281

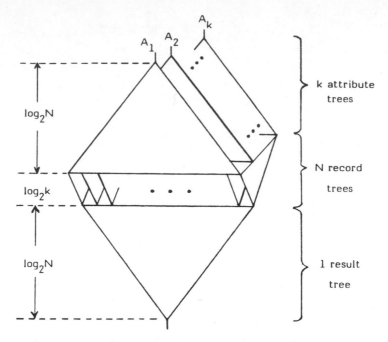

Fig. 2.   Inverted file tree machine (IFTM)
for N records of k attributes each.

3.   THE INVERTED FILE TREE MACHINE

In this section we present the inverted file tree machine (IFTM) as an efficient solution to the multi-key retrieval problem. For records with k attributes the IFTM consists of k attribute trees which are top parts of normal tree machines, each responsible for one attribute (see Fig. 2).

Now the k attribute values of the ith record are stored in the jth square processor of each of the k attribute trees, $1 \leq j \leq N$. Initially we have $j = i$. For each i, $1 \leq i \leq N$, all the ith square processors of the k attribute trees are the leaves of the result tree, a binary tree of triangles identical to the bottom part of the Bentley-Kung machine. Now, given an exact match query, the k specified attribute values are fed into the input roots of the attribute trees in parallel. After arriving at the squares and being compared with the stored attribute values, the record tree of each record reports a one if the record satisfied all k specified values.

The binary tree of triangles then combines the bits by letting each triangle compute the sum of its two inputs, exactly as in the Bentley-Kung machine. Figure 3 shows an IFTM for two attributes storing the records $(a,1)$, $(a,2)$, $(b,3)$, and $(c,5)$.

Let us now compare the IFM and the Bentley-Kung machine storing k-dimensional records with respect to their performance. In the IFTM we can perform m exact match queries with k attributes specified in $2 \log_2 N + \log_2 k + m$ time units, which is an improvement by the factor k. Furthermore, the "inverted file time units" are much shorter because the square processors are much simpler and k times smaller. If exact match queries with at most one answer ask for the location of the satisfying record, the root of the record tree reports the location instead of a one and the location flows

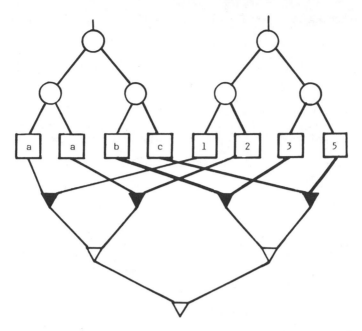

Fig. 3.   IFTM for two attributes storing four records.

through the result tree. The price that has to be paid for the at least
factor k improvement in time is more space needed by the k attribute trees.
However, the IFTM needs less than k times the space of the Bentley-Kung
machine storing k-dimensional records. Similar as in the Bentley-Kung machine, we can perform more types of queries, which will be handled in Sec.
6. The next section shows how the IFTM handles insertions and deletions.

4.   INSERTIONS AND DELETIONS

The only update operation which can be carried out with the present
tools without creating the problem of inconsistency is the non-redundant
insertion. In this section, we will describe how the IFTM can handle non-
redundant and redundant insertions as well as all of the following types
of deletions: non-redundant deletions, redundant deletions and multiple
deletions where one DELETE operation is supposed to remove a list of records,
e.g., specified by a partial match query. The following method for update
operations is based on Song's space allocation scheme [12] which the reader
is assumed to be familiar with. However, Song's scheme is only capable of
handling non-redundant insertions and deletions. To include the more complicated update operations, we introduce two phases for each operation.

4.1.   Deletions

The design of the IFTM implies the problem that records which are supposed to be deleted can be identified only at the root of the record trees.
The records have to be removed, however, from the square processors of the
attribute trees. Thus all operations following a deletion operate on an inconsistent file for a short time, if pipelining is not turned off after a
deletion. Obviously, turning off pipelining is really crucial to the efficiency of VLSI algorithms. In the following, we will suggest a two phase
method for deletions which allows continued pipelining after a DELETE operation by modifying the results of following operations until the record(s)
has (have) been physically removed from the square processors.

Our method necessitates the following extensions to the IFTM:

(i)   Each square processor of the attribute trees accommodates in addi-
        tion to an attribute value a d-bit which is set in order to indi-
        cate that the attribute value will probably be physically removed.

(ii)   It is possible to broadcast information in the record trees from
        the square processors to the roots of the record trees and vice
        versa. Thus the record trees have two strata: top-down logical
        AND is performed, bottom-up we apply "AND FALSE" to partial results
        of following operations on their way top-down in order to negate
        positive partial results and we broadcast an instruction which sets
        the d-bit in the square processors.

(iii)   Each root of a record tree contains in addition to the location of
        a record an f-bit.

(iv)   The result tree has two strata: top-down the results of operations
        are computed, bottom-up physical deletion requests are broadcast
        from the root of the result tree to the roots of the record trees
        using the location as a primary key.

With the extensions (i)-(iv) implemented in the IFTM a DELETE operation is
performed in the following way: each DELETE is carried out in two phases:
(1) the identification phase and (2) the removal phase.

1) The identification phase

        The identification phase identifies the record(s) to be deleted, takes
measures to guarantee consistency of the file and modifies results of operations directly following the DELETE operation. This is accomplished in the
following way. A code for the DELETE operation together with the attribute
value is broadcast down each of the attribute trees. The results of the
comparisons in the square processors are AND - combined in the record trees.
Reaching the roots of the record trees, the record(s) to be deleted is (are)
identified. Each root of a record to be deleted broadcasts two instruction
streams. One stream sends the location of the record down to the root of
the result tree. The other stream traverses the record trees upwards, ne-
gates positive partial results of operations directly following the DELETE
and sets a d-bit in the square processors indicating to following operations
that the marked attribute value is determined to be removed. Using two
different strata in the record trees, the bottom—up stream and the top-down
stream meet at the AND processors. In order to guarantee that partial results of all following operations (not just of every second following operation) are negated, two consecutive "AND FALSE waves" traverse the record
trees upwards. The second wave is initiated by the f-bit in the root of the
record tree which is set at the time the first wave leaves the root on its
way upwards.

2) The removal phase

        In the removal phase the attribute values marked with a set d-bit are
erased. This is done by broadcasting the location of the record upwards
in the result tree. After identifying the root of the appropriate record
tree using the location as a primary key, the square processors in this
record tree are freed using Song's space allocation scheme for the simple
tree machine with one tree controller and one FIRSTFREE count for the IFTM.
In order to guarantee correct behavior of the space allocation scheme, in
the case of a multiple deletion, the removal phases may not start in parallel at the roots of the record trees, but have to be sequentialized at the
root of the result tree.

Thus if b records have been found in the identification phase of one DELETE operation, b consecutive pipelined removal phases are started from the root of the result tree. This ensures that the attribute values of one record are in the square processors with the same number in all the attribute trees and thus they are the leaves of one record tree.

Summarizing, we can state that the suggested method for deletions supports continued pipelining of operations following an arbitrary deletion and it preserves the locality property of the data flow by starting the removal phase where the identification phase ends: at the root of the result tree.

4.2. Insertions

Non-redundant insertions can be easily handled in the IFTM by applying Song's scheme for the simple tree machine to the IFTM using one tree controller for all k attribute trees. The method suggested here additionally supports redundant insertions by making use of two phases: (1) the combined identification and write phase and (2) the removal phase.

1) The combined identification and write phase:

This phase identifies the possibly existing record and treats the record in the same way as the identification phase of the deletion. In parallel, the record is non-redundantly inserted into the IFTM.

2) The removal phase:

The removal phase is identical to the one described for deletions. If the first phase has identified a record to be already existing, the removal phase is applied. If not, we have a do nothing (or inactive) removal phase.

Obviously, the above method handles non-redundant and redundant insertions, and it supports continued pipelining of operations following an arbitrary insertion.

5. THE MODIFIED INVERTED FILE TREE MACHINE

The IFTM contains N record trees each consisting of $\bar{k}$-1 AND processors where $\bar{k}$ is the smallest power of two larger than or equal to k. In the modified IFTM (mIFTM) each record tree is replaced by a systolic linear tuple comparison array similar as suggested by [8]. This array which we call record array combines the ith slightly extended square processors of each attribute tree, $1 \leq i \leq N$, and thus consists of k processors. Figure 4 shows such an extended square processor which in addition to comparing values is able to compute the AND of the comparison results.

More precisely, the jth processor (from left end of the record array) compares the incoming attribute value with the stored attribute value and outputs on its output line t_{out} the AND of this comparison result with the input line t_{in} which is the output of the (j-1)st processor. To make this all work, the inputs to the roots of the attribute trees must be staggered, i.e., the ith component of an operation is input to the ith attribute tree one time step after feeding the (i-1)st component of that operation into the (i-1)st attribute tree. Thus, if the input to the first processor of the record array is one, then the output at the kth processor of the record array will be a bit indicating whether the record fulfills the operation.

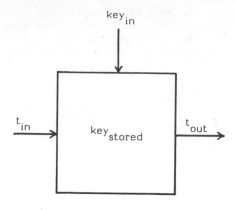

$$t_{out} \leftarrow t_{in} \wedge (key_{in} = key_{stored})$$

Fig. 4.   Square processor in the
modified IFTM.

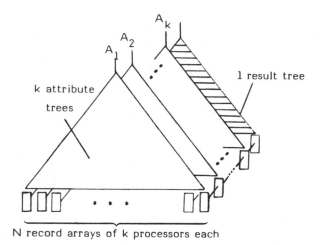

Fig. 5.   Modified IFTM for N records of k
attributes each.

As shown in Fig. 5, the result tree is added in the position of the
(k + 1)st attribute tree.   The leaves of the result tree are square processors which have the function of the roots of the record trees in the IFTM.
The ith result square processor stores the location of the ith record and
inputs the output of the kth square processor of the ith record array.
Thus if the ith record fulfills the present operation, either a one or a
location is broadcast through the result tree.

Obviously, queries can be pipelined in the mIFTM, if they are staggered as discussed above.   With similar two phase methods all types of insertions and deletions which are supported by the IFTM can be handled by
the mIFTM in a pipelined sequence of staggered operations.   Since these
methods are only minor modifications of the methods described for the IFTM,
we will not present them in this paper.

What do we have to pay for saving space in the mIFTM? Let us consider a sequence of m exact match queries with k attributes specified. It takes $2 \log_2 N + k + m$ time units to perform this sequence of queries on the mIFTM. This compares to $2 \log_2 N + \log_2 k + m$ time units for the IFTM. The only difference is that it takes k time units in the mIFTM compared to $\log_2 k$ time units in the IFTM to test the record specified in the first query for equality. Subsequent queries are pipelined and thus can be answered in unit time in both machines.

How many processors and interconnections do we save in the mIFTM? Since we have no record trees, we save roughly $N \cdot \bar{k}$ AND processors, where $\bar{k}$ is the smallest power of two larger than or equal to k. However, the function of these AND processors is integrated in the square processors of the attribute trees. Consequently, the mIFTM consists of 2kN processors in the attribute trees and 2N processors in the result tree. As for the number of interconnections we have $2\bar{k}N$ interconnections in the record trees of the IFTM vs. $k\bar{N}$ interconnections in the record arrays of the mIFTM and thus a saving of $N(2\bar{k} - k)$ interconnections.

6. OTHER TYPES OF QUERIES

In the following, we describe in detail how a large class of queries can be answered using the mIFTM. All different types of queries can be handled by the IFTM in a very similar way which will not be described here.

In a practical application, we want to load the mIFTM with a specified set of records and send a stream of queries and updates through the machine. These queries can be of different types, e.g.,

- exact match query;

- partial match query where values are specified for s < k attributes with the remaining k-s attributes unspecified;

- range query where a range, i.e., an interval $[low_i, up_i]$ is specified for each A_i, $1 \le i \le k$;

- partial range query where a range is specified for s < k attributes;

- best match query where an ideal record and a distance function are specified and we ask for the record closest to the ideal record;

- close match query (also called near neighbor query) where an ideal record, a distance function and a tolerance are specified and we ask for the records (number of records) within the tolerance to the ideal record.

Some of these types of queries may result in more than one answer. In this case, we first restrict to counting the number of answers. The problem of reporting all answers will be treated later.

Different types of queries require different "programs" in the square processors of the record array and in the result tree (min, AND, etc.). Remember the program for exact match queries was

$$t_{out} \leftarrow t_{in} \land (key_{in} = key_{stored})$$

With each attribute value specified in a query we send a code for the query type which selects the adequate program in the square processor. However, there remains the problem that we cannot program the (k + 1)st square processor (of the result tree) without having some irregular data flow. To solve this problem, we insert an extra attribute tree for "attribute A_{k+1}" between the kth attribute tree and the result tree (compare Fig. 5). For each operation we feed its code as well into this tree where it will reach its square processors. This additional tree will also be helpful for queries like close match queries.

For <u>partial match queries</u> with s attributes specified we feed these s values into the corresponding attribute trees and stars into the remaining attribute trees. The comparison star with any attribute value yields true or one in our notation. Thus the program for the first k square processors is

$$t_{out} \leftarrow t_{in} \wedge ((key_{in} = key_{stored}) \vee (key_{in} = *))$$

The program for the (k + 1)st square processor is

$$t_{out} \leftarrow t_{in}$$

The triangle processors add up their two inputs. Therefore for partial match queries we improve query time by the factor s, the number of attributes specified in the query, compared to the simple tree machine.

<u>Range queries</u> are slightly more complicated. Each attribute value now has to be compared with a lower bound low_i and an upper bound up_i which we feed into the corresponding attribute tree one after the other. Thus a wave of lower bounds flows over the squares followed by a wave of upper bounds. When the lower bounds reach the square processors they perform the following program:

$$t_{out} \leftarrow t_{in} \wedge (low_{in} \leq key_{stored})$$

The wave of upper bounds lets the square processors perform the following program:

$$t_{out} \leftarrow t_{in} \wedge (up_{in} \geq key_{stored})$$

The (k + 1)st square processor has to combine the answers from the lower bounds with those from the upper bounds. It receives the collected answers from the lower bounds one time unit before obtaining the collected answers from all upper bounds. Thus it has to save one bit from the lower bounds and AND it with the bit from the upper bounds. The triangle processors again sum up their inputs. Obviously, we speed up query time by the factor k. By providing 2k + 1 attribute trees, one for the lower bound and one for the upper bound of each attribute, we could even achieve a speed up factor of 2k. Since we want to design a general purpose machine, we do not consider this variant any further.

To solve the <u>best match</u> problem, we have to take a closer look at the distance function. The most common distance functions between a record x and the ideal record y are of the form

$$d(x, y) = [\sum_{i=1}^{k} |x_i - y_i|^p]^{1/p}$$

E.g., for p = 1, we have the taxicab or city block distance; for p = 2, we have the Euclidean distance. Since in best match we only compare distance, we can omit the pth root.

Let us consider a best match query with p = 2. We feed the ideal

288

record into the first k attribute trees and let the square processors perform the following program:

$$t_{out} \leftarrow t_{in} + (key_{stored} - key_{in})^2$$

The (k + 1)st square processor outputs its input combined with the memory location. The triangle processors compare the two input distances and pass on the minimum distance together with the corresponding memory location. Best match queries may use more general distance functions, e.g., the Hamming distance. The general form of a distance function is

$$d(x, y) = F(\sum_{i=1}^{k} f_i(x_i, y_i))$$

Obviously, the square processors can be programmed such that this general distance is computed. For this purpose the ith square processor performs

$$t_{out} \leftarrow t_{in} + f_i(key_{stored}, key_{in})$$

and the (k + 1)st square processor computes

$$t_{out} \leftarrow F(t_{in})$$

In a _close match query_, also called _near neighbor query_, a distance function, an ideal record (both as in a best match query) plus additionally a tolerance ε are specified. The tolerance ε is fed into the (k + 1)st attribute tree. The first k square processors have exactly the same function as in the best match query. The (k + 1)st square processor performs

$$t_{out} \leftarrow (F(t_{in}) \leq \varepsilon)$$

Again, the triangle processors just sum up their two inputs.

Summarizing we can say that in all types of queries considered so far the speed up factor is equal to the number of specified attributes.

When the number of answers is more than one, we have to interrupt pipelining to some extent, because the result tree can only output one answer at a time. The problem of waiting until a processor is ready can be solved using well-known techniques from data flow computation. A second problem is to guarantee that answers from different queries do not intermix. This is handled in the following way: the (k + 1)st square processor of each record array broadcasts either the location of the record followed by a bar one time step later (when the record satisfies the query) or only a bar. Each triangle processor in the result tree outputs a location as long as one of the inputs is a location and outputs a bar if both inputs are bars. Thus a wave of answers to a specific query is followed by a wave of bars in the result tree. This implies that in the output of the mIFTM the sequences of answers belonging to queries are separated by one bar. It can be easily shown that a sequence of m queries having a total of s answers can be performed in $2 \log_2 N + k + m + s$ time units of the mIFTM.

Clearly, the mIFTM can be used for queries more general than the ones described above. Since we want to use only one type of square processor for all k attribute trees and we want to restrict ourselves to a regular data flow, the ith square processor should exhibit the following behavior

$$t_{out} \leftarrow t_{in} \ op_i \ f_i(key_{in}, key_{stored})$$

where op_i is an operator and f_i is a function, $1 \leq i \leq k$.

Concerning the (k + 1)st square processor, it can be used for collecting partial answers as we already saw for range queries. For example, consider a query where we want to find the record closest to two given records, i.e., the record with the minimal sum of the distances. This can be performed by two best match queries where the (k + 1)st processor adds up the two partial answers. Obviously, there is a large class of query types with a speed up factor equal to the number of specified attributes over the Bentley-Kung approach. It can be shown that this does not only hold for the modified IFTM, but as well for the IFTM.

7. CONCLUSIONS

In this paper we have presented the IFTM and the modified IFTM as efficient VLSI solutions to a large class of multi-key retrieval problems improving query time of previous machines by the factor number of specified attributes. It is interesting to observe that the inverted file organization as a software structure frequently exhibits poor performance because of the costly intersection operations necessary to answer multi-key queries. In the IFTM and modified IFTM these intersections are performed very efficiently in the record trees and record arrays, respectively. The inverted file idea of using as many single key structures as there are attributes to be supported is exploiting parallelism in an optimal way. We have presented two phase methods which handle non-redundant and redundant insertions and deletions plus multiple deletions while continuously pipelining operations. This is guaranteed because of the fact that operations always access to a consistent file or the version of a file which is kept consistent by additional information. Which of the two machines is the better choice? This will depend on a practical silicon implementation of both machines. Because of the more complex function of the extended square processors, the modified IFTM time units may be longer than the IFTM time units. The increase of the length of a time unit and the average length of a sequence of operations will determine which of the two machines will be the better choice. In contrast to systolic arrays, none of these tree machines has been implemented in silicon yet. Once this will be accomplished, we are ready for a most interesting performance comparison: software versus VLSI hardware.

REMARK

After finishing this paper, the authors became aware of [4] which efficiently supports multi-key retrieval using an architecture very similar to the inverted file tree machine. However, their suggestion does not allow pipelining of arbitrary operations, in particular not pipelining of redundant insertions and deletions. Furthermore no simplification such as the modified inverted file tree machine is suggested for their architecture.

ACKNOWLEDGEMENT

We would like to thank Armin B. Cremers for his criticism concerning the complexity of the inverted file tree machine which motivated us to suggest the modified version.

REFERENCES

1. M. J. Atallah and R. S. Kosaraju, A generalized dictionary machine for VLSI, IEEE Transactions on Computers, Vol. C-34, February 1985.

2. J. L. Bentley and H. T. Kung, A tree machine for searching problems, Proc. 1979 International Conference on Parallel Processing, 257-266.

3. M. Bonuccelli, E. Lodi, F. Luccio, P. Maestrini, and L. Pagli, A VLSI tree machine for relational databases, Proc. 10th Annual International Symposium on Computer Architecture, Stockholm, June 13-17, 1983, 67-75.

4. M. Bonuccelli, E. Lodi, F. Luccio, P. Maestrini, and L. Pagli, VLSI mesh of trees for database processing, Proc. CAAP 83, 8th Colloquium on Trees in Algebra and Programming, Lecture Notes in Computer Science 159, 155-166, Springer (1983).

5. H. P. Kriegel, Performance comparison of index structures for multi-key retrieval, in: Proc. 1984 ACM/SIGMOD International Conference on Management of Data, June 18-21, Boston, 186-196 (1984).

6. H. P. Kriegel and B. Seeger, Multidimensional order preserving linear hashing with partial expansions, in Proc. ICDT '86, International Conference on Database Theory, Rome, September 8-10, 1986, Springer, Lecture Notes in Computer Science, 243, 203-220.

7. H. T. Kung, Let's design algorithms for VLSI, Proc. Conference on Very Large Scale Integration: Architecture, Design, Fabrication, California Institute of Technology, January 1979, 65-90.

8. H. T. Kung and P. L. Lehman, Systolic (VLSI) arrays for relational database operations, Proc. 1980 ACM/SIGMOD International Conference on Management of Data, May 1980, 105-116.

9. J. Nievergelt, H. Hinterberger, and K. C. Sevcik, The grid file: an adaptable, symmetric multi-key file structure, ACM Transactions on Database Systems, 9, 1, 38-71 (1984).

10. T. Ottmann, A. L. Rosenberg, and L. J. Stockmeyer, A dictionary machine for VLSI, IEEE Transactions on Computers, 892-897 (1982).

11. A. K. Somani and V. K. Agarwal, An efficient unsorted VLSI dictionary machine, IEEE Transactions on Computers, 841-852 (1985).

12. S. W. Song, On a high-performance VLSI solution to database problems, Ph.D. Thesis, Department of Computer Science, Carnegie-Mellon University, also available as Technical Report CMU-CS-81-142 (August 1981).

MULTIDIMENSIONAL CLUSTERING TECHIQUES
FOR LARGE RELATIONAL DATABASE MACHINES

Shinya Fushimi*#, Masaru Kitsuregawa**, Hidehiko
Tanaka* and Tohru Moto-oka*

* Dept. of Elec. Eng., Univ. of Tokyo, Tokyo, Japan

** Institute of Industrial Science, Univ. of Tokyo

Mitsubishi Electric Co.
   Kamakura-City, Kanagawa Pref., Japan

ABSTRACT

This paper introduces the formula which gives the performance of the multidimensional clustering techniques. For a couple of access patterns, this formula is evaluated to show the substantial superiority of multidimensional clustering to the classical access method. Based on the theoretical analysis of the formula, the new multidimensional clustering technique for the physical database organization of the large relational database machine is proposed. This technique aims at minimizing the expected number of page accesses by fully exploiting the access pattern to the data items. Its performance and implementation issues are briefly discussed.

1. Introduction

In this paper, the formula which gives the exact performance of the multidimensional clustering technique is introduced. It is computed for some specific query distributions and multidimensional page partitionings. It is shown that, for some distribution of accesses, the multidimensional clustering can reduce the average number of page accesses to $N/3^k$, where N is the number of pages and k the number of clustering attributes. Based on the theoretical analysis of this formula, the new multidimensional clustering algorithm for the physical database organization of the large relational database machine is proposed. The algorithm is essentially the extension of KD-tree algorithm[4,5], but can largely reduce the average number of page accesses of secondary storage devices. This is achieved by adaptive page partitioning with full exploitation of such information as what tuples the relation contains and how and what selection predicates are issued. Its performance, implementation issues, and integration with transpose method are briefly discussed. Although the algorithm was originally designed for the large relational database machine, it can be applied to the physical database organization of the general purpose computer.

This paper is the extended abstract of the parent papers [12, 13], and focuses only the theoretical aspects of the algorithm due to the space limitation. Details on its performance will be found in [12].

2. Background

Since 1979, we have been developing the parallel relational database machine

GRACE[1]. Using the parallel algorithm based on hash and sort, GRACE can perform heavy load operations such as join in the time proportional to the size of operand relations[2]. As a result, all of relational operations, sort, and other functions are executed in $O(N)$ time complexity in GRACE. This makes it possible for GRACE to adopt the data stream oriented processing: all of operations are executed keeping up with the flow of the data stream consisting of the tuples of operand relations. The relational algebra tree compiled from a query is then executed by merging and cascading (pipelining) incoming data streams, which results in the $O(N)$ time execution of the whole query.

GRACE adopts the movable head disks as her secondary memories to accommodate very large database. They are placed in the dedicated disk modules. Thus, while the traditional bottleneck caused by heavy load operations disappears in GRACE, another one appears at the accesses to the disks in disk modules as a compensation of their large capacities. In other words, the performance of GRACE is completely determined by the generation rate of the initial data stream emerging from disks. Considering that tuples in the disk are accessed associatively using the selection predicate specified in the query, we can define the generation rate of the data stream to be the number of tuples satisfying the selection predicate divided by the time interval from the moment the associative access begins to the moment the last tuple of the stream output from disk. To make the denominator small, we should take some means to bunch the necessary tuples into the smallest number of pages and read only such pages from disks. It is clearly impossible to expect such a situation for every query. Instead, we pursue the minimization of the *average number of page accesses* to realize the efficient generation of data streams on the average. For these reasons, we decided to incorporate the index oriented organization of relations with appropriate *clustering* of tuples. But in GRACE, we can put the whole index in the semiconductor memory in disk module, and use specialized hardware to traverse it. Such implementation issues are revisited in section 6.

3. Multidimensional Clustering Techniques for Physical Database Design

Recently, the clustering technique has been widely recognized as the promising method applicable to the physical database design. *Clustering* is the design method for the physical database organization to minimize the expected number of page accesses by accumulating tuples hit by queries into as small number of pages as possible. Therefore, the basic strategy for the clustering can be summarized as "*the more frequently the tuples are accessed, the smaller pages they should be packed together into*".

Define the *base space of the relation* be the Cartesian product of domains of attributes referred to by the relation. Traditional access methods such as hashing, ISAM, and VSAM divide the base space only along the axis of key attribute. *Multidimensional clustering* is the physical database design technique which divides a set of tuples into several pages by partitioning the base space not only along the axis of primary key attribute but also along those of other non-key attributes. The resultant subspaces are formed as hyperrectangles which make altogether the disjoint union of the base space of the relation. The set of tuples in one subspace are stored in one actual page. Hence there exists one-to-one mapping between subspaces and actual pages. The index structure is constructed as page partitioning proceeds recursively.

Multidimensional clustering introduces new degrees of freedom for page partitioning to make the expected number of page accesses much smaller. In addition, range queries and near-neighbor search can be efficiently manipulated[5], and dynamic page split/merge technique can be incorporated in case of page overflow/underflow. Note that the multidimensional clustering does not allow to divide the space into subspaces having any shapes such as circles or triangles. Hence, the multidimensional clustering can generate only a small class of all partitionings the general clustering can produce. But notice that the clustering techniques which can generate the larger class of partitionings cannot associate the efficient access paths. This would become fatal in the physical database design. Taking these facts into consideration, we conclude that the multidimensional clustering occupies the necessary and sufficient position among the whole clustering techniques. Therefore, we will focus and discuss it in the following sections.

4. Analytic Evaluation of Multidimensional Clustering

4.1. Cost Function

We will derive the formula which expresses the average number of page accesses for given query distribution when the base space is partitioned into subspaces of *any* shapes[12]. Then, we apply this formula to the specific query distributions and the multidimensional partitionings, and compare the performance between the multi- and one-dimensional clusterings.

Before we proceed, a couple of ambiguities on terminology should be fixed. *Physical page* is the portion of the actual magnetic device. Each page can store the same volume of data. We call the whole set of physical pages available in the disk the *physical page space of disk*. On the other hand, the *logical page space of the relation* derives from partitioning the base space of the relation. The *logical page* is the partition, i.e., the subset of the base space. The set of tuples in one logical page are stored in one physical page in the disk. The mapping between two spaces gives the *physical allocation of logical pages in the disk*. Note that the number of (physical or logical) pages accessed by the query is independent of the mapping, and exactly depends on how tuples are distributed in the logical page space. Since we focus the number of page accesses, we ignore the overhead time dependent on the page allocation such as seek and ratency time. Hence the difference between these two spaces does not matter here, and we simply use *page space* and *page* for both usages.

Let R be the relation having k attributes $A_1, \ldots, A_k$. And let T and N be the number of tuples of R and the number of pages storing R respectively. D denotes the base space of the relation, i.e., $D = \prod_{i=1}^{k} \mathrm{dom}(A_i)$. We can characterize the distribution of tuples in the base space and the access pattern to them by two distribution functions:

(1) Tuple Distribution Function D(t)
   D(t) is defined so that D(t)=n iff $R(A_1, \cdots, A_k)$ contains n copies of the tuple $t = (v_1, \ldots, v_k)$. Usually, $n=0$ or 1. But n may be more than one when the relation is created by *transpose*. Trivially,

$$\int_{t \in D} D(t)dt = \mathrm{T}.$$

   D(t) affects the possible partitionings of tuples into pages.

(2) Query Distribution Function Q(q)
   Q(q) is the normalized distribution function of the query, that is,

$$Q:Q \rightarrow [0,1]$$

   where

$$Q = \{ \text{ all possible queries } \},$$

and

$$\int_{q \in Q} Q(q)dq = 1.$$

Let $\pi(q)$ denote the number of pages accessed by q. Suppose D is divided into N pages $p_1, \ldots, p_N$. In general, $\overline{\pi}$, the average number of page accesses for given $Q(q)$, can be given as follows:

$$\overline{\pi} = \int_{q \in Q} \pi(q)Q(q)dq. \tag{1}$$

Now, we rewrite $\pi(q)$ to make it easy to manipulate. $\Omega(p_j, q)$ is defined to be true, if the page p_j is accessed by q, and false, otherwise. Define the binary function $\delta(p_j, q)$ as follows:

$$\delta(p_j, q) = \begin{cases} 1 & \text{if } \Omega(p_j, q) = true \\ 0 & \text{if } \Omega(p_j, q) = false \end{cases}$$

Then we have as the concrete form of $\pi(q)$:

$$\pi(q) = \sum_{j=1}^{N} \delta(p_j, q). \tag{2}$$

Further define the following set and function:

$$\Gamma(p_j) = \{ q \in Q \mid \Omega(p_j, q) \}$$

and

$$H(p_j) = \int_{q \in \Gamma(p_j)} Q(q)dq.$$

$\Gamma(p_j)$ denotes the set of queries which access page p_j. $H(p_j)$ represents the sum of the relative frequencies of such queries. Now, we have the following theorem.

[Theorem 1] For given $Q(q)$ and the set of pages $\{p_1, \ldots, p_N\}$, the average number of page accesses is given by

$$\overline{\pi} = \sum_{i=1}^{N} H(p_j). \tag{3}$$

[Proof]

$$\overline{\pi} = \int_{q \in Q} \pi(q)Q(q)dq$$

$$= \int_{q \in Q} (\sum_{j=1}^{N} \delta(p_j, q))Q(q)dq \qquad \text{(by (2))}$$

$$= \sum_{j=1}^{N} \int_{q \in Q} \delta(p_j, q)Q(q)dq$$

$$= \sum_{j=1}^{N} \int_{q \in \Gamma(p_j)} Q(q)dq \qquad \text{(by definition)}$$

$$= \sum_{j=1}^{N} H(p_j). \qquad\qquad \text{(by definition)}$$

Note that the theorem holds independent of the clustering algorithms.

4.2. Queries

Although the formula (1) holds for any set of queries, here we will consider queries formed by conjunctions of ranges. That is, the domain of each attribute assumed to be the ordered set, and the query issued by the user can be put in the the general form:

$$x_1 \leq v_1 \leq y_1 \wedge, \ldots, \wedge x_k \leq v_k \leq y_k,$$

whose meaning is that "select tuples where v_i, the value of its i-th attribute, is greater than or equal to x_i and less than or equal to y_i". All the domains are further assumed to be upper and lower bounded sets, i.e., for any $v_i \in \mathrm{dom}(A_i)$, $\mathrm{MIN}_i \leq v_i \leq \mathrm{MAX}_i$. Then, we can assume without loss of generality that every query refers to *all* of attributes. Considering $x_i \leq y_i$, we can regard this query as the point $\prod_{i=1}^{k} [x_i, y_i]$ in the $2k$-dimensional space, i.e.,

$$Q = \{ \prod_{i=1}^{k} [x_i, y_i] \mid x_i, y_i \in \mathrm{dom}(A_i) \text{ for } all \ i \}.$$

This space is formed as the Cartesian product of k right-angled triangles. The i-th component of the query corresponds to the point in the i-th right-angled triangle Q_i as shown in Fig.1. Three points in Fig.1 show three kinds of possible cases. The point A in the Fig.1 represents the normal range predicate for attribute A_i, whereas the point B and C correspond to the situations that no predicate and exact match predicate are given to A_i respectively. Therefore, this class is general and large enough to include not only the range queries, but exact match and partial match queries.

Since we consider the range queries and multidimensional clustering, the function $\Omega(p_j, q)$ defined above becomes definite. Since q can be regarded as the subset of D, we have:

$$\Omega(p_j, q) \equiv (p_j \cap q) \neq \emptyset.$$

Moreover, every page is of hyperrectangle form in D. Hence, the page $p_j = \prod_{i=1}^{k} [\alpha_{ij}, \beta_{ij}]$ is generally represented as the point Q, as well as queries (Fig.1). Therefore, $\Omega(p_j, q)$ is further reduced to the following form:

$$\Omega(p_j, q) \equiv p_j \cap q \neq \emptyset$$

$$\equiv x_i \leq \beta_{ij} \wedge y_i \geq \alpha_{ij} \text{ for } all \ i.$$

Now, for given page $p_j = \prod_{i=1}^{k} [\alpha_{ij}, \beta_{ij}]$, $\Gamma(p_j)$, the queries which access page p_j, can be represented as the shaded area in Fig.1 for each attribute.

Although it is clearly impossible to compute the right hand side of (3) for general distributions, we can do it for some specific ones and show the substantial superiority of the multidimensional clustering technique to the traditional one-dimensional clustering.

4.3. Examples of Cost Evaluations and Performance Comparison with Traditional Access Method

Here we demonstrate the evaluations of (3) for specific access patterns, and compare the performance of multidimensional clustering with that of traditional access method. As the multidimensional page partitioning, we consider the lattice page partitioning (Fig.2): we suppose the domain of the i-th attribute be evenly divided into n_i intervals of the length $\delta_i = MAX_i/n_i$ as shown in Fig.2. (For simplicity, $MIN_i = 0$ for all i is assumed.) Then clearly $N = \prod_{i=1}^{k} n_i$ holds, and $\beta_{ij} - \alpha_{ij} = \delta_i$ for the i-th component of any page $p_j = \prod_{i=1}^{k} [\alpha_{ij}, \beta_{ij}]$. For this page partitioning, we have the sequence of the values splitting the space for each attribute as:

$$
\begin{cases}
\xi_{ij_i} = \delta_i \cdot (j_i - 1) \\
\eta_{ij_i} = \delta_i \cdot j_i
\end{cases}
\quad (j_i = 1, \ldots, n_i),
$$

and $[\alpha_{ij}, \beta_{ij}]$ as above is equal to $[\xi_{ij_i}, \eta_{ij_i}]$ for some j_i.

4.3.1. Uniform Distribution of One-Dimensional Partial Match query

The *uniform partial match query distribution of degree n*, or n-*dimensional uniform partial match query distribution* accompanies the query distribution function in which the n of k attributes are randomly selected and given the random values from each of n domains to form the predicate. Here evaluated is the distribution where $n = 1$. Although the similar evaluation for this distribution has been already conducted[5], we dare to do it to validate the formula (3) and, in addition, clarify the limitation of the multidimensional clustering.

This distribution can be represented as the overlap of k subdistributions where the i-th subdistribution is given the one k-th total volume (i.e. $1/k$), and denotes the sim-

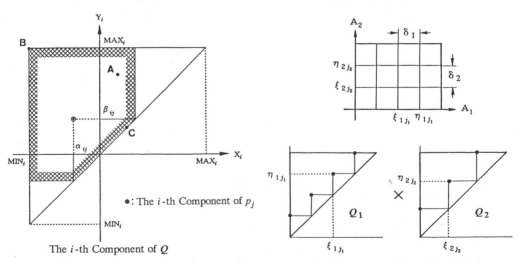

$\bullet$: The i-th Component of p_j

The i-th Component of Q

Fig.1 Representation of Queries and Page

Fig.2 Lattice Page Partitioning and Sequence of Discriminator Values

ple one-dimensional uniform distribution for attribute A_i in Q (Fig.3). Hence, the i-th subdistribution has positive values only on the diagonal edge of the the i-th triangle and value one on the right-angled corners of the others. Here arises the problem on the measure of integration, but we can ignore it by interpreting integration as the simple summation of fully fine granules. Then, for each page $p_j = \prod_{i=1}^{k}[\alpha_{ij},\beta_{ij}]$, we have:

$$H(p_j) = \sum_{i=1}^{k}\frac{1}{k}\cdot\frac{\sqrt{2}(\beta_{ij}-\alpha_{ij})}{\sqrt{2}MAX_i}$$

$$= \frac{1}{k}\cdot\sum_{i=1}^{k}\frac{\delta_i}{MAX_i} = \frac{1}{k}\sum_{i=1}^{k}\frac{1}{n_i},$$

hence,

$$\bar{\pi} = \sum_{j=1}^{N}H(p_j) = \frac{N}{k}\sum_{i=1}^{k}\frac{1}{n_i}. \tag{4}$$

As long as we respect the condition $\prod_{i=1}^{k}n_i=N$, the minimum of (4) can be given by letting $n_1 = \cdots = n_k = {}^k\sqrt{N}$. Then we have,

$$\bar{\pi}_{min} = \frac{N}{k}\sum_{i=1}^{k}\frac{1}{{}^k\sqrt{N}} = N\cdot N^{-\frac{1}{k}} = N^{1-\frac{1}{k}}.$$

On the other hand, when we apply the traditional one-dimensional clustering to this case, we have the average number of page accesses $(\bar{\pi}')$ by letting $n_1 = N, n_2 =, \ldots, n_k = 1$ in (4) as:

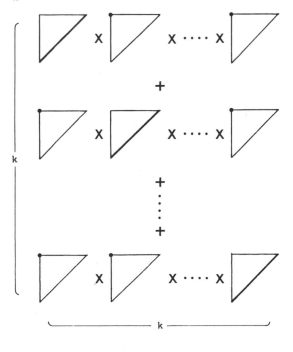

$\bullet, \,\rule[0.3em]{1em}{0.08em}\,$: Queries

Fig.3 Query Distribution of Uniform Partial Match Query of Degree One

$$\overline{\pi}' = \frac{1+(k-1)\cdot N}{k}.$$

Although $\overline{\pi}_{min} < \overline{\pi}'$, we cannot observe the performance improvement as expected by multidimensional clustering. For such a query distribution, only a few number of tuples are retrieved in many cases. But $\overline{\pi}$ reaches, for example, 107.6 for $N=512$ and $k=4$.

This is due to the fact that, for such a distribution, the attributes interfere clustering properties each other when the lattice-like page partitioning is applied. Each of pages retrieved contains a few number of tuples which satisfy the predicate. The desired tuples spread over rather many pages. The principle of the multidimensional clustering technique is that the bunch of tuples accessed frequently should be grouped together in as small pages as possible by exploiting the biases of the distribution both of attributes and values. Then it becomes most effective when a lot of predicates generated from $Q(q)$ refer to many attributes and the query distribution reveals the biases. The uniform one-dimensional partial match distribution is at the worst extreme side from this point of view. Therefore, we cannot expect the large performance improvement in principle. This can be largely relaxed by integrating transpose method (see section 6).

4.3.2. Uniform Range Query

Here we assume the query distribution from which each query is selected uniformly on Q. For the page $p_j = \prod_{i=1}^{k} [\alpha_{ij}, \beta_{ij}]$, let $S(\alpha_{ij}, \beta_{ij})$ be the size of the shaded area for the attribute A_i as in Fig.1. In this case, we have:

$$H(p_j) = \prod_{i=1}^{k} \frac{S(\alpha_{ij}, \beta_{ij})}{S(0, MAX_i)}.$$

Then,

$$\overline{\pi} = \sum_{j=1}^{N} H(p_j) = \sum_{j=1}^{N} \prod_{i=1}^{k} \frac{S(\alpha_{ij}, \beta_{ij})}{S(0, MAX_i)} = \prod_{i=1}^{k} \sum_{j_i=1}^{n_i} \frac{S(\xi_{ij_i}, \eta_{ij_i})}{S(0, MAX_i)}$$

(since D is partitioned in the lattice)

$$= (\prod_{i=1}^{k} (2/MAX_i^2)) \cdot (\prod_{i=1}^{k} \sum_{j_i=1}^{n_i} (\delta_i \cdot j_i \cdot MAX_i -$$

$$\tfrac{1}{2}(\delta_i^2 \cdot (j_i - 1)^2 + \delta_i^2 \cdot j_i^2)))$$

$$= \prod_{i=1}^{k} \left(\frac{n_i}{3} + 1 - \frac{1}{3 \cdot n_i} \right)$$

$$\approx \prod_{i=1}^{k} \left(\frac{n_i}{3} \right) = O(\frac{N}{3^k}).$$

That is, the average number of page accesses is reduced inversely proportional to

3^k. As a comparison, if we apply the traditional one-dimensional clustering, the average number of accesses $\overline{\pi}'$ is given by letting $n_1 = N, n_2 = \cdots = n_k = 1$ as:

$$\overline{\pi}' = \frac{N}{3} + 1 - \frac{1}{3 \cdot N} = O\left(\frac{N}{3}\right).$$

This shows that the multidimensional clustering technique is so effective that it improves the performance in order of the k-th power of 3. Note that in this case, $Q(q)$ does not include any tortion or bias either on attribute or values, but almost always gives predicates to all the attributes. It is expected that $Q(q)$ usually contains biases in the actual environment. We will try to exploiting them to further improve the performance.

5. Algorithm of Generalized KD-tree

5.1. Brief Survey of Multidimensional Clustering Algorithms

Among the multidimensional clustering algorithms proposed so far, KD-tree[4,5] is one of the best-known algorithms. Let $D = \prod_{i=1}^{k} [\text{MIN}_i, \text{MAX}_i]$ be the base space of the relation $R(A_1, \cdots, A_k)$ with k attributes. For given relation, KD-tree method first divide D into two subspaces D_l and D_r using the value of A_1 of the tuple (which is called *discriminator value*) such that it divides evenly the whole set of tuples. Here A_1 is called *discriminator attribute*. Let γ_1 be the discriminator value thus selected. Here we have

$$D_l = [\text{MIN}_1, \gamma_1] \times [\text{MIN}_2, \text{MAX}_2] \times \cdots \times [\text{MIN}_k, \text{MAX}_k],$$

and

$$D_r = [\gamma_1, \text{MAX}_1] \times [\text{MIN}_2, \text{MAX}_2] \times \cdots \times [\text{MIN}_k, \text{MAX}_k].$$

Then this step is recursively applied to D_l and D_r with the discriminator attributes changed in the cyclic order as A_2, A_3, ... , A_k, A_1, ... until the produced subspace contains tuples of the page size or less. The *KD-tree* is created at the same time whose node stores the discriminator attribute and value produced at each recursion step as well as pointers to its left and right sons. The discriminator attribute of the node at level i is thus i modulo $k + 1$. It is obvious that KD-tree algorithm only respects the distribution of tuples and ignores how and what queries are issued. The tuples frequently accessed in the lump may be divided into rather many pages. As a result, relatively many pages are accessed on the average. *Extended KD-tree* [6] tries to resolve such drawback by giving no constraint to the selection of discriminator attributes along the levels of the tree. Another extension of KD-tree[7,8] is also proposed to employ the multi-way structure.

The counterpart of multidimensional tree is the *multidimensional trie* [9,10]. This can be regarded as the trie version of the KD-tree. It differs from the KD-tree technique in that it always selects the discriminator value such that it evenly partitions the given space, not the number of tuples. The corresponding index is constructed as a binary trie, where each node stores the discriminator attribute and the bit position used to partition the space. At the partitioning step, we are free to choose only the discriminator attribute. Once the discriminator attribute is selected, the leftmost unused bit of its bit representation is selected to partition the space and marked "used". In [10], the algorithm of selecting the discriminator attributes is discussed for the partial match queries. As in the pure trie, the very compact representation of the access path is possible. On the other hand, this technique cannot achieve the 100 % load factor of pages, since it always divide the space on the two's power boundaries. As a consequence, the total number of pages storing the tuples might increase. Another drawback is that it cannot be fully adaptive to the tuple and query distributions due to the limited degree of freedom on selection of the discriminator attributes and values.

5.2. Class of Multidimensional Clustering Techniques

The multidimensional clustering algorithms can be classified by the set of partitionings they can create. The cardinality of this set is most important since it directly reflects the degree of freedom on the selection of discriminator values and attributes at each space partitioning step. As for the algorithms mentioned above, for given $D(t)$, those values are given as follows: KD-tree : 1, extended KD-tree : $k^{\log N}$ (when balanced), and multidimensional trie : k^{N-1}. From this point of view, the clustering algorithm can be redefined to be the algorithm which selects one of the possible partitionings it can achieve such that it makes the average number of accesses smallest. In general, the more partitionings the algorithm can produce, the smaller average number of page accesses can be achieved if appropriate discriminator selection algorithm is supplied. Hence the keys to good clustering algorithm are *the large class of partitionings and the adaptive selection of discriminator attributes and values*. The former gives the flexibility and the latter presents the adaptability to the algorithm.

5.3. Generalized KD-tree

KD-tree suffers from relatively large average number of accesses caused by fixing the sequence of discriminator attributes and values. Suppose tuples which fit in N pages are to be partitioned into N pages. There is no choice of discriminator attribute and value in the partitioning step of the KD-tree algorithm. But we can select any of k attributes and, for each of them, any of N-1 candidates of discriminator values. When the procedure to select the discriminator attribute and value is given, the partitioning proceeds as follows. First we select one of $k \cdot (N-1)$ candidates by this procedure as the pair of the discrimination attribute and value, and split the base space by them. Suppose the candidate (A_i, γ_i) is selected and the resultant subspaces contain tuples of n and $N-n$ pages respectively. The new node is created and linked appropriately, and (A_i, γ_i) is stored therein. Then these steps are recursively applied to the subspaces with the number of pages n and $N-n$ until the resulted space contains tuples of one page. The tree thus constructed is called here the *generalized KD-tree* (abbreviated *GKD-tree*).

$C(N)$, the number of page partitionings the GKD-tree can generate, is computed by the recursion:

$$C(N) = \sum_{i=1}^{N-1} k \cdot C(i) \cdot C(N-i)$$

Then we have

$$C(N) = \frac{1}{N} \cdot \binom{2N-2}{N-1} \cdot k^{N-1} \quad (N \geq 1)$$

$$\approx (4k)^{N-1}/N \cdot \sqrt{\pi \cdot (N-1)}.$$

This value is much larger than those of KD-tree, extended KD-tree, and multidimensional trie. This suggests that if such a flexibility of GKD-tree on page partitioning can be fully utilized by the appropriate algorithm, the average number of page accesses will be largely reduced. Note that GKD-tree still enjoys such advantages of KD-tree as good clustering property even for range queries and ability to split/merge pages dynamically in case of page overflow/underflow.

5.4. Algorithm of Selecting Discriminator Attributes and Values

Recall the principle of the clustering that the set of tuples frequently accessed in the lump should be stored in as small number of pages as possible. The same effect would be achieved if we follow the strategy that the set of tuples frequently accessed should *not* be partitioned and the set of tuples rarely accessed should be first partitioned as long as possible. We can identify such tuples by the following theorem.

Suppose the page $p_j = \prod_{i=1}^{k} [\alpha_{ij}, \beta_{ij}]$ is to be divided into two on the discriminator attribute A_i by the discriminator value γ_{ij} ($\alpha_{ij} \leq \gamma_{ij} \leq \beta_{ij}$). Then there produced are two subspaces $p_j{}^l(\gamma_{ij})$ and $p_j{}^r(\gamma_{ij})$, where

$$p_j{}^l(\gamma_{ij}) = [\alpha_{1j}, \beta_{1j}] \times \cdots \times [\alpha_{ij}, \gamma_{ij}] \times \cdots \times [\alpha_{kj}, \beta_{kj}],$$

and

$$p_j{}^r(\gamma_{ij}) = [\alpha_{1j}, \beta_{1j}] \times \cdots \times [\gamma_{ij}, \beta_{ij}] \times \cdots \times [\alpha_{kj}, \beta_{kj}].$$

Let $p_j{}^\Delta(\gamma_{ij})$ be $p_j{}^l(\gamma_{ij}) \cap p_j{}^r(\gamma_{ij})$, which is the hyperplane in the base space.

[Theorem 2]

$$H(p_j) + H(p_j{}^\Delta(\gamma_{ij})) = H(p_j{}^l(\gamma_{ij})) + H(p_j{}^r(\gamma_{ij})), \tag{5}$$

or if we write $H(\alpha_{1j}, \beta_{1j}, \ldots, \alpha_{kj}, \beta_{kj})$ for $H(p_j)$,

$$H(\alpha_{1j}, \beta_{1j}, \ldots, \alpha_{ij}, \beta_{ij}, \ldots, \alpha_{kj}, \beta_{kj})$$
$$+ H(\alpha_{1j}, \beta_{1j}, \ldots, \gamma_{ij}, \gamma_{ij}, \ldots, \alpha_{kj}, \beta_{kj})$$
$$= H(\alpha_{1j}, \beta_{1j}, \ldots, \alpha_{ij}, \gamma_{ij}, \ldots, \alpha_{kj}, \beta_{kj})$$
$$+ H(\alpha_{1j}, \beta_{1j}, \ldots, \gamma_{ij}, \beta_{ij}, \ldots, \alpha_{kj}, \beta_{kj}) \tag{6}$$

$$(\alpha_{ij} \leq \gamma_{ij} \leq \beta_{ij}).$$

[Proof]
Consider Q_i, the projection of Q onto the attribute A_i (Fig.4). The whole shaded area represents the queries accessing p_j in the figure. When this page is partitioned into $p_j{}^l(\gamma_{ij})$ and $p_j{}^r(\gamma_{ij})$ by the value γ_{ij}, $[\alpha_{ij}, \beta_{ij}]$, the i-th component of p_j, is divided into $[\alpha_{ij}, \gamma_{ij}]$ and $[\gamma_{ij}, \beta_{ij}]$. Other components of p_j are left unchanged. In Q_i, the queries accessing these subpages are depicted as vertically and horizontally shaded areas respectively in Fig.4. The cross-shaded area means the overlap of these two queries sets, and can be regarded as the queries accessing the virtual page $p_j{}^\Delta(\gamma_{ij}) = [\alpha_{1j}, \beta_{1j}] \times \cdots \times [\gamma_{ij}, \gamma_{ij}] \times \cdots \times [\alpha_{kj}, \beta_{kj}]$. By the well-known property on the additiveness of the integration regions, clearly the theorem holds.

Suppose that tuples are partitioned into N pages $p_1, \ldots, p_N$. The average number of page accesses is expressed in (3). Now further partition the page p_j on A_i by the value γ_{ij}. By the theorem 2, we have the new average number of accesses as:

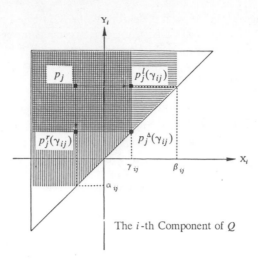

The i-th Component of Q

Fig.4 Relationship Between Page Boundaries and Page Partitioning

$$\overline{\pi}' = H(p_1) + \cdots + H(p_{j-1})$$

$$+ H(p_j{}^l(\gamma_{ij})) + H(p_j{}^r(\gamma_{ij})) + H(p_{j+1}) + \cdots + H(p_N)$$

$$= \sum_{j=1}^{N} H(p_j) + H(p_j{}^\Delta(\gamma_{ij})) = \overline{\pi} + H(p_j{}^\Delta(\gamma_{ij})).$$

Therefore, the average number of accesses increases by $H(p_j{}^\Delta(\gamma_{ij}))$ by this page split. It should be noticed that if the discriminator value is selected so that the set of tuples frequently accessed is partitioned, then $p_j{}^\Delta(\gamma_{ij})$ is virtually accessed by queries frequently issued, hence $H(p_j{}^\Delta(\gamma_{ij}))$ becomes bigger. From this observation, we have the following algorithm:

Input:
　　N, the number of pages the tuples fit into, and D, the space to be partitioned.
Output:
　　The partitioned page space and its associated GKD-tree. The node of GKD-tree consists of four fields: discriminator attribute field, discriminator value field, and two pointers to left and right sons.
Algorithm:
　　First select N-1 candidates for each of k attributes to get all of possible discriminator values. This is done by sorting tuples on each attribute, and then by selecting values which are on the page capacity boundaries. Then compute $H(p_j{}^\Delta(\gamma_{ij}))$ $(i=1,\ldots,k,j=1,\ldots,N-1)$ for each of $k\cdot(N-1)$ candidates. Next, take value and attribute such that they give the minimum value of $H(p_j{}^\Delta(\gamma_{ij}))$, and split the space by them into left and right subspaces as described above. Allocate the node and store this pair as the discriminator attribute and value therein. These form one recursion step. Suppose the base space be divided by this step into left subspace (D_l) and right subspace (D_r) containing tuples of n and N$-n$ pages respectively. Then apply this step recursively to each of D_l and D_r with page number n and N$-n$ respectively, until the given space contains the tuples of the page size or less. At every recursion step, two pointers in the nodes are arranged so that they correctly point to their left and right sons to maintain the recursive structure of page partitioning.

5.5. Time Complexity

Assuming the time complexity of the computation of $H(p_j{}^\Delta(\gamma_{ij}))$ be 1, we show the whole time complexity (denoted by TC(N)) of the algorithm for partitioning into N pages.

The best case is when the tuples are always divided evenly. Then

$$TC(N) = k \cdot (N-1) + 2 \cdot TC(N/2),$$

that is,

$$TC(N) = k \cdot ((logN-1) \cdot N+1) = O(k \cdot N \cdot logN).$$

The GKD-tree is then completely balanced.

The worst case corresponds when one of two resultant space contains only tuples of one page at every recursion step. We have for this case,

$$TC(N) = k \cdot (N-1) + TC(N-1),$$

hence,

$$TC(N) = k \cdot N \cdot (N-1)/2 = O(k \cdot N^2).$$

The GKD-tree for such page partition becomes to reveal the linearly connected link rather than the balanced tree.

5.6. Performance Evaluation

Although we have already evaluated the performance of this algorithm for several distributions of Q(q) and D(t)[12], only one of them is reported here due to the space limitation.

In this evaluation, Q(q) is artificially generated to contain one "mountain" of the distribution at the center of D, while D(t) to contain 2k "mountain"s on the places near boundaries as illustrated in Fig.5, where $k=2$. When KD-tree algorithm is applied, the base space is divided into the lattice with balanced KD-tree. The summit of Q(q) is partitioned into rather many pages. Hence, these pages are frequently accessed along with nearby pages. On the other hand, the GKD-tree algorithm partitions the "skirts" area of Q(q). This is the result from the strategy that the area frequently accessed should not be divided. The performance comparison between KD-tree and GKD-tree is depicted in Fig.6 in which $k=4$. It is clear that the average number of accesses is largely reduced by GKD-tree algorithm. Note that the GKD-tree produced is *not* balanced.

The behavior of the algorithm can be summarized as follows. In general, the query distribution is not uniform. This introduces the undulation of the access pattern in D: there are many "mountains" and, inevitably, many "valleys" (although it is difficult for our intuition to grasp this shape correctly since Q has the dimension two times larger than D). The algorithm first try to divide the base space D so as not to cut across the top of the mountain. In other words, the base space is cut along the valleys. In this phase, the corresponding GKD-tree grows to become balanced. After that, it reluctantly begins to cut the mountain areas, but still obeys the order that the summit area of the mountain should not be cut. So it creeps and goes around the skirts of the mountain to climb and cut. The GKD-tree resulted in this phase has the shape of almost linearly connected link. As a consequence, the GKD-tree is so produced that it is globally balanced but locally grown straight. Note that this shape directly reflects the query distribution: the leaves at lower level point the pages near the top of the moun-

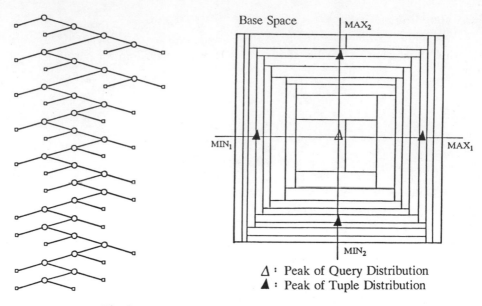

Base Space

MAX_2

MIN_1

MAX_1

MIN_2

Δ : Peak of Query Distribution
$\blacktriangle$: Peak of Tuple Distribution

Fig.5 Example of Page Partitioning and GKD-tree

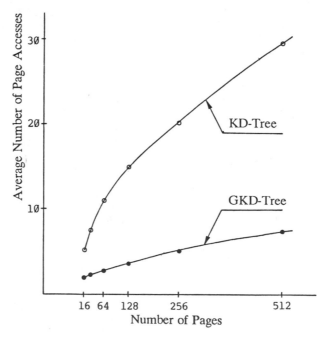

Fig.6 Performance Comparison between
KD-tree Method and GKD-tree Method

tain. Hence, among the leaves in the GKD-tree thus produced, the lower level the leaves exist, the more frequently the corresponding pages are accessed.

6. Discussion

6.1. Implementation Issues

The biggest problem on the implementation is how to maintain query distribution. It is practically infeasible to keep all of the selection predicates since they would become too large. But notice that only the some portion of distribution is required for algorithm to work effectively: it would work well even if the distribution for the queries which are rarely issued is not accurate. We have already developed the algorithm called *gradated statistics* to realize this effectively[12]. The query distribution is implemented by storing selection predicates in the tabular form in the storage of fixed size. One entry consists of $2k$ fields to represent ranges along with the count field expressing how many times the predicate issued. When the storage is filled up, the *gradation* begins: some entry is merged with the other by making lower bits of some attributes "don't care" so that the resulted entry has the smallest value in its count field. By this algorithm, the distribution for queries rarely issued is made "gradated", while that for queries frequently entered is kept still accurate.

Another problem on the implementation is how to compute the candidates for discrimination values. This requires the sorting of a lot of tuples. Our database machine environment resolves this since sorting can be efficiently carried out by using hardware sorter[3] and relatively large semiconductor memory in the disk module.

The index search can be executed efficiently by making use of the property mentioned in the previous section. That is, the pages which are pointed by leaf nodes at lower level are frequently accessed. Then, in many cases, the traverse of the index could be omitted by checking such nodes first. In addition, the index itself can be placed in the semiconductor memory in the disk module, hence the tree need not to be balanced.

6.2. Integration with Transpose Technique

We observed in section 4.2.1 that for uniform partial match query distribution of small degree, the access performance is not improved to our satisfaction even by multidimensional clustering. This reveals the limitation of multidimensional clustering technique. We can further improve the performance by integrating our clustering algorithm with *transposing* of relations[11], i.e., *projective partitioning of the base space*. The limitation of multidimensional clustering can be relaxed at the same time. In general, transpose technique requires to concatenate subtuples to form the result tuple. Therefore, this method is effective only if a few number of concatenations are required. This condition holds in two ways: a few attributes are involved in the query, and/or the number of result tuples are rather small. We notice that the uniform partial match query distribution of rather low degree satisfies at least the latter condition, since such a query is expected to output a few tuples in many cases.

Transposing can be regraded as the *attribute clustering* or *vertical clustering* of the relation, while the multidimensional clustering can be identified as *tuple clustering* or *horizontal clustering*. Our algorithm can be integrated with the transpose method as follows[13]. First, the *attributes* are partitioned so that join operations of the retrieved subtuples using *TID's* are less required (attribute clustering). Then the GKD-tree algorithm is applied to all the subrelations (tuple clustering). The details are beyond the scope of this paper, and will be reported elsewhere.

7. Conclusion

In this paper, we first analyzed multidimensional clustering techniques. We derived the formula representing the average number of page accesses when the multidimensional clustering is applied. This formula was evaluated for two query distributions with lattice-like page partitioning assumed, and superiority of the multidimensional clustering to the traditional access method was shown. Based on the theoretical analysis of this formula, the new multidimensional clustering algorithm was proposed. It was shown that this algorithm can largely improve the access performance by fully exploiting the distribution information of tuples and queries. Implementation techniques and the integration of our algorithm with transposing method were briefly discussed.

References

[1] Kitsuregawa,M.,et. al. *"Relational Algebra Machine GRACE"*, Lecture Notes in Computer Science 147, Springer-Verlag, pp.191-214 (1982)

[2] Kitsuregawa,M.,et. al. *"Application of Hash to Data Base Machine and Its Architecture"*, New Generation Computing, 1, Springer-Verlag, pp.63-74, (1983)

[3] Kitsuregawa,M and Fushimi,S.,et. al. *"The Organization of Pipeline Merge Sorter"*, Trans. of IECE Japan, J66-D, pp.332-339 (1983)

[4] Bentley,J.L., *"Multidimensional Binary Search Trees Used for Associative Searching"*, CACM, Vol. 18, No. 9, pp. 509-517 (1975)

[5] Bentley,J.L., *"Multidimensional Binary Search Trees in Database Applications"*, Trans. on SE, Vol. SE-5, No. 4, pp.333-340 (1979)

[6] Chang,J.M., and Fu,K.S., *"Extended K-D Tree Database Organization: A Dynamic Multi-Attribute Clustering Method"*, Proc. of VLDB, pp.39-43 (1979)

[7] Robinson,J.T., *"The K-D-B-Tree: A Search Structure for Large Multidimensional Dynamic Indexes"*, Proc. of ACM SIGMOD Conf., pp.10-18 (1981)

[8] Ouksel,M. and Scheuermann,P. *"Multidimensional B-Trees: Analysis of Dynamic Behavior"*, BIT 21, pp.401-418 (1981)

[9] Orenstein,J.A., *"Multidimensional Tries Used for Associative Searching"* Information Processing Letters, Vol. 14, No. 4, pp. 150-157 (1981)

[10] Tanaka,Y., *"Adaptive Segmentation Schemes for Large Relational Database Machines"*, in Database Machines, Leilich, H.-O. and Missikoff, M. ed. Springer-Verlag (1983)

[11] Batory,D.S., *"On Searching Transposed Files"*, ACM TODS, Vol.4, No.4, pp.531-544 (1979)

[12] Fushimi,S., et. al., *"Algorithm and Performance Evaluation of Adaptive Multidimensional Clustering Technique"*, to appear in Proc. of 1985 ACM/SIGMOD International Conference on Management of Data, Austin, Texas (1985).

[13] Fushimi,S., *"Design of Secondary Storage System for Relational Database Machine: Physical Database Organization Method Based on Multidimensional Clustering and Transposition"*, Ms. Thesis, Univ. of Tokyo (1983) (In Japanese)

[14] Whang,K.-Y., Wiederhold,G., and Sagalowicz,D., *"Separability - An Approach to Physical Database Design"*, Trans. on Comuters, Vol.C-33, No.3, pp.209-222 (1984)

A METHOD FOR REALISTIC COMPARISONS OF SORTING ALGORITHMS FOR VLSI

Hans-Werner Lang, Manfred Schimmler,
Hartmut Schmeck and Heiko Schröder

Institut für Informatik und Praktische Mathematik
Christian-Albrechts-Universität
D-2300 Kiel,   F. R. Germany

ABSTRACT

A method for comparing the asymptotic performance of different sorting
algorithms for VLSI is proposed. For each algorithm it takes into account the
maximal problem size that is realizable on a single chip under the restrictions imposed by the available technology. This sorting chip is used to perform a sort-split operation on blocks of data in an external merge algorithm
for sorting arbitrarily large sets of data. The performance of the merge
algorithm is determined by the execution time and period of the sorting chip
used. Thus a realistic comparison of the practical feasability of sorting
algorithms for VLSI is obtained.

1. INTRODUCTION

In the past few years a large variety of sorting algorithms for VLSI has
been proposed and analyzed (see e.g. Thompson and Kung, 1977; Nassimi and
Sahni, 1979; Lee et al., 1981; Kumar and Hirschberg, 1983; Miranker et al.,
1983; Nath et al., 1983; Schröder, 1983; Shin et al., 1983; Thompson, 1983;
Bonucelli et al., 1984; Rudolph, 1984; Lang et al., 1985; Akl and Schmeck,
1986). Usually, comparisons of these algorithms are based on their asymptotic VLSI complexity, i.e. their area A, time T, period P, and combinations
of these like AT, AT^2, or ATP (see e.g. Thompson, 1983). However, realistic
comparisons of algorithms for VLSI should take into account the limitations
on the maximum problem size imposed by the available technology. For example,
the large area requirement of the very fast OTN-sort (Nath et al., 1983)
($A=O(n^2 log^2 n)$, $T=O(logn)$) implies that a chip implementing this algorithm
will only be able to sort fairly small sequences, whereas a chip implementing
the relatively slow Ribble sort (Shin et al., 1983) ($A=O(nlogn)$, $T=O(n)$)
should be capable of handling much larger input sizes because of its small
area requirement.

In this paper we propose a method which allows to compare the performance of different sorting algorithms while taking into account the techno-
logical limitations: For every sorting algorithm and with respect to a fixed
technology we use the maximally realizable chip S implementing this algorithm
to perform the basic sort-split operation in a merge algorithm based on
Batcher's bitonic sort (Batcher, 1968) for sorting arbitrarily large

sequences of blocks of data items. Clearly, the time complexity of the merge algorithm depends on the characteristic data of chip S, which is its maximal problem size, its time, and its period. We thus obtain an external sorting algorithm which can be used to realistically compare the practical feasability of sorting algorithms for VLSI.

External sorting algorithms for VLSI have also been proposed by Bonuccelli et al. (1984). But instead of using such an algorithm to compare the performance of different sorting chips they only consider OTN-sort (Nath et al., 1983) which we show to be the worst possible choice. Furthermore, their extension of bitonic sort is unnecessarily complex and not a true external sorting algorithm, since the number of sorting chips needed depends on the problem size.

The paper is organized as follows: In Section 2 we describe the hardware environment necessary to implement the merge algorithm M(S) based on a sorting chip S. The merge algorithm M(S) is described and analyzed in Section 3. In Section 4 the algorithm is used to perform two comparisons of four different sorting algorithms. The first comparison is based on standard assumptions about current technology whereas the second comparison is based on the technology that is assumed to be available at the end of this decade (Mangir, 1983; Thompson and Raghavan, 1984).

2. THE HARDWARE ENVIRONMENT

Let S be a sorting chip capable of sorting sequences of up to n_S data items. Assume that a much larger sequence has to be sorted, which is distributed over a set of storage chips storing several blocks of $n_S/2$ data items each.

Apart from Bonuccelli et al. (1984) this problem has been dealt with by Lee et al. (1981) and Akl and Schmeck (1986). They use a sorting chip with sequential input/output to execute a merge-split operation on blocks of data and assume to have random access to the storage chips via a bus system. Their merge algorithm is based on Batcher's odd-even merge (Batcher, 1968).

Compared with the above the algorithm and hardware structure introduced in this paper has the following major advantages: Since it is based on Batcher's bitonic sort (Batcher, 1968), its control structure is simple. An arbitrary sorting chip can be used to execute the merge-split operation on blocks of data, so there is no limitation to sequential input/output as in Lee et al. (1981) and Akl and Schmeck (1986). A shift memory is used to store the data, so there is no need for a bus system. It is systolic in its nature, i.e. the blocks of data are pumped through the shift registers and the sorting chip at a constant speed. Since this hardware environment for merging can be adapted to the input/output requirements of any kind of sorting chip it enables us to do a fair comparison of VLSI sorting algorithms for any problem size for which the shift memory fits on a single board.

The storage chips are connected to form four FIFO queues as shown in Figure 1. The control unit sends the shift pulses to these four independent shift memories. Initially the two FIFO queues of ST I are the input queues filled with 2^m blocks of data of $n_S/2$ elements each. The two queues in ST II are empty. Throughout the execution of the merge algorithm described below the data blocks are pumped from the input queues through the sorting chip S to the output queues. Whenever the input queues are empty they become the output queues and vice versa.

In this structure the data rates of the shift memory are adapted to the I/O data rates of the sorting chip. This is always possible using the same

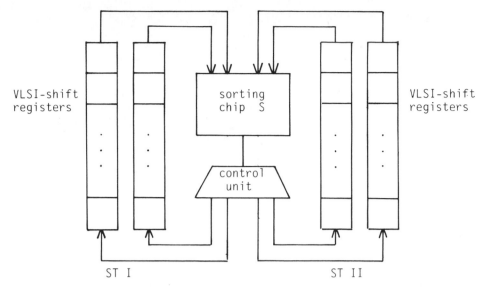

VLSI-shift registers

sorting chip S

VLSI-shift registers

control unit

ST I   ST II

Fig. 1: The hardware structure of the external sorting algorithm.

standard of technology for the memory and the sorting chip. Also the number of pins required for the storage chips is the same as for the sorting chip. So the entire structure is technically feasible whenever chip S is.

3. THE ALGORITHM

The algorithm presented in this section is based on Batcher's bitonic sort (Batcher, 1968), which is a method for sorting 2^m data items in $(m^2+m)/2$ stages of delay, each consisting of 2^{m-1} parallel comparison-exchange operations. If the data items and the comparison-exchanges are replaced by sorted blocks of k data items and merge-split operations, respectively, we obtain a merge algorithm. The correctness of this transformation is proved for any sorting net in Knuth (1973), p. 640. In this paper a sorting chip is used to perform the merge-split operation which thus becomes a sort-split operation. Therefore the merge algorithm can be used to extend the sorting capability of any given sorting chip to arbitrarily large sequences of unsorted blocks of data items.

Let S be a sorting chip which can sort sequences of maximally n_S data items into ascending or descending order in time t_S with period p_S. Let N = $2^{m-1} n_S$ be the number of data items to be sorted.

```
Algorithm M(S)
k_s:=0;
for k:=0 to m-1 do
   for l:=0 to k do
      begin
         if l = k   then   k_s:=k;

         for j:=0 to 2^{m-1}-1 do
            begin
               Shift one block of size n_s/2 from the left input queue and one
               block from the right input queue into S;
               Sort the data on S into ascending order (if bit_k(j) = 0) or into
               descending order (if bit_k(j) = 1);
               Shift the sorted sequence of n_s data items into the left output
```

```
        queue (if bit_{k_S} (j) = 0) or into the right output
        queue (if bit_{k_S} (j) = 1)
      end
    end
```

It can be shown that algorithm M(S) is a realization of Batcher's bitonic sort. The proof has been omitted here for sake of brevity.

4. COMPARISON OF SORTING ALGORITHMS

As mentioned before the time complexity of the merge algorithm M(S) based on the sorting chip S is

$$T_{M(S)}(N) = (N \cdot p_S/n_S + t_S - p_S) \cdot ((\log (2N/n_S))^2 + \log (2N/n_S))/2$$

The asymptotic performance of M(S) is determined almost completely by n_S and p_S, since for large N the term $t_S - p_S$ becomes neglectably small (for many sorting chips it is zero) compared with N/n_S. Therefore the tuple (n_S, p_S) is called the characteristic point of S. Let $T^*_{M(S)}$ denote the function obtained from $T_{M(S)}$ by ignoring the term $t_S - p_S$. In the following $T^*_{M(S)}$ is called the characteristic function of S.

We use a double logarithmic scale to facilitate comparisons of the performance of different sorting chips: In this representation differences in the values of n_S and p_S correspond to simple horizontal and vertical translations of the graph of $T^*_{M(S)}$.

$T^*_{M(S)}$ is used below to compare the performance of several sorting chips. This comparison can be done easily since all characteristic functions have the same shape. For the comparisons we have chosen four sorting algorithms:

- odd-even-transposition sort (OETS) (Knuth, 1973)
- ribble sort (RS) (Shin et al., 1983)
- LS³ sort (Lang et al., 1985)
- OTN sort (Nath et al., 1983)

The version of odd-even transposition sort analyzed here has word parallel and bit serial input and output. It is systolic so that every log n clock pulses a new set of data can be read.

Ribble sort has word serial and bit serial input and output. It has two phases, the reading phase, where the data is input and the writing phase where the data is output. Period and execution time are identical.

LS³-sort is a systolic algorithm sorting bit serially a $\sqrt{n} \times \sqrt{n}$ -array of data items into snake-like ordering. The array is input row by row and pumped through a series of perfect shuffle and OETS networks, iteratively merging increasingly larger sorted subarrays. Every $\sqrt{n}$ log n time units a new data set can be input.

OTN sort uses a mesh of trees as hardware structure. That is an n xn array of processing elements where the elements of each column and each row are leaves of complete binary trees. The n data items are input through the roots of the trees and distributed to all their leaves. The leaf processors compare the data items received from the column and row trees. By summing the results the rank of every data item is obtained at the root of its row tree in time log n. The ranks are used to output the sorted sequence at the roots of the column trees after 2 log n more time units.

Table 1: VLSI complexity of four sorting algorithms for VLSI.

	area	time	period	data rate
OETS	$O(n^2)$	$n + \log n$	$\log n$	n
RS	$O(n \log n)$	$2n \log n$	$2n \log n$	1
LS^3	$O(n \log n)$	$8 \sqrt{n} \log n$	$\sqrt{n} \log n / 2$	$2 \sqrt{n}$
OTN	$O(n^2 \log^2 n)$	$9 \log n$	$9 \log n$	n

The rather different VLSI complexities of the bit-serial versions of these algorithms are summarized in Table 1 assuming that the n data items to be sorted have log n bits each.

Table 2: Characteristic data of four sorting algorithms for VLSI with respect to (a) current and (b) future technology.

	(a) current technology			(b) future technology		
	n_S	t_S	p_S	n_S	t_S	p_S
OETS	25	3	1	775	50	1
RS	200	400	400	200,000	400,000	400,000
LS^3	16	32	2	16,384	512	64
OTN	4	9	9	64	9	9

For every algorithm the characteristic data of the maximal layout realizable in current technology is shown in Table 2a. The time and period are the number of steps needed for sorting a maximal sequence where a step is the time needed by a comparison of two data items. The values listed are estimates based on the assumption that in current technology ($\lambda = 3\mu$) about 100 comparator cells can be integrated on a chip.

The functions $T^*_{M(S)}$ corresponding to the data of Table 2a are shown in Figure 2. Because of the small problem size the merge algorithm based on OTN-sort has the worst asymptotic time complexity. The simple structure and small period of OETS are the reasons why M(OETS) is the asymptotically fastest sorting method.

Assuming that in future technologies the number of logic elements on a chip can be increased by a factor of 1000 (compare Mangir, 1983; Thompson and Raghavan, 1984) we obtain estimates of the characteristic data of layouts that may be realizable at the end of this decade (Table 2b).

Surprisingly even with respect to future technologies M(OETS) is the asymptotically fastest sorting structure (see Figure 3). For OETS the number of pins required might become the limiting factor in the maximal problem size n_S. Only M(RS) changes its relative position and becomes the slowest algorithm.

313

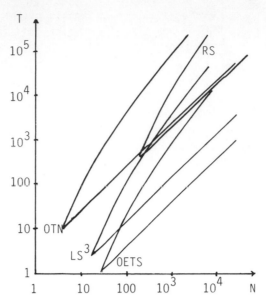

Fig. 2: Characteristic functions of four sorting
algorithms for VLSI with respect to current
technology.

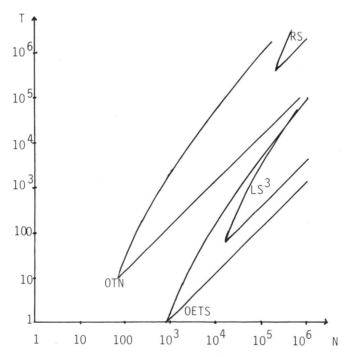

Fig. 3: Characteristic functions of four sorting
algorithms for VLSI with respect to future
technology.

5. CONCLUSION

In this paper we have shown how Batcher's bitonic sort algorithm can be
used to obtain realistic comparisons of the practical feasability of

different sorting algorithms for VLSI. The four sorting algorithms compared in this paper are by no means the only candidates worth being analyzed, but they lead to some interesting results:

Surprisingly, the asymptotically slow odd-even-transposition sort turns out to be the most efficient algorithm for sorting large sequences of data items, even with respect to the much increased integration capabilities of future technologies.

This demonstrates the importance of simple control structures, small area requirements, and small period for the design of VLSI algorithms.

There are several alternatives to this design of a sorting device on one board. Batcher's bitonic sort has been used here mainly because it results in a systolic realization, so that shift registers can be used instead of bus systems and addressable memory, and it is obvious that this design is technically feasible whenever the sorting chip used in this design is. Also it is clear that the data rates between the chips are those determined by the sorting chip. Instead of using just one sorting chip for the merge process several sorting chips can either be used in parallel or, in case of a modular sorting algorithm, can be connected to form a sorting device for a bigger problem size. Further investigations will be done in this direction balancing the I/O rates of the sorting chip and the I/O rates realizable by the environment of the board.

It would also be interesting to extend our method to other classes of algorithms. One necessary requirement would be the decomposability of a problem into a sequence of subproblems of the same type but of substantially smaller size.

REFERENCES

Akl, S.G. and Schmeck, H., 1986, Systolic Sorting in a Sequential
        Input/Output Environment, Parallel Computing 3, pp. 11-23.
Batcher, K.E., 1968, Sorting Networks and their Applications, in: Proc. AFIPS
        1968 SJCC, Vol. 32, AFIPS Press, Montvale, N.J., pp. 307-314.
Bonuccelli, M.A., Lodi, E., and Pagli, L., 1984, External Sorting in VLSI,
        IEEE Trans. Comput. C-33, pp. 931-934.
Chen, T.C., Lum, V.Y., and Tung, C., 1978, The Rebound Sorter: An Efficient
        Sort Engine for Large Files, in: Proc. 4th Int. Conf. on Very Large
        Data Bases, pp. 312-318.
Knuth, D.E., 1973, The Art of Computer Programming, Vol. 3: Sorting and
        Searching, Addison-Wesley.
Kumar, M. and Hirschberg, D.S., 1983, An Efficient Implementation of
        Batcher's Odd-Even Merge Algorithm and its Application in Parallel
        Sorting Schemes, IEEE Trans. Comput. C-32, pp. 254-264.
Lang, H.-W., Schimmler, M., Schmeck, H., and Schröder, H., 1985, Systolic
        Sorting on a Mesh-Connected Network, IEEE Trans. Comput. C-34, pp.
        652-658.
Lee, D.T., Chang, H., and Wong, C.K., 1981, An On-Chip Compare/Steer Bubble
        Sorter, IEEE Trans. Comput. C-30, pp. 396-404.
Mangir, T.E., 1983, Impact and Limitations of Interconnect Technology on VLSI
        and Restructurable VLSI Design, in: Proc. IEEE Int. Conf. on Computer
        Design: VLSI in Computers, pp. 735-739.
Miranker, G.S., Tang, L., and Wong, C.K., 1983, A "Zero Time" VLSI Sorter,
        IBM Journal of Research and Development (2) pp. 140-148.
Nassimi, D. and Sahni, S., 1979, Bitonic Sort on a Mesh-Connected Parallel
        Computer, IEEE Trans. Comput. C-28, pp. 2-7.
Nath, D.D., Maheshwari, S.N., and Bhatt, P.C.P., 1983, Efficient VLSI Net-

works for Parallel Processing Based on Orthogonal Trees, IEEE Trans. Comput. C-32, pp. 569-581.

Preparata, F.P. and Vuillemin, J., 1981, The Cube-Connected Cycles: A Versatile Network for Parallel Computation, Comm. ACM 24, pp. 300-309.

Rudolph, L., 1984, A Robust Sorting Network, in: Proc. Conf. on Advanced Research in VLSI, M.I.T., January 1984, pp. 26-33.

Schröder, H., 1983, Partition Sorts for VLSI, Proc. 13. GI-Jahrestagung, Hamburg, October 1983, Informatik-Fachberichte 73, pp. 101-116.

Shin, H., Welch, A.J., and Malek, M., 1983, I/O Overlapped Sorting Schemes for VLSI, in: Proc. IEEE Int. Conf. on Computer Design: VLSI in Computers, pp. 731-734.

Thompson, C.D., 1983, The VLSI Complexity of Sorting, IEEE Trans. Comput. C-32, pp. 1171-1184.

Thompson, C.D. and Kung, H.T., 1977, Sorting on a Mesh-Connected Parallel Computer, Comm. ACM 20, pp. 263-271.

Thompson, C.D. and Raghavan, P., 1984, On Estimating the Performance of VLSI Circuits, in: Proc. Conf. on Advanced Research in VLSI, M.I.T., January 1984, pp. 34-44.

Database Models

UPDATE PROPAGATION IN THE IFO DATABASE MODEL

Serge Abiteboul* and Richard Hull†

*Institut National de Recherche
 en Informatique et en Automatique
 Rocquencourt, 78153, France

†Computer Science Department
 University of Southern California
 Los Angeles, California 90089-0782, U.S.A.

1. INTRODUCTION

The IFO model [2, 3] is a formal database model which encompasses the
fundamental structural components found in the semantic database modelling
literature [5, 6, 7, 8, 9, 10, 11, 12, 13, 14, 15, 16, 17, 18]. The IFO
model uses a graph-based formalism to represent three basic types of relationships between data: ISA relationships, functional relationships, and relationships arising in the construction of objects from simpler objects (e.g., CONVOYs
are objects built from the simpler objects SHIPs). The presence of these
types of relationships between data objects leads to intricate types of
propagation when updates to the underlying data are made. This extended
abstract reports on a development presented in [3], which formally articu-
lates a coherent semantics for updates and update propagation in the IFO
model.

As will be detailed below, there are three basic types of "local" update propagation in the IFO model. Two of them result from ISA relationships present in a given schema (e.g., if STUDENT ISA PERSON and a given
person moves to another country, then she must be removed from both the
PERSON set and the STUDENT set). The third type of local propagation follows from the presence of structured objects (e.g., if one SHIP in a CONVOY
C is destroyed, then C must be replaced by the set C minus that one ship).
An important principle in the definition of update semantics for the IFO
model is to preserve the identity of objects through such modifications as
much as possible.

In addition to specifying "local" propagational effects in the model,
the paper addresses the "global" aspects of propagation. In particular,
it is shown in [3] that given any "permissible" primitive update to an instance, the propagation of this update on a local level uniquely determines

---

*Part of this work was performed while the author was visiting U.S.C.
†Work by this author supported in part by NSF Grants IST-81-07480 and IST-
 83-06517. Part of this work was performed while he was visiting
 I.N.R.I.A.

exactly one (correct) update for the entire instance. This result follows primarily from the fact that IFO schemata are, formally speaking, directed graphs with a certain acyclicity property. It is clear that the basic approach used here of first analyzing local propagation and then examining its global implications can also be extended to other graph-based database models.

This abstract is organized as follows: In Sec. 2, an informal introduction to the IFO model is given.* And in Sec. 3, an overview of the update semantics for the model is presented. In the interest of brevity, many of the formal definitions are omitted; these definitions, a more complete statement of the update propagation in the IFO model, and proofs of many of the results stated here can all be found in [3].

2. INFORMAL DESCRIPTION OF THE IFO MODEL

In this section the various components of the IFO model are introduced in an informal manner. The emphasis of the discussion will be on the intuition and motivations behind the model, rather than on the precise definitions. (As noted above, precise definitions are presented in [3].)

To begin the discussion, we present a brief overview of the IFO model. An IFO *schema* is (formally speaking) a directed graph with various types of vertices and edges, which will be represented (informally) using various diagrammatic conventions. A family of IFO *instances* is associated with each schema. These correspond to the structured (or formatted) versions of data sets which can be represented using the schema.

To describe the model in more detail we proceed in four steps. We begin by describing "types," which describe the structure of the objects arising in the IFO model. Second, we describe "fragments." These are built from types, and are used to represent functional relationships in the IFO model. Third, we describe how ISA relationships between various types of a schema are incorporated. Finally, we briefly describe how these pieces are combined to form (valid) IFO schemata.

The basis of any IFO schema is the representations of the various object structures that are modelled, called *types*. There are three kinds of *atomic* types in the IFO model, and two constructs for recursively building more complicated types.

One of the three atomic types in the IFO model is called *printable*, and corresponds to objects of predefined types which serve as the basis for input and output. In an IFO schema this category of type is indicated using a square node with the name of the type in it (see Fig. 2-1). The second atomic type in the IFO model is called *abstract*, and is depicted using a diamond. These types correspond to objects in the world which have no underlying structure (at least, relative to the point of view of the database designer or user). The third (and last) atomic type is called *free*, and is depicted using a blank circle. Free types correspond intuitively to entities obtained using one or more ISA relationships, and will be discussed further below. Following [17], the Format Model, and others, it is often desirable to associate a meaningful name with vertices of an IFO schema, as shown in Fig. 2-1a. Such names or labels are called *tokens*,

---

*The reader is forewarned that some of the definitions given in the preliminary report [2] issued on the IFO model have been modified in the final report [3].

(a)   (b)   (c)

Fig. 2-1. Three atomic IFO types.

(a)   (b)

Fig. 2-2. Constructed types.

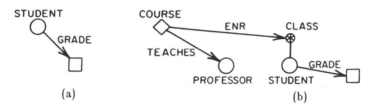

(a)   (b)

Fig. 2-3. Two simple fragments.

and can be specified within the formal model. (Also, the token of a print-able or abstract vertex is often simply the name of the type of that ver-tex.)

The first of the two mechanisms for constructing non-atomic types cor-responds to the procedure of forming (finite) sets of objects of a given structure. Specifically, this is represented by a small circle with a '*' in it, and a large circle below it. For example, Fig. 2-2a shows a type whose objects are finite sets of STUDENTs. The other mechanism for constructing new types out of existing ones in the IFO model is the well-known Cartesian product operator. In the IFO model this is represented by a large circle with a x in it (called a "cross-vertex"), with connecting arcs to one or more vertices below it, which correspond to the object structure types of the various coordinates of the product. For example, ordered pairs with first coordinate a HULL and second coordinate a MOTOR are associated with the vertex MOTOR-BOAT in Fig. 2-2b. The ⊛- and ⊗-constructs can be applied recursively.

Up to this point we have described types, and (speaking informally) have described the *domain*, i.e., the set of "objects," associated with each one of them. An *instance* of a type is defined to be a finite set of objects from the corresponding domain.

The second main structural components of the IFO model are *fragments*, which give the model the power to directly represent functional relation-ships. The representation of functions in the IFO model is very closely related to the representation of functions in the functional data model (e.g., [12, 16]), with one key difference: In the IFO model a sharp dis-tinction is made between vertices serving the role of domain and vertices

321

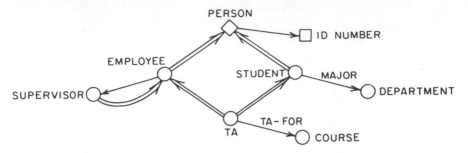

Fig. 2-4.   The TA example.

serving the role of range. As we shall see, this permits a hierarchical
construction of IFO fragments.

   Two fragments are illustrated in Fig. 2-3. The first shows the fragment used to model a set of students (in fact, an instance of the type with
single vertex STUDENT), along with a function mapping students to (printable)
grades. The second fragment is used to store a set of CLASSes and two
functions on that set. The first gives the teacher of each class. The
second gives the enrollment of each class, that is a set of students together with a function giving the grades of the students in that class.
Notationally, if CS101 is one of the classes, then ENR(CS101) denotes the
set of students enrolled in CS101, and ENR(CS101)/GRADE the grade function
for the CS101 class. Also, if Mary is a student in CS101 then ENR(CS101)/
GRADE(Mary) gives her grade in the class. Further nesting of functions is
also permitted in the IFO model. (A construct very similar to this nesting
of functions is introduced in another database model, namely the Verso model
[4, 1].) In the formal model, all functions are partial unless restricted
to be total by an explicit constraint.

   The final structural component of the IFO model is the representation
of ISA relationships. The characteristic feature of an ISA relationship
is that functions (or attributes) are "inherited" downwards along it
(i.e., if one entity type ISA is also a second entity type, then any function
defined on entities of the second type is automatically defined on entities
of the first type). In the IFO model two types of ISA relationship are distinguished, namely *specialization and generalization*. Speaking intuitively
(and briefly), the relationship 'EMPLOYEEs are PERSONs' is an example of
specialization, because an object is essentially a person first, and second-
arily s/he is an employee. On the other hand, the relationships that 'CARs,
PLANEs and BOATs are VEHICLEs' are examples of generalization, because the
basic object-types here are cars, planes, and boats, and they are combined
to form the entity set vehicle for certain purposes. Continuing at an intuitive level, specialization is used when the same object can take on one
or more different roles (e.g., a person might be a student, and then drop
out and become an employee), while generalization is used to group essentially disjoint entity sets. In cases of specialization the structure of
objects is "inherited downwards," in the sense that if people have a given
structure (as objects), then students have the same structure. On the other
hand, in cases of generalization structure is "inherited upwards," in the
sense that if an object is a vehicle then it is either a car, plane, or
boat, and has the structure of one of those. As we shall see in Sec. 3.2
below, the update propagation semantics of these two types of ISA relationship are also different.

   The specialization relationship is depicted using a broad arrow (➡),
whereas the generalization relationship is depicted using a shaded arrow
(➡). Examples of specialization (resp., generalization) are shown in Fig.
2-4 (resp., 2-5).

322

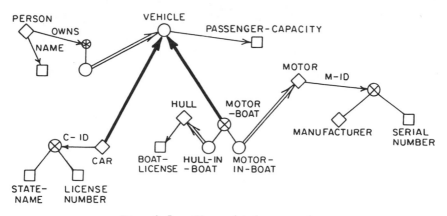

Fig. 2-5.   The vehicle example.

We now briefly describe the "global" restrictions imposed on the ISA relationships occurring in an IFO schema. Each of these five restrictions corresponds to intuitive properties expected of ISA relationships. Together they ensure that each IFO schema has a meaningful family of instances (see [3]), and that update semantics can be defined for IFO schemata in a coherent fashion (as detailed below).

Although the tone of this discussion is largely informal, it is necessary to provide a precise framework so that we may describe the global ISA restrictions in a simple manner. To begin, we give a pseudo-definition of "IFO graphs," which are directed graphs which satisfy all of the properties of IFO schemata, except possibly one or more of the ISA restrictions. (We say 'pseudo-definition' here because we have not given a formal definition of type or fragment.)

Definition: An *IFO graph* is a direct graph $\mathbf{S} = (V, E)$ such that:

1. V is the disjoint union of five sets V_P (*printable vertices*), V_A (*abstract vertices*), V_F (*free vertices*), $V_\otimes$ (*$\otimes$-vertices*), and $V_\circledast$ (*$\circledast$-vertices*);

2. E is the disjoint union of four sets E_O (*object edges*), E_F (*fragment edges*), E_S (*specialization edges*), and E_G (*generalization edges*) (each edge in $E_S \cup E_G$ is also called an *ISA edge*);

3. (V, E_O) is a forest of types; and

4. $(V, E_O \cup E_F)$ is a forest of fragments.

We now consider the five global restrictions placed on ISA relationships in IFO schemata. Each of the restrictions is phrased as a restriction placed on an IFO graph $\mathbf{S} = (V, E)$. The first restriction corresponds to the intuition that each free node inherits its underlying structure from some ISA edge.

ISA1:   Each free vertex is either the tail of at least one specialization edge, or the head of at least one generalization edge, but not both.

The next two ISA rules highlight the importance of some particular vertices, that is the vertices which are fragment roots. These vertices are called *primary* vertices. Primary vertices are the only vertices which

are not buried deeply in a structure and for which the domain does not depend on the value of a function. For that reason, if a new free type is
introduced using ISA relationships, it is required that only primary vertices be used in the formation of this type. More precisely, we have:

> ISA2: The head of each specialization edge and the tail of each gen-
> eralization edge must be a primary vertex; and

> ISA3: The head of each generalization edge must be a primary vertex.

As mentioned earlier, the object structures associated with vertices
in an IFO graph are determined largely by the object edges, and also by
ISA edges. Suppose that the object structure of some vertex p depends
(via some object and ISA edges) on some vertex q. Then one clearly does
not want the object structure of q to be depending on p. This leads to the
restriction:

> ISA4 (informal): There should not be any cycle of structure determina-
> tion.

The last ISA rule concerns specialization, and expresses the intuition
that a given type cannot be a subtype via specialization of two fundamen-
tally different types:

> ISA5: Two directed paths of specialization edges sharing the same
> origin can be extended to a common vertex.

Finally, we define an *IFO schema* to be an IFO graph satisfying the five
ISA rules.

A variety of results on IFO schemas and IFO instances are presented in
[3]. For instance, it is shown there that one can check in linear time
whether an IFO graph is an IFO schema. Also, some properties of the object
structures resulting from the interplay of ISA relationships in an IFO
schema are exhibited, and an algorithm to compute these object structures is
presented.

3.   UPDATE SEMANTICS IN THE IFO MODEL

This section is focused on the semantics of updates in the IFO model.
Because the IFO model is one of the first database models which incorporate essentially arbitrary ISA relationships, functions and object construction, it provides a unique framework in which to study updates. In
particular, it allows us to carefully examine the different ways that a
modification of the data associated with one part of a database schema can
affect data associated with other parts of the schema. The analysis of update propagation presented here highlights many of the subtleties resulting
from these various types of data relationships. In the interest of brevity,
the discussion shall be quite informal, and technical details will be omitted.

Three essentially separate topics will be addressed in order to describe
updates in their full generality. First, the semantics of updates in (non-
atomic) types will be discussed. Second, the relationship between updates
and ISA relationships will be analyzed. As we shall see, the acyclicity of
ISA relationships in IFO schemata (following from Rule ISA4) implies that
when computing the effect of a *permissible* atomic update on an instance of
a given IFO schema, each primary vertex of that schema need be "visited"
at most once. Finally, the impact of updates on fragment instances and general IFO instances will be considered.

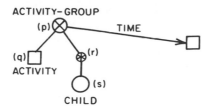

ACTIVITY-GROUP

Fig. 3-1. The kindergarten example.

In the discussion presented here we focus primarily on permissible atomic updates to IFO instances. It will be clear from the discussion that these can be combined to develop convenient "update transactions," so that users could easily specify a group of related primitive updates.

3.1. Updates on Objects

In this subsection we focus exclusively on update semantics for type instances. Two fundamental issues will be discussed, these being (i) updates directed at the primary vertices of types, and (ii) indirect updates caused by updates occurring at the (non-primary) leaves of a type (as might be implied by updates to other types in the schema and ISA relationships).

We begin with an informal discussion which illustrates both of these issues. Consider the fragment shown in Fig. 3-1, which might be used to store information about the daily schedule of activity-groups in a kinder-garten. Specifically, an activity-group is an ordered pair, with first co-ordinate an activity-name (such as 'nap' or 'art') and second coordinate a set of children. For each activity group the function TIME gives the time at which the group will begin the given activity.

The most direct type of update to an instance $\mathbf{I}$ of this type is one which deletes a given activity group (say O) from the instance (yielding the new instance $\mathbf{I} - \{O\}$). In a syntax developed in the full paper [3], such a deletion is denoted by the triple $(p, O, \perp)$. Similarly, a new ac-tivity group (say P) can be inserted into $\mathbf{I}$ (denoted $(p, \perp, P)$). Finally, an existing activity group (say O) might be replaced by a different one (say O'). This type of update is useful, for example, if the value for the new object O' under the function TIME is to be TIME(O). (This might arise, for example, if O' was obtained by adding a child to the group in O.) This notion of modification is also useful in connection with ISA re-lationships, as will be seen later. This update is denoted by (p, O, O').

We now consider how updates at the leaves of the type of Fig. 3-1 affect instances of that type. Intuitively, modifications and deletions to the "active domain" of a leaf of the type will permeate upwards in ob-jects in the active domain of the root of that type. First, suppose that the kindergarten has run out of paint, and that all 'paint' groups are to become 'draw' groups. This modification can be "requested" by the triple $(q,$ 'paint', 'draw'). If ('paint', {Elaine, Jimmy}) is an element of an instance $\mathbf{I}$, then under the update it will be replaced by ('draw', {Elaine, Jimmy}). On the other hand, if the activity 'paint' is to be dropped, this is denoted by $(q,$ 'paint,' $\perp)$, and in the resulting instance ('paint', {Elaine, Jimmy}) will simply be deleted.

The situation is more interesting if objects associated with the ver-tex s are to be modified or deleted. For example, suppose that Jimmy has moved to another town, and so he is to be deleted. (This might be implied by a specialization edge connecting the vertex s with some other vertex modelling the townspeople.) This deletion can be denoted by $(s,$ Jimmy, $\perp)$, and has the effect of replacing ('paint', {Elaine, Jimmy}) by ('paint', {Elaine}) in the overall instance.* And since this is a modification, the value TIME(('paint', {Elaine})) in the new instance will be the value of TIME(('paint', {Elaine, Jimmy})) in the old instance.† Similarly, the modification (s, Jimmy, Johnny) will imply the modification (p, ('paint', {Elaine, Jimmy}), ('paint', {Elaine, Johnny})) in the overall instance (possibly along with others).

We now indicate the formal framework developed in [3] for describing these kinds of update propagation. First, an *update* of a type **R** at a vertex p is defined to be an ordered triple (p, O, O') as indicated in the informal discussion above. A set M of update specs at p is said to be *consistent* if whenever (p, O, P) and (p, O, P') are in M, then O = ⊥. (In other words, a set M of update specs is consistent at p if it does not call for two conflicting actions on an object O.)

Two mechanisms are given for modifying an instance **I** of a type **R**. First, if M is a consistent set of updates at the root of **R**, then M[**I**] is defined to be the "result" of applying M to **I**. Second, if M is a set of updates at the leaves of **R** (which does not contain any insertion), then M[**I**] is defined to be the "result" of applying M to **I**. (This definition is recursive in nature.) Speaking informally, it is shown in [3] that in the latter case, the updates in M can be applied "all at once," or "one leaf at a time."

3.2. Update Propagation Resulting from ISA Edges

In this subsection we focus on updates involving types and ISA relationships. (The extension to fragments will be considered in the next subsection.) Our discussion begins with some informal remarks and examples. After this we outline a mechanism for specifying permissible atomic updates to IFO schemata (in which there are no fragment edges), and describe the *effect* of such permissible updates on IFO instances. It will be shown that this effect of permissible updates is intuitively natural and well-defined, and that it can be computed efficiently.

As noted in the introduction, a fundamental principle in the definition of update semantics in the IFO model is to preserve the identity of individual objects through modifications as much as possible. For example, suppose that **R** is a type with root p in an IFO schema, that O and O' are associated with **R** by an instance **I**, and that the modification (p, O, O') is requested. In this case, the effect at **R** of this update is to simply delete O; however, if there is a specialization edge (q, p) and O is associated with the vertex q, it is appropriate to modify O to O' at q (rather than simply deleting O). Such preservation of identity through modifications is also relevant in the case of generalizations.

In the full paper [3], a *permissible update* is defined to be either an insertion at the primary vertex of some type (satisfying certain conditions), or a modification or deletion at some primary or leaf vertex (again satisfying certain conditions). There are basically three different types of "local" propagations that occur in IFO schemata (without fragments) as a result of a single permissible update. Furthermore these local propagation effects interact on a "global" level in certain ways. We have already

---

*An alternative would be to simply remove ('paint' {Elaine, Jimmy}) from the overall instance if Jimmy is to be deleted. The development of update semantics presented here can easily be modified to satisfy this alternative interpretation of the implications of object deletion in sets.
†Again, this is one of several possible choices of "local" update semantics.

discussed one type of local propagation, namely the type of propagation that results from modifications and deletions at the leaves of types. We now describe the other two types of local propagation (associated with specialization and generalization) and their global interaction on an informal level. The differences in the semantics of update propagation for the two types of ISA relationships give partial justification for distinguishing them as done in the IFO model.

Consider now the IFO schema concerning persons, employees, students, and TAs shown in Fig. 2-4. There are three types of permissible updates involving the vertices at the ends of specialization edges for this schema. First, objects can be deleted, inserted, or modified at the vertex PERSON. If an object is deleted (or modified), then that deletion (modification) is propagated downwards (i.e, from the head of a specialization edge to its tail) in the obvious fashion. On the other hand, if an object is inserted at PERSON, then no other insertions occur. Second, objects associated with any of the vertices EMPLOYEE, STUDENT, or TA can be deleted. In such cases, the deletion again propagates downwards, and it has no impact on vertices above the vertex where the deletion occurred. And third, if an object O is associated with the head of a given specialization edge, then that object can be inserted at the tail of that specialization edge. This insertion has no propagational effects.

The reader will note that objects cannot be modified at specialization tails. Speaking intuitively, this is prohibited because objects associated with the tail of a specialization edge "derive" their existence from the fact that they also occur at the head of that specialization edge. For this reason, all modifications of objects of a given specialization tail p should first be specified at the type at the very "top" of the set of specialization paths emanating from p. Analogously, an arbitrary object cannot be inserted at a specialization tail (unless it is already associated with the associated specialization head). (It is apparent, however, that the set of permissible updates described here could be used to build a transaction which had the effect of modifying or inserting objects at specialization tails.)

We now consider the third type of local propagation, which results from generalization edges. Recall now the example involving the generalization of motorboats and cars which formed the set of vehicles shown in Fig. 2-5. In this situation, objects associated with the vertices MOTORBOAT and CAR can be deleted, inserted, or modified. Such updates have the obvious propagational effect on the set of VEHICLEs. On the other hand, objects associated with the vertex VEHICLE cannot be manipulated directly. Speaking intuitively, this is because objects associated with VEHICLE "derive" their existence exclusively from their presence in some underlying vertex. Thus, database users can manipulate such objects only by manipulating them at their initially defining vertex.

One subtlety of update propagation in connection with generalization relationships occurs if two or more vertices with non-disjoint domains are generalized into some single vertex. In this case the update propagation semantics is rather complex. While it is discussed in more depth in the paper [3], we do not consider this possibility here. In particular, we assume for the present discussion that whenever (p, r) and (q, r) are generalization edges in S, then the set of objects which can be associated with p is disjoint from the set of objects which can be associated with q.

We consider now the global interaction of the three types of local update propagation. Our definitions concerning updates are based on a number of intuitively natural assumptions. One fundamental assumption is that the effect of (the propagation of) an update on a given vertex can be determined from the effect of that update on all of the vertices "immediately

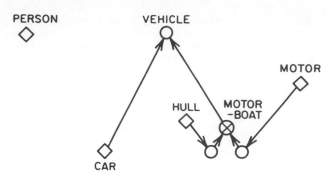

PERSON   VEHICLE

MOTOR

HULL   MOTOR
-BOAT

CAR

Fig. 3-2.   Update propagation graph of domain-
portion of schema in Fig. 2-5.

preceding" it relative to the three types of local update propagation (see
Rules UD 3, 4, and 5). As we shall see shortly, the acyclicity of a given
IFO schema (following from rule ISA4) shall guarantee that this kind of dependence does not lead to "loops" or cycles of update propagation. Furthermore, as we shall see, the total propagational effect of a given permissible update on an IFO instance can be calculated by visiting each
primary vertex of the schema at most once.

We now present a brief outline of the formal development concerning
update propagation for ISA and object relationships in the full paper [3].
First, we consider now only IFO schemata with no fragment edges.

As suggested in the previous discussion, it is possible to define on a
local level the "order" in which the propagation of updates proceeds in a
given IFO schema. This motivates the following:

Definition: Let S = (V, E) be an IFO schema with no fragment edges,
where $V = V_A \cup V_P \cup V_F \cup V_\otimes \cup V_\circledast$, and $E = E_O \cup E_S \cup E_G$. The *update propagation graph* of S, denoted UPG(S), is the graph*($V_{Pr} \cup V_L$, $E_{O'} \cup \rho(E_S) \cup
E_G$) where

    a. V_{Pr} is the set of primary vertices of S;

    b. V_L is the set of non-primary leaf vertices of types of S; and

    c. $E_{O'} = \{(p, q) \mid p$ is the leaf of a type with primary vertex q, and
       $p \neq q\}$.

The update propagation graph of the non-fragment portions of the IFO
schema of Fig. 2-5 is shown in Fig. 3-2. Speaking intuitively, it is easily
verified that if p is a primary vertex of an object schema S or the leaf of
a type of S, then the effect at p of a given update is determined on the
local level by the effects of that update at the set $\{q \mid (q, p)$ is an edge
of UPG(S)$\}$. A fundamental result of the IFO model (which is easily implied
by rule ISA4) is that the update propagation graph of any IFO schema is
necessarily acyclic. Intuitively, this means that when determining the
effect of an update at a given vertex, the update propagation graph can be
traversed from "bottom" to "top," with each vertex being visited at most
once.

---

*For a set of edges F in a directed graph, the <u>reversal</u> of F, denoted
$\rho(F)$, is $\{(q, p) \mid (p, q) \in F\}$.

328

To describe a global modification to an instance **I** of an IFO schema **S** we use the notion of a (*global*) *update,* which is a mapping U which associates a set of updates with each primary and leaf vertex of the schema. The "result" of such an update U on an instance **I** is defined to be the instance obtained by applying the updates in U at the appropriate vertices. (Actually, this application of U to **I** may result in an instance which violates the ISA relationships in the schema; an update is *consistent* if the result does satisfy these relationships.)

Recall now that a permissible update is a single update $u = (r, 0, 0')$ which satisfies certain properties. To define the global impact of u on an instance **I**, we must specify an update U on **I** which, intuitively speaking, corresponds exactly to u and all of the local propagational effects caused by u. In the full paper [3], this update U is characterized by five formally specified "rules." We now present informal statements of each of these.

UD1: $U(r) = \{u\}$, that is the effect of u at r is precisely to perform u.

UD2:   If there is no directed path from r to p in UPG(**S**), then $U(p) = \phi$.

UD3:   If $p \neq r$ is a non-leaf primary vertex of a type **R** with set of leaves Q, then $U(p)$ = the set of updates at p implied by the sets of updates assigned by U at the leaves of **R**.

UD4:   If $p \neq r$ is a specialization tail and $Q = \{q|(p, q) \in V_S\}$, then $U(p) = \{(p, 0, 0')|(q, 0, 0') \in U[Q]$ and 0 is in the active domain of **I**\}.

UD5:* Let $p \neq r$ be a generalization head and $Q = \{q|(q, p) \in E_G\}$. Then $U(p) = \{(p, 0, 0')|(q, 0, 0') \in U[Q]\}$.

Note that rules UD3, UD4, and UD5 express the local propagational effects of object relationships, specialization relationships, and generalization relationships, respectively.

In the paper, an Update Algorithm is given which computes an update U corresponding to a permissible update u in **I**. The following result (stated informally here) is then demonstrated:

Theorem 3.1:   Let **S** be an object schema, **I** an instance of **S**, and u a permissible update for **I**. Then there is exactly one update U corresponding to u for **I**. Furthermore, this update is consistent, and it is computed by the Update Algorithm.

In view of the above theorem, we can unambiguously define the "effect" of a permissible update on an instance of an object schema.

3.3.   Updates on Fragments

In this subsection we briefly describe how the update semantics described above is extended to incorporate fragments, and hence, functional relationships. Two distinct topics are considered, first permissible updates in connection with fragments alone (such as changing a given function value), and then the impact of update propagation elsewhere in the IFO schema on a given fragment instance.

---

*Recall that in this abstract we consider only those schemata for which generalized vertices have disjoint domains.

Consider now the fragment involving courses, professors, and enrollments shown in Fig. 2-3b. A fundamental type of permissible update on instances of this fragment is to change a functional value. For example, if CS101 is a course and currently TEACHES(CS101) = Toto, the new value Lulu can be assigned by TEACHES(CS101) := Lulu (and the value 'undefined' is assigned by TEACHES(CS101) := ⊥). Suppose now that Zaza is a new student to be inserted into the set ENR(CS101). This is requested by: insert Zaza into ENR(CS101). Once Zaza is in place, a grade can be assigned using ENR(CS101)/GRADE(Zaza) := 'A'.

As a result of the ISA rules, the range portion of a fragment can participate in only one type of ISA relationship. Specifically, a vertex of this range portion can be a specialization tail, but not a specialization head or part of a generalization edge. As a result, modifications to the range of a fragment have no impact on the rest of an IFO schema. (A subtlety here is that function assignments must satisfy the specialization relationships called for in the schema. For example, if there is a specialization from the PROFESSOR vertex to a vertex EMPLOYEE elsewhere in the schema, the assignment TEACHES(CS101) := Lulu can be made only if Lulu is already associated with the vertex EMPLOYEE.)

Two types of modification of fragment instances in an IFO schema can result from global propagation elsewhere in the schema. First, if the course CS101 is deleted (by a permissible update, for example), the functional values associated with CS101 are simply dropped. And if a new course is inserted, all function values for it are assigned the value ⊥ (i.e., 'undefined'). (Update semantics are also given in [3] for the case where a domain object is modified.)

The second type of global propagation affecting fragment instances results from specialization edges with tails in the range part of a fragment. For example, suppose that (STUDENT, PERSON) is a specialization edge, that ENR(CS101) = {Mimi, Lolo}, and that Mimi is deleted from PERSON. Then the object {Mimi, Lolo} is modified to become {Lolo}, and the value ENR(CS101) is automatically updated to {Lolo}.

As shown in the full paper [3], these two types of global propagation can be incorporated into the framework given in the previous subsection.

REFERENCES

1. S. Abiteboul and N. Bidoit, Non first normal form relations to represent hierarchically organized data, Journal of Computer and System Sciences 33 (3); 361-393, Dec. 1986.
2. S. Abiteboul and R. Hull, IFO: A Formal Semantic Database Model (Preliminary Report), Proc. ACM SIGACT-SIGMOD Symp. on Principles of Database Systems (1984), pp. 119-132.
3. S. Abiteboul and R. Hull, IFO: A formal semantic database model, To appear in ACM Trans. on Database Systems.
4. F. Banchilhon, et al., Verso: A relational back end data base machine, Proc. Inter. Workshop on Database Machines, San Diego, California (1982).
5. R. Brown and D. S. Parker, LAURA: A formal data model and her logical design methodology, VLDB (1983), pp. 206-218.
6. P. Buneman, R. E. Frankel, and R. Nikhil, An implementation technique for database query languages. ACM Trans. on Database Systems 7 (2): 164-186, 1982.
7. P. P. Chen, "The entity-relationship model - toward a unified view of data," ACM Trans. on Database Systems 1, 1 (1976), 9-36.

8. E. F. Codd, "Extending the database relational model to capture more meaning," ACM Trans. on Database Systems 4, 4 (1979), 397–434.

9. M. Hammer and D. McLeod, "Database description with SDM: A semantic database model," ACM Trans. on Database Systems 6, 3 (1981), 351-386.

10. R. Hull and C. K. Yap, "The format model: A theory of database organization," J. ACM 31, 3 (1984), 518-537.

11. W. Kent, "Limitations of record-based information models," ACM Trans. on Database Systems 4, 1 (1979), 107-131.

12. L. Kerschberg and J. E. S. Pacheco, A functional data base model, Pontificia Universidade Catolica do Rio de Janeiro, Rio de Janeiro, Brazil, February 1976.

13. R. King and D. McLeod, "A methodology and tool for designing office information systems," ACM Trans. on Office Information Systems, 1985.

14. R. King and D. McLeod, Semantic database models, In: Database Design, S. B. Yao, ed., Springer-Verlag, New York (1985), pp. 115–150.

15. D. McLeod and J. M. Smith, Abstraction in databases, Workshop on Data Abstraction, Databases, and Conceptual Modelling, Pingree Park, Colorado (1980), pp. 19-25.

16. D. Shipman, "The functional data model and the data language DAPLEX," ACM Trans. on Database Systems 6, 1 (1981), 140-173.

17. J. M. Smith and D. C. P. Smith, "Database abstractions: Aggregation and generalization," ACM Trans. on Database Systems 2, 2 (1977), 105-133.

18. D. C. Tsichritzis and F. H. Lochovsky, Data Models, Prentice-Hall, Englewood Cliffs, New Jersey (1982).

COMPUTATION-TUPLE SEQUENCES AND OBJECT HISTORIES:

EXTENDED ABSTRACT

Seymour Ginsburg
Computer Science Dept.
University of Southern
California
Los Angeles, CA90089-0782
U.S.A.

Katsumi Tanaka
Dept. of Instrumentation
Engineering,
Kobe University
Rokkodai, Nada, Kobe 657
Japan

INTRODUCTION

Recently, much attention has been focussed on database aspects of office automation and their associated research problems [A,EN,PFK,SIGMOD,SSLKG]. One major component in this area is a facility to manage historical data for many "objects", namely, an object history (management) system. Examples of such systems are those for personal checking account, salary reviews and schedules. However, there has been little widely accepted mathematical formalism which captures major aspects of the historical data for the objects (here called object histories) and permits the identification and analysis of relevant issues. The purpose of the present paper is to introduce a mathematical model allowing such an investigation[1].

The paper consists of three sections. Section 1 introduces the model. In our model, an object history (which frequently includes some computation) is represented by a sequence of tuples, called a *computation-tuple sequence*. A set of "valid" computation-tuple sequences is specified by a *computation-tuple sequence scheme* (abbreviated *CSS*), that includes computation functions, semantic constraints and starting conditions. Section 2 deals with a special type of CSS, called *local*, in which validity-maintenance of a computation-tuple sequence is relatively easy to check. Two characterizations are then presented. The first pertains to the *localness* of semantic constraints, while the second concerns when a given CSS can be replaced by a local CSS with the same set of valid computation-tuple sequences. In Section 3, we introduce the concepts of *bi-simulatability* and *local bi-simulatability* for regarding

---

The first author was supported in part by the National Science Foundation under grant MCS-792-5004. Work done by the second author while he was on leave from Kobe University. The second author was supported in part by the Sakko-Kai Foundation. This paper is an extended abstract version of [GT].

[1]Another such investigation is [CW]. There is very little overlap between the two papers, in either formalism or in questions of concern.

two CSS (over possibly different sequences of attributes) as conveying
the same information. Two major results on local bi-simulatability are
then established. The first discusses (local) bi-simulatability of a
CSS by a CSS (i) which has no computation attributes and (ii) whose
maintenance of computation-tuple sequence validity only involves the
last two computation tuples. The second result shows that local bi-
simulatability preserves the property of having an "equivalent" CSS
which is local.

1. THE MODEL

To motivate what follows, we present a simple object-history example.

Example 1.1 (Checking Account):
Consider the following checking accout plan of a savings and loan
association. There are the usual two actions of DEPOSIT and WITHDRAW
(some AMOUNT), each followed by a computation of the new account
balance. A special type of action pays interest daily on the day's
minimum balance, at the annual (NOW) rate of 5.25% on the first $2000
and a variable higher rate on the rest. Each DATE-value is assumed to
uniquely determine the higher rate, called $TNIR^2$. A typical example of
an object history (here, an individual checking account is considered
as an object) is given in Figure 1.1.

The computation-tuple sequence scheme is a construct whose purpose is
to define a set of valid (for the same object) sequences of computation
tuples. It consists of:

(*) A set of attributes, partitioned into memory, input and
computation attributes;

(**) Appropriate functions which calculates values for the
computation and memory attributes;

(***) Semantic constraints involving the order of the computation
tuples; and

DATE	ACTION	AMOUNT	TNIR	END-D/W BAL	DAILY MIN-BAL	NOW INTEREST	TNI^3	BALANCE
4/1/85	DEPOSIT	2500	.00037	2500	0	0	0	2500
4/1/85	INTEREST	0	.00037	2500	0	0	0	2500
4/2/85	WITHDRAW	200	.00037	2300	2300	0	0	2300
4/2/85	INTEREST	0	.00037	2300	2300	2.87	0.11	2302.98
4/3/85	WITHDRAW	500	.00038	1802.98	1802.98	0	0	1802.98
4/3/85	DEPOSIT	200	.00038	2002.98	1802.98	0	0	2002.98

Figure 1.1: An example of checking account history

---

^{2}TNIR=TREASURY-NOTE-INTEREST-RATE
^{3}TNI=TREASURY-NOTE-INTEREST

(****) Specific sequences (the initialization) with which to start the valid computation-tuple sequence.

Let X be a finite nonempty set of attributes and $A_1,...,A_n$ some fixed listing of the distinct elements in X. Then, $\langle X \rangle$ denotes the sequence $A_1...A_n$, and $Dom(\langle X \rangle)$ the cartesian product[4] $Dom(A_1) \times ... \times Dom(A_n)$. Also, $\langle X|A_i \rangle$ denotes the prefix $A_1...A_{i-1}$, $i \geq 2$.

Definition:
Let $\langle U \rangle$ be a sequence of attributes. A *computation tuple* over $\langle U \rangle$ is an ordered pair $(\langle U \rangle, u)$, or u when $\langle U \rangle$ is understood, where u is an element in $Dom(\langle U \rangle)$. A *computation-tuple sequence* over $\langle U \rangle$ is a sequence **u** of computation tuples over $\langle U \rangle$. By $|u|$ is meant the length of **u**.

To implement (*) and (**), we have:

Definition:
A *computation scheme* (abbreviated *CS*) over $\langle U \rangle$ is a 5-tuple $C = (\langle S \rangle, \langle I \rangle, \langle E \rangle, F_{eval}, F_{state})$, where

1. S, I and E are pairwise disjoint subsets of U (of *state, input,* and *evaluation* attributes, respectively) with S and I nonempty and[5] $\langle U \rangle = \langle S \rangle \langle I \rangle \langle E \rangle$.

2. $F_{eval} = \{e_C|$ for each attribute C in E, e_C is a partial function (called an *evaluation* function) from $Dom(\langle U \rangle)^{P_C} \times Dom(\langle U|C \rangle)$ into $Dom(C)$ for some non-negative integer $\rho_C \}$; and

3. $F_{state} = \{f_A|$ for each attribute A in S, f_A is a partial function (called a *state* function) from $Dom(\langle U \rangle)$ into $Dom(A)\}$.

The integer ρ_C is called the *rank* of e_C ; and $\rho = \{\rho_C, 1| e_C$ in $F_{eval}\}$ is the *rank* of C.

The purpose of a CS is to assist in defining valid computation-tuple sequences. It does this by the notion of consistency. Intuitively, a computation-tuple sequence **u** is consistent if ultimately the values for the state and evaluation attributes are determined by the corresponding state and evaluation functions. More formally, we have:

Definition:
Let $u = u_1...u_m$ ($m \geq 1$) be a computation-tuple sequence over $\langle U \rangle$. Then **u** is said to be *consistent* with a CS $C = (\langle S \rangle, \langle I \rangle, \langle E \rangle, F_{eval}, F_{state})$ over $\langle U \rangle$ if the following conditions hold:

(1) For each $1 \leq h \leq m$ and A in S, if $h \geq 2$ then $u_h(A) = f_A(u_{h-1})$; and

(2) For each $1 \leq h \leq m$ and C in E, if $h > \rho_C$ then[6] $u_h(C) = e_C(u_{h-\rho_C},...,u_{h-1},u_h[\langle U|C \rangle])$.

---

[4]Here, Dom(A) (called the domain of A) is a subset of Dom_∞ of at least two elements, where Dom_∞ is an infinite set of elements.

[5]Given sequences $\langle U_1 \rangle = A'_1...A'_{m1}$ and $\langle U_2 \rangle = B'_1...B'_{m2}$, $\langle U_1 \rangle \langle U_2 \rangle = A'_1...A'_{m1}B'_1...B'_{m2}$.

[6]Let $\langle U \rangle = A_1...A_n$ and $\langle X \rangle$ a sequence of $\langle U \rangle$. For each computation-tuple u over $\langle U \rangle$, the projection of u onto $\langle X \rangle$, denoted $u[\langle X \rangle]$, is the computation-tuple v over $\langle X \rangle$ defined by $v(A) = u(A)$ for each A in X.

Example 1.2:
A computation scheme $C = (\langle S \rangle, \langle I \rangle, \langle E \rangle, F_{eval}, F_{state})$ for the checking account plan is as follows (We omit here the details of F_{eval} and F_{state}):

    $\langle S \rangle$=DATE;
    $\langle I \rangle$=ACTION,AMOUNT,TNIR; and
    $\langle E \rangle$=END-D/W-BAL,DAILY-MIN-BAL,
              NOW-INTEREST,TNI,BALANCE

Note that the current DATE-value u_i(DATE) depends on only one previous computation tuple u_{i-1} since u_i(DATE)=u_{i-1}(DATE)+1 (calendarwise addtion) if u_{i-1}(ACTION)=INTEREST and u_i(DATE)=u_{i-1}(DATE) otherwise. It is readily seen that the values for each C in E depend only on the current and previous computation tuples. Thus, the rank ρ of C is 1.

To discuss (***), we use the well-known notion of a constraint [U]. However, the concept of an arbitrary constraint for computation-tuple sequences[7] is too general for our purposes. We shall restrict our constraints to a special class called *uniform*. These are characterized by the fact that satisfaction holds uniformly throughout a computation-tuple sequence, i.e., holds in every interval of a computation-tuple sequence.

Definition:
A constraint σ over $\langle U \rangle$ is *uniform* if, for each $u=u_1...u_m$ over $\langle U \rangle$, $u \models \sigma$ implies $u_i...u_j \models \sigma$ for all i and j, $1 \leq i \leq j \leq m$.

Clearly, a constraint is uniform iff it is prefix preserving and suffix preserving.

Example 1.3:
A set of constraints for checking account plan of Example 1.1 is $\Sigma = \{\sigma_1, \sigma_2\}$, where σ_1 and σ_2 are functions from the computation-tuple sequences over $\langle U \rangle$ into {true, false} defined for each $u=u_1...u_m$ by

    $\sigma_1(u)$=true iff u_i(DATE)=u_j(DATE) implies u_i(TNIR)=u_j(TNIR) for all $i \leq j$

and

    $\sigma_2(u)$=true iff u_i(ACTION)=INTEREST implies u_i(AMOUNT)=0 for all i.

Obviously, σ_1 and σ_2 are uniform constraints.

The last concept needed is the *initialization*, an appropriate set of computation-tuple sequences with which to start a sequence until all state and evaluation functions can be applied (see (****)). Combining the notions of a CS, uniform constraints and an initialization denoted by *I*, we have:

Definition:
A *computation-tuple sequence scheme* (abbreviated *CSS*) over $\langle U \rangle$ is a triple $T=(C, \ \Sigma, \ I)$, where

    (1) C is a computation scheme over $\langle U \rangle$;

---

[7]A constraint σ over $\langle U \rangle$ is a mapping which assigns to each computation-tuple sequence over $\langle U \rangle$ a value of "true" or "false". If $\sigma(u)$=true, then u is said to satisfy σ, denoted by $u \models \sigma$.

(2) Σ is a finite set of uniform constraints over ⟨U⟩; and

(3) *I* is a prefix-closed subset of {u over ⟨U⟩ | u ⊨ Σ, u is consistent with C, and |u|≦ρ, ρ the rank of C}.

A CSS determines valid computation-tuple sequences as follows:

Definition:
Let T=(C, Σ, *I*) be a CSS. A computation-tuple sequence **u** is said to be *valid* (for T) if

(i) **u** is in *I* or

(ii) **u** ⊨ Σ, **u** is consistent with C and **u**=**vw** for some **v** in *I* of length ρ, where ρ is the rank of C.

Let VSEQ(T) be the set of all valid computation-tuple sequences for T.

It is readily seen that VSEQ(T) is closed under prefixes, but not intervals.

Example 1.4:
For the checking-account plan, let

I= {u over ⟨U⟩ | u=u₁, where u₁(ACTION)=DEPOSIT, u₁(AMOUNT)=u₁(END-D/W-BAL), and u₁(DAILY-MIN-BAL)=0}.

It is obvious that each element in *I* is consistent with C and satisfies Σ.

2. LOCAL COMPUTATION-TUPLE SEQUENCE SCHEMES

In this section, we introduce and note some properties of a special type of CSS called *local*. This type of CSS is characterized by a k_1-*local* ($k_1 \geq 2$) computation scheme and k_2-*local* ($k_2 \geq 1$) uniform constraints.

Definition:
For each integer $k \geq 2$, a computation scheme is said to be *k-local* (abbreviated *k-LCS*) if its rank is less than k.

Clearly, if C is a k-LCS over ⟨U⟩, then for each u=u₁...u_m (m≧k) over ⟨U⟩, **u** is consistent with C if (and only if) u_i u_{i+1}...u_{i+k-1} is consistent with C for all i≦m-k+1. In a k-LCS, to maintain the consistency for adding a computation tuple u to a consistent computation-tuple sequence **u**, we need only check the consistency of the sequence containing the last k computation tuples in **u**, a relatively easy task to do. For a similar reason, it is of interest to consider when satsifaction of a constraint by a computation-tuple sequence under addition of a computation tuple can be maintained by examining satisfaction of the last k-tuples in the new sequence. This leads to the following concept:

Definition:
A *k-local* (1≦k) *constraint* σ over ⟨U⟩ is a constraint over ⟨U⟩ with the following property:

There exists some constraint σ' over $\langle U \rangle$ such that, for all $u = u_1 \ldots u_m$ $(m \geq k)$, $u \models \sigma$ if and only if $u_i u_{i+1} \ldots u_{i+k-1} \models \sigma'$ for all $i \leq m-k+1$.

Given a k-local constraint σ, it is clear that for all $u = u_1 \ldots u_m$ $(m \geq k)$, $u \models \sigma$ iff $u_i \ldots u_{i+k-1} \models \sigma$ for all $i \leq m-k+1$. In addition, it is readily seen that the following properties hold for k-local constraints:

(1) The class of k-local constraints over $\langle U \rangle$, thus k-local uniform constraints over $\langle U \rangle$, is closed under the logical connective $\wedge$.

(2) If σ is a k-local constraint, then σ is k'-local for all $k' > k$.

(3) If σ is a k-local constraint and $u \models \sigma$ for all u, $|u| < k$, then σ is a uniform constraint.

Two characterizations of k-local uniform constraints are next given:

Proposition 2.1: Let σ be a constraint over $\langle U \rangle$ and k a positive integer. Then the following statements are equivalent:

(1) σ is k-local uniform.

(2) There exists a set K of computation-tuple sequences over $\langle U \rangle$ of length at most k such that[8] $Sup(K) = \{u$ over $\langle U \rangle \mid u \models \sigma\}$.

(3) There exists a set K' of computation-tuple sequences over $\langle U \rangle$ of length at most k such that $Sub(K') = \{u$ over $\langle U \rangle \mid u \models \sigma\}$, where $Sub(K')$ denotes the set $\{u$ over $\langle U \rangle \mid$ every interval v of u, $|v| \leq k$, is in $K'\}$.

Combining k_1-LCS and k_2-local uniform constraints, we get:

Definition:
A CSS $T = (C, \Sigma, I)$ is said to be *local* if for some k_1 and k_2, C is k_1-local and Σ is a set of k_2-local uniform constraints. In the above, T is said to be (k_1, k_2)-local.

Consider the CSS $T = (C, \Sigma, I)$ of Example 1.3. Since the rank ρ of C is 1, the computation scheme C is 2-local. Therefore, for all $u = u_1 \ldots u_m$ $(m \geq 2)$, u is consistent with C iff $u_i u_{i+1}$ is consistent with C for all $i < m$. However, there is no k for which σ_1 is k-local. Even though some CSS T over $\langle U \rangle$ is not local, there may exist another CSS T' over $\langle U \rangle$ such that T' is local and $VSEQ(T) = VSEQ(T')$. This situation is of interest for design considerations since validity of computation-tuple sequences for local CSS are relatively easy to maintain. This leads to the following concept:

Definition:
Let $T = (C, \Sigma, I)$ be a CSS over $\langle U \rangle$. Then, T is said to be *locally representable* if there exists a local CSS $T' = (C, \Sigma', I)$ such that $VSEQ(T) = VSEQ(T')$ and Σ' is a set of k-local uniform constraints for some k.
[Note that T' is required to have the same computation scheme and initialization as T.]

---

[8]For each set K of computation-tuple sequences over $\langle U \rangle$, $Sup(K)$ denotes the set $\{u'$ over $\langle U \rangle \mid u$ is an interval of u' for some u in $K\}$.

Example 2.1:
Let $T=(C, \Sigma, I)$ be the CSS in Example 1.2, 1.3, and 1.4, and
$T'=(C, \Sigma', I)$, $\Sigma'=\{o_1', o_2\}$, with $o_1'(u)=$true iff $u_i(DATE)=u_{i+1}(DATE)$
implies $u_i(TNIR)=u_{i+1}(TNIR)$ for all i. Clearly, o_1' is 2-local. The
constraint o_2 is 1-local, and thus 2-local. Hence T' is local.
Furthermore, it is readily seen that VSEQ(T)=VSEQ(T'). Thus, T is
locally representable.

A criterion for a CSS to be locally representable is:

Theorem 2.1:
Let $T=(C, \Sigma, I)$ be a CSS over $\langle U \rangle$ and o the constraint over $\langle U \rangle$
defined by $\{u$ over $\langle U \rangle \mid o(u)=$true$\} =$ Interval(VSEQ(T)), where
Interval(VSEQ(T)) $= \{u$ over $\langle U \rangle \mid u$ an interval of some element in
VSEQ(T)$\}$. Let $T'=(C, \{o\}, I)$. Then, T' is a CSS such that VSEQ(T')
=VSEQ(T), and T is locally representable iff T' is local.

3. BI-SIMULATABILITY

In this section, we consider the question:

> what does it mean for two CSS (over possibly different
> sequences of attributes) to convey the same information?

To examine this, we present two abstractions, *bi-simulatability* and
local bi-simulatability . Bi-simulatability is a natural and basic
concept. Local bi-simulatability combines bi-simulatability with a
local reconstructability property.

Intuitively, the notion of bi-simulatability consists of

> (i) the existence of a one-one onto correspondence between two sets
> of large enough valid computation-tuple sequences, and
> (ii) the preservation of the prefix relationship.

In essence, property (ii) states that when u over $\langle U \rangle$ is bi-simulated
by v over $\langle V \rangle$, then the addition of tuple u to u is bi-simulated by the
addition of some tuple sequence to v, and conversely.

Definition:
Let T_1 and T_2 be CSS over $\langle U \rangle$ and $\langle V \rangle$ resp. Then, T_2 is said to
bi-simulate T_1 if there exist positive integers m_1, m_2 and a one-one
function μ from[9] VSEQ(T_1)$_{m_1}$ onto VSEQ(T_2)$_{m_2}$ such that for all u_1, u_2 in
VSEQ(T_1)$_{m_1}$, u_1 is a prefix of u_2 iff $\mu(u_1)$ is a prefix of $\mu(u_2)$. In
the above, T_2 is said to *bi-simulate* T_1 *via* (m_1, m_2, μ).

Note that if T_2 bi-simulates T_1 via (m_1, m_2, μ), then T_1 bi-simulates T_2
via (m_2, m_1, μ^{-1}), where μ^{-1} is the inverse of μ.

The following establishes a length connection under the bi-
simulatability mapping μ.

---

[9]The symbolism VSEQ(T)$_m$ denotes the set $\{u$ in VSEQ(T) $\mid |u| \geq m\}$.

Proposition 3.1:
Suppose T_2 bi-simulates T_1 via (m_1, m_2, μ). Then, $|\mu(u)| - |u| = m_2 - m_1$ for all u in $VSEQ(T_1)_{m_1}$.

As corollaries of the above proposition, we have:

(1) If T_2 bi-simulates T_1 via (m_1, m_2, μ), then for each nonnegative integer h, there exists a function μ' such that T_2 bi-simulates T_1 via (m_1+h, m_2+h, μ') (one such function μ' is μ restricted to $VSEQ(T_1)_{m_1+h}$), and

(2) Bi-simulatability is an equivalence relation.

If T_2 bi-simulates T_1 via (m_1, m_2, μ) and $\mu(u)$ is known, then u can be obtained by applying μ^{-1} to $\mu(u)$. However, that may not always be feasible. Just as local constraints are an advantage over general constraints, so being able to determine the tail end of u by employing the tail end of $\mu(u)$ is an advantage. This leads to:

Definition:
Let T_1 and T_2 be CSS over $\langle U \rangle$ and $\langle V \rangle$ resp. Then T_2 is said to *locally bi-simulate* T_1 if there exist m_1, m_2, μ, positive integers l_1 and l_2, and functions g_1, g_2 with the following properties:

(1) T_2 bi-simulates T_1 via (m_1, m_2, μ);

(2) g_1 maps $Dom(\langle U \rangle)^{l_1}$ into $Dom(\langle V \rangle)$ such that[10] $tail_1(v) = g_1(tail_{l_1}(\mu^{-1}(v)))$ for each v in $VSEQ(T_2)_{m_2}$; and

(3) g_2 maps $Dom(\langle V \rangle)^{l_2}$ into $Dom(\langle U \rangle)$ such that $tail_1(u) = g_2(tail_{l_2}(\mu^{-1}(u)))$ for each u in $VSEQ(T_1)_{m_1}$.

In the above, T_2 is said to *locally bi-simulate* T_1 via $(m_1, m_2, \mu, g_1, g_2, l_1, l_2)$.

Intuitively, local bi-simulatability means bi-simulatability plus the ability to determine the last tuple of one valid computation-tuple sequence by observing a fixed-length suffix of its correspondent. Local bi-simulatability is also an equivalence relation.

Bi-simulatability and local bi-simulatability are introduced to reflect the notion of two CSS conveying the same information. Thus, it is natural to ask the following questions:

> *To what extent can these two concepts be used to "simplify" the structure of a CSS? In particular, can the evaluation attributes be eliminated?*

We shall show below (Theorem 3.1) that the answer is yes. In order to establish Theorem 3.1, we first prove the following proposition. This result presents a condition on a CSS T_1 under which dropping the evaluation attributes and evaluation functions essentially yields a CSS T_2 which locally bi-simulates T_1.

---

[10]For each computation-tuple sequences $w = w_1 \ldots w_n$ and l ($1 \le l \le n$), $tail_l(w)$ is the suffix $w_{n-l+1} \ldots w_n$ of w.

Proposition 3.2:
Let $T_1=((\langle S\rangle,\langle I\rangle,\langle E\rangle,F_{eval}, F_{state}),\Sigma_1,I_1)$ be a CSS over $\langle U\rangle=\langle S\rangle\langle I\rangle\langle E\rangle$ such that $\rho_C=0$ for each C in E. Then there exists a CSS $T_2 = ((\langle S\rangle, \langle I\rangle, \langle \phi\rangle, \quad F'_{eval}, \quad F'_{state}),\Sigma_2,I_2)$ over $\langle V\rangle=\langle S\rangle\langle I\rangle$ such that

(a) $VSEQ(T_2)= \{u[V] \mid u \text{ in } VSEQ(T_1)\}^{11}$,

(b) Σ_2 is a set of k-local uniform constraints if Σ_1 is, and

(c) T_2 locally bi-simulates T_1.

In the above, part (a) need not hold if $\rho_C \neq 0$ for some C in E.

Theorem 3.1:
Each CSS T_1 is locally bi-simulated by some CSS T_2 having no evaluation attributes. Furthermore, T_2 can be made (2,1)-local if T_1 is local.

Our final result shows that local bi-simulatability preserves local representability. Formally, we have:

Theorem 3.2: Suppose T_1 is locally representable and T_2 locally bi-simulates T_1. Then, T_2 is also locally representable.

On the other hand, in the case of bi-simulatability, we give an example of CSS T_1 and T_2 such that T_2 bi-simulates T_1, T_1 is locally representable, and T_2 is not locally representable. This means that bi-simulatability does not preserve local representability.

4.CONCLUDING REMARKS

In this extended abstract, a model is introduced for describing historical data for *objects* such as checking accounts, credit-card accounts, seminar schedules etc. An object history is represented by a sequence of *computation tuples* over some fixed sequence of attributes. The set of all valid object histories is specified by a CSS. The CSS has three major features (in addition to input data): computation(involving previous computation tuples), *uniform* constraints (involving the sequencing of computation tuples), and specific sequences with which to start the valid computation-tuple sequences. A special type of CSS, vcalled *local,* is then singled out for its particular simplicity in maintaining the consistency of the computation and the satisfaction of the constraints. A necessary and sufficient condition for a CSS to be equivalent to some local CSS is established.

Finally, the restructuring of CSS via the notion of *bi-simulatability* and *local bi-simulatability* are studied. Intuitively, bi-simulatability means the existence of a one-one onto prefix-preserving mapping between two sets of valid computation-tuple sequences. *Local* bi-simulatability further requires the local reconstructability of the

---

[11]Let $\langle X\rangle$ be a subsequence of $\langle U\rangle$. For each computation-tuple sequence $u=u_1...u_m$ over $\langle U\rangle$, the projection of u onto $\langle X\rangle$, denoted by $u[\langle X\rangle]$, is the computation-tuple sequence $x_1...x_m$ over $\langle X\rangle$, where $x_i=u_i[\langle X\rangle]$ for each i.

original computation-tuple sequence by the one locally bi-simulating it. (More precisely, the last computation-tuple of a valid computation-tuple sequence is detrermined from some fixed-length tail end sequence of the corresponding computation-tuple sequence.) Roughly speaking, bi-simulatability and local bi-simulatability are concepts for regarding two differently structured CSS as conveying the same information. A number of basic properties about (local) bisimulatability are established, among which are the following:

(1) Both bi-simulatability and local bi-simulatability are equivalence relations.

(2) Bi-simulatability implies that the addition of one computation tuple to a valid computation-tuple sequence is *bi-simulated* by the addition of exactly one computation tuple to the corresponding computation-tuple sequence.

(3) Each CSS T is locally bi-simulated by a CSS T' which has no evaluation attributes. Furthermore, T' can be made (2,1)-local if T is local.

(4) Local bi-simulatability preserves the property of having an *equivalent* CSS which is local.

The present investigation is just a beginning in the study of object histories. The following are a few of the many problems still to be examined.

(1) The notion of *uniform* and/or *local* constraints discussed here may be too broad. It will be important to identify various classes of meaningful and mathematically tractable constraints which arise in important applications.

(2) There is no clear distinction between the role of the evaluation functions and the role of the uniform constraints. (Indeed, it is possible to express each evaluation function as a special type of uniform constraints.)

(3) It will be important to find more natural methods for specifying the initialization set I, since I is now described as a prefixed-closed set of computation-tuple sequences of length at most ρ.

(4) The purpose of a CSS as introduced here is to specify the set of all valid object histories over a fixed sequence of attributes. In many applications, there are multiple object histories over the same sequence of attributes, e.g., checking-account histories for different individuals or schdules for different seminars. So, it will be necessary to extend our model to include this situation, which will require some method to describe the set of all the valid sets of object histories.

(5) Operations such as projection, join, selection, etc. play a fundamental role in relational database theory. In the same sense, it is important to find such basically important operations on object histories. (One of such important classes is identified in [GT2],[GT3], called interval queries. Given an object history, an interval query returns an interval of the object history as its result. Also, as in relational databases, the projection operation seems to be important on object histories, which is investigated in [GT4].)

342

REFERENCES

[A]   *AFIPS   Office   Automation   Conference   Digests,   1980-
           1983*.

[CW]   J. Clifford and D.S.Warren, "Formal Semantics for Time in
           Databases", *ACM TODS,* Vol.1,   pp.214-254,   1983.

[EN]   C.A.Ellis and G.J.Nutt, "Office Information Systems and
           Computer   Science",   *ACM Computing Surveys,* Vol.12,
           pp.27-60,   1980.

[GT]   S.Ginsburg and K.Tanaka, "Computation-tuple Sequences and
           Object   Histories",   *USC Technical Report*,   TR-83-217,   Nov.
           1983.

[GT2]   S.Ginsburg and K.Tanaka, "Interval Queries on Object
           Histories",   *USC Technical Report*,   TR-84-302,   Feb.   1984.

[GT3]   S.Ginsburg and K.Tanaka, "Interval Queries on Object
           Histories: Extended   Abstract",   *Proc. of 10th
           International   Conference   on   Very   Large   Data   Bases,*
           August 1984, pp.208-217, Singapore.

[GT4]   S.Ginsburg and C.Tang, "Projections on Object Histories",
           USC Technical Report,   TR-84-311,   Oct.1984.

[PFK]   R.Purvy, J.Farrell and P.Klose, "The Design of Star's
           Record   Processing",   *ACM TOOIS,* Vol.1,   pp.3-24,   1983.

[SIGMOD]   *ACM   Databases   for   Business   and   Office   Applications
           Proceedings,   1983.*

[SSLKG]   M.Stonebraker, H.Stettner, N.Lynn, J.Kalash and
           A.Guttman, "Document Processing in a Relational Database
           System", *ACM TOOIS,* Vol.1,   pp.143-158,   1983.

[U]   J.D.Ullman,   *Principles   of   Database   Systems,   2nd
           Ed..,* Computer   Science   Press,   Potomac,   Maryland,   1982.

PROJECTION OF OBJECT HISTORIES

Seymour Ginsburg(*)

Chang-jie Tang(**)

Computer Science Department
University of Southern California
Los Angeles, California 90089-0782

Computer Science Department
Sichuan University
Chengdu, Sichuan
People's Republic of China

INTRODUCTION

In [GT1], the notion of a record-based model for describing historical data (called "object histories") was introduced and some properties noted. The major construct is a "computation-tuple sequence scheme" (CSS), which specifies the set of all possible "valid" object histories for the object. In [GT2], the effect of "interval queries" was examined. The present paper studies the effect of projections on object histories. Specifically, we look at when the projection of the set of all valid object histories described by one CSS is the set of all valid object histories described by some other CSS.

Informally, an object history is a historical record of an object. (Here, each object stands for an individual "thing" or "entity", such as the electricity usage in a specific individual residence or a specific person's checking account, etc.). An object history is viewed as a sequence of occurrences, each occurrence consisting of some input data and, possibly, some calculation. (For example, in an electricity-usage history, one occurrence might be the current meter reading and current price per kilowatt hour, together with the kwh consumption and consumer cost.) In the model, each history is represented as a sequence of tuples (over the same attributes), called a "computation-tuple sequence." A CSS is a construct which defines the set of all possible "valid" computation-tuple sequences (e.g., the set of all possible "valid" electricity-usage histories), and consists of:

(+) A set of attributes, partitioned into state, input and evaluation attributes;

(++) Functions which calculate values for state and evaluation attributes;

(*) This author was supported in part by the National Science Foundation under grants MCS-792-5004 and DCR-8318752.
(**) Research done at the University of Southern California while on leave from Sichuan University.

(+++) Semantic constraints whose satisfaction is to hold uniformly throughout a computation-tuple sequence; and

(++++) A set of specific computation-tuple sequences of some bounded length with which to start a valid computation-tuple sequence.

The paper is divided into six sections. Section 1 reviews the model for object histories. Section 2 concerns a necessary and sufficient condition for the projected set to be described by a CSS. Also, the important notion of membership-W-independence is introduced. This concept, in conjunction with technical conditions, is used several times to guarantee that the projected set be described by a CSS having specific properties. Section 3 deals with two sufficiency hypotheses, one general (Theorem 3.1) and the other special (Theorem 3.2). Section 4 presents four examples . Section 5 extends Theorems 3.1 and 3.2 so that if the original CSS is "local", so is the "end" CSS. The last section exhibits a simple class of CSS to show that the existence of a CSS to define the projected set is recursively unsolvable. Most of the proofs are omitted.

SECTION 1. PRELIMINARIES

Throughout, Dom_∞ is an infinite set of elements (called <u>domain</u> values) and U_∞ is an infinite set of symbols (called <u>attributes</u>). For each A in U_∞, $\mathrm{Dom}(A)$ (called the <u>domain</u> of A) is a subset of Dom_∞ of at least two elements. All attributes considered are assumed to be elements of U_∞. The symbols A, B and C (possibly subscripted) denote attributes and U (possibly subscripted or primed) denotes a nonempty finite set of attributes.

Let X be a nonempty finite set of attributes and $A_1,...,A_n$ some fixed listing of the distinct elements of X. Then $<X>$ denotes the sequence $A_1...A_n$, and $\mathrm{Dom}(<X>)$ the Cartesian product $\mathrm{Dom}(A_1)\times...\times\mathrm{Dom}(A_n)$. For $i \geq 2$, $<X|A_i>$ denotes the prefix $A_1...A_{i-1}$. If Y is a nonempty subset of X, then $<Y>$ denotes the obvious subsequence of $<X>$.

Definition: A <u>computation tuple</u> over $<U>$ is an element in $\mathrm{Dom}(<U>)$. A <u>computation-tuple sequence</u> over $<U>$ is a nonempty sequence $\bar{u}$ of computation tuples over $<U>$. The set of all computation-tuple sequences over $<U>$ is denoted by $\mathrm{SEQ}(<U>)$.

The symbols u, v and w, possibly subscripted or primed, always represent computation tuples; and, $\bar{u}$, $\bar{v}$ and $\bar{w}$ computation-tuple sequences.

With respect to (+) of the Introduction we have:

Definition: An <u>attribute scheme</u> over $<U>$ is a triple $(<S>, <I>, <E>)$, where S, I and E are pairwise disjoint subsets of U (of <u>state</u>, <u>input</u> and <u>evaluation</u> attributes, resp.), with S and I nonempty and[1] and $<U> = <S><I><E>$.

Using the previous notion, we now formalize (++) of the Introduction.

---

[1]For $<U_1> = A_1...A_{m_1}$ and $<U_2> = B_1...B_{m_2}$, $<U_1><U_2> = A_1...A_{m_1} B_1...B_{m_2}$.

Definition: A computation scheme (abbreviated CS) over $<U>$ is a 5-tuple $c = (<S>, <I>, <E>, \mathcal{E}, \mathcal{F})$, where

(1) $(<S>, <I>, <E>)$ is an attribute scheme over $<U>$;

(2) $\mathcal{E} = \{e_C | C \text{ in } E, e_C \text{ a partial function (called an } \underline{\text{evaluation}} \text{ function) from}$

$\text{Dom}(<U>)^{\rho_C} \times \text{Dom}(<U|C>)$ into $\text{Dom}(C)$ for some non-negative integer $\rho_C\}$;
and (3) $\mathcal{F} = \{f_A | A \text{ in } S, f_A \text{ a partial function (called a } \underline{\text{state}} \text{ function) from}$
$\text{Dom}(<U>)$ into $\text{Dom}(A)\}$.
The integer ρ_C is the $\underline{\text{rank}}$ of e_C, and $\rho(c) = \max\{\rho_C, 1 | e_C \text{ in } \mathcal{E}\}$ the $\underline{\text{rank}}$ of c.

The purpose of a computation scheme is to select those computation tuple sequences whose values for the state and evaluation attributes are ultimately determined by the corresponding state and evaluation functions.

Notation: Let $c = (<S>, <I>, <E>, \mathcal{E}, \mathcal{F})$ be a CS over $<U>$. For each A in S and $\emptyset \neq S' \subseteq S$, let

$$\text{VSEQ}(f_A) = \{u_1...u_m | m \geq 1, u_h(A) = f_A(u_{h-1}) \text{ for each h, } 2 \leq h \leq m\}$$

and $\text{VSEQ}(\{f_A | A \text{ in } S'\}) = \underset{A \text{ in } S'}{\cap} \text{VSEQ}(f_A)$.

For each C in E and $\emptyset \neq E' \subseteq E$, let

$$\text{VSEQ}(e_C) = \{u_1...u_m | m \geq 1, u_h(C) = e_C(u_{h-\rho_C},...,u_{h-1}, u_h[<U|C>])$$

$$\text{for}^2 \text{ each h, } \rho_C < h \leq m\} \text{ and}$$

$$\text{VSEQ}(\{e_C | C \text{ in } E'\}) = \underset{C \text{ in } E'}{\cap} \text{VSEQ}(e_C).$$

Let $\text{VSEQ}(\emptyset) = \text{SEQ}(<U>)$ and $\text{VSEQ}(c) = \text{VSEQ}(\mathcal{E}) \cap \text{VSEQ}(\mathcal{F})$.

To formalize (+++) of the Introduction, we have:

Definition: A $\underline{\text{constraint}}$ σ over $\text{SEQ}(<U>)$ is a mapping over $\text{SEQ}(<U>)$ which assigns to each $\bar{u}$ in $\text{SEQ}(<U>)$ a value of "true" or "false". If $\sigma(\bar{u}) = $ true, then $\bar{u}$ is said to $\underline{\text{satisfy}}$ σ. For each set Σ of constraints over $\text{SEQ}(<U>)$, the set $\{\bar{u} \text{ in } \text{SEQ}(<U>) | \bar{u} \text{ satisfies each } \sigma \text{ in } \Sigma\}$ is denoted by $\text{VSEQ}(\Sigma)$.

We shall usually define a constraint σ by just specifying $\text{VSEQ}(\sigma)$.

We restrict our constraints to a special class called "uniform". These are characterized by the fact that satisfaction holds uniformly throughout a computation-tuple sequence, i.e., in every interval of a computation-tuple sequence.

Definition: σ over $\text{SEQ}(<U>)$ is $\underline{\text{uniform}}$ if $u_1...u_m$ in $\text{VSEQ}(\sigma)$ implies $u_i...u_j$ in $\text{VSEQ}(\sigma)$, for all $u_1...u_m$ in $\text{SEQ}(<U>)$ and all i and j, $1 \leq i \leq j \leq m$.

Clearly, $\text{VSEQ}(\Sigma)$ is interval closed for each set Σ of uniform constraints.

---

[2] Let $<U> = A_1...A_n$ and $<V>$ a subsequence of $<U>$. For each computation tuple u over $<U>$, $u[<V>]$ is the computation tuple v over $<V>$ defined by $v(A) = u(A)$ for each A in V.

The last concept needed for a computation-tuple sequence scheme is that of "initialization". (See (++++) of the Introduction.)

Definition: Given a CS C over $<U>$ and a finite set Σ of uniform constraints over SEQ($<U>$), an initialization (with respect to C and Σ) is any prefix-closed subset I of $\{\bar{u}$ in VSEQ(C) $\cap$ VSEQ(Σ)$||\bar{u}| \leq \rho(C)\}$.

Notation: Let S be a prefix-closed subset of SEQ($<U>$) and $\rho \geq 1$ an integer such that $|\bar{u}| \leq \rho$ for each $\bar{u}$ in S. Then VSEQ(S,ρ) $= S \cup \{\bar{u}_1\bar{u}_2|\bar{u}_1$ in $S,|\bar{u}_1| = \rho\}$. When ρ is understood (almost always), VSEQ(S, ρ) is written VSEQ(S).

In particular, if I is an initialization with respect to C and Σ, then

$$\text{VSEQ}(I) = \{\bar{u} \text{ in SEQ}(<U>)|\bar{u} = \bar{u}_1\bar{u}_2 \text{ for some } \bar{u}_1 \text{ in } I \text{ of length } \rho(C)\} \cup I.$$

Clearly, each VSEQ(S), thus each VSEQ(I), is prefix closed.

Definition: A computation-tuple sequence scheme (CSS) over $<U> = <S> <I> <E>$) is a triple (C, Σ, I), where (a) C is a computation scheme over $<U>$; (b) Σ is a finite set of uniform constraints over SEQ($<U>$); and (c) I is an initialization with respect to C and Σ.

A CSS determines valid computation-tuple sequences as follows:

Definition: For each CSS $T = ((<S>, <I>, <E>, \mathcal{E}, \mathcal{F}), \Sigma, I)$, let

$$\text{VSEQ}(T) = \text{VSEQ}(\mathcal{E}) \cap \text{VSEQ}(\mathcal{F}) \cap \text{VSEQ}(\Sigma) \cap \text{VSEQ}(I).$$

A computation-tuple sequence is said to be valid (for T) if it is in VSEQ(T).

Thus, a computation-tuple sequence is valid if it (i) is "consistent" with C, (ii) satisfies each constraint in Σ, and (iii) is either in the initialization, or its prefix of length $\rho(C)$ is in the initialization.

Note that VSEQ(T) is prefix closed.

Notation: Given $<U> = <S><I><E>$, let $<V> = <S'><I'><E'>$, where $\emptyset \neq S' \subseteq S, \emptyset \neq I' \subseteq I$ and $E' \subseteq E$.

The purpose of the present paper is to consider when the projection onto a given V of a given VSEQ(T) is VSEQ(T') for some T', i.e., given $<U>$, $<V>$ and T over $<U>$, when does there exist a T' over $<V>$, of a specified rank[3] such that VSEQ(T') $= \Pi_V$(VSEQ(T)).

We conclude the present section with an example.

Example 1.1: Consider the usage of electricity in an individual residence. Once a month, a meter (ranging from 0 to 99999) is read and the kilowatt hours since the last reading computed. Based on the price per kilowatt hour, the consumer's bill is then calculated. To detect irregularities, such as meter malfunctioning or power tapping, whenever monthly usage exceeds 3000 kilowatt hours there is an investigation by the utility and a bill determination made from the findings. One

---

[3]The projection of $\bar{u} = \bar{u}_1...\bar{u}_m$ onto $<V>$, denoted $\Pi_V(\bar{u})$, is the computation-tuple sequence $v_1...v_m$, where $v_i = u_i[<V>]$ for each i. For $S \subseteq$ SEQ($<U>$), $\Pi_V(S) = \{\Pi_V(\bar{u})|\bar{u} \text{ in } S\}$.

CSS T $= (C, \Sigma, I)$, with $C = (<S>, <I>, <E>, \mathcal{E}, \mathcal{F})$, is as follows:

(a) $<S> = A_1A_2$, where $A_1 =$ year and $A_2 =$ month.

(b) $<I> = B_1B_2$, where $B_1 =$ number (in terms of kilowatt hours, rounded off to the nearest integer) on the meter, and $B_2 =$ current price per kilowatt hour.

(c) $<E> = C_1C_2$, where $C_1 =$ number of kilowatt hours since the last reading, and $C_2 =$ total cost (including 6 1/2% sales tax).

(d) $\mathcal{E} = \{e_{C_1}, e_{C_2}\}$, where e_{C_1} and e_{C_2} are the functions defined for all u_1 and u_2 in Dom($<U>$) by

$e_{C_1}(u_1, u_2[<U|C_1>]) = (u_2(B_1)\text{-}u_1(B_1))$ mod 100,000 and

$e_{C_2}(u_1, u_2[<U|C_2>]) = u_2(B_1)\times u_2(C_1)\times 1.065$. [Obviously, $\rho(C) = 1$.]

(e) $\mathcal{F} = \{f_{A_1}, f_{A_2}\}$, with f_{A_1} and f_{A_2} the functions defined for each u in Dom($<U>$) by $f_{A_1}(u) = u(A_1)+1$ if $u(A_2) = 12$, $f_{A_1}(u) = u(A_1)$ if $1 \leq u(A_2) <$ 12, $f_{A_2}(u) = u(A_2)+1$ if $1 \leq u(A_2) < 12$ and $f_{A_2}(u) = 1$ if $u(A_2) = 12$.

(f) $\Sigma = \{\sigma\}$, where σ is the constraint over SEQ($<U>$) defined by $u_1...u_k$ is in VSEQ(σ) iff $0 \leq (u_{i+1}(B_1)\text{-}u_i(B_1))$mod $100,000 \leq$ 3000 for all i, $1 \leq i \leq$ k-1. [Thus, $u_1...u_k$ is in VSEQ(σ) if in no month is more than 3000 kilowatt hours used.]

(g) $I = \{$u in Dom($<U>$)$|u(C_1) = u(C_2) = 0\}$. [The first time the meter is read, no kilowatt hours have been used, and the bill is zero.]

One valid computation-tuple sequence is given in Figure 1.1 □

	<S>		<I>		<E>	
	A_1 (year)	A_2 (month)	B_1 (KWH reading)	B_2 (price per KWH)	C_1 (KWH consumption)	C_2 (cost)
u_1	1982	12	1033	0.06	0	0
u_2	1983	1	1133	0.06	100	6.39
u_3	1983	2	1223	0.07	90	6.71

Figure 1.1

SECTION 2. A CHARACTERIZATION

In this section we establish a characterization for the existence of a T', of a specified rank, such that VSEQ(T') = Π_V(VSEQ(T)). We need three lemmas.

The first lemma says that for each T with at least one evaluation attribute and each integer $\hat{\rho} \geq \rho(T)$, a $\hat{T}$ of rank $\hat{\rho}$ can be found such that $VSEQ(\hat{T}) = VSEQ(T)$.

Notation: For each CSS $T = ((<S>, <I>, <E>, \mathcal{E}, \mathcal{F}), \Sigma, I)$ and each $\hat{\rho} \geq \rho(T)$, (i) if $E = \emptyset$ let $\hat{T} = ((<S>, <I>, <E>, \overset{\wedge}{\mathcal{E}}, \mathcal{F}), \Sigma, \hat{I}) = T$ and (ii) if $E \neq \emptyset$ let $\hat{T} = ((<S>, <I>, <E>, \overset{\wedge}{\mathcal{E}}, \mathcal{F}), \Sigma, \hat{I})$ be as follows:

(a) $\overset{\wedge}{\mathcal{E}} = \{\hat{e}_C | C \text{ in } E\}$, where $\hat{e}_C$ is the (partial) function from $Dom(<U>)^{\hat{\rho}} \times Dom(<U|C>)$ into $Dom(C)$ defined by

$$\hat{e}_C(\hat{u}_1,...,\hat{u}_{\hat{\rho}-\rho_C}, u_1,...,u_{\rho_C}, u_{\rho_C+1}[<U|C>]) = e_C(u_1,...,u_{\rho_C}, u_{\rho_C+1}[<U|C>])$$

for each $\hat{u}_1,...,\hat{u}_{\hat{\rho}-\rho_C}, u_1,...,u_{\rho_C+1}$ in $Dom(<U>)$.

(b) $\hat{I} = \{\bar{u} \text{ in } VSEQ(T) | \|\bar{u}\| \leq \hat{\rho}\}$.

Lemma 2.1: $\hat{T}$ has rank $\hat{\rho}$ if $E \neq \emptyset$, and $VSEQ(\hat{T}) = VSEQ(T)$. $\square$

The second lemma states that $VSEQ(T)$ is the intersection of $VSEQ(I)$ and an interval-closed set, and the third lemma establishes a necessary condition for the existence of a T' such that $VSEQ(T') = \Pi_V(VSEQ(T))$.

Lemma 2.2: For each CSS $T = ((<S>, <I>, <E>, \mathcal{E}, \mathcal{F}), \Sigma, I)$,

$$VSEQ(T) = VSEQ(I) \cap Interval(\dot{V}SEQ(T)).$$

Lemma 2.3: Let $T = ((<S>, <I>, <E>, \mathcal{E}, \mathcal{F}), \Sigma, I)$ be a CSS over $<U>$ and $\hat{\rho} = \rho(T)$. Suppose there exists a CSS T' over $<V> = <S'> <I'> <E'>$ of rank $\hat{\rho}$ such that $VSEQ(T') = \Pi_V(VSEQ(T))$. Then

(a) $u_2(A) = u_2'(A)$ for each A in S' and each u_1u_2 and $u_1'u_2'$ in $Interval(VSEQ(T))$ satisfying $\Pi_V(u_1) = \Pi_V(u_1')$.

(b) If $E' \neq \emptyset$, then $u_{\hat{\rho}+1}(C) = u_{\hat{\rho}+1}'(C)$ for each C in E' and each $u_1...u_{\hat{\rho}+1}$ and $u_1'...u_{\hat{\rho}+1}'$ in $Interval(VSEQ(T))$ satisfying both $\Pi_V(u_1...u_{\hat{\rho}}) = \Pi_V(u_1'...u_{\hat{\rho}}')$ and[4] $\Pi_V(u_{\hat{\rho}+1}[<U|C>]) = \Pi_V(u_{\hat{\rho}+1}'[<U|C>])$.

We need the concepts of a restriction function and a restriction of $\hat{T}$.

Definition: Let $T = ((<S>, <I>, <E>, \mathcal{E}, \mathcal{F}), \Sigma, I)$ and $<V> = <S'> <I'> <E'>$.

(a) For each C in E', the restriction of e_C on V (with respect to T), denoted e_C^r, is the partial function from $Dom(<V>)^{\rho_C} \times Dom(<V|C>)$ to $Dom(C)$ defined as follows for each $v_1...v_{\rho_C+1}$ in $SEQ(<V>)$: If

---

[4] For each u in $Dom(<U>)$ and C in V, $\Pi_V(u[<U|C>]) = u[<V|C>]$.

$$\{e_C(u_1,...,u_{\rho_C},u_{\rho_C+1}[<U|C>])|u_1...u_{\rho_C+1} \text{ in Interval(VSEQ(T))},$$
$$\Pi_V(u_1...u_{\rho_C}) = v_1...v_{\rho_C} \text{ and } u_{\rho_C+1}[<V|C>] = v_{\rho_C+1}[<V|C>]\}$$

has exactly one element, then $e_C^r(v_1,...,v_{\rho_C}, v_{\rho_C+1}[<V|C>])$ is this element; and is undefined otherwise.

(b) For each A in S', the restriction of f_A on V (with respect to T), denoted f_A^r, is the partial function from $Dom(<V>)$ to $Dom(A)$ defined as follows for each v in $Dom(<V>)$: If

$$\{f_A(u_1)|\Pi_V(u_1) = v, u_1u_2 \text{ in Interval(VSEQ(T)) for some } u_2\}$$

has exactly one element, then $f_A^r(v)$ is this element; and is undefined otherwise.

Using the restriction functions, we have:

Definition: Let $T = ((<S>, <I>, <E>, \varepsilon, \mathcal{F}), \Sigma, I)$ and $<V> = <S'> <I'> <E'>$. The restriction of T on V, T^r, is $((<S'>, <I'>, <E'>, \varepsilon^r, \mathcal{F}^r), \{\sigma^r\}, I^r)$, where $\varepsilon^r = \{e_C^r|e_C \text{ in } \varepsilon, C \text{ in } E'\}$, $\mathcal{F}^r = \{f_A^r|f_A \text{ in } \mathcal{F}, A \text{ in } S'\}$, $VSEQ(\sigma^r) = \Pi_V(\text{Interval(VSEQ(T))})$, and $I^r = \Pi_V(I)$ if $E' \neq \emptyset$ and[5] $I^r = Prefix_1(\Pi_V(I))$ if $E' = \emptyset$.

$\rho(T^r) = \max\{1, \rho_C|C \text{ in } E'\}$. Thus, $\rho(\hat{T}^r) = \rho(\hat{T})$ if $E' \neq \emptyset$.

Using Lemma 2.1-2.3, one can show the following characterization Theorem.

Theorem 2.1: For $T = ((<S>, <I>, <E>, \varepsilon, \mathcal{F}), \Sigma, I)$ over $<U>$, $\hat{\rho} \geq \rho(T)$ and $<V> = <S'><I'><E'>$, the following two conditions are equivalent:

(a) There exists a CSS $T' = ((<S'>, <I'>, <E'>, \varepsilon', \mathcal{F}'), \Sigma', I')$, of rank $\hat{\rho}$ if $E' \neq \emptyset$ and of rank 1 if $E' = \emptyset$, such that $VSEQ(T') = \Pi_V(VSEQ(T))$.

(b) $VSEQ(\hat{T}^r) = \Pi_V(VSEQ(T))$.

Proof: The key steps to prove that (a) implies (b) are:
$$\Pi_V(VSEQ(T)) = VSEQ(\Pi_V(\hat{I})) \cap VSEQ(\hat{\sigma}^r),$$
$$\supseteq VSEQ(\Pi_V(\hat{I})) \cap VSEQ(\hat{\sigma}^r) \cap VSEQ(\hat{\varepsilon}^r) \cap VSEQ(\hat{\mathcal{F}}^r)$$
$$= VSEQ(\hat{T}^r).$$

The proof of the reverse containment depends on the following points:
$$\Pi_V(VSEQ(T)) \subseteq VSEQ(\hat{\sigma}^r), \Pi_V(VSEQ(T)) \subseteq VSEQ(\Pi_V(\hat{I})),$$
$$\Pi_V(VSEQ(T)) \subseteq Interval(\Pi_V(VSEQ(T))) \subseteq VSEQ(\hat{\mathcal{F}}^r), \text{ and}$$
$$\Pi_V(VSEQ(T)) \subseteq VSEQ(\hat{\varepsilon}^r). \quad \square$$

---

[5]For each $S \subseteq SEQ(<U>)$, $Prefix_1(S) = \{u'|u' \text{ a prefix of (length 1) of some } u \text{ in } S\}$.

We now introduce a concept, membership-W-independence, which is utilized in a corollary below, as well as in sections 3 and 5.

Definition: Let W be a subset of U such that $U-W \neq \emptyset$, and S_1 and S_2 subsets of $SEQ(<U>)$. Then S_1 is said to be membership-W-independent with respect to S_2 if $S_1 = \{\bar{u} \text{ in } S_1 \cup S_2 | \Pi_{U-W}(\bar{u}) = \Pi_{U-W}(\bar{u'}) \text{ for some } \bar{u'} \text{ in } S_1\}$. If $S_2 = SEQ(<U>)$, the phrase "with respect to S_2" is omitted.

An obvious equivalent formulation for S_1 to be membership-W-independent with respect to S_2 is that $S_1 \supseteq \{\bar{u} \text{ in } S_2 | \Pi_{U-W}(\bar{u}) = \Pi_{U-W}(\bar{u'} \text{ form some } \bar{u'} \text{ in } S_1\}$.

Corollary: Let $T_i = ((<S>, <I>, <E>, \mathcal{E}_i, \mathcal{F}_i), \Sigma_i, I_i)$, $i = 1,2$, be CSS over $<U>$ of rank ρ, with $VSEQ(T_1) \subseteq VSEQ(T_2)$. Let I_1 be membership-(U-V)-independent with respect to at least one set in the family

$$\{VSEQ(e_C), VSEQ(f_A), VSEQ(\sigma)|e_C \text{ in } \mathcal{E}_1, f_A \text{ in } \mathcal{F}_1, \sigma \text{ in } \Sigma_1\}.$$

Suppose that there exists a CSS T_2' over $<V>$, of rank ρ, such that $VSEQ(T_2') = \Pi_V(VSEQ(T_2))$. Then there exist a CSS T_1' over $<V>$, of rank ρ, such that $VSEQ(T_1') = \Pi_V(VSEQ(T_1))$.

SECTION 3. SUFFICIENCY

We now consider two sufficiency results dealing with the existence of a T' such that $VSEQ(T') = \Pi_V(VSEQ(T))$.

Using Theorem 2.1 and Lemma 2.3, one can prove the following:

Lemma 3.1: Let $T = ((<S>, <I>, <E>, \mathcal{E}, \mathcal{F}), \Sigma, I)$ and $\hat{\rho} \geq \rho(T)$. Then there exists a CSS T', of rank $\hat{\rho}$ if $E' \neq \emptyset$ and rank 1 if $E' = \emptyset$, such that $VSEQ(T') = \Pi_V(VSEQ(T))$ iff (a) and (b) of Lemma 2.3 hold and

(*) $\Pi_V(VSEQ(\hat{T})) = \Pi_V(VSEQ(\hat{I})) \cap \Pi_V(\text{Interval}(VSEQ(\hat{T})))$.

Using Lemma 2.3 and 3.1, we can prove the following theorem:

Theorem 3.1: Let $T = ((<S>, <I>, <E>, \mathcal{E}, \mathcal{F}), \Sigma, I)$, $\hat{\rho} \geq \rho(T)$ and $<V> = <S'> <I'> <E'>$. Suppose that (a) f_A and e_C are total functions for each A in S-S' and C in E-E', (b) $VSEQ(\Sigma)$ and $VSEQ(f_A)$ are membership-(U-V)-independent for each A in S', and (c) if $E' \neq \emptyset$, then $VSEQ(e_C)$ is membership-(U-V)-independent for each C in E'. Then there exists a CSS T', of rank $\hat{\rho}$ if $E' \neq \emptyset$ and rank 1 if $E' = \emptyset$, over $<V>$ such that $VSEQ(T') = \Pi_V(VSEQ(T))$.

Proof: It suffices to show that the hypotheses of Lemma 3.1 hold.

Let $\hat{T} = ((<S>, <I>, <E>, \hat{\mathcal{E}}, \mathcal{F}), \Sigma, \hat{I})$. The key step is a consideration of

(*) of Lemma 3.1. Clearly, the left side is a subset of the right side. Let

$$S_V = VSEQ(\{\hat{e}_C | C \text{ in } E'\}) \cap VSEQ(\{f_A | A \text{ in } S'\}) \cap VSEQ(\Sigma) \text{ and}$$

$$S_{U-V} = VSEQ(\{\hat{e}_C | C \text{ in } E-E'\}) \cap VSEQ(\{f_A | A \text{ in } S-S'\}).$$

It can be shown that the right side of (*) of Lemma 3.1 is contained in $VSEQ(\hat{I}) \cap$

$S_V \cap S_{U-V} = VSEQ(\hat{T})$, namely the left side of (*). $\square$

Our second sufficiency result also uses Lemma 3.1.

Theorem 3.2: Let $T = ((<S>, <I>, <E>, \mathcal{E}, \mathcal{F}), \Sigma, I), <V> = <S> <I>$ $<E'>$, and $\hat{\rho} \geq \rho(T)$. Suppose that e_C is of rank 0 for each C in E-E'. Then there exists a CSS T' over $<V>$, of rank $\hat{\rho}$ if $E' \neq \emptyset$ and of rank 1 if $E' = \emptyset$, such that $VSEQ(T') = \Pi_V(VSEQ(T))$.

Proof: By Lemma 3.1, it suffices to show that (a) and (b) of Lemma 2.3, as well as (*) of Lemma 3.1, hold. The key step is in the consideration of (*) of Lemma 3.1, namely to show that $\Pi_V(VSEQ(\hat{I})) \subseteq \Pi_V(\text{Interval}(VSEQ(\hat{T}))) \subseteq \Pi_V(VSEQ(\hat{T}))$. $\square$

SECTION 4. EXAMPLES

We now present four examples relating to Theorems 2.1, 3.1, and 3.2.

Example 4.1: Let $T = ((<S>, <I>, <E>, \mathcal{E}, \mathcal{F}), \Sigma, I)$ be the CSS of rank 1 given in Example 1.1. Let $<V> = <A_1A_2> <B_1B_2> <C_2>$ and $\hat{\rho} = 1$. It is easily seen that conditions (a) and (b) of Theorem 3.1 hold, but (c) does not (e_{C_2}

depends on C_1). Now let $\overline{T} = ((<S>, <I>, <E>, \overline{\mathcal{E}}, \mathcal{F}), \Sigma, I)$, where $\overline{e}_{C_1} = e_{C_1}$ and $\overline{e}_{C_2}$ is obtained by substituting $e_{C_1}(u_1, u_2[<U|C_1>])$ for $u_2(C_1)$ in e_{C_2}, i.e.,

$$\overline{e}_{C_2}(u_1,u_2[<U|C_2>])=u_2(B_2)\times(u_2(B_2)\times(u_2(B_1)-u_1(B_1))\bmod 100,000\times 1.065.$$

Clearly, $VSEQ(\overline{T}) = VSEQ(T)$ and $\overline{T}$ satisfies (a), (b) and (c) of Theorem 3.1.

Thus, there exists a CSS T' such that $VSEQ(T') = \Pi_V(VSEQ(\overline{T})) = \Pi_V(VSEQ(T))$.

By Theorem 2.1, one such T' is $\hat{T}^r = ((<A_1A_2>, <B_1B_2>, <C_2>, \hat{\mathcal{E}}^r, \mathcal{F}^r), \{\hat{\sigma}^r\},$

$\hat{I}^r)$, where:

(a) $\hat{\mathcal{E}}^r = \{\hat{e}_{C_2}^r\}$, $\hat{e}_{C_2}^r$ being defined for all v_1 and v_2 in Dom($<V>$) by

$$\hat{e}_{C_2}^r(v_1,v_2[<V|C_2>]) = 1.065 \ v_2(B_2)((v_2(B_1)-v_1(B_1)) \bmod 100,000).$$

[Indeed, note that $\hat{e}_{C_2} = e_{C_2}$. Then for u_1u_2 in Interval($VSEQ(T_1)$) $\subseteq VSEQ(e_{C_1})$ such that $\Pi_V(u_1u_2) = v_1v_2$,

$$\hat{e}_{C_2}(u_1,u_2[<U|C_2>]) = 1.065 \ u_2(B_2)u_2(C_1)$$

353

$$= 1.065 \ u_2(B_2)((u_2(B_1)\text{-}u_1(B_1)) \ \text{mod} \ 100,000), \ \text{since} \ u_1u_2 \ \text{is in VSEQ}(e_{C_1})$$

$$= 1.065 \ v_2(B_2)((v_2(B_1)\text{-}v_1(B_1)) \ \text{mod} \ 100,000), \ \text{since} \ \Pi_V(u_1u_2) = v_1v_2.$$

Thus, $\{\hat{e}_{C_2}(u_1,u_2[<U|C_2>])|\Pi_V(u_1u_2) = v_1v_2, \ u_1u_2 \ \text{in Interval(VSEQ}(T))\}$ has exactly one element, namely $1.065 \ v_2(B_2)((v_2(B_1)\text{-}v_1(B_1)) \ \text{mod} \ 100,000).]$

(b) $\mathcal{F}^r = \{f^r_{A_1}, \ f^r_{A_2}\}$, $f^r_{A_1}$ and $f^r_{A_2}$ being the functions on Dom($<V>$) defined for each v in Dom($<V>$) by $f^r_{A_1}(v) = v(A_1)+1$ if $v(A_2) = 12$, $f^r_{A_1}(v) = v(A_1)$ if $1 \le v(A_2)<12$, $f^r_{A_2}(v) = v(A_2)+1$ if $1 \le v(A_2)<12$ and $f^r_{A_2}(v) = 1$ if $v(A_2) = 12$.

(c) $v_1...v_k$ is in VSEQ($\hat{\sigma}^r$) if $0 \le (v_{i+1}(B_1)\text{-}v_i(B_1)\text{mod} \ 100,000) \le 3000$ for all i.

(d) $\hat{I}^r = \{v \ \text{in Dom}(<V>)|v(C_2) = 0\}.$ $\square$

An illustration of Theorems 2.1 and 3.2 is the following:

Example 4.2: Consider a CSS for the monthly sales record of a California discount store for a year. Let $T = ((<A_1A_2>, <B_1B_2>, <C_1C_2C_3>, \ \mathcal{E}, \ \mathcal{F}), \ \emptyset, \ I)$, where:

(a) A_1 is the number of the month, A_2 is the item name, B_1 is the monthly total sales, which includes the 6.5% sales tax collected, B_2 is the cost of the item purchased by the store, C_1 is the tax collected, C_2 is the monthly sales less monthly purchased cost and tax, and C_3 is the year's sum to date of the C_2-entries.

(b) $\mathcal{E} = \{e_{C_1}, \ e_{C_2}, \ e_{C_3}\}$, where $e_{C_1}(u) = ^6 \ .061 \ u(B_1)$, $e_{C_2}(u) = (u(B_1)\text{-}u(B_2))\text{-}u(C_1)$, and $e_{C_3}(u_1,u_2[<U|C_3>]) = u_1(C_3)+u_2(C_2)$ for all $u, \ u_1, \ u_2$ in Dom($<U>$). [Thus, $\rho_{C_1} = \rho_{C_2} = 0$ and $\rho_{C_3} = 1$. Hence, $\mathcal{P}(T) = 1.]$

(c) $\mathcal{F} = \{f_{A_1}, f_{A_2}\}$, where $f_{A_1}(u) = u(A_1)+1$ if $u(A_1) < 12$ and undefined otherwise, and $f_{A_2}(u) = u(A_2)$ for all u in Dom($<U>$).

(d) $I = \{(1, \ \text{item}, \ b_1, \ b_2, \ .061b_1, \ .939b_1\text{-}b_2, \ .939 \ b_1\text{-}b_2) \ |\text{item in Dom}(B_1), \ b_1 \ge 0, \ b_2 \ge 0\}.$

Now let $<V> = <A_1A_2B_1B_2C_3>$. Then $U\text{-}V = \{C_1,C_2\}$. By Theorem 3.2 (since $\rho_{C_1} = \rho_{C_2} = 0$), there exists a CSS T' over $<V>$ of rank 1, such that VSEQ(T') $= \Pi_V(\text{VSEQ}(T))$. By Theorem 2.1, one such T' is $\hat{T}^r = ((<A_1A_2>, <B_1B_2>, <C_3>, \hat{\mathcal{E}}^r, \mathcal{F}^r), \{\hat{\sigma}^r\}, \hat{I}^r).$

Example 4.3: Let $T = ((<A>, <B>, <C_1C_2C_3>, \ \mathcal{E}, \ \mathcal{F}), \ \emptyset, \ I)$ be as follows:

(a) The domain of each attribute is the integers.

(b) $\mathcal{E} = \{e_{C_1}, \ e_{C_2}, \ e_{C_3}\}$, with $e_{C_1}, \ e_{C_2}$, and e_{C_3} defined for each u_1u_2 in

---

[6]The tax rate on the total sales (which includes the 6.5% tax) is $6.5/106.5 = .061$ approximately.

SEQ($<U>$) by $e_{C_1}(u_1,u_2[<U|C_1>]) = u_2(B)$, $e_{C_2}(u_1,u_2[<U|C_2>]) = u_2(B)+u_1(C_2)$ if $u_2(B)+u_1(C_2) > 0$ and undefined otherwise, and $e_{C_3}(u_1,u_2[<U|C_3>]) = u_2(C_2)$.

(c) $\mathcal{F} = \{f_A\}$, where $f_A(u) = u(A)$ for each u in Dom($<U>$).

(d) $I = \{(a,b,b,b,b)|a$ arbitrary, b positive$\}$.

[Such a CSS formalizes the situation where a department of a company has two resources, each starting at the same level, and increasing or decreasing at the same amount, e.g., one resource might be the number of employees and another the number of telephones.]

Let $<V> = <A><B><C_1>$. VSEQ(e_{C_1}) and VSEQ(f_A) are membership-C_2C_3-independent, but neither e_{C_2} nor e_{C_3} is a total function. Thus, condition (a) of Theorem 3.1 is violated, while conditions (b) and (c) hold. There is no CSS T' over $<V>$ (of any rank) such that VSEQ(T') $= \Pi_V$(VSEQ(T)). □

Our fourth example illustrates the role played by $\hat{\rho}$ in Theorems 2.1 and 3.1, namely, that for some CSS T and integer ρ, there may be no CSS T' of rank ρ such that VSEQ(T') $= \Pi_V$(VSEQ(T)), but there may be a CSS T'' of rank $\hat{\rho} > \rho$ such that VSEQ(T'') $= \Pi_V$(VSEQ(T)).

Example 4.4: Let $<S> = <S'> = <A>$, $<I> = <I'> = <B>$, $<E> = <C_1C_2>$, $<E'> = <C_1>$, $<U> = <S><I><E>$, $<V> = <S'><I'><E'>$, and Dom(A) = Dom(B) = Dom(C_1) = Dom(C_2) = N $\cup$ {-1}, N being the set of nonnegative integers. We now exhibit CSS T_∞, T_1,...,T_i,... with the following properties (for each positive integer m):

(1) T_m and T_∞ are of rank 1.

(2) There is no CSS T', of rank at most m, over $<V>$ such that VSEQ(T') $= \Pi_V$(VSEQ(T_m)).

(3) There is a T'_m of rank m+1 such that VSEQ(T'_m) $= \Pi_V$(VSEQ(T_m)).

(4) There exists a CSS T_0, of rank 1, over $<S'><I'>$ such that VSEQ(T_0) $= \Pi_{<S'><I'>}$(VSEQ(T_∞)) (so that the state function is uniquely defined), but there is no T'' over $<V>$ such that VSEQ(T'') $= \Pi_V$(VSEQ(T_∞)).

Let $T_m = ((<S>, <I>, <E>, \mathcal{E}, \mathcal{F}), \emptyset, I_m)$ and $T_\infty = ((<S>, <I>, <E>, \mathcal{E}, \mathcal{F}), \emptyset, I_\infty)$, where:

(5) $\mathcal{E} = \{e_{C_1}, e_{C_2}\}$, with e_{C_1} and e_{C_2} defined for each u_1u_2 in SEQ($<U>$) by $e_{C_1}(u_1,u_2[<U|C_1>]) = \text{sign}(u_1(A)-u_1(C_2))^7$ and $e_{C_2}(u_1,u_2[<U|C_2>]) = u_1(C_2)$;

(6) $\mathcal{F} = \{f_A\}$, with f_A defined by $f_A(u) = u(A)+1$ for each u in Dom($<U>$);

---

[7]By definition, sign(x) = 1 if x $\geq$ 0 and sign(x) = -1 otherwise.

(7) $I_m = \{(0,0,-1,n)|0 \le n \le m\}$ and $I_\infty = \{(0,0,-1,n)|n \ge 0\}$.

It can be shown that (1)-(4) hold. □

SECTION 5. LOCALNESS

An important class of CSS, called "local", was discussed in [GT1] and [GT2]. In the present section, we show that "local representability" is not preserved under projection but is under the hypotheses of Theorems 3.1 and 3.2.

Definition: $T = (C, \Sigma, I)$ is $\underline{(k_1,k_2)\text{-local}}$ if (1) $k_1 \ge 2$ and $k_2 \ge 1$, (2) for each $\bar{u}$, $|\bar{u}| \ge k$, $\bar{u}$ is in $VSEQ(C)$ iff $\{\bar{w}| \; |\bar{w}| = k_1$, $\bar{w}$ an interval of $\bar{u}\} \subseteq VSEQ(C)$, and (3) for each $\bar{u}$, $|\bar{u}| \ge k_2$, $\bar{u}$ is in $VSEQ(\Sigma)$ iff $\{\bar{w}| \; |\bar{w}| = k_2$, $\bar{w}$ an interval of $\bar{u}\} \subseteq VSEQ(\Sigma)$. T is $\underline{local}$ if it is (k_1,k_2)-local for some k_1 and k_2. T is said to be $\underline{locally}$ $\underline{representable}$ if there exists a local CSS $\bar{T}$ such that $VSEQ(\bar{T}) = VSEQ(T)$.

If $T = (C, \Sigma, I)$ is (k_1,k_2)-local, then the maintenance of a computation-tuple sequence being in $VSEQ(C)$ just involves checking the last k_1 computation tuples, and the maintenance of being in $VSEQ(\Sigma)$ the last k_2.

If T over $<U>$ is local and there exists a CSS T' over $<V>$ such that $VSEQ(T') = \Pi_V(VSEQ(T))$, is T' locally representable? The answer is no.

Example 5.1: Let $T = ((<A_1A_2>, <B>, <C>, \mathcal{E}, \mathcal{F}), \Sigma, I)$ be the CSS where:

(a) $Dom(A_1) = Dom(A_2) = \{1,2\}$ and $Dom(B) = Dom(C)$ is the set of integers.

(b) $\mathcal{E} = \{e_C\}$, where $e_C(u_1,u_2[<U|C>]) = u_2(B)$ for each u_1u_2.

(c) $\mathcal{F} = \{f_{A_1}, f_{A_2}\}$, where $f_{A_1}(u) = u(A_1)$ and $f_{A_2}(u) = u(A_2)$ for each u.

(d) $\Sigma = \{\sigma\}$, where

$$VSEQ(\sigma) = \{u_1...u_m|u_i(A_1) = 1 \text{ and } u_i(C) \le 10 \text{ for each } i\}$$
$$\cup \{u_1...u_m|u_i(A_1) = 2 \text{ and } u_i(C) \ge -10 \text{ for each } i\}.$$

[$VSEQ(\sigma)$ is not membership-A_1-independent.]

(e) $I = \{(1,a_2,b,b)|a_2 \text{ in } \{1,2\}, b \le 10\} \cup \{(2,a_2,b,b)|a_2 \text{ in } \{1,2\}, b \ge -10\}$.

Clearly, T is $(2,1)$-local. Let $<V> = <A_2><B><C>$. Then

$$\Pi_V(VSEQ(T)) = \{v_1...v_m|v_i(A_2) = v_{i+1}(A_2),$$
$$v_j(B) = v_j(C) \le 10, \; 1 \le i \le m\text{-}1, \; 1 \le j \le m\}$$
$$\cup \{v_1...v_m|v_i(A_2) = v_{i+1}(A_2), \; v_j(B) =$$
$$v_j(C) \ge -10, \; 1 \le i \le m\text{-}1, \; 1 \le j \le m\} \text{ and}$$

$Interval(\Pi_V(VSEQ(T))) = \Pi_V(VSEQ(T))$.

Obviously, there exists a CSS T_1' of rank 1, thus T_ρ' of rank $\hat{\rho}$ for each $\hat{\rho} \ge 1$, such that $VSEQ(T'_\rho) = \Pi_V(VSEQ(T))$. It can be shown that there is no local CSS T' over $<V>$ such that $VSEQ(T') = \Pi_V(VSEQ(T))$. □

The above example shows that Theorem 2.1 cannot be extended to include localness. However, as is noted below, the sufficiency conditions of Theorems 3.1 and 3.2 permit local representability to pass through projection.

<u>Theorem 5.1</u>: Given the hypotheses and notation in Theorem 3.1, suppose that T is (k_1,k_2)-local and $k = \max\{k_1,k_2,\hat\rho+1\}$. Then $\Pi_V(VSEQ(T)) = VSEQ(\hat{T}^r)$ and $\hat{T}^r$ is (k,k)-local. Expressed otherwise, if T is locally representable and satisfies the hypotheses of Theorem 3.1, then there exists a locally representable $\overline{T}$ over $<V>$ such that $VSEQ(\overline{T}) = \Pi_V(VSEQ(T))$.

<u>Proof</u>: The key step is to show that for each $\bar{v}$ of length at least k, $\bar{v}$ is in $VSEQ(\hat\sigma^r)$ iff $\{\overline{v'}|\ |\overline{v'}| = k,\ \overline{v'}$ is an interval of $\bar{v}\} \subseteq VSEQ(\hat\sigma^r)$. Since $\hat\sigma^r$ is uniform, the "only if" is clear. The proof of the "if" is long. □

The analogous result for Theorem 3.2 is:

<u>Theorem 5.2</u>: Let $T = ((<S>, <I>, <E>, \varepsilon, \mathcal{F}), \Sigma, I)$ be a (k_1,k_2)-local CSS, $\hat\rho \geq \rho(T)$ and $k = \max\{k_1,k_2,\ \hat\rho+1\}$. Suppose that e_C is of rank 0 for each C in E-E'. Then $\Pi_V(VSEQ(T)) = VSEQ(\hat{T}^r)$ and $\hat{T}^r$ is (k,k)-local.

SECTION 6. DECIDABILITY

Since evaluation functions, state functions, constraints and initializations can be general, it is to be expected that for arbitrary T the existence of a T' over $<V>$ (of some specified or unspecified rank) such that $VSEQ(T') = \Pi_V(VSEQ(T))$ is recursively unsolvable. We now present a very simple class $\mathcal{T}$ of CSS such that the projection problem is recursively unsolvable for arbitrary T in $\mathcal{T}$.

<u>Notation</u>: Let Δ_∞ be a countably infinite set of abstract symbols. For each $n \geq 1$ and pair of tuples $x = (x_1,...,x_n)$ and $y = (y_1,...,y_n)$ of nonempty words over some finite subset Δ of Δ_∞; let $T(x,y) = ((<A_1A_2A_3>, <B>, <C_1C_2>, \varepsilon^{(x,y)}, \mathcal{F}^{(x,y)}), \emptyset, I^{(x,y)})$ be the CSS of rank 1 defined as follows:

(a) $Dom(A_1)$ is the set of positive integers, $Dom(A_2) = Dom(A_3) = \{0,1\}$, $Dom(B) = \{1,...,n\}$ and $Dom(C_1) = Dom(C_2) = \Delta^+$.

(b) $\varepsilon^{(x,y)} = \{e_{C_1}^{(x,y)}, e_{C_2}^{(x,y)}\}$, abbreviated $\{e_{C_1}, e_{C_2}\}$, where, for each u_1u_2 in $SEQ(<U>)$, $e_{C_1}(u_1,\ u_2[<U|C_1>]) = u_1(C_1)x_{u_2(B)}$ and $e_{C_2}(u_1,\ u_2[<U|C_2>]) = u_1(C_2)y_{u_2(B)}$. [$e_{C_1}$ concatenates $x_{u_2(B)}$ to $u_1(C_1)$ and e_{C_2} appends $y_{u_2(B)}$ to $u_1(C_2)$.]

(c) $\mathcal{F}^{(x,y)} = \{f_{A_1}^{(x,y)}, f_{A_2}^{(x,y)}, f_{A_3}^{(x,y)}\}$, written $\{f_{A_1}, f_{A_2}, f_{A_3}\}$, where for each u,

$f_{A_1}(u) = u(A_1) + 1$, $f_{A_2}(u) = u(A_2)$, $f_{A_3}(u) = 0$ if $u(A_2) = 0$ and $u(C_1) = u(C_2)$, $f_{A_3}(u) = 1$ otherwise. [f_{A_1} "numbers" the computation tuple, f_{A_2} keeps $u(A_2)$ constant, and f_{A_3} tests for whether or not $u(C_1) = u(C_2)$ when $u(A_2) = 0$.]

(d) $I^{(x,y)} = \{(1,0,1,i,x_i,y_i), (1,1,1,i,x_i,y_i) | 1 \leq i \leq n\}$.

Let τ be the set of all $T^{(x,y)}$.

<u>Theorem 6.1</u>: Let $<V> = <A_1 A_3>$ $<B>$ $<C_1 C_2>$ and $\hat{\rho} \geq 1$. It is recursively unsolvable to determine, for an arbitrary $T^{(x,y)}$ in τ. (a) Whether or not there exists a CSS T' over $<V>$ such that $VSEQ(T') = \Pi_V(VSEQ(T^{(x,y)}))$, and (b) Whether or not there exists a CSS T' over $<V>$, of any rank, such that $VSEQ(T') = \Pi_V(VSEQ(T))$.

<u>Proof</u>: It can be shown that the following three statements are equivalent:

(1) There exists an integer $k \geq 1$ and sequence $i_1, ..., i_k$, $1 \leq i_j \leq n$ for each j, such that $x_{i_1} ... x_{i_k} = y_{i_1} ... y_{i_k}$.

(2) There is no CSS T' over $<V>$, of rank $\hat{\rho}$, such that $VSEQ(T') = \Pi_V(VSEQ(T^{(x,y)}))$.

(3) There is no CSS T' over $<V>$ such that $VSEQ(T') = \Pi_V(VSEQ(T^{(x,y)}))$.

Whether or not (1) holds is the Post Correspondence Problem, which is well known [P] to be recursively unsolvable. Hence, whether or not (2) holds or (3) holds is also recursively unsolvable. This implies the recursive unsolvability of (a) and (b). □

REFERENCES

[GT1]   S. Ginsburg and K. Tanaka, "Computation-tuple Sequences and Object Histories," to appear in the ACM Transactions on Database Systems.

[GT2]   S. Ginsburg and K. Tanaka, "Interval Queries on Object Histories." USC Computer Science Department, TR-84-302, Feb. 1984.

[P]   E.L. Post, "A Variant of a Recursively Unsolvable Problem," Bull. Amer. Math. Soc., Vol. 52 (1946), pp. 264-268.

FUNCTIONAL ENTITY RELATIONSHIP MODEL AND UPDATE OPERATIONS

(EXTENDED ABSTRACT)

Masanobu Matsuo[1] and Laurian M. Chirica[2]

[1]Sumitomo Electric Industries Ltd.
[2]California Polytechnic State University
San Luis Obispo, California

1. INTRODUCTION

A data model called FERM (Functional Entity Relationship Model) is
introduced in order to improve the semantic modeling capability of database systems [Matsuo84]. FERM is an integration and extension of the
Functional Data Model and the Entity Relationship Model [Chen76]. FERM
consists of a small number of primitive elements such as entity sets,
attribute types, and database storage functions together with a set of
integrity constraints called Domain-Range Constraints. Functional relationships between entities and attributes of an entity are represented
in FERM by database storage functions. High-level semantics such as ISA
relationships, existence dependencies, mandatory relationships, case and
disjointness relationships are all expressed in FERM in terms of Domain-
Range Constraints on database storage functions. Thus, consistent database storage states with respect to such high-level semantics can be
studied by analyzing the effect of database update operations on Domain-
Range Constraints.

A formal definition of update (insertion, deletion, replacement)
operations in FERM is given. This remedies two major shortcomings of
previous work in the Functional and Entity Relationship Model area: (1)
the lack of a formal definition; (2) the absence of update operations in
these models.

2. IMPROVEMENT IN THE SEMATIC MODELING CAPABILITY

Data model is recognized as a key concept in DBMS and database application development [Bracchi83, Borkin80, Biller77, Brodie81, Codd80].

A new data model called FERM (Functional Entity Relationship model)
has been proposed based on entities, attributes and functional relationships. The database storage set is introduced in order to model a time-
varying database. Update operations and integrity constraints are defined in terms of the database storage set.

A FERM schema is an 11-tuple
    M = (DT, AS, type, ES, FE, eset, FS, argtype, restype, C)
where:
(1) DT is a finite family of data types. Each data type T has a value
     set T_V and a set T_O of operations and predicates on that value
     set;
(2) AS is a finite set of symbols called attribute symbols;
(3) type is a function type: AS→DT. Given A ε AS, type(A) in the at-

tribute type of A. Each A ε AS has the value set and the opera-
tions/predicates set of type(A);
(4) ES is a finite set of symbols called entity symbols;
(5) FE is a finite family of sets called entity sets;
(6) eset is a function eset: ES→FE. Given E ε ES eset(E) is the
entity set of E;
(7) FS is finite set of symbols called database storage function
symbols;
(8) argtype is a function argtype: FS → ES$^+$. Given f ε FS, argtype(f)
is a sequence of entity symbols representing the argument type of
database storage function f;
(9) restype is a function restype: FS→ES ∪ AS. Given f ε FS,
restype(f) is an entity symbol or an attribute symbol representing
the result type of database storage function f;
(10) key is a function key: ES→FS$^+$. Given EεES, key(E) is a sequence
of database storage function symbols representing a key which can
identify individual entities in eset(E);
(11) C is a finite set of semantic integrity constraints.

Attribute types
    location = string(30)
    department_name = string(10)
    soc_sec_number = 0..999999999
    employee_name = string(30)
    phone_number = string(7)
    position = (secretary, coder, typist)
    speciality = (EE, CS, PHY, CHEM)
    start_year = 1980..2000
    project_name = string(30)

Entity sets and database storage functions
Entity: employee
    attribute functions
       SSN: employee → soc_sec_number
       PHN: employee → phone_number
       ENM: employee → employee_name
    constraints
       SSN is a proper key.
Entity: department
    attribute functions
       DNM: department → department_name
       LOC: department → location
    constraints
       DNM is a proper key.
Entity: spouse
    attribute functions
       SS: spouse → soc_sec_number
    constraints
       SS is a proper key.
Entity: engineer
    attribute functions
       SPC: engineer → speciality
    constraints
       isaE*SSN is an inherited key.
Entity: staff
    attribute functions
       POS: staff → position
    constraints
       isaS*SSN is inherited key.
Entity: manager
    attribute functions

360

```
      STY: manager → start_year
   constraints
      isaP*isaE*SSN is an inherited key.
Entity: project
   attribute functions
      JNM: project → project_name
   constraints
      JNM is a proper key.


Relationship functions
      WK:    employee → department
      MRRG: employee → spouse
      MGR:  project → manager
      isaE:  engineer → employee
      isaS:  staff → employee
      isaP:  manager → engineer
      PARTD: project × engineer → start_year
      PART:  project × staff → Bool
```

The functional diagram of this database schema is shown in Figure 1.

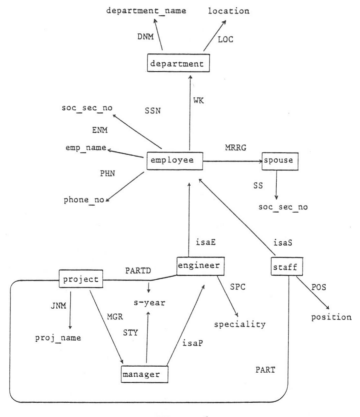

Figure 1

The interpretation of a FERM schema is assured to contain a storage set ST. A formal definition of ST is given in Appendix I. The reason for introducing ST is to model changes of database states with time, reflecting changes of real-world situations. Moreover, database users can use this abstracted database state in order to specify semantic constraints. The initial storage state of a database represents the database state which contains no information.

A storage state is called <u>consistent</u> if it satisfies all semantic constraints. Otherwise the state is said to be <u>inconsistent</u>.

Database storage functions are used to model time-varying functional relationships between entities or between entities and attribute values. For example, the fact that an employee has a unique social security number in the database, is modeled by the database storage function.

SSN: employee × ST → the value set of soc_sec_number.

For simplicity, such a function will be specified as

SSN: employee → soc_sec_number.

Given an employee entity e ε employee, and storage state s ε ST of the database, the value SSN(e, s) ε soc_sec_number gives the employee's social security number. Also, if the value SSN(e, s) is null, this indicates that the social security number of the employee is not defined in the database state s. The fact that an employee works for a department is modeled by the database storage function,

WK: employee × ST → department.

Given an employee entity e ε employee, and two different state s, s'ε ST, WK(e, s) may be different from WK(e, s'). By having a ST component in the argument list of any database storage function, changes of working departments of employees are modeled directly and naturally.

By introducing boolean attribute type Bool, a database storage function f such that restype(f) = Bool and the arity of argtype(f) is k, can represent the k-ary relationship. For example, a many-to-many relationships between projects and staff are represented by the database storage function

PART: project × staff → Bool

The attribute of a relationship is also represented by a database storage function. For example, the attribute grade of the relationship TAKE between students and courses is represented by the database storage function

TAKE: student × course → grade

The domain of a database storage function SSN with respect to a given storage state s is the set,

DOM(s, SSN) = {a ε employee |SSN(a, s) ≠ null}.

DOM(s, SSN) represents the set of employees having known social security numbers in the storage state s. The range of a database storage function MRRG with respect to a given storage state s is the set

RAN(s, MRRG) = {MRRG(a, s) |aε DOM(s, MRRG)}.

RAN(s, MRRG) represents the set of all spouses of all employees in the database state s.

3. UNIFIED APPROACH TO REPRESENTATION OF HIGH-LEVEL SEMANTICS

Representation of high-level semantics such as ISA relationships, existence dependencies, mandatory relationships, case relationships, disjointness relationships, and aggregations can be done in terms of a class of integrity constraints, called Domain-Range Constraints (DRCs) (See Appendix II). This makes it possible to analyze correct update operations w.r.t. high-level semantics in a unified way, by studying the correctness conditions w.r.t. DRCs, rather than considering individual high-level semantics on a case by case basis.

Consider the database schema shown in Figure 1. A possible integrity constraint may be placed on this schema, asserting that each employee in the database must have both a name and a social security number.

This constraint may or may not be satisfied, depending upon the given state of the database. A database state s satisfies the above constraint, if and only if

(1) DOM(s, SSN) = DOM(s, ENM).

Such a condition, referred to as an equality domain constraint, is a predicate over database storage set ST. This condition is abbreviated as D(SSN) = D(ENM) and is viewed as an unary predicate over ST defined by formula (1).

Similarly the following constraint may be placed on this schema: "Each employee in the database must have a soc_sec_number but may have no phone."

Such a constraint, referred to as an Inclusion Domain Constraint, is abbreviated as $D(PHN) \subseteq D(SSN)$.

4. DEFINITION AND ANALYSIS OF UPDATE OPERATIONS

The major contribution of our work, is a formal framework for constructing correct update operations w.r.t. DRCs. By giving only correct update operations to the DBMS user, in some instances integrity checking can be done at design-time or compile-time. For every database storage function, there exists an assignment function. For example, consider WK. assign-WK: employee × department × ST→ST is defined as follows:
(1) For all b ε department, and s ε ST, assign-WK (null, b, s) = s.
(2) For all a ε employee- {null}, b ε department and s ε ST,
    assign-WK (a, b, s) = s' such that,
    for all x ε employee,
    (2 a) WK(x,s') = if x = a then b else WK(x,s)
    (2 b) g(y,s') = g(y,s) for all g such that g ≠ f and all y.
The fact that the employee e stops working for the department d is represented by assign-WK(e, null, s) assuming that WK(e, s) = d. The fact that the employee is transferred from WK(e, s) to the department d is represented by assign-WK (e, d, s) assuming that WK(e, s) ≠ null. Therefore assign-WK (e, d, s) has the following three semantic variations:
(1) Insertion. Assume that e ≠ null and WK(e, s) = null.
(2) Deletion. Assume that e ≠ null and WK(e, s) ≠ null and d = null.
(3) Replacement. Assume that e ≠ null and WK(e, s) ≠ null and d ≠ null.
Since one assignment function represents the above three types of semantics depending on its argument values, we can introduce the following types of update operation for WK. Properties of assignment functions are shown in Appendix III.

Let STM = ST × {OK, error}.

(a) The insertion operation of type 0 into WK is the function $I<WK^0>$: employee × department × ST→STM defined as: for all a ε employee, b ε department, and s ε ST,
    $I<WK^0>$(a, b, s) = if b∉RAN(s, WK) and WK(a, s)=null and a≠null and b≠null then <assign-WK(a, b, s), OK> else <s, error>.
    Let e ε employee, d ε department, and s ε ST. The fact that the first employee e is assigned to a new department d is represented by $I<WK^0>$(e, d, s).

(b) The insertion operation of type 1 into WK is the function $I<WK^1>$: employee × department × ST→STM defined as:
    $I<WK^1>$(a, b, s) = if b∈RAN(s, WK) and WK(a, s) = null and a≠null then <assign-WK(a, b, s), OK> else <s, error>.

    The fact that an employee e is assigned to a department d where some employees are working is represented by $I<WK^1>$(e, d, s).

The multiple insertion operation in SSN, ENM of types(0, 0) respectively $I<SSN^0, ENM^0>$ is a transaction which inserts a unique social security number of one employee and unique name of another employee into a database. Note that the employee whose social security number is inserted may be different from the employee whose name is inserted in this operation.

Recall the insertion operation $I<SSN^0, ENM^0>$. In this operation,

the employee whose social security number is inserted may be different
from the employee whose name is inserted. Then how can we define an insertion operation which inserts a social security number and a name for
the same employee to a database? We must impose an equality constraint
on an argument of $I<SSN^0, ENM^0>$. For example, we can define the following operation

u: employee $\times$ soc_sec_number $\times$ name $\times$ ST $\to$ STM such that:
for all e ε employee, sn ε soc_sec_number, nm ε name and s ε ST,
$u(e, sn, nm, s) = I<SSN^0, ENM^0>(e, sn, e, nm, s)$. Consider D(SSN)=D(ENM).
Although this update operation u is correct with respect to D(SSN)=D(ENM),
$I<SSN^0, ENM^0>$ is not correct. Similarly, deletion, replacement operation, are defined.

From the above framework, what kinds of update operations (or transactions) are correct w.r.t. DRC, can be analyzed formally. In fact, how
to generate such correct update operations and how to impose equality
constraints on an arguments of an update function are shown in [Matsuo84]

5. SYNTHETIC APPROACH TO GENERATION CORRECT UPDATE OPERATIONS W.R.T. DRCS

In existing and proposed data models the fixed set of primitive update operations is given and integrity constraints are checked at run-
time in order to preserve constraints [Stonebraker76]. At best, in
RDM, normalization methods try to eliminate update anomalies which are
side-effects of FDs (functional dependencies). In our work, we present
an approach to reducing some of run-time integrity checkings in update
operations by introducing a transaction composition of more than one
operation and contraction mechanisms [Matsuo84]. In this approach,
since correct update operations w.r.t. DRCs are synthesized at design-
time, DBA makes available correct transactions to the user by composing
synthesized update operations. This approach is therefore viewed
directly opposite to normalization of database schema in RDM, since the
normalization process involves constructing a "good" database schema
based on a given set of update operations in order to eliminate anomalies.

An algorithm generating a minimal and complete set of correct update operations w.r.t. DRCs is given and correctness of this algorithm
is proved [Matsuo84]. As long as synthesized update operations are used
to construct transactions, such transactions always preserve DRCs. This
result reduces the development process of large database applications,
and improves the reliability of applications.

REFERENCE

[Biller77]   Biller, H., Neuhold, E.J. Semantics of data bases: the
               semantics of data models, Information systems 1977.
[Bracchi83]   Bracchi, G., et al. A set of integrated tools for the
               conceptual design of database schemas and transactions,
               Methodology and tools for database design, North-Holland
               publishing company, 1983, 181-204.
[Brodie81]   Brodie, L.M. Association: A database abstraction for
               semantic modelling, Information modeling and analysis, ER
               institute, 1981, 583-608.
[Borkin80]   Borkin, S.A. Date models, MIT press, 1980.
[Codd80]   Codd, E.F. Data models in database management, Proc. of
               the workshop on Data Abstraction, Databases and Concep-
               tual Modelling, 1980.
[Chen76]   Chen, P.P. The entity-relationship model-Toward a unified
               view of data, ACM TODS, Vol. 1, No.1, 1976.
[Matsuo84]   Matsuo M., Functional Entity Relationship Model, Ph.D.
               thesis, University of California, Santa Barbara, 1984.
[Stonebraker 75] Stonebraker, M. Implementation of Integrity

constraints and views by query modification, ACM SIGMOD
International Symposium on Management of Data, 1975, 65-
78.

Appendix I

Database storage set
     Let f be a database storage function symbol in FS, argtype(f) =
$<E_1,\ldots,E_n>$ where $E_1,\ldots, E_n \in$ ES, and restype(f) = D where $D \in ES \cup AS$.
Argtype(f) is interpreted as the Cartesian product eset(E_1) x ..x
eset(E_n) which is denoted as arg(f). Restype(f) is interpreted as
eset(D) if $D \in$ ES otherwise type (D)$_V$, which is denoted as res(f).
     A function s_f; A → B is called finite, if and only if the set
domain(s_f) = {a $\in$ A|s_f(a) $\neq$ null$_B$} is finite. A function s_f: A → B is
called strict, if and only if s_f(null$_A$) = null$_B$. [A → B] denotes the
set of all finite and strict functions from A to B.
Given a set FS of database storage functions symbols with argtype(f) and
restype(f) for each f in FS, the set ST of database storage states is
defined as
          ST = X$_{f \in FS}$[arg(f) → res(f)]
i.e., the Cartesian product of [arg(f) → res(f)] for all f$\in$FS.☐
     For example, let FS = { f, g, h }. A storage state s$\in$ST is defined
as a sequence $<s_f, s_g, s_h>$ where
          $s_{f\varepsilon}$ [arg(f) → res(f)],
          $s_{g\varepsilon}$ [arg(g) → res(g)],
          $s_{h\varepsilon}$ [arg(h) → res(h)].
     For each f $\in$ FS there exists a function null$_f$ $\in$ [arg(f) → res(f)]
defined as null$_f$(a) = null for all a$\in$arg(f). The database storage state
null$_{ST}$ = $<null_f>_{f \in FS}$ is referred to as the initial database state. For
example, let FS = { f, g, h }. The initial storage state is
          null$_{ST}$ = $<null_f, null_g, null_h>$

Appendix II

(1) Assume that in this database example (Fig. 1) we want to express the
fact engineers and staff members are at the same time employees in an
organization. Additionally, we want to express the fact that managers
are engineers and thus, by transitivity, managers are employees. These
semantic facts are expressed by ISA-type relationships in many high-
level database models.
     In FERM, an ISA-relationship is a database storage characterized by
a set of Domain-Range Constraints. For example, the fact that
isaP: manager→engineer is an ISA relationship is expressed by the
constraints.
     (a) D(isaP) = D(manager)
     (b) |D(isaP)|=|R(isaP)|, i.e., isaP is injective.

(2) Consider the relationship between an employee and his/her spouse
          MRRG: employee → spouse.
A personnel database may want keep information about a spouse entity
only as long as it is related to an employee entity in the database.
This type of relationship is referred to as an existence dependency
relationship since the existence of the spouse entity depends upon a
related employee divorces his/her spouse or leaves the company, the corresponding spouse, i.e., MRRG(employee, s), should be deleted from the
database (s is the database state before the spouse entity is deleted
from the database). The existence dependency as described above is
characterized by the following constraint:
          R(MRRG) = D(spouse).

(3) Assume that an employee may be an engineer or a staff member but

not both. This semantic constraint may be expressed in FERM by the following predicate on ST

   RAN(s, isaE) $\cap$ RAN(s, isaS) = $\emptyset$

or, by abstracting on s ε ST,

   R(isaE) $\cap$ R(isaS) = $\emptyset$.

This condition (Disjointness relationship) simply states that the set of employees who are engineers is disjoint from the set of employees who are in staff positions.

Appendix III

The function assign-f has the following properties.

Property 1
For all a, a' ε arg(f) - { null }, b, b' ε res(f), and s ε ST:
assign-f(a, b, assign-f(a', b', s)) =
if a = a' then assign-f(a, b, s)
else assign-f(a', b', assign-f(a, b, s'))

Property 2
For all a, a' ε arg(f) - { null }, b ε res(f), and s ε ST,
f(a, assign-f(a', b, s)) = if a = a' then b else f(a, s)

Property 3
For all a ε arg(f) - { null },
assign-f(a, null, null$_{ST}$) = null$_{ST}$

Property 4
For all s ε ST, f(null, s) = null

Property 5
For all f, g ε FS such that f$\neq$g, a ε arg(f) - { null },
b ε res(f), c ε arg(g) - { null }, d ε res(g), and all s ε ST,
assign-f(a, b, assign-g(c, d, s)) = assign-g(c, d, assign-f(a, b, s))

Property 6
For all a ε arg(f), f(a, null$_{ST}$) = null.

Property 7
For all s ε ST, a ε arg(f) - { null }, and b ε res(f) - { null },
DOM(assign-f(a, b, s), f) = DOM(s, f) $\cup$ { a },

Property 8
For all s ε ST, a ε arg(f), b ε res(f), and g ε FS, g $\neq$ f,
DOM(assign-f(a, b, s), g) = DOM(s, g).

Property 9
For all s ε ST, a ε arg(f) - { null }, and b ε res(f) - { null },
RAN(assign-f(a, b, s), f) = RAN(s, f) $\cup$ { b }.

Property 10
For all s ε ST, a ε arg(f), b ε res(f) - { null }, and g ε FS, g $\neq$ f,
RAN(assign-f(a, b, s), g) = RAN(s, g).

Property 11
For all s ε ST, arg(f) - { null },
DOM(assign-f(a, null, s), f) = DOM(s, f) - { a }.

Property 12
For all s ε ST, a ε arg(f) - { null },
let s' = assign-f(a, null, s),
if a ε DOM(s, f) and there exist no a' ε DOM(s, f) where a $\neq$ a' and
f(a, s) = f(a' s)
then RAN (s', f) = RAN(s, f) - { f(a, s) } else RAN(s', f) = RAN(s, f).

AN ALGEBRA FOR AN ENTITY-RELATIONSHIP MODEL

AND ITS APPLICATION TO GRAPHICAL QUERY PROCESSING

Bogdan Czejdo

David W. Embley

Department of Computer Science
University of Houston
Houston, Texas

Department of Computer Science
Brigham Young University
Provo, Utah

INTRODUCTION

Algebras for entity-relationship (ER) models have recently begun to emerge. Chen proposed an algebra for a binary ER model with directional relationships [2]. Markowitz and Raz [5] and Parent and Spaccapietra [7] have proposed algebras for ER models that are derived from or patterned after relational algebra. Markowitz and Raz have also proposed a query language, ERROL, based on their algebra [6].

The ER algebra defined in this paper is for a binary ER model. The operators of the algebra transform ER diagrams to ER diagrams, and every diagram represents a possible user query. Thus, the ER algebra has immediate application as a query language in which users graphically manipulate ER diagrams to formulate queries.

THE BINARY ENTITY-RELATIONSHIP MODEL

The binary entity-relationship model used for the algebra defined in this paper is based on the binary ER model of Chen's taxonomy [1]. Our model is initially limited to binary relationships between distinct entity sets; in particular no entity set can be related to itself. We can extend the ER algebra to a binary model that allows an entity set to be related to itself and to other models in the taxonomy, but the chosen model provides an appropriate vehicle in which to initially explore the properties and applications of the algebra.

An important property of our algebra is the invariance of entity identity under the operations of the algebra. To maintain the identity of entities, a *surrogate key* is introduced for every entity set in the model [3]. Each entity set in the database is defined in terms of a set of attributes in the usual way. To this set of attributes, a surrogate key attribute is added. In the derived database scheme, each surrogate key is a key for the stored relation representing its entity set. Surrogate keys and surrogate key values, however, are not directly available to a user of the database. Users cannot view them or make use of them to formulate queries. Whenever a new entity is inserted, the system generates a unique surrogate key value for the entity. Surrogate key values cannot be modified and cannot be deleted except when the entity is removed from the database.

Logically, a surrogate key value represents an entity. In addition to providing a means to maintain the identity of entities, surrogate keys are also useful for representing relationships. Surrogate keys are foreign keys in relationship relations. Thus, they obviate the need for designating primary keys to be used as foreign keys.

Basic Model Definition

Let U be the set of all attribute names, S be the set of all surrogate key names, and N be the set of all names of entity and relationship sets.

An *entity set descriptor* is a 4-tuple (EN, EA, ES, EK) where EN is an element of N, EA is a (possibly empty) subset of U, ES is an element of S, and EK is a (possibly empty) set of nonempty subsets of EA. Elements of EK are keys for the entity. We demand that the names EN be unique so that no two entity sets have the same name. We further demand that the surrogate key names ES be unique. Since both the entity name and the surrogate key of an entity set descriptor are unique, we could omit one or the other. Both are retained, however, and used in the discussion that ensues.

A *relationship set descriptor* is a 4-tuple (RN, RA, RK, RE) where RN is an element of N, RA is a (possibly empty) subset of U, RK is a (possibly empty) set of nonempty subsets of RA, and RE is a pair with several restrictions described in the following paragraph. Each element of RK is a key component of the relationship set. We demand that the relationship set names RN be unique, so that no two relationship sets have the same name. We also demand that RN $\cap$ EN be empty so that a name in RN $\cup$ EN uniquely identifies an entity or a relationship set.

RE defines the connections between a relationship set and its related entity sets. Since binary relationships can be many-many, many-one, one-many, or one-one, we define an entity cardinality designator EC. EC is either the symbol * or the symbol 1 where * stands for "many" and 1 stands for "one". We define RE to be the pair (REA, REB) where REA and REB are both pairs (ES, EC). There are two pairs because the model is binary. Each pair defines a link to an entity set. ES is a surrogate key and thus identifies the entity set, and EC gives the cardinality of the link. Since our model requires that no entity set can be related to itself, the surrogate key ES in REA must be distinct from the surrogate key ES in REB.

A *binary entity-relationship model* (BERM) is a pair (E, R) where E is a set of entity set descriptors and R is a set of relationship set descriptors.

The graphical representation of a BERM is an ER diagram. Figure 1 shows the ER diagram that is used by all the examples in this paper. Key attributes of the sets EK and RK are underlined. Recall that EK and RK are sets of sets; the elements in the sets that are members of EK and RK are joined together with arcs. Since surrogate keys have a one-to-one correspondence with entity set names, only the entity set name is displayed in the ER diagram.

Let (E, R) be a BERM with n entity set descriptors in E and m relationship set descriptors in R. The *derived database scheme* is a set of n+m relation schemes, one for each entity set descriptor in E and relationship set descriptor in R. The derived relational database scheme for the ER diagram of Figure 1 is shown in Figure 2. In Figure 2, keys are underlined and surrogate key attributes are identified by the prefix "e" concatenated to an entity set name.

The Extended Binary Entity-Relationship Model

In this Section a BERM is extended so that each entity and relationship set descriptor contains a relational algebra expression for the relation that corresponds to its entity or relationship set. There is no need for such an extension if the model remains static, because each entity and relationship set is trivially represented by a stored relation in the derived database scheme. The ER algebra to be defined, however, manipulates the model so that the correspondence is no longer trivial.

In the extended model, an *entity set descriptor* is redefined to be a 5-tuple (EN, EA, ES, EK, EX) where EN, EA, ES, and EK are as above, and EX is a well-formed relational algebra expressions over the derived database scheme.

Similarly, in the extended model a *relationship set descriptor* is redefined to be a 5-tuple (RN, RA, RK, RE, RX) where RN, RA, RK, and RE are as above, and RX is a well-formed relational algebra expressions over the derived database scheme. We adopt the relational algebra and the notation of [4] for the expressions in EX and RX.

The extended binary entity-relationship model (EBERM) is a pair (E, R) where E and R are respectively sets of entity and relationship set descriptors as redefined.

368

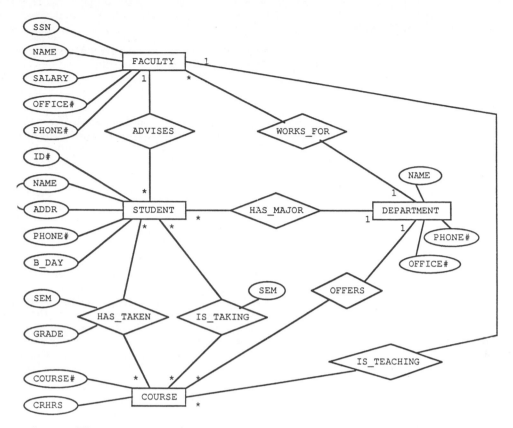

Figure 1. Entity-Relationship Diagram for Sample Database.

```
FACULTY (eFACULTY, SSN, NAME, SALARY, OFFICE#, PHONE#)

STUDENT (eSTUDENT, ID#, NAME, ADDR, PHONE#, B_DAY)

COURSE (eCOURSE, COURSE#, CRHRS)

DEPARTMENT (eDEPARTMENT, NAME, PHONE#, OFFICE#)

ADVISES (eSTUDENT, eFACULTY)

HAS_TAKEN (eSTUDENT, eCOURSE, SEM, GRADE)

IS_TAKING (eSTUDENT, eCOURSE, SEM)

IS_TEACHING (eCOURSE, eFACULTY)

WORKS_FOR (eFACULTY, eDEPARTMENT)

HAS_MAJOR (eSTUDENT, eDEPARTMENT)

OFFERS (eCOURSE, eDEPARTMENT)
```

Figure 2. A Database Scheme for the ER Diagram in Figure 1.

THE ENTITY-RELATIONSHIP ALGEBRA

The entity-relationship algebra defined in this paper is called ERA and is a pair (C, O) where C is the carrier of the algebra and O is the set of operators. The carrier C is a set of EBERMs, and the operators of O are given below.

Delete Relationship Set. This operator removes a specified relationship set descriptor from set R. The set E is not affected.

Delete Entity Set. This operator removes a specified entity set descriptor from set E. All relationship set descriptors associated with the entity set removed are also removed.

Relationship Set Projection. Attributes of relationship sets may be removed by invoking this operator. If attributes that are key components are removed, the set of key components for the relationship set must be adjusted. If, for example, the attribute SEM is removed from the HAS_TAKEN relationship set in Figure 1, GRADE would have to become a key component. This is because it is possible that a student has taken the same course twice and received different grades so that {eSTUDENT, eCOURSE} alone would not be a key. Observe that it is also possible that a student may have taken the same course twice and received the same grade. A consequence of obeying the request to remove SEM causes these two facts to become indistinguishable.

Entity Set Projection. Attributes of entity sets may be removed by invoking this operator. Since surrogate key names may not be removed, even when all attributes are removed from an entity set, an entity still maintains its identity. An entity in an entity set with no attributes, however, can only be identified through its relationships. For example, if all the attributes were removed from the FACULTY entity set in Figure 1, a particular faculty member could still be identified as the one who advises the student with some given ID#.

To exemplify the style of formal definition used for all the ERA operators, we give a formal definition for entity set projection. Let Z be a set of attributes, then project(E_1, Z) is defined as follows:

enabling conditions	action
1. $E_1 \in E$	/* project on the attributes Z */
2. $Z \subset E_1.EA$	$E_1.EA \leftarrow Z$
	let P be $Z \cup \{E_1.ES\}$
	$E_1.EX \leftarrow \pi_P E_1.EX$
	/* eliminate keys including deleted attributes */
	for each set $S \in E_1.EK$
	if $S \not\subset Z$ then $E_1.EK \leftarrow E_1.EK - \{S\}$

Relationship Set Selection. Selection operators do not alter the ER diagram. Instead they alter the expressions in the descriptors so entity and relationship sets are restricted according to some conditional expression. The attributes in conditional expressions are limited to those in the entity or relationship set.

As an example, the selection operator select(HAS_TAKEN, SEM='Fall 83') would restrict the relationship set HAS_TAKEN to include only those tuples that have a semester value of 'Fall 83'. The effect is to restrict the relationship to include only relations with students who took courses during the Fall Semester of 1983.

Entity Set Selection. Entity selection is more complex than relationship selection. Not only is it necessary to update the entity set descriptor, but it is also necessary to update the relationship set descriptors that are associated with the entity set. The relationship descriptor must be changed so that their expressions remove relations that involve removed entities.

As an example, consider the operator select(STUDENT, ADDR='β House' **or** PHONE#=2991) applied to the ER Diagram of Figure 1. This operator limits the entities in STUDENT to those that either live in the β House or have phone number 2991. In addition, the relationship sets ADVISES, HAS_TAKEN, IS_TAKING, and HAS_MAJOR are restricted to involve only those entities remaining in STUDENT.

Relationship Set Union, Intersection, and Difference. The relationship union, inter-section, and difference operators permit compatible relationship sets between the same two entity sets to be combined into one. Relationship sets are compatible if they have the same set of attributes. They may be combined through union, intersection, or set difference.

As an example if the operators select(STUDENT, ID#=3784), project(HAS_TAKEN, {SEM}), and union(HAS_TAKEN, IS_TAKING) were applied to the ER Diagram in Figure 1, the resulting relationship (HAS_TAKEN-OR-IS_TAKING) would contain a list of all the courses that the student with ID# 3784 has or is taking.

Reduce Entity Set. This operator reduces an entity set E_1 into new relationship sets that are dependent on E_1. After the new relationship sets are created, E_1 and its original associated relationships are removed. If there are n relationship sets associated with E_1, then as many as $(n-1)n/2$ new relationship sets are created, one for every pair except for pairs that create a relationship set that relates an entity set to itself. The new relationship sets are created by joining the pairs and E_1. An example of this operator is provided below.

Rename Attribute. Attribute renaming operators give the user control over the natural join generated by the reduction operator. For example, it may be necessary to keep names of a faculty member separate from names of students during a reduction of the ER Diagram of Figure 1.

Rename Entity or Relationship Set. Because some of the operators generate names for relationship or entity sets that may be long or unnatural, the user may wish to rename them.

AN ER ALGEBRAIC QUERY LANGUAGE

A natural application for ERA, the entity-relationship algebra just described, is an ER algebraic query language. In this section we define an ERA-based query language. In the next section we explain how to implement the language as an interactive, two-dimensional, graphical query language.

A query of the ER algebraic query language has two parts: (1) a diagram transformation part and (2) a display part. In the diagram transformation part the user may apply the ERA operators as desired. In the display part a relational algebra expression is created by joining with natural join the expressions of each entity or relationship set descriptor in the transformed diagram and then projecting on the set of attributes in the diagram. Since join is commutative and associative, there is no concern about the order in which the descriptor expressions are joined. The relational algebra expression is then applied to the underlying database resulting in a relation, which is displayed as a table.

Sample Query #1: "List the ID#'s and names of students majoring in Computer Science."

        delete(FACULTY)
        delete(COURSE) /* See Figure 3a. */
        select(DEPARTMENT, NAME='Computer Science')
        project(STUDENT, {ID#, NAME})
        project(DEPARTMENT, φ) /* See Figure 3b. */

As the diagram is transformed, the system builds expressions, retaining them in the expression components of the entity and relationship set descriptors. After the above operators have been invoked, the expression in the EX component of STUDENT is

$$\pi_{\text{eSTUDENT,ID\#,NAME}}\text{student} \ ,$$

the expression in the RX component of HAS_MAJOR is

$$\pi_{\text{eSTUDENT,eDEPARTMENT}}\sigma_{\text{NAME='Computer Science'}}(\text{department} \bowtie \text{has\_major}) \ ,$$

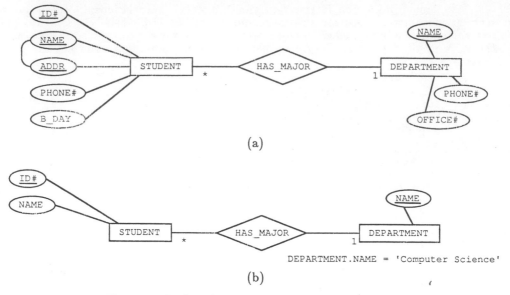

(a)

DEPARTMENT.NAME = 'Computer Science'

(b)

Figure 3. Reduced Diagrams for Sample Query #1.

and the expression in the EX component of DEPARTMENT is

$$\pi_{\text{eDEPARTMENT}}(\sigma_{\text{NAME}='\text{Computer Science}'}\text{department}) \ .$$

When this query is displayed, a single expression is created by joining these expressions and projecting out the surrogate keys. The final expression can then be optimized and applied to the database to return a result.

A GRAPHICAL QUERY LANGUAGE

The operators of the entity-relationship algebra can be easily implemented using a graphical interface. A possible screen layout is to list all the operator names (delete, project, select, union, intersect, difference, reduce, and rename) at the top of the screen, reserve an area at the bottom for messages from the system and for user input, and leave the large remaining area in the center for ER diagrams. Large diagrams can be viewed using a window mechanism.

A user invokes operations to reduce a diagram by first using a pointing device to move a cursor to an operator. Then, depending on the operator, the user either moves the cursor to select the operator's arguments or enters text in the message area. If an operator is to be applied two or more times in succession, it need not be repeatedly selected. Thus, for example, several entity and relationship sets could be deleted by just selecting the delete operator and then pointing successively at the entity and relationship sets on the diagram to be deleted. After the diagram has been transformed, the results can be displayed or printed.

Sample Query #2: "List names of faculty who are either advising or teaching the student with ID# 7932." Assuming that the screen initially contains the ER diagram in Figure 1, this query can be done graphically as follows:

1. Choose delete, and point at DEPARTMENT and HAS_TAKEN.
2. Choose select, point at STUDENT, and enter ID#=7932.
3. Choose project, point at STUDENT, point at COURSE, point at IS_TAKING, and point at FACULTY followed by the attribute NAME (in FACULTY).

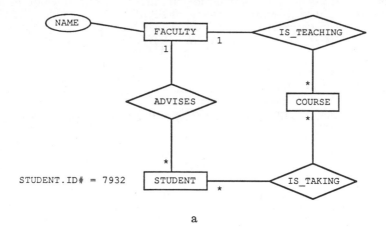

a

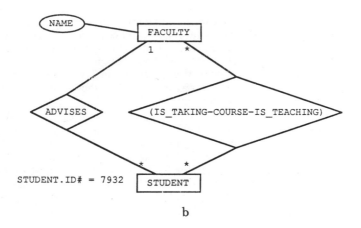

b

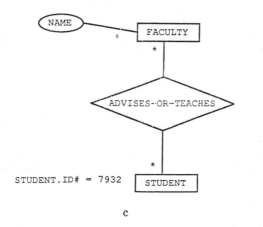

c

Figure 4. Reduced Diagrams for Sample Query #2.

At this point in the query formulation the diagram on the screen would be as in Figure 4a, and the ERA operators delete(DEPARTMENT), delete(HAS_TAKEN), select(STUDENT, ID#=7932), project(STUDENT, φ), project(COURSE, φ), project(IS_TAKING, φ), and project(FACULTY, {Name}) would have been generated.

4. Choose reduce, and point at COURSE.

After this step, the diagram would be as in Figure 4b, and the operator reduce(COURSE) would have been generated.

5. Choose union, and point at ADVISES and (IS_TAKING-COURSE-IS_TEACHING).

6. Choose rename, point at (ADVISES-OR-(IS_TAKING-COURSE-IS_TEACHING)), and enter 'ADVISES-OR-TEACHES'.

7. Display result.

The final diagram is as in Figure 4c and represents the user's query. The generated relational algebra expression is complex, but when executed, does yield the desired result.

SUMMARY

We have defined an entity-relationship algebra for a particular ER model, the EBERM. An EBERM extends the binary ER model to include relational algebraic expressions defined over the database scheme derived from a given instance of the model. A set of algebraic operators, each of which maps an EBERM to an EBERM, was suggested. An important property of algebra is the invariance of entity set identity under the operations. Rigorous definitions for the model were provided so that the semantics of the algebraic operations could be defined.

Based on the algebra, we defined two query languages: the first directly employs the operators of the algebra to arrive at a result, and the second is a graphical query language in which an ER diagram is manipulated interactively. In the latter, a pointing device is used to select operators and point at the parts of the diagram to be altered. Both languages obtain a result for a query by forming and executing a relational algebra expression obtained from the model as transformed by the ER algebra.

REFERENCES

1. P.P. Chen, "A preliminary framework for entity-relationship models," in *Entity-Relationship Approach to Information Modeling and Analysis*, ed. P.P. Chen, pp. 19-28, North-Holland, Amsterdam, The Netherlands, 1983.

2. P.P. Chen, "An algebra for a directional binary entity-relationship model," *Proceedings of the International Conference on Data Engineering*, pp. 37-40, Los Angeles, California, April 1984.

3. E.F. Codd, "Extending the database relational model to capture more meaning," *ACM Transactions on Database Systems*, vol. 4, no. 4, pp. 397-434, December 1979.

4. D. Maier, *The Theory of Relational Databases*, Computer Science Press, Rockville, Maryland, 1983.

5. V.M. Markowitz and Y. Raz, "A modified relational algebra and its use in an entity-relationship environment," in *Entity-Relationship Approach to Software Engineering*, ed. C. Davis, S. Jajodia, P. Ng & R. Yeh, North-Holland, Amsterdam, 1983.

6. V.M. Markowitz and Y. Raz, "ERROL: an entity-relationship, role-oriented, query language," in *Entity-Relationship Approach to Software Engineering*, ed. C. Davis, S. Jajodia, P. Ng & R. Yeh, North-Holland, Amsterdam, 1983.

7. C. Parent and S. Spaccapietra, "An entity-relationship algebra," *Proceedings of the International Conference on Data Engineering*, pp. 500-507, Los Angeles, California, April 1984.

Structures and Performance of Physical Database Models

RECORD-TO-AREA MAPPING IN THE CODASYL ENVIRONMENT

Leszek A. Maciaszek

Department of Computing Science
University of Wollongong
Wollongong, NSW 2500, Australia

ABSTRACT

A physical level design problem of record-to-area mapping
is addressed in terms of an integrated database design
methodology. A problem of gathering record usage statistics,
as a prerequisite of the record-to-area mapping, is discussed.
A stepwise heuristic algorithm for the record-to-area mapping
in the CODASYL database environment is proposed and
illustrated by an example. The iterative character of the
algorithm and the necessity for the design refinements and
feedbacks are exposed.

1. INTRODUCTION

A problem of physical database design (called thereafter
materialization) deals with "... finding an optimal
configuration of physical files and access structures - given
the logical access paths that represent the interconnection
among objects in the data model, the usage pattern of those
paths, the organizational characteristics of stored data, and
the various features provided by particular database
management system (DBMS)" (Whang *et al.*, 1982).

This paper refers to the initial stages of the
materialization phase. We present a method of gathering record
usage statistics, that in turn are utilized in the process of
assigning records to areas. We believe that our paper makes
the contribution of presenting a novel and creative algorithm
to this aim. Though we emphasize the iterative character of
the algorithm, we clearly define the initial design conditions
and decline as input those parameters that must not be known
to the Database Administrator (DBA) at these initial
materialization stages.

2. TERMINOLOGY

We distinguish between the physical storage structure and the physical allocation structure. The storage structure definition is a responsibility of the DBMS and its Data Storage Definition Language (DSDL), whereas the allocation structure is determined according to the device and the file management provisions of an Operating System (OS) and its Command Language (CL). We presuppose the reader's knowledge about the terminology relevant to physical allocation structure (file, block, etc.) and – on this understanding – we proceed to define the terms specific to physical storage structure in the scope determined by this paper.

An *area* is a largest named subdivision of the stored database. Its allocation structure equivalent is called a *file* – a portion of external storage media space controlled as an entity by the OS. In some DBMS-s (e.g. IDMS) the areas can be mapped onto the files in a variety of ways, in other systems (e.g. DMS-1100) the areas are related to the files in strictly 1:1 correspondence. At run time, one or more areas are "readied" in groups called *realms*. Thus, the realm is a subschema counterpart of the schema area. An area is either initialized (created) by the DBMS when it is opened for initial load or preinitialized by means of a utility routine of the DBMS. Its size is specified in the DSDL in number of pages. DSDL provides also for possible extensions with the specified step up to a maximum size.

A *page* is a fixed-length portion of an area. Ultimately, a page becomes a unit for transfers between external storage medium and DBMS buffer space, and it is called then a *block* (physical record). (However, exceptions are possible. For instance, a page in VAX DBMS consists of one or more disk blocks of 512 bytes each.) Thus, all pages have the same size in a given area. No doubt, this is a severe restriction on part of the DSDL in the light of complex distribution of record occurrences belonging to many different record types and connected by truly network blend of CODASYL's sets. This hinders the optimization of performance, especially when several record types are calced (hashed) within the same area.

A *storage record* is a direct result of transforming a logical record to an actual storage format. It includes a subset (not necessarily proper, and possibly empty) of the data items from the logical record. It also contains all necessary pointers, record length information, and other control data. This implies that a given logical record type may be mapped onto more than one type of storage record. In this case, the storage records may also have overlapping (redundant) data items from the logical record. All storage records related to a certain logical record are linked together, either by direct or indirect (i.e. through an index) pointers. The storage records can only be addressed by the DBMS, not by application programs. The DBMS is solely responsible for resolving space allocation problems caused by the growth of storage records within a page (and, possibly, for further splitting of storage records into smaller units). It is essential to recognize that for the process of record-to-area mapping the notion of logical record is sufficient and, therefore, we do not make further distinctions between logical and storage records.

3. GATHERING RECORD USAGE STATISTICS

The physical design process commences with the stage in which the record usage statistics are calculated. At this initial stage, the choice of the record as a measure for usage statistics is the only practical alternative despite that the notion of record has got here more logical than physical flavor. No physical notion, for example a storage record or a page, can be exercised because neither is defined in this design stage yet. From this point of view our understanding of record usage statistics, i.e. as a sort of interface between the logical and physical view of the database, verges upon the notion of access path as defined by Katz and Wong (1983); it also resembles a logical level traversal type of Effelsberg and Loomis (1984), a transaction definition language of Staniszkis et al. (1982), and search strategies of Batory and Gotlieb (1982). However, and regretably, none of the above notions could have been used in our model directly at this stage since the technicality of the problem and its purpose are slightly different and not that of performance prediction. We merely aim at calculating a coefficient called a relative design importance of a record in the physical design considerations.

The relative design importance (RDI) of a record is inferred from the logical record access counts that can in turn be determined by means of analysing the function specifications of conceptualization (Maciaszek, 1986). Note that since the intra-record structures were already determined during the logical design (formalization), it is easy to define access patterns of functions (transactions) to logical records. A simple abstraction process can do this job by modifying (in fact simplifying) the function specifications in such a way that they traverse now the logical records rather than attributes. Figure 1 is an example of Petri-net-like access structures of three function specifications.

However, the access patterns of functions are not sufficient to get the logical record access counts. Traversal techniques (search strategies) to individual records are also required to be known. Fortunately, this task is taken care of earlier during formalization. Though these strategies are likely to be modified later in the materialization process, they give a sound basis for access counts considerations at these initial materialization stages. Four traversal techniques, i.e. the ways of reaching a logical record in a database, are recognized here:

(1) CALC/DBKEY – uses a CALC key, database key, or currency indicator value;
(2) INDEXED – uses an index defined either for a set (e.g. an ordinary or index-pointer array) or for a record type (e.g. a B-tree).
(3) SCAN – accesses all the record occurrences by means of pointers within a set or by means of physical contiguity within an area.
(4) PARTIAL SCAN – similar to SCAN but terminates as soon as

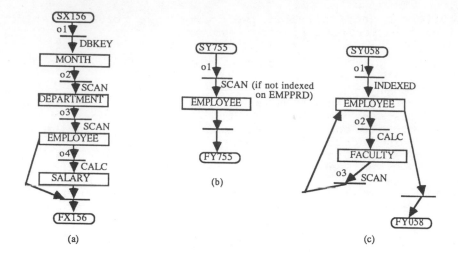

(a) EMPSLR01 - Determine the global number of employees and the total salary in a given month for each department and then list surnames, first names and monthly salaries of the employees.

(b) EMPSTF01 - Get surnames, first names, number of publications, periods of employment and ranks of the university staff members.

(c) EMPSTF02 - Enter doctors on position of a section chairman into the Faculty Council and list members of the Faculty Council according to the surnames, first names, degrees, ranks, and positions.

Figure 1. Access structures of the three functions: (a) EMPSLR01, (b) EMPSTF01, (c) EMPSTF02.

the criteria of searching (response set) have been satisfied.

In an integrated and shared database system a record can be retrieved in a variety of ways. SCAN and PARTIAL SCAN, when based on the physical contiguity within an area, can be applied to all database records. Member records of the sets can be always reclaimed by either SCAN or PARTIAL SCAN or INDEXED applied to those sets. If a record has been located as CALC or DIRECT, it can be accessed randomly. Moreover, the secondary search can be provided on some data items of the record (usually by means of a B-tree).

In general, the CALC/DBKEY and INDEXED techniques are oriented toward the entry point search. Thus, they mainly apply to the problem of finding an individual record fast and only rarely are used for the set traversal (i.e. as a navigational support). On the other hand, the SCAN and PARTIAL SCAN techniques are almost exclusively applied as the tools of a navigational search and are concerned with the efficient processing of a subset of member records within the set, such that the subset involved satisfies the search criteria.

Having known how an occurrence (or occurrences) of record type R_i is going to be retrieved by the function F_j, the DBA is in a position to calculate the expected cardinality (i.e. a number of record occurrences of R_i) in processing F_j. We shall call this measure a search length SL_j. In case of the CALC/DBKEY and INDEXED techniques, the search lengths are determined for both the entry point and the navigational support. In case of the SCAN and PARTIAL SCAN we define the search lengths in terms of navigation through either the set or the area. However, we note an important difference between the SCAN and PARTIAL SCAN – the cardinality of the subset records satisfying the search criteria for SCAN can be and most commonly is greater than 1, whereas for PARTIAL SCAN the cardinality is almost always equal to 1 (this is caused by the very nature of PARTIAL SCAN that is normally used to scan for a record occurrence responding to the search criteria and perhaps even being the entry point in the next step of the function processing – when no more efficient technique, i.e. CALC/DBKEY or INDEXED, can be applied).

The following are the search length formulas for the four traversal techniques (in case of the CALC/DBKEY and INDEXED technique, the length of the entry point search is given on the left-hand side of the expression and the length of the navigational search is shown on the right-hand side):

(1) CALC/DBKEY:

$$1 + p_o - p_b \cong SL_j \cong C_{R(i)} / 2, \text{ where:}$$

$C_{R(i)}$ – cardinality of records in areas in which R_i can be placed,

p_o – probability of overflow of calc-chain to another page (zero for DBKEY),

p_b – probability of finding the record occurrence in a buffer.

(2) INDEXED:

$$H_I \geq SL_j \cong C_{R(i)} / 2, \text{ where:}$$

$C_{R(i)}$ – cardinality of R_i in the database,

H_I – height of the index or the number of secondary storage accesses required to search the index (in fact, an assumption is made here that a master index will not require a disk access since it will have been placed in main memory when the area is opened; however, this gain of one access is neutralized later by a need to access the data area after scanning the whole index area); in general, the height of an index is on the order of $O(\log_b C_{R(i)})$, where b is the branch factor, i.e. the number of key values in a node (or, in the case of B-trees, one greater than the number of key values).

(3) SCAN:

$$SL_j \cong C_{R(i)} * (1 - p_r), \text{ where:}$$

$C_{R(i)}$ – cardinality of R_i in the set or in the database,

p_r – probability that two or more record occurrences R_i are placed in one page.

(4) PARTIAL SCAN:

$$SL_j \cong (C_{R(i)} * (1 - p_r)) / 2, \text{ where:}$$

$c_{R(i)}$ - cardinality of R_i in the set or in the database,
P_r - probability that two or more record occurrences R_i are placed in one page.

The stage of record usage statistics is concluded by finding the relative design importance of record types RDI_{ij}. This coefficient has a profound influence on the further materialization stages. The materialization as a whole is centered around the record design (let us recall that even sets are materialized in records). Thus, attaching design priorities to the records cannot be overemphasized in the process of considering design alternatives. Figure 2 presents an example in which a tabular form is used to calculate the RDI_{ij} of five record types R_i in the pre-canned function environment consisting of five functions F_j. The formulas necessary to arrive at the RDI_{ij} are also shown. RDR_i stands for the relative design rank of a function and expresses the relative importance of a function (transaction) in the set of functions under consideration (Σ RDR_j = 1). The RDR_j are calculated during the conceptualization (Maciaszek, 1986). They lend themselves as important factors to identify and supress the effect of interferences among user views on the design process. They also represent a direct response to the requirements of the theory of separability and by leading to the vertical partitioning of a design scope make the problem tractable.

4. ASSIGNING RECORDS TO AREAS

The record-to-area mapping consists the second stage of our physical design methodology. No doubt, it has not been treated with due attention by the researchers. Most often this topic is either neglected or only pointed out in the methodologies (e.g. cp. Staniszkis et al. (1983)). As an exception we can mention the discussion in Teorey and Fry (1982) where a simplified heuristic algorithm is investigated.

The task of record-to-area mapping is clearly related to the problem of record placement strategy. It follows that the knowledge about record placement is inherent in the mapping problem and vice versa. To resolve the conflict an iterative reasoning must be applied. At the outset it should be noticed that the preliminary decisions on placement strategies have already been taken during the phase of formalization. Those decisions were based on the access patterns of the set of functions to the records and we can safely recognize them as a starting-point.

According to the 1978 Draft Specification of DSDL (CODASYL, 1978) three placement strategies are available to DBA:
1. CALC
2. CLUSTERED
3. SEQUENTIAL

Functions / RR(j)		0.2	0.1	0.3	0.25	0.15	RDI(ij)
Record Types		F1	F2	F3	F4	F5	
R1	SL(j)	20	40	40		5	
	RP(j)	0.3448	0.4444	0.3809		0.2500	
	RI(j)	0.0690	0.0444	0.1143		0.0375	0.2652
R2	SL(j)	1	15	3	4		
	RP(j)	0.0174	0.3333	0.0286	0.5714		
	RI(j)	0.0035	0.0333	0.0086	0.1428		0.1882
R3	SL(j)	5		7		8	
	RP(j)	0.0682		0.0667		0.4000	
	RI(j)	0.0172		0.0200		0.0600	0.0972
R4	SL(j)	30		50	3		
	RP(j)	0.5172		0.4762	0.4386		
	RI(j)	0.1034		0.1429	0.1071		0.3534
R5	SL(j)	2	10	5		7	
	RP(j)	0.0345	0.2222	0.0476		0.3500	
	RI(j)	0.0069	0.0222	0.0143		0.0525	0.0959
Σ SL(j)		58	45	105	7	20	Σ = 1
Σ RP(j)		1	1	1	1	1	

SL(j) - average search length for R(i) in processing F(j)
RP(j) - relative priority of retrieval of R(i) with respect to F(j)
RL(j) - relative importance of R(i) with respect to F(j)
RDI(ij) - relative design importance of R(i) with respect to all functions F(j)

$RP(j) = SL(j) / \Sigma SL(j)$

$RI(j) = RR(j) * RP(j)$

$RDI(ij) = \Sigma RI(j)$

Figure 2. Tabular aid to calculate the relative design importance of record types (example).

Since it is allowed for an area to contain occurrences of several record types and for occurrences of a record type to span several areas, a potential solution space is enormous. In fact, the number of design alternatives is the product of power sets of the set of record types and the set of areas minus a negligable constant to eliminate some repeating or useless combinations (such as no record types in the area). Thus, the number of possible mappings is proportional to $O(2^n * 2^m)$, where n and m represent the number of record types and areas, respectively.

Moreover, the problem is not manageable by a well-known class of linear optimization techniques or other noniterative algorithms because of some conflicting conditions in the implementation alternatives (e.g. a member having two owners in two sets can only be clustered with respect to one set) and since some alternatives can override the effects of the other design choices. In these circumstances, a solution to the problem lies in a stepwise heuristic algorithm or at the best in techniques of integer linear programming such as branch-and-bound algorithms (originally conceived as back-track programming) and perhaps also in techniques of dynamic programming (v. Reuter and Kinzinger (1984) for some conclusions from experiences of using heuristic and analytic methods in physical database design).

The following more or less quantitative rules-of-thumb are formulated for our heuristic algorithm (note that most of them tighten the solution space):

383

(1) Consider the relative design importance RDI of records in the process of record-to-area mapping (Figure 2).

(2) A cluster of record occurrences consisting of an owner and member CLUSTERED NEAR OWNER is stored within one area (though not necessarily all clusters of the pertinent set type have to be mapped into one and the same area).

(3) Record occurrencess of a particular type with PLACEMENT SEQUENTIAL are stored in one area.

(4) It follows from (2) and (3) that only records located PLACEMENT CALC are candidates, though still unlikely, to be spread over more than one area.

(5) An area does not run over more than one subschema (this ensures that only a subset of applications is affected by a failure in an area).

(6) A number of areas linked by sets should be minimized (to facilitate recovery if not for other reasons).

(7) While keeping a number of areas as small as possible (according to our experiences more than 15-20 areas per subschema introduces an inadmissible overhead on the part of the I/O transfers and memory requirements), the areas must not extend over more than one disk volume and perhaps with the load factor not exceeding 70% (for system availability and recovery reasons).

(8) By definition, the areas of MODE INDEX or POINTER are separate from DATA areas and as such they can influence the rule (7).

(9) A subschema includes owner record types if the pertinent member record types are of INSERTION AUTOMATIC.

(10) An owner of singular sets is located in an area where the member record type (of one of these singular sets) having the highest relative design rand RDR is stored.

From what has been said one can conclude that the following sets of objects are involved in the process of record-to-area mapping:

(1) Set of record types:
$R = \{R_1, \ldots, R_i, \ldots, R_r\}$

(2) Set of areas:
$A = \{A_1, \ldots, A_i, \ldots, A_a\}$

(3) Set of subschemata:
$X = \{X_1, \ldots, X_i, \ldots, X_x\}$

(4) Set of set types:
$S = \{S_1, \ldots, S_i, \ldots, S_s\}$.

Obviously, it has to be assumed that the cardinalities of sets R, X, S as well as the mutual connections and overlappings among those sets are known a priori. The cardinality of A can only be determined in the process of relating records to areas. Thus, an initial design situation can be expressed in a way exemplified in Figure 3. The example comprises nine record types, $n(R) = 9$, twelve set types, $n(S) = 12$, and seven subschemata, $n(X) = 7$. The RDI of record types and their placements are also shown - C stands for CALC, V - CLUSTERED VIA, S- SEQUENTIAL. Moreover the record type R_1 is named SYSTEM, i.e. it is the owner of singular sets, and the record R_2 is the AUTOMATIC member in the set type S_4.

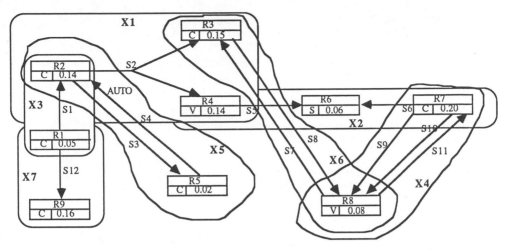

Figure 3. Initial design situation for a record-to-area
          mapping (example).

It is our belief that the example of Figure 3 can be
considered as typical for a real-life, though scaled, design
problem. Therefore the example can serve as a validation
vehicle of plusability of our rules-of-thumb. Perhaps the
first observation should concern the rule (5) - it can be seen
from the diagram that this rule, though otherwise justified,
cannot be complied with in the mapping process. As shown, in
order to be consistent with that rule a separate area would be
created for each of the nine record types involved. Evidently
this would be contrary to the very idea of a database as
opposed to a conventional file system.

However, we do not decline the rule (5) entirely. The
subschemata express collections of user functions
(transactions) and in our approach to database design the user
needs are of highest priority. A possible solution to this
deadlock is to combine the rule (5) with the rule (1). To this
end we determine subschema-driven relative design importance
of records XRDI by simply adding RDI-s of all record types
embraced by a pertinent subschema X_i (Figure 4).

Subschema	Records in the Subschema	XRDI
X1	R2, R3, R4	0.43
X2	R4, R6, R7	0.40
X3	R1, R2	0.30
X4	R7, R8	0.28
X5	R2, R5	0.27
X6	R3, R8	0.23
X7	R1, R9	0.21

Figure 4. Subschema-driven relative design importance of
          record types (example).

As a consequence the process of record-to-area mapping
commences with the record types belonging to subschemata of

XRDI = max, unless there is a contradiction accruing from the rule (9). In our example the highest XRDI, equal 0.43, is attached to the subschema X_1. The rule (9) is relevant for X_1 because the record type R_2 is the AUTOMATIC member of the set type S_4. Thus the scope of X_1 would need to be extended by including a pertinent owner R_5 or else precautions have to be taken not to run programs that store, delete or update in keys the record R_2.

We are now in a position to formulate a stepwise and recursive heuristic algorithm for a record-to-area mapping in the CODASYL environment:

STEP1. Draw a diagram of an initial design situation as shown in Figure 4.

STEP2. Determine the subschema-driven relative design importance of record types XRDI and sort them in descending order $XRDI_1$, ... $XRDI_x$, where x stands for total number of subschemata and $XRDI_i \rightarrow X_i$ (i=1,2,..., x).

STEP3. For any X_i, i=1,2, ..., x, determine the cardinalities of record types with placements:
(a) CALC - n(C),
(b) CLUSTERED - n(V),
(c) SEQUENTIAL - n(S).

STEP4. If for X_i the cardinality of CALC record types n(C) $\leq$ 1, then define an area A_i as being consistent with the boundaries of X_i, i.e. $A_i = X_i = \{R_i: R_i \in X_i\}$.

STEP5. If for X_i the cardinality of CALC record types n(C) > 1, and all CALC record types are owners of CLUSTERED VIA or SEQUENTIAL record types invoked within the same subschema, then define an area A_i as being consistent with the boundaries of X_i, i.e. $A_i = X_i = \{R_i: R_i \in X_i\}$.

Special cases of STEP4 and STEP5 and possible iterations:

(a) If there is a reason to assign more than one area to a CALC record type as mentioned in the rule-of-thumb (4) (e.g. record STUDENT could be scattered over two areas in order to process separately the male and female students). Note, however, that such situations are considered exceptional - if each record type is assigned only to one area then the complexity of the design problem is reduced to the order of $O(2^n)$, where n stands for the number of record types.

(b) If SEQUENTIAL record types $R_s \in X_i$ are independently enclosed by another subschema X_j, perhaps as the only record types in this subschema, that it is justifiable to separate R_s out and to locate R_s in a distinct area $A_j = X_j = \{R_s\}$. The above action is relinquished if XRDI of X_j is lesser than XRDI of X_i by an order of magnitude.

(c)   If VIA record types $R_v \in X_i$ can be CLUSTERED with
        more than one owner record type R_o, then consider
        the relative design ranks RDR of all the owners
        involved $\{R_o\}$ and cluster R_S with respect to the
        owner of highest RDR even though this owner could be
        outside of the X_i scope. Then apply the
        rule-of-thumb (2) and define the area $A_j \supseteq \{R_v, R_o\}$.
        If the owner R_o is from outside of the X_i scope then
        define $A_i = X_i - R_v$.

STEP 6.   After each decision that relates a record type to an
          area modify the design situation accordingly and
          repeat the steps 2 - 6. Refine and verify those
          steps with regard to the rules-of-thumb (6) - (10).

We believe that the disciplined approach presented above
allows to map record types to areas in the least awkward and
the most fair manner with respect to the whole database user
community. The following shows how this algorithm arrives at
the set of areas for our example of Figures 3 - 4.

1.   The Subschema X_1 has the largest $XRDI_1 = 0.43$. There are
      three record types embraced by this subschema $X_1 = \{R_2,
      R_3, R_4\}$, and $n(C) = 2$, $n(V) = 1$. Moreover, the
      rule-of-thumb (9) is relevant to this subschema since R_2
      is the AUTOMATIC member in the set S_3. After applying
      STEP 5 and the rule (9) to the subschema X_1 the defined
      area is $A_1 = \{R_2, R_3, R_5\}$. This situation is illustrated
      in Figure 5.

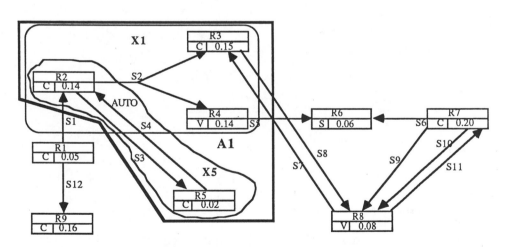

Figure 5. Design situation after the first iteration (area A1
          is defined).

2.   The next subschema to be considered is $X_2 = \{R_4, R_6, R_7\}$
      with $XRDI_2 = 0.40$. However the record type R_4 has already
      been assigned to the area A_1 and a new subschema-driven

relative design importance of the remaining record types R_6 and R_7 is equal $XRDI_2' = 0.06 + 0.20 = 0.26$. On this basis we rather proceed now to the subschema with the bigger XRDI (that is the subschema X_3) and defer a bit considerations concerning X_2.

3. The subschema $X_3 = \{R_1, R_2\}$ with XRDI = 0.30. However, again the record type R_2 has already been assigned to the area A_1, thus diminishing the value of the subschema-driven relative design importance of the record type to $XRDI_3' = 0.05$. We also note that R_1 is the owner of the two singular set types S_1 and S_{12} and the only other subschema involved is X_7. Thus, per the rule-of-thumb (10), we define a new area $A_2 = \{R_1, R_9\}$. This decision is even further motivated by the SEQUENTIAL placement of R_9. The design situation is visualized in Figure 6.

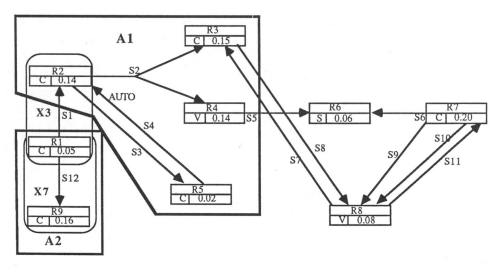

Figure 6. Design situation after the second iteration (area A2 is defined).

4. The next subschema of interest is $X_4 = \{R_7, R_8\}$ with $XRDI_4 = 0.28$ and $n(C) = 1$, $n(V) = 1$. When applying STEP 4 and considering the special case (c) another area is determined $A_3 = \{R_7, R_8\}$ (Figure 7).

5. As a result, the only record type left, i.e. not included in an area, is R_6. As being of SEQUENTIAL placement the record type R_6 is subjected to the special case (b) and, therefore, the last area defined is $A_4 = \{R_6\}$. Figure 7 shows the final design solution of the record-to-area mapping for our example.

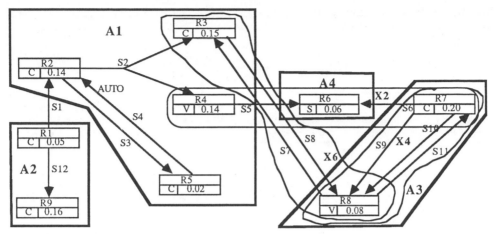

Figure 7. Final design solution after the third and fourth
iteration (areas A3 and A4 are defined).

5. CONCLUSION

Our approach to the physical database design in general,
and to the record-to-area mapping in particular, is an
outgrowth of a wider database design methodology and as such,
we believe, it is globally consistent, integrated and
iterative. We claim that the presented algorithm leads to the
feasible design solution that can become semioptimal as the
design process progresses and further feedbacks are applied.
It is also our belief that the algorithm is accurate and
specific enough to be used as the design technique in any
production environment based on a CODASYL DBMS. The algorithm
is meant to serve as a preliminary specification for a
computer-assisted design tool, which - when integrated with
the CAD tools for conceptual and logical designs - will be an
initial physical design component of the Database Design
Workbench (DDW).

REFERENCES

BATORY, D.S. and GOTLIEB, C.C. (1982): A Unifying Model of
Physical Databases, *ACM Trans. Database Syst.*, 4, pp.509-539.

BATORY, D.S. (1984): *Modelling the Storage Architectures of
Commercial Database Systems*, Tech. Report TR-83-21, University
of Texas at Austin, 58pp.

BEIGHTLER, C.S. PHILLIPS, D.T. and WILDE, D.J. (1979):
Foundations of Optimization, Prentice-Hall, 487pp.

CODASYL (1978): Report of the CODASYL Data Description
Language Committee, *Inform. Syst.*, 4, pp.247-320.

EFFELSBERG, W. and LOOMIS, M.E.S. (1984): Logical, Internal,
and Physical Reference Behaviour in CODASYL Database Systems,
ACM Trans. Database Syst., 2, pp.187-213.

KATZ, R.H. and WONG, E. (1983): Resolving Conflicts in Global Storage Design Through Replication, *ACM Trans. Database Syst.*, 1, pp.110-135.

MACIASZEK, L.A. (1986): *An Enhanced Conceptual Structure Derivation,* University of Wollongong, Department of Computing Science, Preprint 86-1, 44pp.

REUTER, A. and KINZINGER, H. (1984): Automatic Design of the Internal Schema for a CODASYL Data Base System, *IEEE Trans. on Soft. Eng.*, 4, pp.358-375.

STANISZKIS, W SACCA', D. MANFREDI, F. and MECHIA, A. (1983): Physical Data Base Design for CODASYL DBMS, in: *Methodology and Tools for Data Base Design*, ed. S. Ceri, North-Holland, pp.119-148.

TEOREY, T.J. and FRY, J.P. (1982): *Design of Database Structures, Prentice-Hall*, 495pp.

WHANG, K.-Y. WIEDERHOLD, G. and SAGALOWICZ, D. (1981): Separability - an Approach to Physical Database Design, *Proc. 7th Int. Conf. Very Large Data Bases*, Cannes, France, pp.320-332.

WHANG, K.-Y. WIEDERHOLD, G. and SAGALOWICZ, D. (1982): Physical Design of Network Model Database Using the Property of Separability, *Proc. 8th Int. Conf. Very Large Data Bases,* Mexico City, Mexico, pp.98-107.

AN OPTIMAL TRIE CONSTRUCTION ALGORITHM

FOR PARTIAL - MATCH QUERIES

Narao Nakatsu

Center for Educational Technology
Aichi University of Education
Kariya, Aichi 448, Japan

Abstract

This paper presents design algorithms of a trie file to answer partial-match queries. The author assumes in this paper that each field of a record is mapped into several bits which construct a bucket address into which the record is stored.

At each level of a trie, at most 2^d-way branching is possible according to predetermined d bits. The problem is the allocation and the selection of the d bits at each level.

Major results are as follows;
1) When data is uniform, an optimal trie is obtained.
2) When queries are uniform and data is not uniform, three factors which affect the efficiency are found. A new unbalance measure is defined and a brief description of a trie construction algorithm is presented.
3) When neither queries nor data is uniform, a heuristic algorithm which is based on four observations is given.

1.Introduction

Efficient retrieval algorithm by secondary keys is required in information retrieval systems and database systems. Index files will be usually created for fields which are frequently specified in a query. Many valuable file organization techniques such as the B-tree or ordered lists have been presented for an index file. Though an index file gives fast access, it requires much space and it is impractical to make index files for many fields. The partial-match retrieval is related to the retrieval by secondary keys without any index files. In this paper, a file design for answering partial-match queries will be presented.

Let A_1, ..., A_k be finite set of fields, where A_i takes values from a domain D_i ($1 \leq i \leq k$). A record r is an ordered k-tuple (a_1, a_2,..., a_k), where $a_i \in D_i$. A file F is a collection of records and is a subset of $D_1 x D_2 x...x D_k$. A partial-match query is a specification of values for zero or more fields of a record. Formally a partial-match query is an element of $(D_1 \cup \{*\}) x (D_2 \cup \{*\}) x ...x (D_k \cup \{*\})$, where * indicates the unspecification of the field. The answer of a partial-match query is a set of all records in the file whose fields match the specified values.

[Example 1] Consider a file of suppliers whose fields are (name, address, item, price) in this order. A partial-match query might specify an item name and a price such as "display all records where item equals ball and price equals 1.5 dollars."

When a file is small, to search the entire file may be tolerable. In this case, such techniques as the pattern matching machines or preprocessing information are useful[1,2]. However an efficient file organization is required usually from a standpoint of response time. We assume that a file is divided into a set of buckets which are stored on a directly accessible secondary device such as a magnetic disk. A bucket can contain a fixed number of records at most which depends on the computer system. An address is computed from a record and the record will be stored in the bucket of the address. The number of buckets accessed to answer a query will predominate in the retrieval time. The average number of buckets which are accessed to answer one query is considered for the measure of optimum.

Many results have been obtained for the purpose of good partial-match retrieval. A hash coding is one of the most useful technique. As Rivest suggests, however, it is very space consuming unless data is uniform[3]. The trie structure stands efficiently in such a case. For a trie, Rivest discussed the average case behavior for uniform data and uniform queries. Burkhard has investigated a trie construction method with good worst case performance[4]. Alager and Soochan discussed a trie design for non-uniform query distribution[7]. All of them discussed a binary attributed file where $D_i=\{0,1\}$. The goal of this paper is to present an elaborate algorithm for an optimal trie construction in general cases.

2.**Basic** concepts

We assume that a bucket is addressed by an n-bit address which will be determined by values of fields of a record. There are 2^n buckets at most and n will be determined by a system designer according to the number of records and the capacity of a bucket.

How to allocate each bit of a bucket address to each field of a record is one of interesting issues. Aho and Ullman have given an optimal solution to this issue under the assumption that data is uniform, a query distribution is given beforehand and any bucket can be accessed in one read operation[8]. By the term "uniformity of data", the author means that there is no empty buckets.

The selection of hash functions is another important problem. A hash function which maps a record to a bucket address directly is studied by Rivest and Burkhard[3-6], which is a different approach from us and is useful only for uniform data. This paper concerns about the construction of a trie index file. We define a trie which is slightly different from an original trie[9].

Assume that each record of a file has k fields. Each record is stored in a bucket whose address is specified by an n-bit sequence. For each field A_i of a record, a hash function h_i is defined, where h_i is a mapping from the domain D_i to bit sequences of length w_i ($\sum_i w_i = n$). We write $h(D_i)=b_{i1}b_{i2}...b_{iw_i}$ where $b_{ij}=0$ or 1. Assume that an address of a record is obtained by $h_1(a_1)h_2(a_2) ... h_k(a_k)$, the concatenation of bit sequences, where a_i is the value of the i-th field. The symbol b_{ij} is used to specify the j-th bit of $h_i(a_i)$ and is called an address bit. When an address is given, we can access the bucket of the address by going along the trie down from the root to a leaf.

[Definition 1] A 2^d-way trie is a tree where a bucket is stored in a leaf node. Each internal node is labeled by d different address bits and 2^d branching based on the specified d bits is made at most. We insist that 1) every node on a path from the root to a leaf is labeled by different address bits; 2) when address bits $b_{i_1j_1}$, $b_{i_2j_2}$,..., $b_{i_dj_d}$ are labeled at an internal node, all buckets in one subtrie of that node have the same values at the bit positions $b_{i_1j_1}$, $b_{i_2j_2}$, ..., $b_{i_dj_d}$ ($1\leq i_m\leq k$, $1\leq j_m\leq w_{i_m}$, $1\leq m\leq d$).

For special cases, we define two classes of a trie. A trie is a pruned trie iff each internal node has at least 2 non-null sons. A trie is called fixed order when all internal nodes of the same level are labeled by the same address bits.

[Example 2] Consider the following file where the corresponding bucket address is shown for each record. This file has 2 fields and $w_1=3$ and $w_2=2$. This file may be stored as shown in Figure 1, where $d=1$.

	b_{11}	b_{12}	b_{13}	b_{21}	b_{22}
r_1	0	0	1	0	0
r_2	0	1	0	0	1
r_3	0	1	0	1	1
r_4	0	1	1	1	0

For the optimum, the number of internal nodes as well as the number of buckets (leaf nodes) to be accessed should be taken into account.

One interesting issue is the order of address bits to be tested. Again consider the file in Example 2. A trie in Figure 2 is also feasible. Which one is better? and what is the best?

The problems to construct an optimal trie are summarized as follows:
1)When a bucket address is represented by an n-bit sequence, how many bits should be allocated to each field?
2)What is the order of testing the bits for branching?
3)How to choose a hash function for each field?

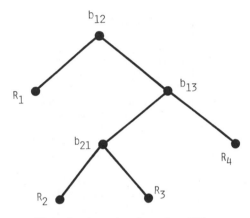

Fig. 1- A trie for the file
of Example 2.

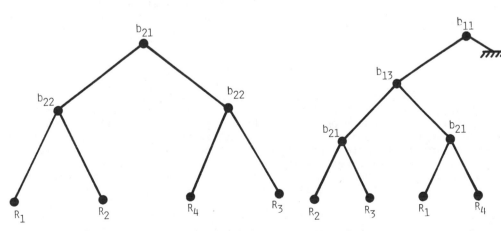

Fig. 2- Another trie for the file   Fig. 3- A trie suitable for
of Example 2.   unsuccessful search.

The last problem is the unsuccessful search. Consider the next example.

[**Example** 3] The trie of Figure 3 is also possible for the records of Example 2. Fig. 3 is not pruned. When only the first field A_1 is specified in a query and its hashed value equals 100, we know there is no record which matches the query by reading the root node only. On the other hand, 4 leaf nodes should be accessed in the trie of Fig. 2. Fig. 3 is better if the probability of unsuccessful search is high. We can not unconditionally conclude which is better.

Many researchers have studied on binary attributed files where each field is defined on {0,1}. They claim non-binary attributed files can be treated as binary ones. However the author is afraid that the problem 1) is omitted (one bit is always assigned to each field) and the correspondence between a field and a block of address bits disappears. Problems 3) and 4) are new problems.

3.**Uniform** data

Throughout this section, the uniformity of data is assumed. Let p_i ($1 \leq i \leq k$) be the probability that the i-th field is not specified in a query. We assume p_i is independent from one another. In general, p_i will be obtained by observing a set of queries. Our goal is to show the construction algorithm of an optimal trie in the sense of minimizing the number of nodes to be accessed.

For the simplicity of analysis, a binary (2-way) fixed order trie is considered. Let n_v be an internal node of level v (the root node has level 0) and $f(n_v)$ be the father node of n_v. Let l_i ($1 \leq i \leq k$) be the number of address bits of the field A_i labeled on the nodes on the path from the root to $f(n_v)$. Note that $\sum_i l_i = v$ and $l_i \leq w_i$ because each internal node is labeled by only one bit.

Let b_{i1}, b_{i2}, ..., b_{il_i} be address bits of A_i labeled on the path from the root to $f(n_v)$. Because the trie is binary, n_v is specified by an n-dimensional vector $(\mathbf{c}_1, \mathbf{c}_2, ..., \mathbf{c}_k) = (c_{11}, ..., c_{1w_1}, c_{21}, ..., c_{2w_2}, ..., c_{kw_k})$, v elements of which are 0 or 1 and the remaining n-v elements are don't care. The probability $p(n_v)$ that n_v is accessed for a query is given as follows:

$p(n_v) = \Pi_i \{p_i + (1-p_i)q_i(\mathbf{c}_i)\}$ -----(1), where $q_i(\mathbf{c}_i)$ is the probability that the field A_i is specified in the query and its hashed sequence equals $\mathbf{c}_i$.

Sum of all combinations of $\mathbf{c}_i$ ($1 \leq i \leq k$) will give the average number of internal nodes of level v accessed by a query, which is represented by $N(v)$. By the fact $\sum_{\mathbf{c}_i} \{p_i + (1-p_i)q_i(\mathbf{c}_i)\} = 2^{-l_i}p_i + 1 - p_i$, we have $N(v) = \Pi_i (2^{-l_i}p_i + 1 - p_i)$ where $\sum_i l_i = v$. ------(2)

To my great surprise, (2) is similar to the one proposed by Aho et al[8]. Note that no bucket should be empty so that $\sum_{\mathbf{c}_i} p_i + (1-p_i)q_i(\mathbf{c}_i) = 2^{-l_i} + 1 - p_i$ may hold. To construct an optimal trie, $N(v)$ should be minimized at every level v. We have the following theorem.

[**Theorem** 1] $N(v)$ is minimum iff $l_i = v/k - \log\{p_i/(1-p_i)\} + 1/k \sum_i \log\{p_i/(1-p_i)\}$ ($1 \leq i \leq k$).

(Proof) As in Aho, the Lagrangian multiplier method is utilized[8].

Considering that a node has 2^d sons and d bits are tested at each level, Theorem 1 holds for a 2^d-way trie by replacing v with dv. Theorem 1 is useful only to minimize the number of nodes of level v accessed. We must try to minimize total number of nodes to be accessed. Fortunately l_i will be incremented by 1/k (one at most) when level is incremented by one. By this observation, we can obtain an optimal construction algorithm of a trie.

Following corollaries are immediately obtained from Theorem 1.

[**Corollary** 1] When queries are uniform, $l_i = v/k$. This means that each field should be tested as equally as possible at each level.

[**Corollary** 2] When $p_i > p_j$, $l_i < l_j$ holds.

Corollary 2 means that more bits should be allocated to a field which is specified with higher probability in a query and address bits of the field should be tested at a lower level.

The following two points should be taken into account for applications.

i) Each l_i is an integer and the integer solution that minimizes $N(v)$ is required.

ii) Each w_i should be determined with reference to the size of the domain D_i. If domain $D_i=\{male,female\}$, then $w_i=1$ in spite of Theorem 1.

By the above discussion, Algorithm 1 is obtained for an optimal trie.

[**Algorithm** 1] Construction of an optimal trie.

Assumption: Each record has k fields. The probability of unspecification of each field is given by $(p_1, p_2, \ldots, p_k)$. The number of address bits is assumed n. $\log\{p_i/(1-p_i)\}$ is denoted by α_i for simplicity.

1: /* w_i is the number of address bits allocated to the field A_i. m_i is the number of address bits of A_i which are already tested. */

$m_i\text{<--}0$, $w_i\text{<--}0$ for all i in $[1,k]$.

2: /* consider the size of domains. */

$j\text{<--}0$, $s\text{<--}k$.

for i=1 until k do

    calculate $l_i=n/s-\alpha_i +1/s \sum_j \alpha_j$.

    if $|D_i|<2^{l_i}$ then begin

                    $w_i\text{<--} \lceil\log|D_i|\rceil$, j=j+1, n=n-$w_i$, s=s-1.

                  end

end

3: calculate $w_i=n/s-\alpha_i +1/s \sum_{w_j=0} \alpha_j$ for each i such that $w_i=0$.

4: for i=1 until k do

    Find a hash function h_i which maps the domain D_i to w_i bits.

    end

5:if there is no empty bucket then goto 6

        else give up the optimal trie,apply Algorithm 2 and stop.

6: /* v is a level of a trie */

for v=1 until $\lceil n/d \rceil$ do

    calculate $l_i=dv/k-\alpha_i +1/k \sum_j \alpha_j$ for each i $(1\leq i\leq k)$.

7:    for j=1 until Min(d, n-(v-1)*d) do

        Sort the set $\{l_i-m_i|l_i>m_i,m_i\leq w_i,1\leq i\leq k\}$ in the descending order. Select the first element of the set. Let the suffix of the element be t.

        One bit of the hashed value of the field A_t is used for branching at the level v.

        /* Note: the order of address bits of A_t to be tested is not significant because the trie is perfect. */

        $m_t\text{<--}m_t+1$.

    end

end

                                                  end of Algorithm 1

The critical point of Algorithm 1 is the selection of hash functions. For the optimum, it seems desirable that all records are stored in as less buckets as possible and no bucket overflows. However this is not always true as can be seen in Example 4.

The goodness of a hash function heavily depends on a set of records. A set of records will be changed by insertion and deletion of records. By this fact, it may be safe to select random hash functions if insertions occur frequently and empty bucket disappears soon. Section 4 is the study of non-uniform data, where empty buckets exist and the shape of a trie is not the perfect 2^d-way tree.

4. **Non-uniform** data

We also consider a binary trie for simplicity. Assume that there are N non-empty buckets $(N<2^n)$ and any element of a domain D_i is equally selected when the attribute A_i is specified in a query.

4.1 **Uniform** queries

Consider the case of uniform query first, where the probability of unspecification of a field is the same for all fields. In this case,the probability that an internal node of level v is accessed is given by the formula (1) by replacing p_i by p. By the same way as the case of uniform data, the sum of all combinations of c_i $(1 \leq i \leq k)$, $N(v) = \sum_{c_i} \cdots \sum_{c_k} p(n_v)$, will give the average number of nodes of level v which are accessed by a query. However, the formula $N(v)$ cannot be simplified because the trie is not a perfect trie yet. For the efficient retrieval, it is desirable to make $N(v)$ smaller. By the formula (1), we have the following observation which is presented by Rivest[3] and Alager[7].

[Observation 1] It is better to construct such a trie that has less nodes at less level, i.e., the most unbalanced trie is the best.

Alager and Soochan presented a measure of unbalance, $s_{ij}/(N-s_{ij})$, where N is the number of buckets and s_{ij} is the number of buckets whose address bit b_{ij} has 0.

Whether unbalance is caused by the non-uniformity of data or by the non-randomness of a hash function should be remarked. Observation 1 holds only if the probability that an address bit b_{ij} has 0 equals 1/2. Consider the following example.

[Example 4] A file is given in Figure 4, where $D_1=[1,100]$ and $D_2=[1,1000]$. Hash functions h_1 and h_2 are given in Fig. 5, where $w_1=w_2=2$. By the measure of Alager[7], the address bit b_{21} will be tested at the root. But almost 80 percent of queries have 1 at b_{21} and it is not so good.

By Example 4, non-randomness of hash functions must be taken into account to have an unbalance measure. Let s_{ij} be the number of buckets whose address bit b_{ij} has 0. Let t_{ij} be the probability that b_{ij} has 0 when A_i is specified in a query. t_{ij} is determined by h_i. We can define a measure as $u_{ij}=s_{ij}t_{ij}+(N-s_{ij})(1-t_{ij})$. An address bit b_{ij} such that u_{ij} is the least will be tested at the root. In Example 4, b_{22} should be tested at the root by our measure. Once a bit b_{ij} is selected for a label at the root, the set of records is divided into two subsets, a set of records which have 0 in b_{ij} (left subtrie) and a set of records which have 1 in b_{ij} (right subtrie). The same procedure is applicable to each subtrie. Formal algorithm of the construction of unbalanced trie is trivial and is omitted.

Once hash functions are fixed, above discussions are applicable. Now consider the determination of hash functions. Suppose the number of address bits allocated to each attribute is already known. Intuitively the following observation is obtained.

[Observation 2] It is better that buckets which are accessed by high probability are empty.

For the records of Example 4, hash functions of Figure 6 will be more suitable. By our unbalance measure, the trie of Fig. 7 is obtained.

Now we consider the last problem of pruning. Again we begin with an example.

[Example 5] Consider the trie of Figure 7. When both A_1 and A_2 are unspecified, all 12 nodes are accessed. When only A_1 is specified, 3 nodes of level 0 or 1 are always accessed. The nodes of level 2 are accessed at the probability of 41/100, because the probability that b_{11} is 0 is 41/100 under the assumption of a random query. Nodes of level 3

	A_1	A_2
R_1	50	700
R_2	30	950
R_3	40	930
R_4	70	200
R_5	60	550

Fig. 4- A file considered in Example 4.

[1, 25]	00	[1, 100]	00
[26, 50]	01	[101, 200]	01
[51, 75]	10	[201, 900]	10
[76,100]	11	[901,1000]	11

Fig. 5- Hash functions of the file in Fig.4.

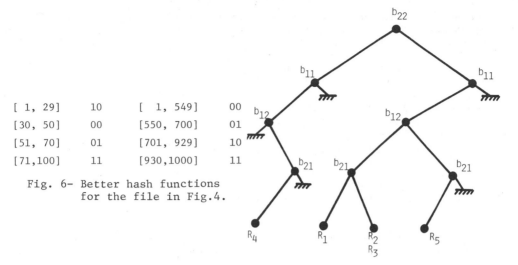

[1, 29]	10	[1, 549]	00
[30, 50]	00	[550, 700]	01
[51, 70]	01	[701, 929]	10
[71,100]	11	[930,1000]	11

Fig. 6- Better hash functions for the file in Fig.4.

Fig. 7- The better trie obtained by Observations 1 and 2 for the file in Fig.4.

are accessed at the probabilities 20/100, 21/100, 20/100, respectively. By the same way, the average numbers of nodes accessed are $2.843+4.864p +4.293p^2$ in the trie of Fig. 7 and $2.333+3p+1.667p^2$ in the pruned trie of Figure 7. The pruned trie of Fig. 7 is always superior to that of Fig.7.

For the file of Example 4, the pruned trie of Fig. 7 seems the best. We must answer the question that a trie should be pruned or not. The following observation is easily recognized.

[Observation 3] A pruned trie is better when the probability of unspecification is close to 1.

The proof of Observation 3 is trivial because the average number of nodes accessed approaches the number of nodes in the trie as p approaches 1. The pruned trie has the least nodes. In generally, it depends on p whether an edge had better be pruned or not. We will show a brief heuristic algorithm for a construction of better trie as follows.
1) Select hash functions such that empty buckets have high probabilities.
2) Decide the order of address bits tested according to the unbalance measure.
3) Prune edges if it decreases the average number of nodes to be accessed.

4.2 Non-uniform queries

The probability that a field is specified in a query will have a great influence on the order of test of address bits. Again we consider a fixed order pruned trie. We concentrate on the selection of an address bit which is tested at the root. Once such a bit is found, we can apply the same procedure to the left and right subtries, respectively.

Consider the two tries, a trie where an address bit b_{ij} is tested at the root and an address bit b_{mn} is tested at level 1 and the other trie where b_{mn} is tested at the root and b_{ij} is tested at level 1. There are 16 cases according as subtries are null or not. Considering each case, we will have the following observation which is presented by Alager[7].

[Observation 4] An attribute whose probability of unspecification is low had better be tested near the root.

By above four observations, we know such a trie that is most unbalanced and the probability of unspecification of the attribute tested at the root is low is efficient. We can propose a heuristic algorithm which depends on the hash functions, unbalance measure, pruning and the probability of unspecification.

The measure of unbalance u_{ij} $(1 \leq i \leq k,\ 1 \leq j \leq w_i)$ is obtained in 4.1.

[**Algorithm** 2] A heuristic algorithm to construct a good trie.

/* Each w_i is already known by Algorithm 1. Let the address bits be $b_{11}, \ldots, b_{1w_1} b_{21} \ldots b_{2w_2} \ldots b_{k1} \ldots b_{kw_k}$ */

1: Determine hash functions h_i heuristically by Observation 2.

2: For each address bit $b_{ij}(1 \leq i \leq k, 1 \leq j \leq w_i)$, compute the unbalance measure u_{ij}.

3: for i=1 until k do v_i <-Min$\{u_{ij}\}$ end

4: Consider the set C=$\{(v_i, p_i) \mid 1 \leq i \leq k\}$.

5: Exclude an element of C which is greater than any other element of C.

6: Elements of C suggest candidates of bit which will be tested at the root.

7: Select one element of C by some criterion.

8: The address bit selected in 7 is labeled at the root.

   Apply Algorithm 2 to both subtries.

9: Try to prune the trie if pruning brings better trie.

                                        end of Algorithm 2

   To have an optimal trie, tries must be made for all candidates and the best one should be selected, which will take intolerable time. To bound the computing time, a simple criterion which estimates roughly the average number of nodes accessed will be adopted at the step 7.

5. **Concluding** remarks

   There have been many good file design such as ABD or PMF for partial-match retrieval[3-6]. The shortcoming of these studies is the assumption that any bucket into which a record is stored is not empty. Every method which uses hash functions to compute bucket addresses directly degrades efficiency as the number of empty bucket increases. Tries seem most suitable for partial-match retrieval.

   In this paper, a construction algorithm of an optimal trie is discussed. For uniform data, an optimal algorithm is obtained. For non-uniform data, two observations, 1) unbalanced tries are better than balanced tries, and 2) address bits of attributes whose probabilities of unspecification are low should be tested at low level, are reconfirmed. Other two factors which affect the optimum are newly introduced and two additional observations are presented. Based on these observations, Algorithm 2 is obtained, in which optimum is thrown away for practical use.

   One of remaining problems is to investigate the efficiency of Algorithm 2, which will be done by a simulation. Another remaining problem is the dynamic maintenance of tries as B-trees[9].

References

[1] Aho, A.V. and Corasick, M.J., "Efficient string matching :An aid to bibliographic search", CACM, 18, 6, 333-340, June 1975.

[2] Nakatsu, N., "Studies on string manipulation problems ánd their applications", Ph.D. thesis, Kyoto Univ., 1983.

[3] Rivest, L.R., "Partial-match retrieval algorithms", SIAM J. of Comput., 5, 1, 19-50, March 1976.

[4] Burkhard, W.A., "Partial match retrieval", BIT, 16, 13-31, 1976.

[5] Burkhard, W.A., "Hashing and trie algorithms for partial match retrieval", ACM TODS 1, 2, 175-187, June 1976.

[6] Burkhard, W.A., "Non-uniform partial-match file designs", Theoretical Computer Science, 5, 1-23, 1977.

[7] Alager, V.S. and Soochan, C., "Partial match retrieval for non-uniform query distributions", Proc. of NCC, 775-780, 1979.

[8] Aho, A.V. and Ullman, J.D., "Optimal partial-match retrieval when fields are independently specified", ACM TODS, 4, 2, 168-179, June 1979.

[9] Knuth,D.E., "The art of computer programming", 3, sorting and searching, Addison-Wesley, 1973.

FOPES: FILE ORGANIZATION PERFORMANCE ESTIMATION SYSTEM

Pavel Zezula and Jan Žižka

Computing Center
Technical University
Brno, Czechoslovakia

ABSTRACT

A new file organization performance prediction system is presented.
Even though many useful ideas from previous works have been accepted, the
system has a new architecture and it comprises many novel modeling possibilities.

INTRODUCTION

A large body of work has been done on the field of data base performance prediction. Interesting surveys with extensive reference lists are
/Schkolnick, 1978; Yao and Navathe, 1981; Sevcik, 1981/. However, there is
also a study /Chilson, 1983/ aimed at investigating methodologies, techniques and tools which are currently available to assist users in the important performance decision-making function. The results of the study are
not very satisfactory because they show that the usage of those tools in
practice is still mostly on the experimental level. There are many reasons
for this situation. For example, the absence of a good physical data base
conceptual framework, information requirements specification, workload specification, etc. Those topics are still waiting for new good ideas which
would solve them. But even such essential problem as the file organization
modeling for performance estimation has not been adequately solved.

MOTIVATIONS

Though the reluctance of the application area to use the tools available could be the justification for developing a new file organization
performance prediction system, let us analyse some of the details. It is
a common experience that the performance of different types of file organizations differ substantially even if a constant workload is supposed.
Nevertheless, whenever a new model dealing with some more complex performance prediction problems appears /index selection, record segmentation,
query optimization, etc./, the file organization access cost is exceptionally supposed to be provided by any of the independently developed file
organization models. Instead, a new very simplified solution concerning
the file organization access cost is suggested to provide the desired
estimation. Unfortunately, this is usually too superficial to be able

to reflect all the important features of the file organization considered. In the consequence of this, this kind of models is independent of the file organization type which is far from the reality.

Those are our objectives for developing a new system which could serve a large range of applications. We have accepted many useful ideas from previous works especially /Yao, 1977/. But we established a new architecture, refined the model specification parameters and increased the computational accuracy. As a result of this, we have reached a system enabling its users not only to model more different types of file organizations but offering them extended possibilities to reflect variety of implementation techniques as well.

Though the emphasis on the automation of performance estimation aids is essential we have implemented our analytical File Organization Performance Estimation System /FOPES/ as a program module in Pascal language. The system has a single input and output and it can be used both independently, giving the file organization performance characteristics, and within some other perfomance evaluation models /for case selection or optimization/ embedded using the simple input-output interface.

Even though the performance complexity of databases stems from the interactions among the individual logical objects /files/, the importance of the file organization performance modeling has increased since the theory of Separability was introduced by Whang /Whang et al., 1981/. According to this theory, the optimal access configuration of a database is decomposed into the tasks of designing the optimal configurations of individual logical objects independently of one another. In such case, the required performance estimations for individual logical objects can be provided by FOPES, which is the system we are going to introduce here now.

FOPES GENERAL DESIGN

In the following, the main features of FOPES will be presented. We speak about system rather than about model which is considered here as a special case of the system, determined by a specification of the system´s external parameters.

From the logical point of view we shall suppose that the file is a collection of homogeneous records, including either single or repeated items for individual record instances. Stating that an attribute-name:value pair is a keyword, it means that one or several different keywords containing the same attribute-name constitute an item. Records of a file are homogeneous if all of them include items specified over the same set of attribute-names. A query consists of a number of clauses of the form:
        attribute-name:list of values
connected by the operators $\wedge$ and $\vee$, structured as the disjunctive form. The series of the clauses connected by the boolean operators $\wedge$ will be called the conjunct. The subconjunct will be that part of conjunct, the clauses of which are specified by the same attribute-name. The response set in general is a subset of the file. Depending on the specification of the subset, we can distinguish among the keyword response sets, the query clause response sets, the subconjunct response sets, etc. The query class is a sequence of queries, performed serially on a specific file during a finite time interval so that some other more complex transaction /eg. JOIN/ can be accomplished.

When modeling using FOPES, the data search tree is considered to be a tree structure reflecting all the possibilities how to access data elements of the file organization. Remember that not only the records but also many other data /keywords, pointers etc./ can be utilized while searching in the file. Moreover, in FOPES each of such data elements must appear in the data search tree as many times as many paths lead to it.

On the other hand, the process, steering the navigation through the

data search tree so that the query response set can be located, is designated as the search strategy. Then the union of the data search tree and the
search strategy can be called the access mechanism.

From the performance point of view, to execute a query means to access
and evaluate some of the data search elements. The number of accessed
elements depends on the query. But even if a specific query is considered,
there are usually many ways how to get the required query response set.
The specific choice of the query evaluation process determining the number
of accessed elements is for each file organization given by its search strategy. According to this, that part of the data search tree which is accessed
while evaluating a query will be denoted as the access path.

Since we are interested in large volumes of data organized on secondary
storage, this fact cannot be neglected in models of these file organizations.
In FOPES, the secondary storage is supposed to be an area on a disc /or discs/
with moving heads and the implementation of a file on this storage is modeled
as a set of storage buckets. Each storage bucket contains a disjoined subset
of the data search tree elements satisfying the following conditions:

    1. all elements of a storage bucket are of the same average size;
    2. whenever a subset of storage bucket elements is required this trans-
       action is processed in a time interval during which no single
       element from any other storage bucket is required.

From the hardware point of view, each storage bucket occupies a part,
possibly mutually overlapping part of the secondary memory which is devoted
to the file organization.

Having explained all the necessary terminology, we can make clear to
understand the term file organization model. In FOPES, the file organization
model is a specification of the access mechanism as well as its physical
representation on the secondary storage by means of the storage buckets.
Since this specification is made in FOPES explicitly it is usually called
the file organization model parameters.

FOPES is a system enabling its users to develop models of file organizations by specifying two groups of parameters. The first group, conven-
tionally called the logical modeling level is addressed to the access mechanism features. The second group, also called the physical modeling level,
is devoted to the specification of the implementation on the secondary
storage by menas of the storage buckets. The inner interface between the two
levels is the access path as shown in Figure 1. The models obtained are
evaluation models providing the following performance related characteristics for performance analysis, comparison and prediction:

    1. query class response time;
    2. space requirements;
    3. for all the queries from the query class:
       a/ query /conjunct, subconjunct, query clause/ response set size,
       b/ access path,
       c/ query response time,
       d/ response times for the individual storage buckets.

The display of the characteristics in the output is optional.

ACCESS MECHANISM

Two major components of the access mechanism were recognized in the
previous chapter, namely the data search tree and the search strategy. In
the following we shall briefly describe the modeling possibilities for both
the components. Since the detailed description of the access mechanism is
beyond the scope of this short paper we shall concentrate on the important
features deliberately leaving out the implementation details.

The data search tree in general is a tree structure with multiway
branching nodes. A node contains a set of data elements each of which is
associated with a pointer entering a descendent node except the case when
the node is the leaf node of the tree. The root of the data search tree is

a node which has no branches entering it and it lies on the first level of the tree. A node which lies at the ned of a path of length j-1 from a root is on the j-th level of the tree. As shown in Figure 2, three kinds ofelements can be recognized in a node of the data search tree.

A convenient structure how to form an idea of a file is a table. Each row of the table represents a record, the columns stand for the individual attributes. Items can be found on the intersections of rows and columns. Though items are usually too small to be stored separately on the computer memory, the natural way is to store them record by record. But there are also some applications for which grouping items of a specific attribute guarantees a better performance. Such files are known as the transposed files. In FOPES both types of files, the record oriented as well as the transposed can be modeled by specifying the data search tree in a convenient manner.

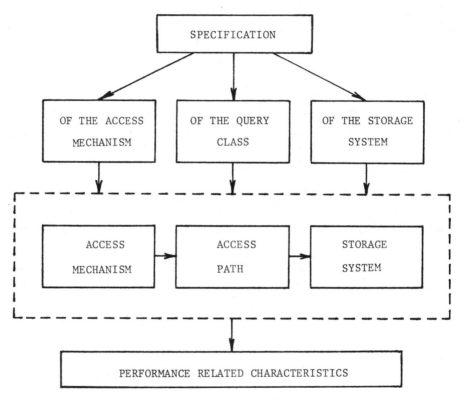

Figure 1.   FOPES architecture.

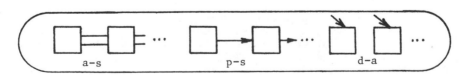

Figure 2. Data search tree node.
a-s address-sequential elements
p-s pointer-sequential elements
d-a directly accessible elements

The auxiliary search structures in file organizations are known as directories which provide for an efficient search for selected groups of attributes. This kind of structure, supporting a straight navigation to key-sets determined by keywords of an attribute or a combination of attributes, is considered to be the partial directory in FOPES. In case the partial directory is used the root node of the data search tree contains attribute identifications for all the partial directories involved. Then the general structure of the directory data search tree can be thought as it is in Figure 3.

Each partial directory consists of a subtree representing a search structure making it possible to access keywords of a specific attribute. If the multiple attribute partial directory is considered, the leaves of the first attribute subtree are also the roots of a forest of the second attribute keyword subtrees, etc.

Any of the partial directories ends with accession lists. They serve as an interface to the file but they reflect the keyset search implementation as well. In general, the data search tree of FOPES offers the cellular structure to model this search. It should be noticed that it includes the inverted and the list organizations as its two special cases. For more details see /Hsiao, 1971; Yao, 1977/.

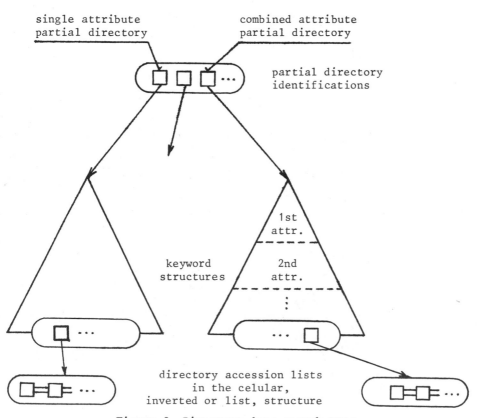

Figure 3. Directory data search tree.

Though the directory and the file are two recognizable parts of file organizations, it is the accession list pointers which connect the parts and create an integrated file organization structure. Nevertheless, there are two ways how to implement the pointers. The direct pointers indicate the desired keyset elements by using the physical address of the secondary storage location on which the objects are stored. On the other hand, the value of a relative pointer is the object identifier within the file structure. The effects of the two options on constructing the data search tree can be seen in Figure 4.

However, the data search tree of any file organization model is just a search tree structure. The other major component of the access mechanism is the search strategy which determines the access path for a specific query. An implicit search strategy is employed in our models unless it is staded explicitly otherwise. The imlicit search respects the data search tree and the nature of the data elements appearing on the individual levels. Notice, that all the processing assumptions will be summarized later. At this place, two important explicit specifications for the search strategy are to be mentioned:

1. clustering – the uniform distribution of data elements is supposed in the data search trees of our models. In case the data elements of a real organization are clustered according to keywords of a specific attribute, this attribute can be declared in the model as the clustering attribute. Then the search startegy will consider the records of the file as clustered.
2. overlapping – since even a query clause can be specified by several values, each of them identifying a keyset, not necessarily disjoined to the other keysets of the query clause, some of the paths leading to the required records might be exactly the same, duplicated paths. If there is a level from which this kind of duplicated paths is omitted, this level should be declared as the overlapping level. According to the structure of the query, we can distinguish among the query clause, the subconjunct, the conjunct and the query overlapping levels.

STORAGE SYSTEM

The storage system in our models can be described as a set of storage buckets. Specifically, the storage bucket represents a certain area of the secondary memory and it contains some of the data search tree elements.

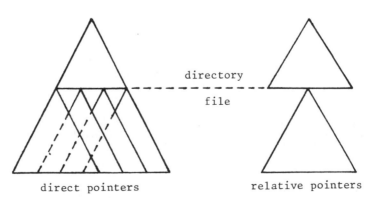

directory

file

direct pointers   relative pointers

Figure 4.   Directory – file connections.

In addition to what has already been said in the previous chapters, here are
the following assumtions concerning the storage bucket contents. One of the
following two types of the data search tree elements is supposed to be in a
single storage bucket:
1. the pointer-sequential or the directly-accessible elements, or
2. the sequences of the address-sequential elements.
Those elements of the storage bucket which are calculated as the objects for
access are supposed to be accessed serially, one after another. Even though
the uniform, respectively clustered, distribution of accessed elements within
the bucket is always supposed, the desired elements can be retrieved /swept
out from cylinders and blocks of the bucket/ either in the random or the
batch sweeping modes. More precisely, the batch sweeping mode is based on
the assumption that, whenever a block /cylinder/ is accessed to retrieve
a desired element, all the elements which are to be accessed there are retrieved at one time, so that no single block /cylinder/ can be accessed more
than once. On the contrary, whenever the random sweeping mode is specified,
the requests for reading blocks /cylinders/ are determined solely by the
distribution of desired elements in the bucket. That is why there is usually
a high probability for a block /cylinder/ of the bucket to be read as many
times as many elements are requested in it. In FOPES, the percentage of elements accessed in both the random and the batch sweeping modes can be determined by specifying the block /cylinder/ sweeping mode ratio.
    All the storage bucket parameters can be summarized as the following:
- data search tree and its access path allocation;
- number of blocks and cylinders;
- number of buffers devoted to this bucket;
- probability for a buffer to be filled with data from the bucket before
  the search starts;
- type of access to the first element of the bucket /random, sequential,
  without access/;
- cylinder and block sweeping mode ratioes;
- hardware performance related parameters /CPU-time consumed to evaluate
  an accessed element; transfer time per unit of storage; block size; number of blocks in a cylinder; block and cylinder random access times;
  block and cylinder sequential access times/.

MODELING CAPABILITY AND VERIFICATION OF THE FOPES

    To demonstrate the modeling possibilities of our system, we present
a list of different types of the file organizations which as we have proved
can be modeled as special cases of the system. They are:
- sequential organizations;
- primary key file organizations /hashing including the structures for
  dynamic files, index sequential, tree organizations eg. B-tree/;
- secondary multiple key file organizations /inverted, cellular, multilist,
  transposed, multiring, attribute-based tree/.
Once a file organization performance prediction system is established, it can
serve for various practical applications, eg. file organization determination,
secondary access structure optimization, clustering analyses, query optimization, query access cost estimation, reorganization planning and others.
    During verification of FOPES we concentrated our attention upon demon-
strating and proving the correct behavior of our system for some widespread
file organizations, used usually in database management systems. The details
were presented in /Zezula and Žižka, 1985/. Here we would like only to
mention a comparison of a real DBMS based on modification of the B-tree and
its adequate model. Having known real characteristics of a specific file
and the number of actually accessed cylinders and blocks for different primary key searches, we have found the deviation of the actual and the estimated values to be between 1 and 8 per cent, typically around 3 per cent.

Like many other analytical models, FOPES is based on the probability theory and two-valued logic. In our models, all values of an attribute are supposed to appear equifrequently in the file. Then, the probability that attribute 1 has the value V1 and attribute 2 has the value V2 in the same record is equal to the probability that attribute 1 has the value V1 times the probability that attribute 2 takes the value V2. When the queries are specified, the attribute values are supposed to be reffered equifrequently. Each block of the storage bucket contains the same number of elements. Moreover, since each storage bucket element has the same probability to qualify in a query, its placement among the blocks /cylinders/ of secondary storage is independent except for the case when the storage bucket elements are clustered.

Even though these assumptions approximately satisfy in many cases, it is important to understand their impact on the file organization performance when they are not actually satisfied. It has been shown in /Christodoulakis, 1984/ that these assumptions often result in predicting only an upper bound of the expected system cost. The imlications of nonrandom placement, non-uniformity and dependencies of attribute values on databse design and perfor-mance estimation are also discussed in that paper.

REFERENCES

Chilson, D. W., and Kudlac, M. E., 1983, A survey of logical and physical design techniques, in: "Proceedings of Databases for Business and Office Applications," Database Week, San Jose, 70:84.

Christodoulakis, S., 1984, Implications of certain assumptions in database performance evaluation, ACM TODS 9,2, 163:186.

Hsiao, D. K., 1971, A generalized record organization, IEEE Trans. on Comp., 1490:1495.

Schkolnick, M., 1978, A survey of physical database design methodology and techniques, in: "Proceedings of 4-th VLDB Conf.," Berlin, 474:486.

Sevcik, K. C., 1981, Data base system performance prediction using an analytical model, in: "Proceedings of 7-th VLDB Conf.," 182:198.

Whang, K. Y., Wiederhold, G., and Sagalowicz, D., 1981, Separability as an approach to physical database design, in: "Technical report STAN-CS-81-898," Stanford University.

Yao, S. B., 1977, An attribute based model for database access cost analysis, ACM TODS 2,1, 45:67.

Yao, S. B., and Navathe, S. B., 1981, in: "Data base design techniques," Springer - Verlag.

Zezula, P., and Žižka, J., 1985, An application of the analytical model in the physical database design, in: "Proceedings of 8-th International Seminar on Database Management Systems," Piešťany, Czechoslovakia, 236:246.

EMPIRICAL COMPARISON OF ASSOCIATIVE

FILE STRUCTURES

D. A. Beckley*, M. W. Evens†,
and V. K. Raman‡

*AT&T Bell Laboratories
 Naperville, Illinois 60566

†Computer Science Department
 Illinois Institute of Technology
 Chicago, Illinois 60616

‡Northeastern Illinois University
 Chicago, Illinois 60625

ASSOCIATIVE VS. PRIMARY KEY FILE STRUCTURES

This paper describes an experimental comparison of three associative
file structures (balanced K-d trees, unbalanced K-d trees, and inverted
lists) with a flat file structure. Performance is compared for different
classes of retrieval queries: exact match, partial match, range search,
nearest neighbor, and best match. The data base used for these experiments
contains half a million characters of medical data, and consists of static,
patient history information about stroke victims. The experimental design
procedure for the file structure comparison is described in detail.

Knowledge-based systems and expert systems are often viewed as data
base systems with added intelligence to augment the user interface. B-trees
have been a dominant file structure for data base systems, but they are not
particularly effective for knowledge based systems needing multikey retrieval query capacity. With the expanding growth of artificial intelligence applications, associative file structures have found many uses, particularly in areas such as image processing, robotics, cartography, pattern
recognition, statistics, and information retrieval.

Primary key file structures typically do not support multiple key retrieval, only "GET" and "GET_NEXT" functions. Associative file structures,
on the other hand, support a richer set of retrieval functions. "EXACT"
match, "PARTIAL" match, "RANGE" search, "NEAREST_NEIGHBOR," and "BEST"
match queries are examples. These functions give the illusion of retrieving
data based on the content of the record rather than the record's address location or file index and involve more substantive relations among the keys
being matched. B-trees have probably been the most popular primary key
structure (Comer, 1979; Vandendorpe, 1980). This structure is efficient,
has low maintenance, and provides reasonable retrieval for both sequential
and random access of primary keys. There is no standard for associative
file structures, and in fact one may never evolve owing to the complexity
of the issues. The combination of multiple keys and multiple query types

covers a very broad scope of applications for which it would be hard to generalize and optimize retrieval functions.

FILE STRUCTURE PERFORMANCE FACTORS

The critical parameters of file structure performance include a measure of user perceived responsiveness and a measure of imposed system workload. Wall clock time (total elapsed time) can be used to quantify user responsiveness. Measurements of finer granularity are often assumed necessary to filter out factors that might distort experimental results. However, the control of noise factors rather than their elimination was the approach used for our measurements.

The second measure of performance quantifies the imposed system workload. Two factors commonly used to derive cost for work include CPU time and I/O counts. For each candidate file structure, the CPU cost (units of CPU time consumed times CPU cost rate) is added to the I/O cost (number of I/O requests times I/O cost rate) to obtain the cost of resources used. Assuming equal responsiveness (wall clock time), the file structure with the smallest cost is considered the best. But these conclusions are applicable only to the measured query class. With associative queries the same file structure may not be the best for all query classes.

Mathematical performance analyses generally predict best, worst, and average performances in terms of the file size "n" and number of keys "k". Environmental and implementation factors are normally not modelled. In addition, the assumed distributions for data and queries do not always accurately reflect real distributions. Only carefully controlled experiments on actual machines with real data can show whether the published mathematical analyses are valid in the real environment. Cardenas and Sagamang (1977) have modelled and analyzed implementation-oriented factors for doubly chained trees, and have shown the performance sensitivity to implementation strategies. The influence of implementation alternatives motivated the undertaking of experiments with a number of different file structures using a large medical data file on a common commercial machine, the IBM 3081K running the JES3 job entry system and OS/MVS.

COMPARISON OF FILE STRUCTURES

The comparison of file structures is difficult because of the complexity of factors used in even a simple direct access storage device (DASD) subsystem. Both quantitative factors such as record length and blocksize and qualitative factors such as queuing algorithms and dynamic reconnection must be considered. In comparing file structures, however, these factors are all "secondary" factors whose influence must be controlled. The "primary" factors are separated into controllable primary factors and observable primary factors. Controllable primary factors are the query classes and file structures; "levels" (e.g., partial match and range search with flat files and inverted lists) for these factors are chosen by the experimenter. The observable factors are the factors that are measured or observed from the experiment. Since performance can be quantified by response (wall clock time) and cost (CPU and I/O), the wall clock time, CPU time, and I/O overhead are the observable primary factors.

Selection of the "levels" of interest for each factor determines the cost of a comparison experiment. In general if an experiment has l_i levels for the "ith" factor and there are q factors, then the number of combinations or sessions for the experiment is the product of the $l_1...l_q$ levels for the q factors. Experimental cost varies linearly with the number of sessions.

The first test of experimental data is the "test of statistical significance" to determine if there is any significant difference in performance
among any of the file structures. If no significant difference can be sub-
stantiated, a ranking of the file structures cannot be inferred from the
experiment. If a difference is detected, the file structures are ranked
into groups with measurable differences. This performance information is
then combined with query usage and weights of importance of the metrics to
obtain the "most appropriate" file structure.

Experimental Controls

In order to control noise factors in this experiment, the environment
conditions had to be as similar as possible for all file structures. The
underlying access method, PL/1 Regional (1) direct access, was used for all
retrieval functions; thus there was no favoritism for the low level access
method used. Furthermore, with direct access, the time consuming overhead
spent with file open processing was eliminated. By setting the number of
buffers to 1 and the blocking factor to 1, the number of records read by
all runs of a given session were the same and could be precisely measured.
Furthermore, any advantage of the default buffering for one file structure
over another was eliminated.

CPU time and wall clock time are not so precisely controllable. Software probes were implanted at the beginning and end of algorithm processing
to eliminate operating system overhead in query interpretation. The probe
at the end of each algorithm generated an instrumentation record for each
query. These instrumentation records were then processed to analyze the
file structure performance.

Interference from system workload because of CPU activity and I/O contention could impact time measurements. In order to decrease system impact
as much as possible, the experiment sessions were executed at times of the
lightest system workload. Furthermore, the data for all file structures
resided on the same physical device, an IBM 3380 volume of user data which
had very little competing activity during the light workload periods.

Associative Query Types

A medical application for associative retrieval was used to exemplify
the steps of file structure comparison. The data consisted of patient
history information for stroke victims. All five query classes were relevant to this application. The file is static and contains 328 fields, but
only the following seven fields were chosen as retrieval keys: stroke type,
race, age, diabetes, systolic blood pressure, and diastolic blood pressure.
The following five queries exemplify requests from each of the query
classes.

Exact Match. In an exact match query, a single key value is specified
for each key.

```
SELECT (STROKE_TYPE  = E,
        RACE         = W,
        AGE          = 070,
        SEX          = F,
        DIABETES     = N,
        SYSTOLIC_BP  = 140,
        DIASTOLIC_BP = 080);
```

(List any embolism case of a white, 70-year-old female without diabetes, with systolic blood pressure of 140 and diastolic blood pressure of 80)

Applications needing only exact match are typically implemented with primary structures rather than associative structures. Of the exact match queries collected from the survey, no successful matches were found. To test the exact match algorithms, a control set of queries were generated. This set of queries has been denoted EXAC*.

Partial Match. In a partial match query, a single key value is specified for one or more of the keys, but not all of the keys.

SELECT (STROKE TYPE = I,

```
        RACE        = W, ·
        SEX         = M);
```

(List any intracerebral hemmorrhage case for a white male)

Range Search. A range search query may involve a full or partial subset of the retrieval keys, but both a lower and an upper bound value are given for each key specified.

SELECT (141 <= SYSTOLIC BP <= 169,

```
        090 <= DIASTOLIC_BP <= 121,
        075 <= AGE          <= 080);
```

(List any case of a 75- to 80-year-old patient with systolic blood pressure between 141 and 169, and diastolic blood pressure between 90 and 121)

Nearest Neighbor. The nearest neighbor query uses a distance function to quantify the nearness of one record to another. The function used in this experiment is the square root of the sum of normalized distance values. The absolute value of the difference in AGE values is divided by 150 to obtain the normalized AGE value; the absolute value of the difference in SYSTOLIC blood pressures and the absolute value of the difference in DIASTOLIC blood pressures are both divided by 250 to obtain the respective normalized values. The distance measurement between two records is defined as the square root of the sum of the squares of these normalized values. Only the numeric keys are used in nearest neighbor queries.

SELECT (NEAREST (AGE   = 060),

```
        NEAREST (SYSTOLIC_BP  = 140),
        NEAREST (DIASTOLIC_BP = 090));
```

(List the nearest case(s) with age around 60, systolic blood pressure around 140, and diastolic blood pressure around 90)

Best Match. Best match queries do not have a distance function. The best match query assumes all keys are of character type or coded values. For each key in the query, an ordered list of "preference" values is specified from the most desired value to the least acceptable. The dominance among keys is determined by the ordering of the preference lists in the query. The search begins for the most preferred key value in the preference list of the most dominant key; preference lists of less dominant keys will then be used successively to identify the best record(s).

SELECT (PREF (STROKE_TYPE = T, E, L),
```
        PREF (DIABETES   = I, O, D));
```

(List the case(s) with the highest ranking for stroke type and diabetes:

```
                                1 1 1 1 1 1
Ranking             1 2 3 4 5 6 7 8 9 0 1 2 3 4 5
                    ---------------------------

Stroke_type..  T T T T E E E E L L L L x x x
Diabetes.....  I O D x I O D x I O D x I O D)
```

FLAT FILES VS. K-D TREES VS. INVERTED LISTS

Four file structures were considered as candidate file structures for
the application. These included the flat file (FL), the balanced K-d tree
(BKD) (Bentley, 1975; Bently, 1979; Friedman, et al., 1977), the unbalanced
K-d tree (UKD), and the inverted list (IL). All structures were loaded in
case number order, independent of any of the retrieval keys. The balanced
K-d tree was the only structure for which there was any file optimization
before the query processing. The balancing algorithm produced a K-d tree
for which the average depth was 8.2 and the maximum depth was 11. The unbalanced K-d tree had an average depth of 10.1 and a maximum depth of 18
(Beckley, Evens, and Raman, 1985).

The motivation for this experiment was to obtain data with which to
empirically compare the performance of each file structure for each query
class. The four file structures and the five query classes (two levels for
exact match) were the primary factors; the number of sessions was thus 24.
Performance was measured in terms of I/O requests and CPU time against wall
clock time. Counts of I/O requests were repeatable for repeated runs, but
CPU time and wall clock time were not. The first task was to determine the
number of runs necessary for each session of the experiment. This test was
independent of file structures. A sample set of queries was processed 30
times and the total CPU and wall clock times for each run were calculated.
A normal distribution could be assumed for the 30 sets of totals, and the
standard deviation of these totals could be assumed as the standard deviation of the population of runs executed with the existing probes. Using
error estimates of 4% of the means and a 95% confidence level for the flat
file structure, the worst case query class (BEST match) required almost 30
runs. The same number of runs was used for all query classes.

In a similar manner, the number of queries for each session was determined. Sample queries were collected via a questionnaire sent to Dr.
Daniel B. Hier of the Neurology Department at Michael Reese Hospital and
to students attending an Information Storage and Retrieval class at IIT.
A run of all samples was made for each file structure to determine the variance within a session. Using these standard deviations, error estimates
of 10% of the means, and 95% confidence level, the sample size was chosen
to be 50 queries.

The results of the experiment are summarized in Table 1. The means
and standard deviations are calculated for the 1500 query executions of
each level (file structure-query class combination). Comparisons of the
CPU time, wall clock time, and I/O counts between the balanced K-d tree
and the unbalanced K-d tree were similar with the exception of wall clock
times for range search, nearest neighbor, and best match query classes.
Most analyses of K-d trees have been with balanced K-d trees, but this experiment seems to indicate that balancing overhead is not always necessary.
The wall clock time differences for the range search, nearest neighbor, and
best match queries in favor of the unbalanced K-d tree were very surprising.
CPU time was an order of magnitude less than the wall clock time so it did
not account for the difference. Neither did the number of I/O's differ a
significant amount to account for the difference in wall clock time. The
experiment was designed to negate system workload impact; furthermore, if
workload were not filtered out, its impact would have been observed in all
query classes.

The CPU times, wall clock times, and certainly the I/O counts were nearly the same for all query classes of the flat file. The motivation for including this file structure in the experiment was to verify the results of the query algorithms and have a baseline with which to compare the other file structures.

The I/O count for the inverted list file structure included the counts to both the inverted lists as well as the uninverted list of the structure. The performance dependency of this file structure on the number of query keys is amplified by its poor performance for the exact match queries which include each key. Keys for which there are only a few key values, for example SEX (male or female), require retrieval of a significant portion of the entries in the corresponding inverted lists for those keys. The implementation of the inverted list used for this experiment kept the key values of each inverted list sorted, but not the associated pointers to the uninverted list.

Each of the 18 combinations of observable factors (CPU time, Wall clock time, and I/O counts) with a query class (EXAC, EXAC*, PART, RANGE, NEAR, and BEST) gives a null hypothesis to be tested. The hypotheses state that for a given observable factor and query class, the means of the observable factor for all file structures are equal.

The means for each observable factor of each level were computed and an analysis of variance was used to either accept or reject the null hypotheses. The analysis of variance predicts the CPU time, wall clock time, and number of I/O's for each observations. The difference between the actual and predicted response is due in part to the file structure and in part to random error. The result of this test is an F value that gives the ratio of these two categories of errors. When the "F" value is near 1, the hypotheses hold; when the value is large, the means differ substantially. The analysis of variance does not rank the file structures, nor does it resolve which file structures are significantly different when there are more than two candidate structures. A large "F" value merely implies that the means of all file structues for the session are not equal; a small "F" value does not prove that the hypotheses are true, it only means that there is no significant reason to doubt it. Table 2 gives the F ratios. Since the "F" values were all very large, the hypotheses were all rejected. The PR value gives the significance probability value associated with the F value. This is the probability of getting an even bigger F value if the null hypothesis is true.

The Student-Newman-Keuls test was used to detect differences between means and to provide a ranking of the groups for each metric that had significant differences. File structures among which there is no significant difference are equally ranked. Table 2 also gives the Student-Newman-Keuls ranking for this experiment and shows that the K-d trees are the superior structures for almost all primary factors of each query class level. The flat file used less CPU time for all query classes than the inverted list, but at the same time the wall clock time for the inverted list was less than that for the flat file. It is interesting to note that the rankings between the flat file and the inverted list for wall clock times do not mirror the rankings for the number of I/O's. As Cardenas (1975) pointed out, perhaps the most important performance factor is the manner in which lists or indices are actually organized and managed.

No file structure is uniformly best for all query classes of each performance metric; the relative ranking between the two K-d tree structures is not the same. The next objective was to obtain an overall ranking of the file structures for each metric. Table 2 gives the means of each SNK group; there is statistical evidence to suspect differences among these groups of file structures. Normalized SNK factors were then computed for

Table 1

QUERY CLASS	FACTOR	MEAN (milli.sec) (or count)	STANDARD DEVIATION	MEAN (milli.sec) (or count)	STANDARD DEVIATION
		------ FILEST=BKD -----		------ FILEST=UKD -----	
EXAC*	CPU	4.556341	.510363	6.004426	1.236799
EXAC	CPU	4.463728	.344283	5.155953	1.140000
PART	CPU	65.239951	33.906399	80.479466	36.904977
RANG	CPU	171.227311	65.795805	172.066678	77.855711
NEAR	CPU	251.071432	93.897889	222.990773	118.405812
BEST	CPU	189.793357	42.547883	183.214101	42.770615
EXAC*	Wall	131.131921	36.145879	167.681612	41.492480
EXAC	Wall	132.174059	28.733901	139.475778	37.799410
PART	Wall	2143.803596	1080.565503	1755.740003	750.877311
RANG	Wall	5309.605192	1706.941078	3524.287155	1489.451182
NEAR	Wall	5599.895268	2360.365654	3335.738412	1954.117855
BEST	Wall	6245.380843	1445.875051	4018.518333	931.497733
EXAC*	I/O	9.280	0.567	12.320	2.438
EXAC	I/O	9.220	0.414	10.760	2.338
PART	I/O	125.640	59.943	160.320	68.540
RANG	I/O	313.620	96.543	322.520	128.066
NEAR	I/O	335.700	141.268	303.180	177.483
BEST	I/O	371.980	77.778	370.800	83.605
		------ FILEST=FL ------		------ FILEST=IL ------	
EXAC*	CPU	235.242987	5.778073	1508.546163	338.891321
EXAC	CPU	234.411475	6.801802	1387.768126	396.173147
PART	CPU	248.597109	11.250889	657.996097	358.277020
RANG	CPU	258.995029	17.339745	641.936521	342.096591
NEAR	CPU	279.199689	10.445615	371.962595	445.640812
BEST	CPU	239.550823	6.794660	284.511715	172.375303
EXAC*	Wall	9443.984710	985.228531	5005.904715	3092.966473
EXAC	Wall	9063.587458	1976.609515	5847.836355	2881.461092
PART	Wall	9206.685641	1777.080575	5142.206209	2902.323583
RANG	Wall	9209.356403	1714.526587	4807.826030	2635.452104
NEAR	Wall	8890.100953	2244.266953	4000.386936	2911.067052
BEST	Wall	8998.028585	2076.297676	4713.742040	2886.587313
EXAC*	I/O	567.000	0.000	1317.540	255.979
EXAC	I/O	567.000	0.000	1215.080	298.277
PART	I/O	567.000	0.000	612.100	282.580
RANG	I/O	567.000	0.000	647.740	317.215
NEAR	I/O	567.000	0.000	503.660	520.961
BEST	I/O	567.000	0.000	542.280	289.766

Table 2

CPU

SNK Group	Exact*	Exact	Partial	Range	Nearest	Best
A	-BKD UKD- 5.28038 (1.000)	-BKD UKD- 4.80984 (1.000)	-BKD- 65.23995 (1.000)	-BKD UKD- 171.64699 (1.000)	-UKD- 222.99077 (1.000)	-UKD- 183.21410 (1.000)
B	-FL- 235.24299 (0.874)	-FL- 234.41148 (0.834)	-UKD- 80.47947 (0.974)	-FL- 258.99503 (0.814)	-BKD- 251.07143 (0.812)	-BKD- 189.79336 (0.935)
C	-IL- 1508.54616 (0.000)	-IL- 1387.76813 (0.000)	-FL- 248.59711 (0.691)	-IL- 641.93652 (0.000)	-FL- 279.19969 (0.623)	-FL- 239.55082 (0.444)
D			-IL- 657.99610 (0.000)		-IL- 371.96260 (0.000)	-IL- 284.51172 (0.000)

Wall Clock Time

SNK Group	Exact*	Exact	Partial	Range	Nearest	Best
A	-BKD UKD- 149.40677 (1.000)	-BKD UKD- 135.82492 (1.000)	-UKD- 2143.80360 (1.000)	-UKD- 3524.28715 (1.000)	-UKD- 3335.73841 (1.000)	-UKD- 4018.51833 (1.000)
B	-IL- 5005.90471 (0.477)	-IL- 5847.83635 (0.360)	-BKD- 1755.74000 (0.948)	-IL- 4807.82603 (0.774)	-IL- 4000.38694 (0.880)	-IL- 4713.74204 (0.860)
C	-FL- 9443.98471 (0.000)	-FL- 9063.58746 (0.000)	-IL- 5142.20621 (0.545)	-BKD- 5309.60519 (0.686)	-BKD- 5599.89527 (0.592)	-BKD- 6245.38084 (0.553)
D			-FL- 9206.68564 (0.000)	-FL- 9209.35640 (0.000)	-FL- 8890.10095 (0.000)	-FL- 8998.02859 (0.000)

I/O Counts

SNK Group	Exact*	Exact	Partial	Range	Nearest	Best
A	-BKD UKD- 10.80 (1.000)	-BKD UKD- 9.99 (1.000)	-BKD- 125.64 (1.000)	-BKD UKD- 318.07 (1.000)	-UKD- 335.70 (1.000)	-BKD UKD- 371.39 (1.000)
B	-FL- 567.00 (0.574)	-FL- 567.00 (0.538)	-UKD- 160.32 (0.929)	-FL- 567.00 (0.245)	-BKD- 303.18 (0.877)	-IL- 542.28 (0.126)
C	-IL- 1317.54 (0.000)	-IL- 1215.08 (0.000)	-FL- 567.00 (0.093)	-IL- 647.74 (0.000)	-IL- 503.66 (0.240)	-FL- 567.00 (0.000)
D			-IL- 612.10 (0.000)		-FL- 567.00 (0.000)	

Table 3

FILEST	CPU	WALL	I/O	OVERALL
UKD	.995	1.000	.986	.996
BKD	.975	.747	.988	.864
IL	.000	.712	.037	.363
FL	.737	.000	.197	.261

each SNK group to obtain a measure of relative performance among the groups (Table 2). Assuming input from the data base designer that assigns weights of importance to each query class (EXAC = .05, EXAC* =.05, PART = .20, RANG = .50, NEAR = .10, and BEST = .10), a sum of products was computed for each metric of each file structure. Each query class weight factor was multiplied by the SNK normalized group factor of the file structure for the corresponding query class; the sum of these products was a measure to rank the file structures relative to each other for each metric (Table 3). Assuming additional input from the data base designer that assigns weights of importance to each metric (CPU = .3, wall clock time = .5, and I/O count = .2), another sum of products was computed for each file structure, this one being a measure for the overall ranking of the file structures.

SUMMARY

This paper has attempted to demonstrate a useful approach for comparing file structures. Typically the performance of no file structure is dominant for all metrics and all query classes; the choice of a file structure for a particular application must consider data, queries, and the environment.

ACKNOWLEDGEMENTS

The kind efforts of Dr. Daniel B. Hier, M.D. of the Neurology Department at Michael Reese Hospital are greatly appreciated. He provided the data and was also a source for test queries. Helpful comments from Jim Vandendorpe and Jon Bentley must also be acknowledged. Finally, we are grateful to AT&T Bell Laboratories for use of the computing resources at the Indian Hill Computation Center for the experiment.

REFERENCES

Beckley, D. A., Evens, M. W., and Raman, V. K., An Experiment with Balanced and Unbalanced K-d Trees for Associative Retrieval, 8th Int. COMPSAC Proceedings, IEEE Computer Society, IEEE Catalog No. 84CH2096-6:256-262 (1984).

Beckley, D. A., Evens, M. W., and Raman, V. K., Multikey Retrieval from K-d Trees and Quad-trees, Int. Conference on Management of Data, ACM, Austin, Texas, May 1985 (1985).

Bentley, J. L., Multidimensional Binary Search Trees used for Associative Searching, CACM, Vol. 18, No. 9:509-517 (1975).

Bentley, J. L., Multidimensional Binary Search Trees used in Database Applications, IEEE Trans. Soft. Eng., Vol. SE-5, No. 4:333-340 (1979).

Cardenas, A. F., Analysis and Performance of Inverted Data Base Structures, CACM, Vol. 18, No. 5:253-263 (1975).

Cardenas, A. F., and Sagamang, J. P., Doubly-Chained Tree Data Base Organization - Analysis and Design Strategies, The Computer Journal, Vol. 20, No. 1:15-26 (1977).

Comer, D., The Ubiquitous B-tree, Computing Surveys, Vol. 11, No. 2:121-137 (1979).

Friedman, J. H., Bentley, J. L., and Finkel, R. A., An Algorithm for Finding Best Matches in Logarithmic Expected Time, ACM TOMS, Vol. 3, No. 3:209-226 (1977).

Vandendorpe, J., A Crash Tolerant B-tree Data Structure for Data Base Retrieval Systems, Ph.D. Thesis, Illinois Institute of Technology (1980).

HYBRID SORTING TECHNIQUES IN GRID STRUCTURES

K.P. Tan* and H.W. Leong**

*Dept. of Information Systems and Computer Science
Nat. Univ. of Singapore, Kent Ridge, Singapore 0511, Singapore

**Dept. of Comp. Sci., Univ. of Illinois at Champaign-Urbana
1304 Springfield Ave., Urbana, Illinois 61801, USA

ABSTRACT

This paper introduces two sorting algorithms based on two grid structures, the square grid and the upper triangle grid. The n elements to be
sorted are distributed over the grid points. The hybrid sorting technique
in each structure comprises a bubblesort (or quicksort) with a grid heapsort.
The computational complexity for both algorithms in the worst case achieves
$O(n^{3/2})$. Explicit pointers or stacks are not used. No extra storage is
required. The algorithm has some obvious properties to fit the basic concept
of parallel processing. A pascal program for the square grid algorithm is
attached.

1. INTRODUCTION

There are many different well-known fundamental sorting algorithms [Kn],
[Lo], and a few dozens of improved methods combining two to three fundamental
sorts. Some sorting techniques require extra storage (e.g., hashsort) or a
lot of data moves such as insertion sort in [Kn]. Some issues are particularly characterized for merging problems [HL]. Some are designed for nearly
sorted lists [CK]. Let n be the number of elements to be sorted. Most of
the unstructural sorts are of $O(n^2)$ in the worst case such as bubblesort in
[Kn], shellsort [Bo] and quicksort [Ho], etc. In general, structured sorts
have better performance. Heapsort [Wi] is a binary tree structure and has
worst case performance of $O(n \log n)$. Munro and Suwanda [MS] considered the
upper triangle of a matrix with i data elements stored at the i-th diagonal,
called the rotated sort list. They achieved the insert/deletion and search
time of $O(n^{1/3} \log n)$. These basic operations on the implicit data structures
were also worked out by Frederickson [Fr]. In this paper, we consider a grid
structure from which two grid patterns are generated, namely the square grid
and the left upper triangle grid. Each grid point represents a sorting element and has two sons at the lower level. First, the elements in the grid
are preprocessed so as to satisfy the heap property. Then the elements are
sorted in row-wise order. At each step, the smallest unsorted element is
found and placed in its correct position. The grid structure and the heap
property allows these steps to be done efficiently. Furthermore, the heap
property is preserved after each step. Both sorting methods achieve $O(n^{3/2})$
performance in the worst case. The relative ordering of the data elements
is implicit in the pattern, rather than explicit in pointers. Stack is also
unnecessary. No extra storage is needed.

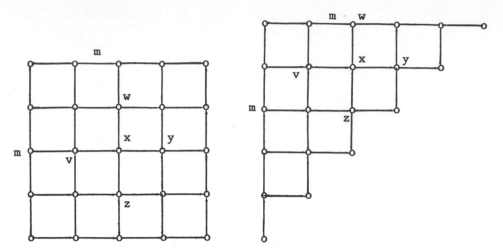

Figure 1. A square grid G.   Figure 2. A left upper triangle grid T

2. DEFINITION

Consider the square grid structure G in Fig. 1, and the left upper triangle grid structure T in Fig. 2, where each grid point represents an element to be sorted. The actual data are stored in a one-dimensional array A.
However, the mapping from one storage scheme to the other is straightforward.
For example, the (i,j) element in the square grid G corresponds to the ((i-
1)*m+j)-th element in the one-dimensional array. Thus, henceforth we shall
only refer to the elements in the grid width which refers to the grid coordinates. We shall adopt the convention that the <u>root</u> is the element at the
top left corner of the grid, and the grid grows to the right and down. Thus,
for any grid point x, the sons of x, if they exist, are the two grid points
y and z that lie to the right of x, and below x respectively. Similarly the
parents of x, if they exist, are the two grid points v and w, lying to the
left and above x respectively. Thus all the grid points not lying on the
boundary have exactly two sons and two parents (hence, the name biparent).
For any grid point x, the subgrid with root at x consists of x and all its
descendants, i.e., all the grid points lying to the right of it and below it.
A grid is to satisfy the heap property if every grid point x satisfies the
conditions, $x \geq v$ and $x \geq w$, where v and w, if they exist, are the parents
of x. We shall call such a grid a biparental heap. It is easy to see that
for any biparental heap, the root is the smallest element. Furthermore,
every row is sorted in increasing order and so are the columns.

For an a x b rectangular subgrid S in Fig. 3, where a=j-1 and b=m-i,
the restore operation takes at most 2(a+b-2)-1 comparisons. The longest path
taken to move the new element y to its proper position has length equal to
(a+b-2) moves. Furthermore, each move requires two comparisons, except for
the last move which requires only one comparison since there is only one son.

For the triangular grid in Fig. 6, each subgrid is also triangular.
Furthermore, for a subgrid at row i, the restore operation of each element
requires 2(m-i) comparisons in the worst case.

3. SQUARE GRID

Consider a square grid structure with side m as shown in Fig. 3. Each
grid point denotes a sorting element. Let $n=m^2$, where n is the total number
of sorting elements distributed over the entire square grid. See Fig. 4(a)

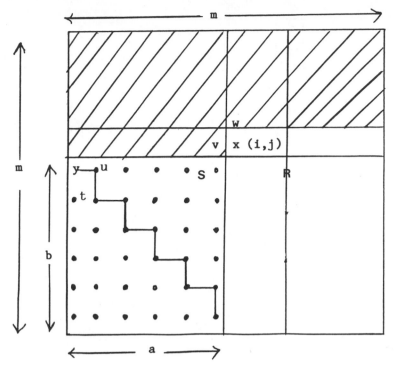

Fig. 3 A square grid of side m. The solid step line denotes the possible
       path to be taken by element y.

when n=25. The objective of this square grid is to sort the n elements in
the row-wise order from the left to the right. See the example in Fig. 4(c).

3.1 Preprocess

Sort every column of m elements in ascending order, then sort every row
in increasing order from the left to the right. Consequently, the grid
satisfies the heap property. See Fig. 4(b).

3.2 Algorithm Description

Assume the elements in the shaded area in Fig. 3 are already sorted in
row-wise order. Now, we want to put the next smallest element x at position
(i,j) where $1 \leq i, j \leq m$. Thus, we need the following:

3.2.1 Find the smallest.   In Fig. 3, the smallest element among the un-
sorted elements can only be either x at (i,j) or y at $(i+1,1)$ since x is the
root of R and y is the root of S. Thus the smaller of the two is the desired
element. In the special case when j=1, then clearly x is the smallest element and is in its proper position.

1	2	17	10	7
21	32	26	23	14
30	29	11	3	19
6	20	8	18	33
13	4	31	25	5

1	2	3	5	8
4	6	7	10	11
13	14	17	18	20
19	21	23	26	29
25	30	31	32	33

1	2	3	4	5
6	7	8	10	11
13	14	17	18	19
20	21	23	25	26
29	30	31	32	33

(a) Randomly distributed   (b) After column sort   (c) After sort, row-wise
                                and row sort   increasing

Fig. 4 A square grid of n elements, where n=25 and m=5.

417

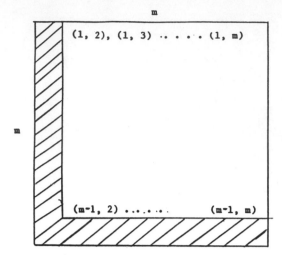

Fig. 5 Elements in the shaded area are in their right positions

Also for the final row, all the elements are sorted and in their proper position. That is, elements in the shaded area in Fig. 5 are in their right positions. Thus, our algorithm should only apply to those elements whose coordinates fall in the ranges, where $2 \leq j \leq m$ and $1 \leq i \leq m-1$ as shown in Fig. 5

3.2.2 <u>Interchange</u>.   In Fig. 3, if $x \leq y$, no interchange is necessary since element x is the desired smallest element. If $x > y$, x and y should be interchanged.

3.2.2 <u>Restore the heap property</u>.   After the interchange, the heap property may be destroyed, then the restore operation must be done. We claim, however, that the new y can only move to points within the block S. The restore step requires $2(a+b-2)-1$ comparisons in the worst case. The solid step line denotes the possible path to be taken by element y.

Finally, we give the sorting algorithm in detail.

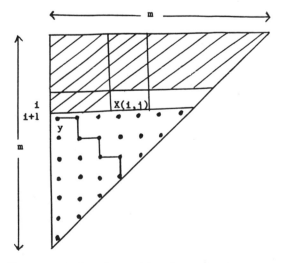

Fig. 6 In the left upper triangle grid, elements in the shaded area are
       sorted in row-wise order. The solid step line denotes the possible
       path to be taken by element y.

```
BPHEAPSORT1
  begin
    for j := 1 to m do sort column j
    for i := 1 to m do sort row i
    for i := 1 to m-1 do
    for j := 2 to m do
      if G[i,j] > G[i+1,1] then
        begin
          interchange G[i,j] and G[i+1,1]
          restore subgrid rooted at G[i+1,1]
        end
  end
```

3.2.3 <u>Algorithm analysis.</u>   Let $P(m)$ be the number of comparisons required for the preprocessing step and $S(m)$ be the number of comparisons needed for the subsequent sort.   We have

$$P(m) = 2m \times m(m-1)/2$$

and

$$S(m) = \sum_{i=1}^{m-1} \sum_{j=2}^{m} (1+2(a+b-2)-1)$$

$$= 2 \sum_{i=1}^{m-1} \sum_{j=2}^{m} (m+j-i-3)$$

$$\sim 2m^3$$

The total number of comparisons becomes

$$T(m) = P(m) + S(m)$$
$$\sim 3m^3$$

In terms of n, the total number of elements, we have

$$T(n) \sim 3n^{3/2}$$

That is, the performance of this sort in the worst case is of $O(n^{3/2})$.

4.   LEFT UPPER TRIANGLE GRID

Consider the triangular grid with side m and total number of grid points n, where $n = m(m+1)/2$ or $m \sqrt{2n}$ (see Fig. 6).   Similar to the square grid, the objective is to sort the n elements in the row-wise order from the left to the right.   See example in Fig. 7(c).

4.1 <u>Preprocess</u>

Similar to the square grid, perform column sort and row sort as the pre-process.   Consequently, this grid pattern satisfies the heap property as shown in Fig. 7(b).   The number of comparisons for this step is

$$P(m) = 2 \sum_{i=1}^{m} i(i-1)/2$$

$$= \frac{1}{3} m^3 - \frac{1}{3} m$$

1	2	17	10	7
21	32	26	23	
30	29	11		
6	20			
13				

1	2	7	10	11
6	17	20	23	
13	26	29		
21	32			
30				

1	2	6	7	10
11	13	17	20	
21	23	26		
29	30			
32				

(a) Randomly, distributed   (b) After column sort   (c) After sort,
                                  and row sort   row-wise increasing

Fig. 7 A left upper triangle grid of n elements, where n=15 and m=5

4.2 Algorithm description

Again, assume that the shaded area in Figure 6 has already been sorted in row-wise order and we need to fill the grid point (i,j) with the smallest unsorted element. The method is similar to that for the square grid. However, the significant feature in this case is that while processing the i-th row, each restore operation requires the same number of comparisons in the worst case, namely 2(m-i-1). The reason is that the longest path in the triangular subgrid rooted at (i+1,1) has (m-i-1) moves. The solid step line denotes the possible path to be taken by element y at (i+1,1). We give the entire algorithm below:

```
BPHEAPSORT2
    begin
        for j := 1 to m-1 do sort column j
        for i := 1 to m-1 do sort row i
        for i := 1 to m-1 do
        for j := 2 to m-i+1 do
    if T[i,j] > T[i+1,1] then
        begin
            interchange T[i,j] and T [i+1,1]
            restore subgrid rooted at T[i+1,1]
        end
    end
```

4.3 Algorithm analysis

For the first row, the restore operation is done at most (m-1) times. Furthermore, each of these operations takes 2(m-2) comparisons in the worst case. For the second row, the restore operations is done at most (m-2) times, each requiring in the worst case 2(m-3) operations and so on. Thus,

$$S(m) = \sum_{i=1}^{m-1} i(1+2(i-1))$$

$$= \frac{2}{3} m^3 - \frac{3}{2} m^2 + \frac{5}{6} m$$

Taking into account the cost of preprocessing, the overall performance is

$$T(m) = P(m) + S(m)$$

$$= m^3 - \frac{3}{2} m^2 + \frac{1}{2} m$$

In terms of n, we have

$$T(n) \sim 2 \sqrt{2} \ n^{3/2}$$

which is slightly better than the result obtained in the square grid.

5. DISCUSSION

For the square grid, the side length is $m \sim \sqrt{n}$. If the given number of data elements n cannot form a perfect square, dummy elements are required to fill the vacancy for the sake of making the implementation simple. Suppose the sorting is in ascending order and in a top-down and left-to-right approach, the unoccupied grid points at the lower part of the square grid should be filled with dummy elements of large values. Similar treatment is applied to the upper triangle grid where $m \sim \sqrt{2n}$. In general, the square grid sorting consists of two parts, the preprocess and the grid heap sort. In the case of reverse ordered array (i.e. the worst case), the array is already sorted consequently after the preprocess on row sorts and column sorts. No grid heap sort is needed. This is a characteristics of the heap property in a square grid structure. However, it is not true for the upper triangle grid. As a result, the square grid sort in the worst case is faster. The pascal program for the square grid sort is illustrated in the Appendix. The program for the upper triangle grid is also lengthy.

420

Therefore, it is not included in the Appendix. The following table shows the sorting time of the algorithms for the two grid structures, the quicksort and the heapsort for different values of n. The sort time is measured in msec.

Table 1 Sorting Time (msec)

Array	n	Square Grid	Upper Triangle Grid	Bubble	Quick	Heap
Random ordered	1000	1328	1136	6430	350	350
	2000	3504	3052	24106	1048	770
	4000	10048	6310	95320	3326	1686
	8000	28064	23648	–	11600	3734
Reverse ordered	4000	4028	7288	95236	45214	1556
	8000	10814	20220	–	189238	3350

For a random ordered array, the sort times of both the square grid and the upper triangle grid are of $O(n^{3/2})$ while that of quicksort and heapsort are of $O(n \log n)$. Therefore, the two grid sorts seem slower in the table. Between the square grid sort and the upper triangle grid sort, the latter is slightly faster. However, for the reverse ordered (or worst) case, the square grid sort boosts to its credential in speed and is even better than the quicksort, which is of $O(n^2)$.

6. CONCLUSION

In this paper, two grid patterns are considered: the square grid and the upper triangle grid. To each structure, an embedded sort is done as the preprocess, $P(n)$. It is followed by a grid heap sort, $S(n)$. The comparison of their performance in the worst case is shown in Table 2.

Table 2 Grid Sorting Performance in the Worst Case

Grid structure	$P(n)$	$S(n)$	$T(n)$
Square	$n^{3/2}$	$2n^{3/2}$	$3n^{3/2}$
Upper Triangle	$\frac{2\sqrt{2}}{3} n^{3/2}$	$\frac{4\sqrt{2}}{3} n^{3/2}$	$2\sqrt{2}\, n^{3/2}$

Both the two grid sorts achieve the consequence of $O(n^{3/2})$ in the worst case, which lies between the well-known results, $O(n^2)$ for the quicksort and $O(n \log n)$ for the heapsort. The upper triangle grid sort performs slightly better than the square grid sort. Because of the characteristics of decomposing an array into many individual sublists, all the row sorts, followed by all the column sorts, in the preprocess can be handled simultaneously in a parallel processing machine. On the other hand, heapsort has no such favourable property. It is obvious how speed improvement is possible with the simple parallelism for this hybrid sorting of a grid structure.

421

ACKNOWLEDGEMENT

The authors would like to acknowledge the valuable discussion with C.L. Liu and to thank G.H. Ong for his assistance in implementing pascal program for all the sorting algorithms used in this paper.

REFERENCES

[Bo]   J. Boothroyd, Algorithm 201: Shellsort, Comm. ACM 6 8, August 1963, pp. 445.
[CK]   C.R. Cook and Do Jin Kim, Best Sorting Algorithms for Nearly Sorted Lists, Comm. ACM 23 11, November 1980, pp. 620-624.
[Fr]   Greg N. Frederickson, Implicit Data Structures for the Dictionary Problem, JACM 30 1, January 1983, pp. 80-94.
[Ho]   C.A.R. Hoare, Algorithm 63 (Partition) - 64 (Quicksort) - 65 (Find), Comm. ACM 4 7, July 1961, pp. 321-322.
[HL]   F.K. Hwang and S. Lin, A Simple Algorithm for Merging Two Disjoint Linearly-Ordered Sets, SIAM Journal of Computing 1 1, 1972.
[Kn]   D.E. Knuth, "The Art of Computer Programming", Vol. 3, Sorting and Searching, 2nd ed., Addison-Wesley, Reading, Massachusetts, 1973.
[Lo]   H. Lorin, "Sorting and Sort Systems", Addison-Wesley, Reading, Massachusetts, 1975.
[MS]   J. Ian Munro and Hendra Sawanda, Implicit Data Structures for Fast Search and Update, Journal of Computer and System Sciences 21 2, October 1980, pp. 236-250.
[Si]   R.C. Singleton, Algorithm 347 (An Efficient Algorithm for Sorting with Minimal Storage), Comm. ACM 12 3, March 1969, pp. 185-187.
[Wi]   J.W.J. Williams, Algorithm 232 (Heapsort), Comm. ACM 7 6, June 1964, pp. 347-348.

APPENDIX : PASCAL PROGRAM FOR SQUARE GRID SORT

```
(* =========================================== *)

PROCEDURE RCSort( VAR x : ArvType;
                      L, U, Step : INTEGER

VAR
    i, J, Temp : INTEGER;

BEGIN
  i := L;

  REPEAT
  J := i + Step;

    REPEAT
    IF x[ i ] > x[ J ]
      THEN
      BEGIN
        Temp    := x[ J ];
        x[ J ] := x[ i ];
        x[ i ] := Temp;
      END;
    J := J + Step;
    UNTIL J > U;

  i := i + Step;
  UNTIL i > ( U - 1 );

END (* of PROCEDURE RCSort *);

(* =========================================== *)

PROCEDURE SqGridSort( VAR x : ArvType;
                          N : INTEGER );

VAR
    nMinus1, nMinus2,
    nPlus1,
    i, J,
    Size,
    L, U,
    Base,
    Root,
    SubRoot,
    GridPoint,
    Temp,
    Son1,
    Son2,
    MinSon        : INTEGER;
    Done          : BOOLEAN;
    nn            : REAL;

BEGIN
( Create square grid )
nn := SQRT( N );
n  := TRUNC( nn );
IF ( nn - n ) > 0

  THEN n := n + 1;
Size := n * n;

IF N < Size
  THEN FOR i := ( N + 1 ) TO Size DO
       x[ i ] := MaxInt;

( Sort row )
L := 0;

  FOR i := 1 TO n DO
  BEGIN
    U := L + n;
    L := L + 1;
    RCSort( x, L, U, 1 );
    L := U;
  END;

( Sort column )
nMinus1 := n - 1;
Base    := nMinus1 * n;

  FOR J := 1 TO n DO
  RCSort( x, J, J + Base, n );

( Sorting )
nMinus2 := n - 2;
nPlus1  := n + 1;

  FOR i := 1 TO nMinus2 DO
  BEGIN
    Base := ( i - 1 ) * n;
    Root := Base + nPlus1;
```

```
    FOR J := 2 TO n DO
    BEGIN
    GridPoint := Base + J;
    IF x[ GridPoint ] > x[ Root ]
      THEN
      BEGIN ( exchange )
        Temp          := x[ Root ];
        x[ Root ]     := x[ GridPoint ];
        x[ GridPoint ] := Temp;

      ( restore )
      SubRoot := Root;
      Done := FALSE;

        WHILE ( NOT Done ) DO
        BEGIN
          Son1 := SubRoot + 1;
          Son2 := SubRoot + n;
          IF x[ Son1 ] < x[ Son2 ]
            THEN MinSon := Son1
            ELSE MinSon := Son2;
          IF x[ SubRoot ] <= x[ MinSon ]
            THEN Done := TRUE

            ELSE
            BEGIN
              Temp           := x[ MinSon ];
              x[ MinSon ]    := x[ SubRoot ];
              x[ SubRoot ]   := Temp;
              SubRoot := MinSon;
            END;

          IF ( ( SubRoot ) DIV n ) = nMinus1
            THEN
            WHILE ( NOT Done ) DO
            BEGIN
              Son1 := SubRoot + 1;
              IF x[ SubRoot ] <= x[ Son1 ]
                THEN Done := TRUE

                ELSE
                BEGIN
                  Temp          := x[ Son1 ];
                  x[ Son1 ]     := x[ SubRoot ];
                  x[ SubRoot ]  := Temp;
                  SubRoot := Son1;
                END;

            END (* of WHILE *);

        END (* of WHILE *);

      END (* of IF *);

    END (* of FOR J *);

  END (* of FOR i *);

Base := nMinus2 * n;
Root := Base + nPlus1;

  FOR J := 2 TO n DO
  BEGIN
    GridPoint := Base + J;
    IF x[ GridPoint ] > x[ Root ]
      THEN
      BEGIN
        Temp          := x[ Root ];
        x[ Root ]     := x[ GridPoint ];
        x[ GridPoint ] := Temp;

      SubRoot := Root;
      Done := FALSE;

        WHILE ( NOT Done ) DO
        BEGIN
          Son1 := SubRoot + 1;
          IF x[ SubRoot ] <= x[ Son1 ]
            THEN Done := TRUE

            ELSE
            BEGIN
              Temp          := x[ Son1 ];
              x[ Son1 ]     := x[ SubRoot ];
              x[ SubRoot ]  := Temp;
              SubRoot := Son1;
            END;

        END (* of WHILE *);

      END (* of IF *);
    END (* of FOR J *);

  END (* of FOR J *);

END (* of SqGridSort *);
(* ================================================= *)
```

423

CAD/VLSI Databases

A MODEL AND STORAGE TECHNIQUE FOR

VERSIONS OF VLSI CAD OBJECTS

Won Kim[1] and D.S. Batory[2]

MCC, Austin, Texas[1]
Dept. of Computer Sciences, University of Texas, Austin[2]

ABSTRACT

VLSI CAD ojbects involve several representations. For each representation, designers experiment with multiple versions. In this paper, we first outline a logical framework that captures the dimensionalities in CAD objects and their versions, and then outline a storage technique that balances the requirements for reducing storage redundancy and fast access to different versions of the same design object.

INTRODUCTION

Many researchers have addressed aspects of the problem of version control, both for CAD applications and software development (Katz and Lehman [1984], Kaiser and Habermann [1982], McLeod, et al [1983], Plouffe, et al [1983], and Haynie and Gohl [1984]). However, we believe that most major issues are not well-understood or clearly defined. Indeed, it is not even entirely clear what they mean by *versions,* and how much of version control should be supported by the system and how much by users.

In VLSI CAD, designers typically experiment with various implementation alternatives, such as using different technologies, different design algorithms, and different design constraints. Within each alternative, a design can have several design representations (Boolean, logic, circuit, geometry, etc.), with the design moving from abstract specifications to progressively more concrete ones at succeeding representations. Designers usually generate multiple versions of a given representation of a design, for example, to fix bugs or to optimize performance. A model is needed that will capture these dimensionalities in CAD objects and their versions.

Versions are typically generated by making relatively small changes to other versions. Some form of differential file techniques can be used to promote storage sharing among versions of the same design (Katz and Lehman [1984]). However, differential file-based techniques allow fast access to only one version (usually the most current), which is not necessarily appropriate. A storage structure for versions is needed that will optimize storage sharing and yet allow fast access to any specific version as well as respond to requests that span multiple versions.

This paper reports the preliminary results of our work on these two aspects of version control, namely the definition of a logical framework for capturing the various dimensions in CAD objects and their versions, and a proposal for a storage structure for versions that balances the conflicting requirements of a high degree of data sharing among versions and fast access to any specific versions.

A VLSI DESIGN EXAMPLE

The properties of a CAD object outlined below are applicable to any of the representations. For concreteness, however, we will illustrate the concepts in the context of circuit representations. The description of a circuit consists of two parts: its interface and implementation (EDIF [1984], and McLeod, et al. [1983]). The *interface* of a circuit specifies the function of the circuit and lists its inputs and outputs. The *implementation* is usually defined as connections among less complex circuits, each of which is assigned its own interface and implementation. Thus circuit description, as well as the design itself, is often hierarchical; it enables implementation details to be developed or revealed in a progressive manner.

Consider a circuit for a 4-bit adder. Figure 1 shows its interface specification: a pair of 4-bit numbers (X,Y) are input and a 5-bit number representing their sum (Z) is output. Details about the circuit's internals are not shown, but are specified in the circuit's implementation.

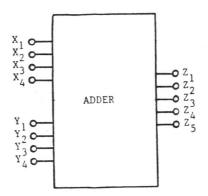

Figure 1. Interface of a 4-bit Adder

One possible implementation of an adder is shown in Figure 2. It is realized as a ripple-carry through four adder-slice circuits. According to its interface, an adder-slice takes a pair of 1-bit numbers (X_i, Y_i) and the carry from a previous slice (C_{i-1}) and produces their 1-bit sum (Z_i) and carry (C_i).

An implementation of an adder-slice is shown in Figure 3. It consists of two half-adders and an OR-gate. The interface of a half-adder has two inputs and two outputs; its implementation consists of an AND-gate and an XOR-gate (Figure 4). Similarly, implementation circuits for 2-input AND-, OR-, and XOR-gates could also be given. However, we will assume for the purposes of this paper that such gates are primitives that do not require further decomposition.

428

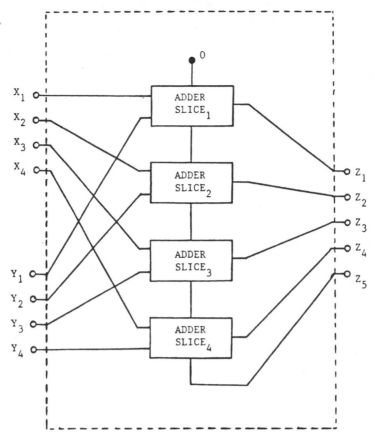

Figure 2. An Implementation of a 4-bit Adder

MODELING CONCEPTS

Databases must be stratified in order to distinguish the levels of detail that are present in design objects. Stratification may be accomplished through the use of *molecular objects* (Batory and Buchmann [1984]), objects which are represented by two sets of heterogeneous tuples (i.e., tuples from different relations). One set describes the object's interface; the other describes its implementation.

Implementations that realize a particular interface are typically not unique. Figure 5, for example, shows an alternative implementation of an adder-slice; there are many others. Objects that share the same interface but have different implementations are called *versions*. (Another common name for version is *design alternative.* Although some authors choose to give these terms different meanings, we make no distinction). Versioned objects are said to be occurrences of a single *object type* (or interface). Figure 6 illustrates the general relationship between object types and object versions. Three object types are shown. Type A has three implementations (versions), type B has none (no implementation for it has yet been specified), and type C has two. An example of object type C is the adder-slice, where Figure 5a shows the object type and Figures. 3 and 5b are its versions.

The relationship between object types and their versions, called *type-version generalization,* has special properties. First, there is the notion of attribute inheritance; all attributes (i.e., interface properties) of an object type are inherited by its versions. Second, object types have update restrictions. Modifying an object type (interface) results in a new object type; the original object type remains unchanged. In addition, versions of the original object type are not considered versions of the new type. As an example, suppose the adder-slice type is modified so that it takes a pair of 2-bit inputs and produces a 2-bit sum and a carry. This new adder-slice type is a different object type from that of the adder-slice of Figure 5a.

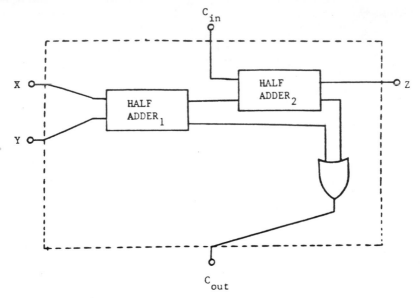

Figure 3. An Implementation of an Adder-Slice

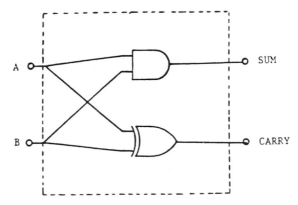

Figure 4. An Implementation of a Half-Adder

The connection between molecular objects and type-version generalization is rather simple. Every object version is a molecular object. The implementation part of a molecular object remains as is. However, its interface is factored into object type data (e.g., circuit type, input and output parameters) and object version data (e.g., version creation date, name of circuit designer). That is, shared interface data is separated from version specific interface data.

Another concept which is essential for modeling objects is *instantiation*. Figure 2 is an example. An adder uses four copies, or *instances,* of the adder-slice circuit. Each instance is distinct (i.e., each has its own separate set of inputs, outputs, and coordinate positions on a circuit diagram), yet each shares the same adder-slice features. Instantiation distinguishes an object from its copies. Copies are simply reproductions of a master; the copies themselves are *not* versions.

Instantiation, like type-version generalization, involves attribute inheritance. An instance of an object inherits all attributes of that object. Instances also have ·attributes that are not inherited (e.g., coordinate position of the instance on a circuit diagram, input and outputs that are specific to the instance). It is the non-inherited attributes which enable different instances to be distinguished. An example will be considered shortly.

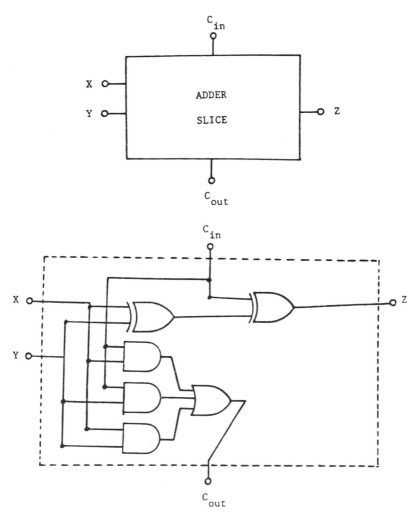

Figure 5. Interface and Another Implementation of an Adder-Slice

Instantiation is applicable to object versions and object types. If an object type is instantiated, no implementation of the object is specified; only the interface is copied. If an object version is instantiated, both the interface and its implementation are copied. An instance of an object version inherits the attributes of both the object type and its version; an instance of an object type just inherits the attributes of the type. Figure 7 shows the attribute inheritance graph relating object types, versions, and their instances.

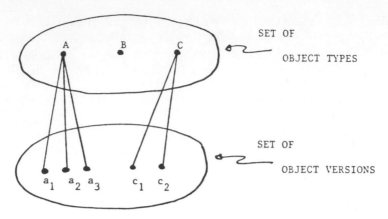

Figure 6. Relationship Between Object Types and Object Versions

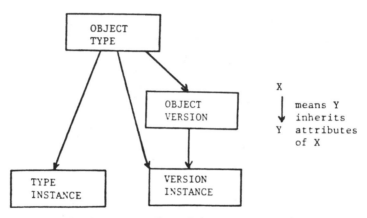

Figure 7. Attribute Inheritance Graph

As an example of attribute inheritance, consider the adder-slice circuit. Its object type has attribute CIRCUIT_TYPE (which specifies that it is of type adder-slice) and attributes that describe each of its three input pins and two output pins. Attributes of an object version which are not inherited from the object type are CREATION_DATE (the date that the version was defined) and DESIGNER (the name of the version's creator). Attributes of an instance which are not inherited from the object type or object version are X-COORD and Y-COORD (the coordinate positions of the instance in a circuit diagram) and I# (instance number). Thus an instance of an object type would inherit the CIRCUIT_TYPE attribute; if the instance were of an object version, it would also inherit attributes CREATION_DATE and DESIGNER.

There is an important semantic difference between instances of object types and instances of object versions. Let V be a version of object type T and let X be a version of some other object type. Whenever an instance of V is used in the implementation of X, V is an inherent or fixed part of X's design. However, when an instance of T is used instead, this creates a 'socket' or 'template' in X in which *any* version of type T may be placed. This gives rise to the concept of *parameterized versions.*[1]

Parameterized versions can be understood as templates. Each socket can be plugged with versions (parameterized or non-parameterized) of a specified type. Parameterization naturally supports hierarchical relationships among version instances. Figure 8 shows an instance of object version A which has two parameters; one is filled with object version B (which itself is parameterized) and the other contains object version C (which is not parameterized). B has three parameters; the first two are filled with non-parameterized versions D and E, and the remaining parameter is unplugged.

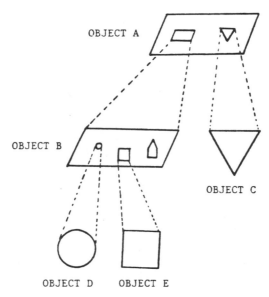

Figure 8. Hierarchical Relationship Among Object Instances

A *plug and socket diagram* is used to indicate particular relationships among version instances. Figure 9a gives a notation for a non-parameterized version (a box). Figure 9b shows the notation for a version with three parameters (a circle with three emanating lines). Figure 9c shows the notation used to describe the structure of Figure 8.

We say that a version is *complete* if it has no parameters; otherwise it is *incomplete*. Suppose t1 and t2(y) are versions of type T. Version t1 is nonparameterized; t2(y) is. Let x(t) be a parameterized version of type X, where t is a parameter of type T. x(t) and x(t2(y)) are incomplete versions; x(t1) and t1 are complete.

It is worth noting that x(t), x(t2(y)), and x(t1) represent different implementations of type X. Each is recognized in our framework as distinct versions of X.

---

[1] Object types might also be parameterized. Conceptually this may be handled in our framework by defining object supertypes. Versions of these supertypes are object types. These types could then be parameterized or nonparameterized. The details remain to be resolved.

Consider a more concrete example. Figure 2 shows an implementation of an adder which utilizes four instances of an adder-slice type. Let A (as_1, as_2, as_3, as_4) denote this version, where $as_1 \dots as_4$ are parameters of type adder-slice. Suppose Figures. 3 and 5b define two unparameterized versions of an adder-slice.[2] Call them S1 and S2. By plugging in S1 or S2 for each of the parameters of A, we can define a spectrum of complete versions {A(S1,S1,S1,S1) ... A(S2,S2,S2,S2)} of an adder. Leaving any parameters unplugged (e.g., A(S1,as_2,S2,S1)) results in an incomplete version.

Versions are often derived from other versions. This may occur by plugging the parameters of a parameterized version (as illustrated above), or it can occur by replacing portions of a circuit with a new design. In either case, tracking the evolution of versions is an important requirement in CAD environments. The generation of version x(t2(y)) from version x(t), for example, could be modeled as a binary relationship <x(t), x(t2(y))> and could be recorded automatically by the DBMS in a version-history relation. This relation should be an implicit part of the DBMS CAD support facilities and can be queried by users. It is presently believed that the history relation itself need not be an explicit part of CAD database schemas.

The concepts explained above also give insight into some implementation issues of object versions. An object version is *basic* if it is not parameterized or none of its parameters are plugged. An object version is *composite* if it is parameterized and at least one of its parameters has been plugged.

From a database user's perspective, basic versions are indistinguishable from composite versions. However, their internal representations are quite different. Basic versions have direct internal representations (e.g., one way is clustering the tuples that define the implementation with those that define the interface). Composite versions have indirect internal representations: only an encoding of their plug and socket diagram is stored. Thus composite versions are implemented as a tree of references to basic versions and possibly other composite versions.

(a) unparameterized   (b) object with   (c) plug and socket
     object   3 parameters   diagram

Figure 9. Plug and Socket Diagram Concepts

A STORAGE TECHNIQUE FOR VERSIONS

A version is typically derived by making relatively limited changes to the implementation portion of another version. (Katz and Lehman [1984]) proposes a variation of differential file techniques to promote sharing of portions of a design common to different versions. The technique is tailored to allow fast access to only the most current version.

---

[2] Actually, Figure 3 defines an adder-slice in terms of two half-adder object type instances. If we were to be consistent with our notation, Figure 3 represents an object version with two parameters of type half-adder. These parameters would be plugged with the non-parameterized half-adder object of Figure 4 in order for the adder-slice to be complete.

However, adopting a storage structure that favors access to one version over all other versions is not necessarily a good idea. For example, a designer may be interested in the most recent version of a design, while a tester may be validating a previous version. Further, storage sharing among versions at the level of records tends to complicate the storage structure and access to the data. These considerations led to the proposal in (Plouffe, et al [1983]) not to support storage sharing among versions.

Neither of these approaches is entirely satisfactory. In this section, we present a new storage structure for versions that can satisfy the conflicting requirements of storage sharing of records common to different versions and fast access to any specific versions.

In general, a new version of a design is created by making changes to one or more existing versions. For expository simplicity, however, we will assume that a version is derived from only one existing version. Then the history of version creations is captured in a directed tree we will call a *version-derivation hierarchy*, such as that shown in Figure 10. In the figure, versions vs-1, vs-2 and vs-3 were derived from vs-0; vs-4 and vs-5 from vs-2, and so on. When a version vs-j is derived from vs-i, we say vs-i is the *parent version* of vs-j, and vs-j is a *child version* of vs-i. The meaning of *ancestor versions*, *descendant versions*, *leaf versions*, and *non-leaf versions* follows naturally.

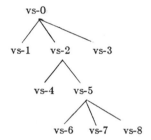

Figure 10. A version-derivation hierarchy

Our technique is based on the following rules.

1. When a leaf version is created, it *inherits* records from its parent version.

   For example, when version vs-4 was created, it inherited all records of vs-2.

2. When a record is inserted into a leaf version, the leaf version *owns* the record. The record is visible only to that version and other versions which may be derived from that version.

   For example, when a new record is inserted into vs-4, it is not visible to any other versions currently on the version-derivation hierarchy.

3. When a record is inserted into a non-leaf version on the version-derivation hierarchy, the record is visible only to that version, and not to any descendant versions.

   For example, when a new record is inserted into vs-2, the record is only visible to vs-2.

4. When a record is deleted from a non-leaf version, the record must continue to be visible to descendant versions that inherited the record.

   For example, when a record is deleted from vs-2 that had been inherited by vs-5, the record is still visible to vs-5.

We consider an update to be a combination of a deletion and an insertion, and as such updates are only indirectly considered in this paper.

Further, we assume that records owned by a version will be clustered (i.e., stored physically close together). This means that access to a specific version will be accomplished by visiting only the clusters of records belonging to the versions on the chain of ancestors from the root version to the version of interest.

To support the above requirements, we propose to augment each sharable record of a version with a bit map, and a field that indicates the version owning the record, as shown in Figure 11. For simplicity, we will assume that the bit map has a fixed size, which is equal to the maximum number of versions of a single object interface the database system will allow. Each bit in the bit map corresponds to a version on the version-derivation hierarchy. A zero-bit means that the record is visible to the corresponding version, and a one-bit means that it is not.

Algorithms for inserting, deleting, and retrieving records follow. A more detailed treatment of the algorithms, including a proof of correctness, will appear in a sequel to this paper.

Algorithm–Insert:

Assume that record r is to be inserted into version v-i.

1. [initialize bitmap] Set the bitmap entries for all versions on the version-derivation hierarchy (except v-i) to 1. Bit v-i and all other bits (that do not correspond to existing versions) are 0.

2. [initialize owner] Set v-i as the owner of record r.

   Note: Step 1 means that record r is visible only to v-i and to no other currently existing versions.

   Step 2 ensures that record r is automatically inherited by versions that are subsequently derived from v-i.

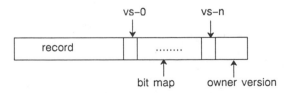

Figure 11.   An augmented record

Algorithm–Delete:

Assume that record r is to be deleted from version v-i.

1. [logical deletion] Set bit v-i in the bitmap of r to 1.

2. [replication] Let D be the set of child versions of v-i that inherit r. Insert a copy of r (with the bitmap as updated in Step 1) into each version in D, with that version as the owner.

3. [physical deletion]. If v-i is the owner of r, physically delete the record.

   Note: Step 1 makes r invisible to v-i. r is also made invisible to the descendants of v-i.

   Step 2 replicates r for each child version of v-i that inherited r from v-i. A copy of r will be owned by each version in D. All instances (the original and its copies) will share the same bitmap, and thus preserve the inheritance of r by descendant versions.

   Step 3 physically deletes r if no version inherits r.

Algorithm–Retrieve:

Assume the records of version v-i are to be retrieved.

1. [version scan] Let C be the set of versions that form a chain from the root version to v-i. Using the version directory, scan the set of records owned by each version in C.

2. [qualification] A record r belongs to version v-i if:

   1. the v-i bit in the bitmap of r is 0, and

   2. if v-i does not own the record, then the bits of the versions that form the path beginning with the owner version to version v-i must be all 0.

   Note: Step 1 eliminates the examination of versions whose records v-i could not have inherited.

   Step 2 eliminates a record if it is known that an ancestor of v-i (and a descendant of the owner version) deleted the record.

   We illustrate these algorithms below. Suppose that the maximum number of versions on a version-derivation hierarchy is 4.

   Version v-0 is the first version created, and it owns the three records shown in Figure 12a; that is, these records are inserted into v-0. Note that bitmap entries for each of these records are all set to 0, indicating that versions to be derived from v-0 can inherit these records.

   Next version v-1 is derived from v-0 by inheriting Record 1 and Record 2, and inserting a new record Record 4. This requires deletion of Record 3 with respect to version v-1, and insertion of Record 4 into v-1. The resulting state of bitmaps in the 4 records is shown in Figure 12b.

Then version v-0 inserts a new record Record 5. Since v-1 has already been derived from v-0, the insertion of this new record should be transparent to v-1. However, versions yet to be derived from v-0 should be able to inherit this record. Therefore, only the bit corresponding to v-1 is set to 1 in Record 5, as shown in Figure 12c.

Now the designer of v-0 deletes Record 2, which v-1 had inherited. The record should continue to be visible to v-1; however, it should not be visible to v-0. The resulting bitmap for Record 2 is shown in Figure 12d. Note that Record 2 was physically deleted and re-inserted with v-1 as its owner.

Next, a new version v-2 is derived from v-1. The records it inherits are Record 1, Record 2, and Record 4. Now suppose Record 1 is deleted from v-1. Record 1 should now be visible to v-0 and v-2. Record 1 is now duplicated, as shown in Figure 12e. The bitmap in Record 1 indicates that versions other than the current descendants of v-0 may inherit the record. Record 1' (a copy of Record 1) is owned by v-2, and versions to be derived from v-2 may inherit it.

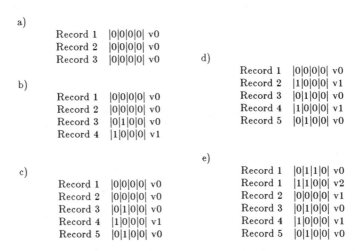

a)
Record 1	|0|0|0|0| v0
Record 2	|0|0|0|0| v0
Record 3	|0|0|0|0| v0

b)
Record 1	|0|0|0|0| v0
Record 2	|0|0|0|0| v0
Record 3	|0|1|0|0| v0
Record 4	|1|0|0|0| v1

c)
Record 1	|0|0|0|0| v0
Record 2	|0|0|0|0| v0
Record 3	|0|1|0|0| v0
Record 4	|1|0|0|0| v1
Record 5	|0|1|0|0| v0

d)
Record 1	|0|0|0|0| v0
Record 2	|1|0|0|0| v1
Record 3	|0|1|0|0| v0
Record 4	|1|0|0|0| v1
Record 5	|0|1|0|0| v0

e)
Record 1	|0|1|1|0| v0
Record 1	|1|1|0|0| v2
Record 2	|0|0|0|0| v1
Record 3	|0|1|0|0| v0
Record 4	|1|0|0|0| v1
Record 5	|0|1|0|0| v0

Figure 12. A scenario for insertion and deletion of records

CONCLUDING REMARKS

In this paper, we reported the preliminary results of our work on two of the three major research areas in version control for VLSI CAD. First, we presented a logical framework for describing the various dimensionalities that exist in VLSI CAD objects. Then we described a technique for storing records of related versions that will allow sharing of common portions of the versions and fast access to any specific versions.

The third research area, which we did not address, is that of monitoring changes in versions and notifying affected designs. Some changes at one level of a design may impact the validity of designs at other levels (Widerhold, et al. [1982]). Further, the validity of a version vs-i of a design may be affected by changes in another version vs-j, when vs-i was derived from vs-j or vs-i instantiated vs-j.

REFERENCES

1. Batory, D. and A. Buchmann. "Molecular Objects, Abstract Data Types, and Data Models: A Framework," in Proc. Intl. Conf. on Very Large Databases, August 1984.
2. Electronic Design Interchange Format, preliminary specification, version 0.8
3. Haynie, M. and C. Gohl. "Revision Relations: Maintaining Revision History Information," IEEE Database Engineering bulletin, vol. 7, no. 2, June 1984, pp. 26-33.
4. Kaiser, G. and A. Habermann. "An Environment for System Version Control," Tech Report, Dept. of Computer Science, Carnegie-Mellon University, November 1982.
5. Katz, R. and T. Lehman. "Database Support for Versions and Alternatives of Large Design Files," IEEE Trans. on Software Engineering, vol. SE-10, no. 2, March 1984, pp. 191-200.
6. McLeod, D., K. Narayanaswamy, and K. Bapa Rao. "An Approach to Information Management for CAD/VLSI Applications," in Proc. Databases for Engineering Applications, Database Week 1983 (ACM), May 1983, pp. 39-50.
7. Plouffe, W., W. Kim, R. Lorie, and D. McNabb. "Versions in an Engineering Database System," IBM Research Report: RJ4085, IBM Research, Calif., October 1983.
8. Wiederhold, G., A. Beetem, and G. Short. "A Database Approach to Communications in VLSI Design," IEEE Trans. on Computer-Aided Design of Integrated Circuits and Systems, vol. CAD-1, no. 2, April 1982, pp. 57-63.

STORAGE AND ACCESS STRUCTURES FOR GEOMETRIC DATA BASES

J. Nievergelt and K. Hinrichs

Department of Computer Science
University of North Carolina
Chapel Hill, NC 27514, USA

ABSTRACT

Geometric computation and data bases, two hitherto unrelated computing technologies, have begun to influence each other in response to the growing use of graphics and computer-aided design. CAD imposes a new challenge to data base implementers. A data base system for CAD must manage "in designer real time" large collections not of points, but of spatial objects, in such a way that proximity queries (such as intersection, contact, minimal tolerances) are answered efficiently. There are many techniques for reducing the problem of storing spatial objects to storing (sets of) points. Common to all of them is the problem that simple queries on objects turn into complex queries on points — much more complex than orthogonal range queries.

We describe a technique which is particularly suitable for storing geometric objects built up from simple primitives, as they commonly occur in CAD. Proximity queries are handled efficiently as part of the accessing mechanism to disk. This technique is based on a transformation of spatial objects into points in higher-dimensional parameter spaces, and on the data structure grid file that answers region queries of complex shape efficiently. The grid file is designed to store higly dynamic sets of multi-dimensional data in such a way that common queries are answered using few disk accesses: a point query requires two disk accesses, a region query requires at most two disk accesses per data bucket retrieved. We describe a software package that implements the grid file and some of its applications.

1   GEOMETRIC COMPUTATION AND DATA BASES

Geometric computation and data bases, two hitherto unrelated computing technologies, have begun to influence each other in response to the growing use of graphics and computer-aided design (CAD). We recall their origins, goals, typical techniques, and explain the difficulties each of them has in handling the requirements of the other.

1.1   THREE GENERATIONS OF COMPUTING APPLICATIONS

The types of computer applications dominant at different times may be classified into three generations according to their influence on the development of computing.

The first generation, characterized by numerical computing, led to the development of many new algorithms. It transformed numerical analysis from a craft to be practiced by every applied mathematician into a field for specialists. It soon became obvious that writing good (efficient, robust) numerical software requires so much knowledge and effort that this task cannot be left to the applications programmer. The development of large portable numerical libraries became one of the major tasks for professional numerical analysts.

The second generation, hatched by the needs of commercial data processing, led to the development of many new data structures. It focused attention on the problem of efficient management of large, dynamic data collections, initially under batch processing conditions. Searching and sorting were recognized as basic operations whose time requirements turned out to be the bottleneck for many applications. Data base technology emerged to shield the end-user from the details of implementation (storage techniques, features dependent on hardware and operating system), by presenting the data in the form of logical models that highlight relationships among data items rather than their internal representation, and by introducing the abstraction of access path to hide detailed access algorithms of underlying data structures.

We are now on the threshold of a third generation of applications, dominated by computing with pictorial and geometric objects. This change of emphasis is triggered by today's ubiquitous interactive use of personal computers, and their increasing graphics capabilities. It is a simple fact that people absorb information fastest when it is presented in pictorial form, hence computer graphics and the underlying processing of geometric objects will play a role in the majority of computer applications. The field of computational geometry has emerged as a scientific discipline during this past decade in response to the growing importance of processing pictorial and geometric objects. It has already created novel and interesting algorithms and data structures, and is beginning to impact data base technology under the label (hopefully temporary) of non-standard database applications. In order to understand how geometric computation is likely to affect data bases, it is useful to survey some milestones in this rapidly developing field which is replacing the traditional areas of numeric computation and of data management as the major research topic in algorithm analysis.

1.2 COMPUTATIONAL GEOMETRY - THEORY AND PRACTICE

During the seventies geometric problems caught the attention of researchers in concrete complexity theory. They brought to bear the finely honed tools of algorithm analysis and achieved rapid progress. Elementary problems (e.g. determining intersections of simple objects such as line segments, aligned rectangles, polygons) yielded elegantly to general algorithmic principles such as divide-and-conquer or plane-sweep. But in many instances a surprisingly large increase of difficulty showed up in going from two to three dimensions: for example, intersection of polyhedra is still an active research topic where major efficiency gains are to be expected. The theory of computational geometry, although well underway, has as yet explored only a fraction of its potential territory.

The practice of computational geometry is even less well understood. Many important geometric problems in computer- aided design, in geographical data processing, in graphics do not lend themselves to being studied and evaluated by the asymptotic performance formulas that the algorithm analyst cherishes. For example, asymptotics does not help in answering the question whether we can access an object in one disk access or two, thus being able to display it "instantaneously" on the designer's screen - realistic assumptions about the size of today's central memories are needed. Nor will asymptotics settle the argument raging in the CAD community between proponents of boundary representations and adherents of constructive solid geometry - taste, experience, and type of application are the relevant parameters. And below the highly visible issues of object representation, data structures and algorithms hide the tantalizing details of the numerics of computational geometry, such as the problems caused by "braiding straight lines", which may intersect repeatedly.

Commercially available software in computer graphics and CAD has not yet taken into account the results of computational geometry. Straightforward algorithms are being used whose theoretical efficiency is poor as compared to known results. Perhaps the straightforward algorithms are better in practice than theoretically optimal ones, but such difficult questions have hardly been investigated, as CAD systems development today is so labor intensive that all resources are absorbed by just getting the system to work, and algorithm analysis has so far been limited to theoretically measurable performance.

We know by analogy with numerical analysis what the next step should be in the maturing process of computational geometry: The development of efficient, portable, robust program libraries for the most basic, frequent geometric operations on standard representations of geometric objects. In other words, we must develop the geometric subroutine library of CAD, thus exposing theoretical results to stringent practical tests.

1.3 THE CONVENTIONAL DATA BASE APPROACH TO "NON-STANDARD" DATA

Data base technology has developed over the past two decades in response to the needs of commercial data processing. The key concepts introduced and supported by data base software mirror the reality that used to be handled manually by office clerks. Large quantities of records of a few different types, identified by a small number of attributes, mostly retrieved in response to relatively simple queries: point queries that ask for the presence or absence of one particular record, interval or range queries that ask for all records whose attribute values lie within given lower and upper bounds. More complex queries tend to be reduced to these basic types.

Data base software has yet to take into account the specific requirements of geometric computation, as can be seen from the terminology used: Geometric objects are lumped into the amorphous pool of "non-standard" applications. The sharp distinction between the logical view presented to the user and the physical aspects that the implementer sees has been possible in conventional data base applications because data structures that allow efficient handling of point sets are well understood. The same distinction is premature for geometric data bases: in interactive applications such as CAD efficiency is the real issue, and until we understand geometric storage techniques better we may not be able to afford the luxury of studying geometric modeling divorced from physical storage. Consider the following example. A set of polyhedra might be stored in a relational data base by using the boundary representation (BR) approach: a polyhedron p is given by its faces, a face f by its bounding edges, an edge e by its endpoints s_1 and s_2. Four relations polyhedra, faces, edges and points might have the following structure:

A tuple in the relation
- polyhedra is a pair (p_i, f_k) of identifiers for a polyhedron and a face: f_k is a face of polyhedron p_i.
- faces is a pair (f_k, e_j) of identifiers for a face and an edge: e_j is a bounding edge of face f_k.
- edges is a triple (e_j, s_m, s_n) of identifiers for one edge and two points: s_m and s_n are the endpoints of edge e_j.
- points is a triple (s_n, x, y): s_n is the identifier of a point, x and y are its coordinates.

This representation smashes an object into parts which are spread over different relations and therefore over the storage medium. The question whether a polyhedron P intersects a given line L is answered by intersecting each face f_k of a polyhedron p_i with L. If the tuple (p_i, f_k) in the relation polyhedra contains the equation of the corresponding plane, the intersection point of the plane and the line L can be computed without accessing other relations. But in order to determine whether this intersection

point lies inside or outside the face f_k requires accessing
tuples of edges and points, i.e. accessing different blocks of
storage, resulting in many more disk accesses than the geometric
problem requires.

1.4 GEOMETRIC MODELING SEPARATED FROM STORAGE CONSIDERATIONS

In this early stage of development of geometric data base
technology, we cannot afford to focus on modeling to the
exclusion of implementation aspects. In graphics and CAD the real
issue is efficiency: 1/10-th of a second is limit of human time
resolution, and a designer works at maximal efficiency when
"trivial" requests are displayed "instantaneously". This allows a
couple of disk accesses only, which means that geometric and
other spatial attributes must be part of the retrieval mechanism
if common geometric queries (intersection, inclusion, point
queries) are to be handled efficiently.

A key problem that affects efficiency is how to reduce
complex objects to simpler ones chosen from predefined
primitives. Among the standard techniques known we have already
discussed how boundary representations stored in a relational
data base prevent efficient access based on geometric queries.
The problem of an object being torn apart happens also in another
standard modeling technique, constructive solid geometry (CSG).
Let us briefly discuss the consequences of basing the physical
storage structure directly on such modeling techniques.

In constructive solid modeling a complex object is
constructed from simple primitives, such as cubes or spheres, by
means of Boolean operations union, intersection and difference.
The construction process is represented by a tree. Each leaf of a
CSG tree contains a simple object, each internal node contains a
Boolean operation. To each node a geometric transformation such
as scaling, translation and rotation may be assigned. The Boolean
operation is applied to the objects represented by the left and
right subtree of the node. A geometric transformation assigned to
a leaf is applied to the simple object stored in the leaf, a
geometric transformation assigned to an internal node is applied
to the object resulting from the Boolean operation stored in this
node. Now consider the query whether a solid in CSG
representation intersects a simple object such as a point or a
line segment. Even if the solid and the query are far apart, all
the components of the solid must be examined in a tree traversal
to detect this. What is lacking is some concisely stated
geometric information that describes global properties of the
solid and its location in space.

2   AN APPROACH TO COMBINED GEOMETRIC MODELING AND STORING:
APPROXIMATION, TRANSFORMATION TO PARAMETER SPACE,
GRID FILE

The technique we now present for modeling and storing
spatial objects is based on:

- approximation of complex spatial objects by simple shapes,
  e.g. containers;
- transformation of simple spatial objects into points in
  higher-dimensional parameter spaces;
- the grid file for point storage.

Complex, irregularly shaped spatial objects can be represented or approximated by simpler ones in a variety of ways, for example: decomposition, as in a quad tree tessellation of a figure into disjoint raster squares of size as large as possible; representation as a cover of overlapping simple shapes; enclosing it in a container chosen from a class of simple shapes. The container technique allows efficient processing of proximity queries because it preserves the most important properties for proximity-based access to spatial objects, in particular: it does not break up the object into components that must be processed separately, and it eliminates many potential tests quickly (if two containers don't intersect, the objects within won't either). As an example, consider finding all polygons that intersect a given query polygon, given that each of them is enclosed in a simple container such as a circle or an aligned rectangle. Testing two polygons for intersection is an expensive operation as compared to testing their containers for intersection. The cheap container test excludes most of the polygons from an expensive, detailed intersection check.

Any approximation technique limits the primitive shapes that must be stored to one or a few types, for example aligned rectangles or boxes. An instance of such a type is determined by a few parameters, such as coordinates of its center and its extension, and can be considered to be a point in a (higher-dimensional) parameter space. This transformation reduces object storage to point storage, increasing the dimensionality of the problem but without loss of information. Combined with an efficient multidimensional data structure for point storage it is the basis for an effective implementation of data bases for spatial objects.

2.1 TRANSFORMATION TO PARAMETER SPACE

Consider a class of simple spatial objects, such as aligned rectangles in the plane (i.e. with sides parallel to the axes). Within its class, each object is defined by a small number of parameters. For example, an aligned rectangle is determined by its center (cx, cy) and the half-length of each side, dx and dy.

An object defined within its class Ω by k parameters, can be considered to be a point in a k-dimensional parameter space H assigned to Ω. For example, an aligned rectangle becomes a point in 4-dimensional space. All of the geometric and topological properties of an object can be deduced from the class it belongs to and from the coordinates of its corresponding point in parameter space.

Different choices of the parameter space H for the same class Ω of objects are appropriate, depending on characteristics of the data to be processed. Some considerations that may determine the choice of parameters are:

- Distinction between location parameters and extension parameters. For some classes of simple objects it is reasonable to distinguish location parameters, such as the center (cx, cy) of an aligned rectangle, from extension parameters, such as the half-sides dx and dy. This distinction is always possible for objects that can be described as Cartesian products of spheres of various dimensions. For example, a rectangle is the product of two 1-dimensional spheres, a cylinder the product of a

1-dimensional and a 2-dimensional sphere. Whenever this
distinction can be made, cone-shaped search regions
generated by proximity queries as described in section 2.3
have a simple intuitive interpretation: The subspace of the
location parameters acts as a "mirror" that reflects a
query.

Independence of parameters, uniform distribution. As an
example, consider the class of all intervals on a straight
line (Fig. 1). If intervals are represented by their left
and right endpoints, lx and rx, the constraint lx ≤ rx
restricts all representations of these intervals by points
(lx, rx) to the triangle above the diagonal. Any data
structure that organizes the embedding space of the data
points, as opposed to the particular set of points that
must be stored, will pay some overhead for representing the
unpopulated half of the embedding space. A coordinate
transformation that distributes data all over the embedding
space leads to more efficient storage. The phenomenon of
non-uniform data distribution can be worse than this. In
most applications, the building blocks from which complex
objects are built are much smaller than the space in which
they are embedded, as the size of a brick is small compared
to the size of a house. If so, parameters such as lx, rx
that locate boundaries of an object, are highly dependent
on each other. Fig. 1 shows short intervals on a long line
clustering along the diagonal, leaving large regions of a
large embedding space unpopulated; whereas the same set of
intervals represented by a location parameter cx and an
extension parameter dx, fills a smaller embedding space in
a much more uniform way. With the assumption of bounded d
this data distribution is easier to handle.

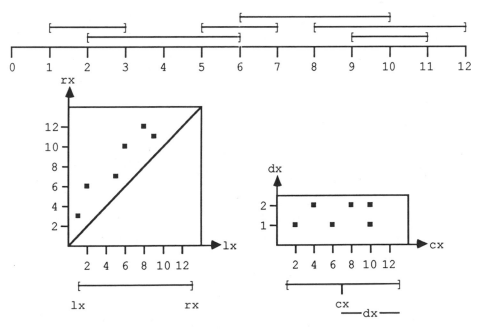

Fig. 1: Intervals on a straight line.

2.2 INTERSECTION QUERIES LEAD TO CONE-SHAPED SEARCH REGIONS

Intersection is a basic component of other proximity queries, and thus deserves special attention. CAD design rules, for example, often require different objects to be separated by some minimal distance. This is equivalent to requiring that objects surrounded by a rim do not intersect. Given a class Ω of simple spatial objects with parameter space H, and a set $\Gamma \subset \Omega$ of simple objects represented as points in H, we consider three types of queries:

- point query: given a query point q, find all objects $A \in \Gamma$ for which $q \in A$.
- point set query: given a set Q of points, find all objects $A \in \Gamma$ which intersect Q.
- geometric join query: given another class Ω' of spatial objects with parameter space H', and a set $\Gamma' \subset \Omega'$, find all pairs $(A, A') \in \Gamma \times \Gamma'$ of intersecting objects.

Point query

For a query point q compute the region in H that contains all points representing objects in that overlap q.

1) Let Ω be the class of intervals on a straight line. An interval given by its center cx and its half length dx overlaps a point q with coordinate qx if and only if cx - dx ≤ qx ≤ cx + dx.

2) The class Ω of aligned rectangles in the plane (with parameters cx, cy, dx, dy) can be treated as the Cartesian product of two classes of intervals, one along the x-axis, the other along the y-axis. All rectangles which contain a given point q are represented by points in 4-dimensional space lying in the Cartesian product of two point-in-interval query regions (Fig. 2). The region is shown by its projections into the cx-dx-plane and the cy-dy-plane.

3) Let Ω be the class of circles in the plane. As parameters for the representation of a circle as a point in 3-dimensional space we choose the coordinates of its center (cx, cy) and its radius r. All circles which overlap a point q are represented in the corresponding 3-dimensional space by points lying in the cone with vertex q shown in Fig. 3. The axis of the cone is parallel to the r-axis (the extension parameter), its vertex is q considered as a point in the cx-cy-plane (the subspace of the location parameters).

Point set query

Given a set Q of points, the region in H that contains all points representing objects $A \in \Gamma$ which intersect Q is the union of the regions in H that result from the point queries for each point in Q. The union of cones is a particularly simple region in H if the query set Q is a simple spatial object.

448

1) Let Ω be the class of intervals on a straight line. An interval $I = (cx, dx)$ intersects a query interval $Q = (cq, dq)$ if and only if its representing point lies in the shaded region shown in Fig. 4; this region is given by the inequalities $cx - dx \leq cq + dq$ and $cx + dx \geq cq - dq$.

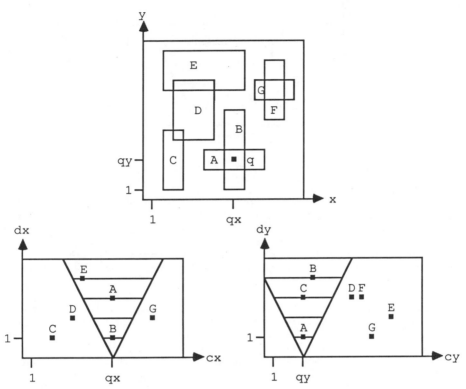

Fig. 2: Search region for a point query in the class of aligned rectangles in the plane.

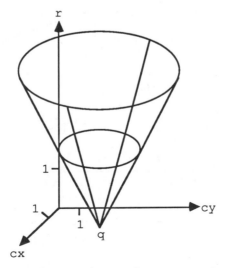

Fig. 3: Search region for a point query in the class of circles in the plane.

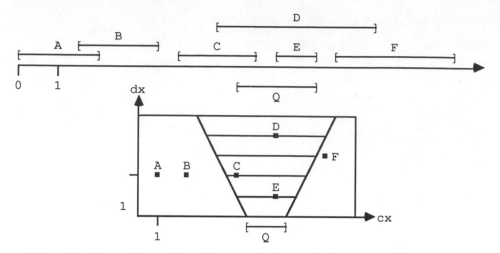

Fig. 4: Search region for an interval intersection query.

2) Let Ω be the class of aligned rectangles in the plane. If Q is also an aligned rectangle then Ω is again treated as the Cartesian product of two classes of intervals, one along the x-axis, the other along the y-axis. All rectangles which intersect Q are represented by points in 4-dimensional space lying in the Cartesian product of two interval intersection query regions.

3) Let Ω be the class of circles in the plane. All circles which intersect a line segment L are represented by points lying in the cone-shaped solid shown in Fig. 5. This solid is obtained by embedding L in the cx-cy-plane, the subspace of the location parameters, and moving the cone with vertex at q along L.

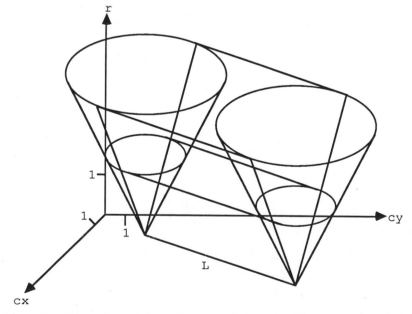

Fig. 5: Search region for an intersection query with a line L.

Geometric join query

Let Ω' be another class of simple spatial objects with parameter
space H', and $\Gamma' \subset \Omega'$. For every $A \in \Omega$ let $H'_A \subset H'$ be the set
of all points in H' representing $A' \in \Omega'$ such that A and A'
intersect. Denote by P_A the point in H representing a spatial
object $A \in \Omega$. The region in the Cartesian product H $\times$ H' that
contains all points representing pairs (A, A') $\in \Gamma \times \Gamma'$ of
intersecting objects is the union of the sets $\{P_A\} \times H'_A$ for all
$A \in \Omega$; this region is particularly simple for the different
classes of simple spatial objects.

Let Ω be the class of points, Ω' the class of intervals on a
straight line. Then H $\times$ H' is the 3-dimensional space. All pairs
(p, I) of points p with coordinate x and intervals I = (cx, dx)
such that p $\in$ I are represented by points lying in the solid
shown in Fig. 6. This solid is obtained by moving the search
region for a point-in-interval query along the bisector in the
x-cx-plane.

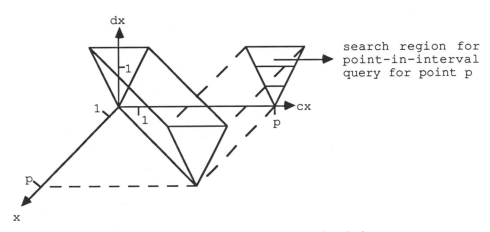

Fig. 6: Search region for a geometric join query.

2.3 EVALUATING REGION QUERIES WITH A GRID FILE

We have seen that proximity queries on spatial objects lead
to search regions significantly more complex than orthogonal
range queries. The grid file [NHS 84] is a structure for storing
multidimensional point data designed to allow the evaluation of
irregularly shaped search regions in such a way that the
complexity of the region affects CPU time but not disk accesses.
The latter limit the performance of a data base implementation.

The grid file partitions space into raster cells and assigns
data buckets to cells. The partition information is kept in
scales, one for each axis of space; the assignment is recorded in
an array called grid directory. The directory is likely to be
large and must therefore be kept on disk, but the scales are
small and can be kept in central memory. Therefore, the grid file
realizes the two-disk-access principle for single point retrieval

(exact match query): by searching the scales, the k coordinates
of a data point are converted into interval indices without any
disk accesses; these indices provide direct access to the correct
element of the grid directory on disk, where the bucket address
is located. In a second access the correct data bucket (i.e. the
bucket that contains the data point to be searched for, if it
exists) is read from disk. A query region Q is matched against
the scales and converted into a set I of index tuples that refer
to entries in the directory. Only after this preprocessing do we
access disk to retrieve the correct pages of the directory and
the correct data buckets whose regions intersects Q (Fig. 7).

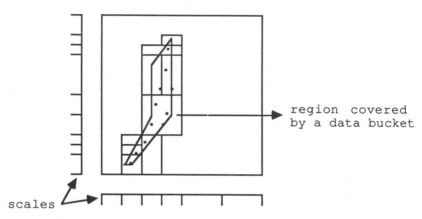

region covered
by a data bucket

scales

Fig. 7: Data buckets retrieved in answer to a region query.

A geometric join query is answered in an analogous way. Let
f and f' be the two grid files involved, and let H and H' be the
underlying higher-dimensional spaces. The scales of f and f'
define a grid on the Cartesian product H x H'. The cells of this
grid which intersect the search region in

H x H' are determined by matching the scales of the two grid
files against the search region. As in the case of proximity
queries on a single grid file this computation needs no access to
disk. If a cell intersects the search region the corresponding
pair of buckets $(B_f , B_{f'})$ is accessed from disk via the grid
directories of f and f'. If the Cartesian product of the bucket
regions of B_f and $B_{f'}$ is completely contained in the search
region all pairs of objects corresponding to pairs of points
$(p_f , p_{f'})$ with $p_f \in B_f$ and $p_{f'} \in B_{f'}$ fulfill the join condition.
If the Cartesian product of the bucket regions of B_f and $B_{f'}$ is
not completely contained in the search region all pairs of points
$(p_f , p_{f'})$ with $p_f \in B_f$ and $p_{f'} \in B_{f'}$ must be checked in order to
see whether they lie inside or outside the search region, i.e.
whether the corresponding pairs of objects fulfill the join
condition. A buffer of minimal size of two pages receives pairs
of data buckets $(B_f , B_{f'})$ according to a scheduling policy
similar to the one mentioned in [MKY 81].

452

2.4 THE GRID FILE SYSTEM

The grid file system [Hin 85b] is one of the building blocks of the Smart Data Interaction package [BHMN 85]. This package includes a Prolog interpreter that gives the user interactive access to the data stored in grid files, and serves as a powerful query language that permits deduction.

The package is implemented in Modula-2 and built on top of the virtual machine OSSI, a set of library modules which hides the machine-dependent details from Modula-2 application programs. OSSI, for Operating System Standard Interface [BHHM 86], has been implemented on several computers and operating systems, and allows us to port large Modula-2 programs without any changes to their source code. OSSI standardizes I/O operations, memory management, and utilities such as string handlers and large-set operations.

The grid file system consists of about 6000 lines of source code. Its interface towards client programs provides utility and query procedures for:
- creating, deleting, opening and closing a grid file.
- inserting and deleting records in a grid file.
- changing non-key information in a record.
- point query: find all records with given key values $x_1, \ldots, x_k$ (if keys are unique at most one record will be found).
- range query: find all records whose key values xi lie in given intervals $[l_i, u_i]$ $(1 \leq i \leq k)$.
- user defined region query: the user has to write a procedure which is called by the grid file system and determines whether a grid cell (given by intervals $[l_i, u_i)$ $(1 \leq i \leq k))$ intersects the search region defined by the user.
- nextabove, nextbelow: given key i with key value x_i, find the records with key values above or below x_i and next to x_i ; this gives the user the possibility to process the records sequentially with respect to one key.
- join query: the join query is a generalization of the join operator known from relational data bases. The user has to write some procedures which are called by the grid file system and guide the join query.
- counting: the above queries can be performed by only counting the records, but not transferring them to the user program.

2.5 CASE STUDIES OF APPLICATIONS

The grid file software package has been used to store and process geometric objects in the following applications [Hin 85a].

Producing layout masks for integrated circuits. Mask generated by a CAD system for chip design (David Mann Format) are presented as a set of aligned rectangles. Fabrication requires that a mask is represented as the set of connected components generated by rectangle overlap, i.e. a set of aligned polygons (all edges parallel to the coordinate axes, Manhattan geometry).

This transformation program was implemented by processing rectangles in a 4-dimensional grid file and computing the connected components by intersection queries.

Preprocessing plotter files. In a CAD system for mechanical engineering plotter files are preprocessed in order to reduce the total distance along which the raised pen has to be moved. The task of finding an optimal solution to this problem is equivalent to the traveling salesman problem and therefore NP-complete. The plotter files contain line segments and arcs which have to be drawn. The end points of the line segments and the arcs are stored in a 2-dimensional grid file. A reduction of the total pen plotting time is achieved by nearest neighbor queries on this grid file. A similar method using quad trees is presented in [And 83].

Analyzing photographic satellite data. A photograph obtained by a satellite consists of 512 * 512 pixels. Each pixel is assigned four color values in the range from 0 to 255. These pixels are stored in a 4-dimensional grid file. The ground imaged by these pixels is then classified into water, forest, fields, residential and metropolitan areas etc., by range queries on this grid file.

Managing simple spatial objects. An interactive program manages large sets of simple spatial objects, e.g. rectangles, circles and segments. These objects, each of which is defined by a fixed number of parameters, are stored in different grid files, one for each type of object. The program allows the user to insert and delete simple spatial objects and to perform proximity queries (e.g. intersection, containment) on the stored data.

Processing geographic data. The Swiss Federal Office for Statistics made available to us a file which contains raster information about Switzerland. Each record in this file represents a square of 100 meters by 100 meters (1 hectare). Switzerland covers about 4'000'000 hectares, but only about 100'000 hectares have beeen registered. Besides the coordinates of the corresponding hectare there are other attributes stored in each record, for instance the identification number of the municipality the hectare belongs to or the type of ground cover of the hectare; since these attributes are not used as keys for performing queries they are neglected. The records have been inserted into a two-dimensional grid file using as keys the two coordinates of the corresponding hectare. Typically, these records are accessed by range queries to find all the hectares that belong to a rectangular region.

REFERENCES

[And 83]
D. P. Anderson: Techniques for reducing pen plotting time, ACM Trans. Graphics 2, 3 (1983), 197-212.

[BHHM 86]
E. S. Biagioni, G. Heiser, K. Hinrichs and C. Muller: OSSI - a portable operating system interface and utility library for Modula-2, to appear in IEEE Software.

[BHMN 85]
E. S. Biagioni, K. Hinrichs, C. Muller, and J. Nievergelt:
Interactive deductive data management - the Smart Data
Interaction package, Wissensbasierte Systeme, GI - Kongress 1985,
München, Informatik - Fachbericht 112, Springer Verlag, Berlin,
Heidelberg, New York, Tokyo, 1985, 208 - 220.

[Hin 85a]
K. Hinrichs: The grid file system: implementation and case
studies of applications, Diss. ETH No. 7734, Swiss Federal
Institute of Technology, 1985.

[Hin 85b]
K. Hinrichs: Implementation of the grid file: design concepts and
experience, BIT 25 (1985), 569 - 592.

[MKY 81]
T. H. Merrett, Y. Kambayashi, H. Yasuura: Scheduling of
page-fetches in join operations, Proc. 7th Intern. Conf. on Very
Large Data Bases, Cannes, France (1981), 488 - 497.

[NHS 84]
J. Nievergelt, H. Hinterberger, K. C. Sevcik: The grid file: an
adaptable, symmetric multikey file structure, ACM Trans. on
Database Systems 9, 1 (1984), 38 - 71.

A CONCEPTUAL BASIS FOR GRAPHICS-BASED

DATA MANAGEMENT

(Extended Abstract)

Daniel Bryce and Richard Hull*

Computer Science Department
University of Southern California
Los Angeles, California 90089-0782

ABSTRACT

   This paper introduces the data structures and graphics-based user interface to the Engineering Support Environment (ESE). This proposed system combines
and extends principles of database management, configuration management, and
graphics interfaces to support the engineering life cycle activities in
connection with the design, analysis, and manufacture of physical objects.
ESE can be broken down to three major components: the design-related information storage component, the design data storage component, and a
graphics-based user interface. Design-related information is represented
in ESE using the semantically oriented IFO database model. Design data in
this system is represented using AND/OR DAGs, which are extended in ESE to
store historical design information and to provide sophisticated configuration management capabilities. Access to these components is provided
through a multi-frame, graphics-based user environment which supports rich
browsing capabilities, including interactive data updates during browsing
sessions.

1. INTRODUCTION

   The past couple of decades have witnessed the development of a number
of important principles for simplifying user interfaces to general-purpose
database management systems, including for example the notions of data independence and data sublanguages [7, 38]. Significant recent developments
have been the introduction of object-oriented semantic database models,
which permit the representation of data relationships more directly than
is typically possible using the classical record-based models [1, 12, 17,
36], and the design and implementation of a number of techniques for providing interactive, graphics-based user access to computer systems in general [10] and to databases in particular [13, 39]. In spite of this work
focusing on general-purpose databases, there have been few applications of
these principles to provide natural interfaces to data management systems

*Work by this author supported in part by the National Science Foundation
 Grant IST-83-06517.

457

supporting specific, complex, data-intensive activities, such as engineering design or software development. The purpose of this paper is to introduce the conceptual framework of the Engineering Support Environment (ESE), which provides a graphics-based interface for data management associated with engineering life cycle activities. A portion of the proposed ESE system, which incorporates some of the ESE data management capabilities in a graphical interface, has been prototyped in the SNAP system [2].

To accomplish its objectives, ESE combines and extends principles and techniques already developed in two broad areas: configuration management and graphics interfaces. For storing actual design information, including alternative versions, the ESE system uses a variant of the AND/OR DAG introduced in [27] and applied to supporting VLSI design and software engineering in [25, 34], respectively. The ESE system extends the AND/OR DAG model of these other investigations in two ways. First, the ESE system incorporates mechanisms into the AND/OR DAG model at a fundamental level for retaining past versions of proposed designs, and supports the construction of configurations involving portions of the design from both the present and the past. The second extension of the AND/OR DAG model is that the ESE system can be used to store information about the individual components of an instantiation of a given design configuration (e.g., a prototype or manufactured version), a capability which is not typically needed in VLSI design or software engineering. In the area of graphics interfaces, the ESE system provides sophisticated browsing access into the database which is similar in spirit to the GUIDE system [39], but which extends it in several ways. First, it provides graphics access to IFO database schemas, which are considerably richer than the entity-relationship schemas of GUIDE. Furthermore, the IFO model provides a natural method for depicting meaningful database subschemas, while avoiding the spaghetti-like subschema representations that other semantic data models often lead to. Second, the ESE system provides graphics access to the AND/OR DAG data structure, and hence to design information. And third, it permits interactive updating of data during browsing sessions. The ESE system also extends previous work on database modelling by incorporating historical information into the semantic IFO database model at a fundamental level.

The ESE system as overviewed here provides a basis for managing computer-aided design data which represents a substantial improvement over existing CAD/CAM database support systems, such as TORNADO [37] and IPAD [8, 9, 15]. In these and related systems, design data is generally stored using one or more of the conventional, record-based database models (relational, hierarchical, or network). These models do not provide natural representations for a variety of data relationships arising in the design process, including the part-subpart hierarchy and the correspondence between requirements specification and designed components. In contrast, the ESE system provides data structures, such as AND/OR DAGs and directly modelled functional relationships, which closely parallel the types of data relationships and restrictions which arise. As a result, ESE is easier for designers and engineers to use than existing CAD/CAM systems. A second benefit of the ESE data structures is in the area of supporting software packages, which analyze proposed designs. In order to apply an analysis package using a CAD/CAM database support system, design data must be translated from the system description into the format required by the analysis package. In the ESE system, this translation is facilitated by the fact that design data is modelled in an essentially direct manner, whereas in existing systems the data is modelled within a less natural record-based framework.

Although directed towards the engineering of physical, three-dimensional objects, various aspects of the ESE system are also relevant to other fields of engineering, and to database applications in general. For example, the extended AND/OR DAG model used in ESE, which incorporated his-

458

torical data and sophisticated configuration construction, can also be used
to support virtually all other engineering design life cycles. Furthermore, the paradigm used in ESE to incorporate historical data into the IFO
model is applicable to a wide variety of database applications. Also,
many of the principles of the graphics interface to ESE database schemas
and design data can be applied in other database applications. In particular, the approach for representing meaningful portions of database schemas
in ESE is a significant improvement over existing techniques for representing database subschemas.

This extended abstract presents an overview of the basic components
of the ESE system. A more complete description of the data structures and
the browsing interface for design-related information is given in [3] (see
also [2]). The remainder of this abstract is divided into three sections.
Section 2 describes the functional capabilities of the ESE; Sec. 3 compares
ESE with related systems; and concluding remarks are made in Sec. 4.

2. FUNCTIONAL CAPABILITIES OF ESE

ESE is an integrated environment which will support the full range
of engineering design, analysis and implementation activities. It provides
uniform, coherent interfaces for supporting tools required throughout the
various phases of the system life cycle. The main ESE capabilities can be
roughly divided into three broad areas, these being: an object-based
semantic data manager, the graphical interfaces for data access and for
schema design, and an integrated subsystem to provide version control and
configuration management of design data. The discussion which follows
briefly overviews the three main functional capabilities and then presents
each of them in more detail.

The first component of ESE is the database model used for representing
design-related information. The basic framework for this is the IFO database model as presented in [1]. IFO is a simple yet powerful data model
which serves as an excellent basis for the ESE user interface. Some specific advantages of IFO are:

1. IFO is an object-based semantic model. It is now widely accepted
   that the object-based semantic database models [5, 12, 17, 30] provide
   a better basis for user support than record-based models [16, 36].
   Also, as suggested by [26], object-based models lend themselves much
   more easily to browsing interfaces than record-based models. Finally,
   IFO provides mechanisms and constructs which mirror the prevalent
   kinds of relationships naturally arising between data stored in a
   database, specifically: ISA relationships, functional relationships,
   and relationships arising from the construction of complex objects from
   simpler objects.

2. It is easy to build an effective graphics-based interface for IFO.
   IFO is a graph-based model, where schemas are described as directed
   graphs with certain properties. Also as shown in [2], features of
   IFO make it possible to display meaningful subschemas in an orderly
   fashion.

3. It is simple and natural to add history to the IFO model. This is
   because IFO is object-based, and is clearly and concisely defined.
   A fundamental requirement for managing historical information is
   keeping track of which objects maintain their identities through time.
   These objects correspond to the <u>primary objects</u> in IFO.

Also, IFO provides a natural foundation upon which additional structures
can easily be built.

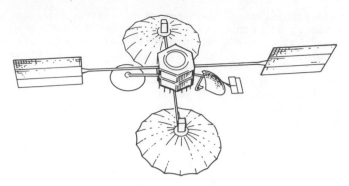

Fig. 1.   Model of a spacecraft.

The second basic capability of ESE is that it provides a convenient and efficient interface to design data, primarily through an extension of IFO called the MAC (MAnager of Configurations). The MAC stores the part-subpart hierarchy of design components, keeps track of alternative proposals for given parts of a design, records the evolution of a design through time, and provides sophisticated configuration management capabilities. Because the MAC is integrated into the ESE schema in a coherent fashion, it is quite simple to store information associated with parts in a designed system, and to relate it to other information in the database. It is based on a variant of AND/OR DAGs, as introduced in [27], and applied to textual data in [34], and more recently to VLSI design in [25]. MAC extends this previous work by permitting users to combine components of the proposed design from past and present in order to compare and analyze related configurations. Also, the MAC distinguishes between design information (in which multiple occurring components can be represented once) and instantiation information (where information concerning each occurrence of such components must be stored). Finally, ESE provides a graphical interface into MAC data which permits users to create, browse, and update design proposals. Furthermore, this interface gives users multilevel (i.e., layered) views of the data, thereby providing a convenient mechanism for hiding detailed information until it is requested.

The third basic capability of ESE is to provide the user with a coherent interactive graphical interface for virtually all activities. A number of recent systems have addressed various aspects of this problem. Several systems use graphics to present data, state queries, or browse through stored data [4, 11, 13, 39]. On the other hand, some systems [22, 33] with less sophisticated graphics interfaces provide browsing and update facilities. Thus, while techniques have been developed to provide interactive graphical interfaces to isolated portions of a database, there is currently no system which allows database users to perform all of their interactions through a graphics-based interface. In contrast, ESE provides a unified, coherent graphical interface to both schema and data (see also [2, 18]). ESE presents the user with a representation of the schema and various types of data, and the user can freely browse and update this data by using any of these representations.

The discussion of the three main ESE capabilities in the remainder of this section is based on a simple example concerning the design of a spacecraft, such as the one shown in Fig. 1. For this example, design data to be maintained include: the part names, alternative choices, and the part-subpart relationship between parts of the spacecraft. Peripheral information includes: the surface thermal properties of the components, and information about the designers responsible for each component (including

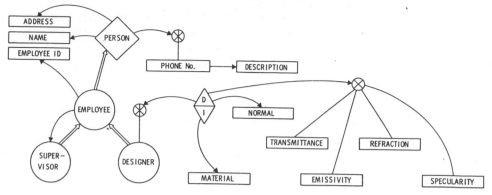

Fig. 2.   An ESE schema.

employee-id, name, address, supervisor, and phone numbers). An ESE schema
for representing this information is shown in Fig. 2, and an AND/OR DAG depicting an actual design for the spacecraft is shown in Fig. 3.

2.1.   Representation of Design-Related Information in ESE

The ESE schema shown in Fig. 2 represents both the spacecraft design
and the institutional framework within which the design is performed. This
subsection focuses on the representation of information aside from the
actual design data. As will be shown, design related information is represented using object types, fragments, and ISA relationships. The exposition
of these components is brief, since it is taken directly from the IFO model
as described in [1]. This subsection concludes by indicating how the IFO
model is extended in ESE to incorporate historical information.

Various objects, such as the surface properties, the designers, employees, employee information, people and associated information, are represented in an ESE schema with object types. There are three atomic object
types and two constructs for recursively building more complicated object
types. Examples of atomic types are:

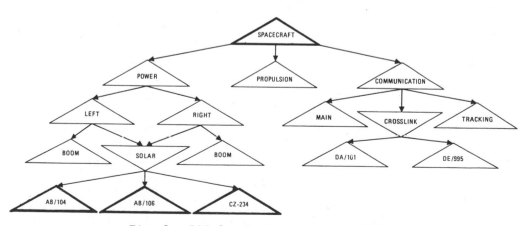

Fig. 3.   DAG for design portion of MAC.

461

1. Printable information, such as a person's name, an employee's id, a phone number, the description of a phone number, and surface properties, is modeled with representable object types. Representable object types correspond to objects of predefined types which serve as the basis for input and output. A representable object type is depicted in a schema as a square with an identifier and a type.

2. Objects like people are represented using abstract object types. We cannot print a person, we can only input attributes of a person. In an ESE schema, an abstract object type is depicted as a diamond with an identifier.

3. To represent a subtype (e.g., employee) of an already existing type (e.g., person), we use a free object type. A free object type is depicted as a circle with an identifier. The underlying structure and some associated attributes of a free object are inherited via ISA relationships. (Free object types also arise from the construction of supertypes as detailed below.)

Examples of the two constructs for creating complex objects are:

1. Objects like the set of phone numbers associated with a person are represented using a ⊛-vertex. This construct is represented by a circle with a star in it with a connecting arc to one child object type.

2. The four surface properties can be viewed as a single unit by using a ⊗-vertex. This corresponds to the well known Cartesian product operator. A cross-vertex is represented as a circle with a cross in it with connecting arcs to one or more child object types.

One type of relationship commonly arising between objects is that one object is an attribute of another. For example, a person has-a name, a person has-a set of phone numbers, an employee has-an employee-id, and an employee has-a supervisor. Fragments give the power to directly represent these functional relationships in ESE. An example of a simple fragment is the tree with root node person and associated subtrees corresponding to attributes name, address, set of phone numbers, and descriptions of the phone numbers as shown in Fig. 2. A second example is the tree with root employee and subtrees employee-id and supervisor. Object types at the roots of fragment types typically represent important entity sets in the database, and are called primary object types. Object types in the range of functions are not used for anything else, except possibly to state that the range objects form a subset of a primary object set (i.e., supervisor is an employee). This makes it easy to hide details irrelevant to a given task; it is therefore possible to provide users with comprehensible diagrams of ESE subschemas.

Fragments can be used to represent nested functions in a recursive fashion. Consider a situation where a person has a set of phone numbers, and each phone number has a description (which could be different for two different people). In particular, suppose that Judy has a set of phone numbers, one of which is her work phone (614-1342), and further suppose that Jack, Judy's secretary, answers Judy's phone and takes messages if Judy is out of the office. Relative to Judy, 614-1342 is described as her main work phone number, however relative to Jack, 614-1342 is described as a backup phone number. Therefore, there is not one description for a phone number, but rather a description relative to each person who has that phone number. Figure 2 shows how to represent this information in an ESE schema.

The other type of relationship prevalent between objects is that objects of one type are automatically of another type. In the running example, a designer is an employee, a supervisor is an employee, and an employee is a person. ESE provides two types of ISA relationships, namely specialization and generalization. The specialization relationship is depicted using a broad arrow (➡). The schema in Fig. 2 shows how to represent the designer, employee, person specialization hierarchy. The type of an object and its attributes are inherited downward in a specialization hierarchy. Therefore, all attributes of persons are attributes of employees and all attributes of employees are attributes of designers. In particular, a designer has an employee id, a name, an address, and a set of phone numbers (each of which has a description). The second type of ISA relationship (although not used in this paper) is generalization, represented by using a shaded arrow (➡). Generalization is used if a common supertype is to be formed from two or more separate entity sets. For example, the supertype legal entity might be represented as the generalization of persons and corporations. In this case, legal entities would be represented using a derived object type. This illustrates the fundamental difference between specialization and generalization: in specialization object type is inherited downwards, whereas in generalization object type is inherited upwards.

This subsection concludes by briefly indicating how the ESE system extends the IFO model to incorporate historical information. To do this, it first describes in conceptual terms how historical data is represented and stored by an ESE database, and then describes the kinds of data access commands that are supported. Users can think of an ESE database as containing the states of the database for each time from its inception up to the present. To accomplish this, the system need only remember the contents of the database at each time immediately following an update. In this manner, the historical contents of a database can be specified in a finite manner, but users can view the database as having existed continuously in time. In terms of physical implementation, the database does not store the entire contents of the database after each update, but rather it keeps track of changes. Also, instead of storing the impact of each update, a user may specify how frequently the database should remember the database contents. For example, a user may instruct the system to remember the state after every update for the past two days, at the midnight of each day for the past week, on Sunday midnight for the past month, and finally on specific days prior to that.

Using the conceptual basis for historical data given here, it is easy to see how the notions of object insertion and deletion (or creation and destruction) at a time t can be defined, and also such notions as a function f having a value continuously during the time interval, or the function f changing its value on a given object at time t. As a result, ESE allows users to ask questions such as "What are the previous addresses of a person named Johnson (and on what dates did she move)?", "When did Johnson become an employee?", and "Was Johnson associated with the Design Department during the entire month of May, 1983?". More complete details are presented in [3].

2.2. Representation of Design Information Using the MAC

MAC is an essential feature of ESE which forms the basis for managing design data. A MAC stores the family of part-subpart relationships of a given application, provides structures for version control and version selection, facilitates configuration management, and manages data for both generic and instantiated components. From the user's point of view, a MAC consists of two components, which provide access to design information and instantiation information respectively. The discussion which follows introduces the MAC system in conceptual terms.

The primary underlying data structure of the MAC is a type of the AND/ OR DAG, closely related to the AND/OR DAGs of [27, 34, 25]. Figure 3 shows such a DAG for our current example. Each node in the graph corresponds to a part or an alternative in the design, the name of which is included as a label for the node.

Parts are modelled in a MAC by using AND nodes. The part labeled spacecraft is the root of this DAG. The spacecraft has three subparts: power, propulsion and communications subsystems. An AND node in the AND/OR DAG is depicted in a MAC expansion as a triangle with a horizontal base on the bottom (and thus corresponds to the propositional connective ∧). The part-subpart relationship is shown as a directed arc from an AND node to an AND or an OR node. An AND node may represent a physical part (i.e., solar panel AB/104) or grouping (i.e., power subsystem or spacecraft). It is also possible for certain AND nodes to serve both functions.

Design alternatives are modelled in a MAC by using OR nodes. In the running example there are three possible interchangeable solar panels. This information is represented in a MAC by using an OR node for the solar panel node. An OR node is depicted in a MAC expansion as a triangle with a horizontal base on the top (corresponding to ∨). The has-version relationship is represented as an arc from an OR node to an AND or an OR node.

An interesting feature of the design structure of the MAC is shown in the power subsystem in Fig. 3. The power system is decomposed into the left and right sides, each of which has a boom and a solar panel. However, since the booms are different nodes, they could actually be different types of booms. On the other hand, since the left and the right side point to the same solar panel selection node, the same selected solar panel design must be used on both sides.

A user may distinguish between major and minor components in a MAC. Major components are parts with special significance, typically because they are the roots of design structures or they appear many times in a design. A user may combine major components from various times to build a special DAG for creating new components or configurations. Another significance of major compnonets is that in the browsing interface for a MAC, a user may scroll through the major components by requesting the next or previous major component. Major vertices are shown with broad borders in a MAC expansion. For the running example, the spacecraft and the solar panels are distinguished as major components.

An important aspect of the MAC design system is that it maintains the history of modifications, using a historical MAC DAG. Users can browse around this structure by moving their window (i.e., current view) into the AND/OR DAG, both within the relevant time frame, and also forwards and backwards in time. A central tool for utilizing data in the historical MAC DAG is a DerIved MAc DaG for I/O (DIMADGIO - pronounced DiMaggio). A DIMADGIO is a type of view into the historical MAC DAG. In particular a DIMADGIO is a (non-historical) AND/OR DAG formed by selecting design components from various times. Typically, a DIMADGIO will contain one or more closely related design proposals. DIMADGIOs can be used to restrict attention to a particular set of components, and to combine components (possibly from different times) to create new components.

A DIMADGIO of particular interest is a configuration. A configuration is essentially a subgraph of a DIMADGIO for which each OR node has only one child. A configuration is built relative to a DIMADGIO, and therefore it can include portions of the current design proposal, and design proposals of previous times. In this manner, users can configure various proposed alternatives and compare their effectiveness. For example, a user may con-

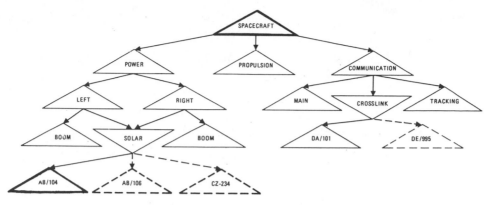

Fig. 4. Configuration C-SAMPLE-1.

figure one spacecraft design, called C-SAMPLE-1, with solar panel AB/104
and crosslink antenna DA/101, and a second configuration, called HWW-014a,
which consists of solar panel AB/106 and crosslink antenna DA/101. These
two proposals can be output and analyzed to find advantages of one proposal over the other. MAC browsing and update facilites allow a user to
define configurations and give them names. Figure 4 shows configuration
C-SAMPLE-1, which was built from the current DIMADGIO. Nodes and arcs
which are not directly included in the configuration are shown with dotted
lines.

Many aspects of the ESE approach to configuration management are novel.
In particular, the MAC subsystem distinguishes between alternatives in the
current design and alternatives which existed in some past version of the
design. Also, actual configurations are built using what may be viewed as
a two-phase approach: first building a DIMADGIO, which incorporates design
components from different times and may still include alternative versions
(to facilitate comparison); and second making specific choices for the alternatives. A fundamental reason for this approach is that DIMADGIOs permit users to combine design data from past and present, but structure this
data as a flat AND/OR DAG with no time dimension. DIMADGIOs serve as a
convenient middle ground between the rich historical MAC DAG data structure and the restrictive notion of a configuration, and thus serve as a
convenient mechanism for comparing related design proposals. While this
approach to configuration management appears promising, its true propriety
will be determined through experimentation with the ESE prototype.

In the engineering life cycle, the design information often drives
the manufacturing effort. One may wish to build and experiment with a
prototype, while identifying and tracking each individual physical part
included in this prototype. For example, in the design phase it is desirable to say that the same type of solar panel is used in two places, but
for analysis (and manufacturing) each part must be individually addressed.
Indeed, this is a fundamental difference between the management of textual
and physical information. Typically, software systems can be managed entirely within the design and the configuration portions of a MAC. However,
when dealing with physical systems configurations are used to create instantiated components.

The steps for building an _instantiation_ from configuration C-SAMPLE-1
are as follows:

465

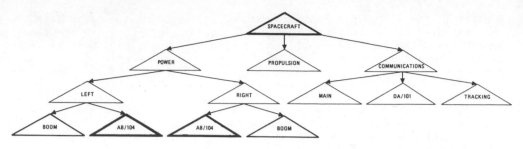

Fig. 5. Instantiation of configuration C-SAMPLE-1.

1. Collapse all OR nodes. The resulting DAG contains only AND nodes.
   In reference to configuration C-SAMPLE-1, for instance, the solar OR
   node is replaced with the AB/104 node. AB/104 is now a subpart of the
   left and right parts of the power subsystem. Similarly, DA/101 is replaced as the subpart of the crosslink antenna of the communication
   subsystem.

2. Duplicate all part components with indegree greater than one. For example, the left and the right side of the power system both reference
   an instance of solar panel AB/104. Since they do not really both use
   the same panel (there are two panels of the same type), two separate
   panel AB/104 nodes must be created, one being a subpart of the left
   and one a subpart of the right side.

   Upon completion of these steps, the system produces an instantiation
tree as shown in Fig. 5. However, the label given to each instantiated
node is the same as the label of the corresponding node in the design DAG,
and therefore some nodes may have the same label. The user can use the
interactive browser to display and update the instantiation tree (or distinct names can be created through some automatic or semi-automatic procedure).

   The discussion of the MAC data structures concludes by indicating how
MACS are incorporated into an ESE schema. As shown in Fig. 2, a MAC abstraction is depicted in an ESE schema as a diamond with a line through the
center, a D in the upper triangle, and an I in the lower triangle. Attributes can be associated with parts in the design portion of a MAC by
placing a function arrow from the upper half of the MAC diamond to an object type. Similarly, attributes can be associated with parts in the instantiation component by placing a function arrow from the lower half of
the MAC diamond to an object type. For example, the schema in Fig. 2 specifies that a design part has surface properties, a surface normal, and a
set of designers. An instantiated part has one attribute, a description
of the material it is made from.

2.3. The Graphical Interface of ESE

   A central aspect of the entire ESE system is the presence of a unified
graphical interface which supports virtually all data access and manipulation activities. User interaction with the interface occurs primarily
within a framework of browsing, that is of interactively exploring and possibly updating data. ESE provides a general browser for design-related information, and an extended browser for MAC expansion DAGs.

   The ESE user environment consists of an α-frame, a graphics schema
frame and a graphics data frame (possibly distributed over more than one

monitor). The α-frame is used for command input, printable query output, and system error or prompt messages. Printable information may be presented (requested) to (from) the user either by using predefined templates or by using the system defaults. Predefined templates can be used to either present objects one at a time or a number of objects at the same time. The graphics schema frame displays the <u>active schema</u> and the MAC expansions. The active schema is the portion of the schema which the user has identified to be of current relevance. The graphics data frame displays data which has geometric information associated with it.

Users are given direct access to the ESE schema and allowed to hide portions of the schema, move around the schema, and reposition schema components. Data is requested by enabling nodes and arcs, and issuing commands to browse through data. When browsing a set, the user may restrict and order it to display a subset or to show the information sorted on some attribute. Mechanisms for browsing and updating MAC DAGs, building DIMADGIOs, configurations, and instantiations; and for correcting historical errors are also provided. Some of these capabilities have been implemented in [2].

3. COMPARISON WITH EXTANT SYSTEMS

ESE spans many disciplines and draws from previous work in data modelling [1; 12, 14, 17, 20, 23, 30, 31, 32], databases for CAD/CAM [8, 9, 15, 18, 21, 29, 37], browsing and graphical interfaces [6, 11, 13, 24, 26, 33, 39], and version control/configuration management [25, 28, 34, 35]. The following is a comparison of ESE with representative systems from the latter three topics, whose objectives or characteristics are most closely related to ESE. This discussion assumes the reader is familiar with these systems.

An essential characteristic of databases for CAD/CAM is representing data in a format natural for the designer, and easily translated to analysis packages. Most extant systems extend the conventional record-based models. For example, TORNADO [37] is a CODASYL-type network database system which includes a number of desirable characteristics for CAD applications, such as variable length records, aditional types, and handling of compound object classes. Perhaps the most comprehensive study on the data requirements for CAD/CAM systems was performed for the IPAD (Integrated Programs for Aerospace-Vehicle Design) project [8, 9, 15]. As a result of this study, a data manager called IPIP was developed. It is based on a generalized record-based model which combines aspects of the relational, hierarchical, and network models. In particular, it allows the same data to be represented in one or more of the conventional models. Furthermore, the same design information can be represented in a different way in the external views. In contrast, the ESE data model is not record-based, and allows designers to think in terms of designed components in a direct manner. Also, since the structures in the data model are closely related to the kinds of relationships arising in engineering, it appears that it will be straightforward to develop interfaces from the ESE system to various application packages.

Relative to query specification, ESE provides many of the same features as GUIDE [39]. In both ESE and GUIDE, the database schema is displayed, and the user expresses queries by identifying portions of the schema. Multiple levels of detail in schemas provide assistance in locating relevant information. GUIDE also includes a hierarchical subject directory to help acquaint users with the type of information contained in the schema. However, the data model GUIDE operates on, a simple version of the Entity-Relationship model [5], does not include ISA relationships, and does not permit complex objects in the range of functions. ESE extends the general principles applied in GUIDE by providing a uniform graphical interface to the richer ESE data model, and by permitting interactive updating during browsing sessions. In particular, ESE provides graphical interfaces for query specification and browsing (incuding updating) to the design related information through an ESE schema, and to design related activities, such as configuration management and instantiation, through the MAC subsystem. Additionally, ESE presents data with geometric representations to the user, and allows identification and simple modification via a graphical interface.

The IFO portion of the ESE interface is similar in functional capability to the recent SKI interface for semantic data models [18]. Specifically, the SKI system permits users to browse schemas in a variety of ways, and to express selection-type queries. A fundamental difference between the SKI and ESE approaches is that in SKI, displayed portions of a database schema are represented using a regimen of horizontal bands and are constructed dynamically during each user session. In ESE, on the other hand, the user is given much greater freedom in the arrangement of subschemas.

Finally, the most promising work to date on version control and configuration management tools has been based on the AND/OR graph as introduced in [27]. In particular, [34] applied to the model to programming support environments, and [25] applied the model to VLSI design. In [25] a VLSI design database is regarded as consisting of alternating levels of OR nodes and AND nodes. The paper shows how to represent the database using the Event model [19], but does not consider historical design changes or mechanisms for creating consistent configurations. Tichy [34] permits AND/OR nodes to be intermixed freely. He also points out that the variety of interpretations of nodes in the graphs are actually overloading the simple AND/OR model. Therefore, additional node types are required to build intelligent software tools. In ESE a similar view is taken. Nodes may be freely intermixed, and since the MAC subsystem is embedded into a general data model in the ESE system, the user (or application program) may define the meaning of the nodes in the graph by associating functions (i.e., attributes) with these nodes. ESE also provides a historical dimension, supports views into the design structure with DIMADGIOs, and distinguishes between design and instantiation information. Finally, whereas [34] outlines a simple method to check for well-formed configurations, ESE provides tools for automatic detection and correction of incompatible configuration units.

4. CONCLUSION

This extended abstract introduced a general framework for an engineering support environment. In particular, it introduced the MAC subsystem for storing design information and showed how to embed it within the semantically oriented IFO model. The MAC subsystem supports sophisticated configuration management and permits data management for instantiated components. ESE provides a uniform graphical interface of virtually all activities. This interface includes mechanisms for viewing portions of the database, and for updating in the midst of a browsing session. This extended abstract is based on [3], where a more complete description of the ESE system may be found.

A prototype, called SNAP, which focuses on the graphics-based browsing to IFO schemas system has been developed at the University of Southern California [2]. This prototype provides a basis for testing much of the underlying philosophy of the ESE system. SNAP, which is written in Lisp on a Symbolics 3600, includes an implementation of the IFO data model, and a graphical interface for designing, browsing, and querying of IFO schemas. Future development will permit us to study the ease of using the configuration management system of MAC and aspects of data sharing.

REFERENCES

1. S. Abiteboul and R. Hull, IFO: A Formal Semantic Database Model, ACM Trans. on Database Systems (in press).
2. D. Bryce and R. Hull, SNAP: A graphics-based schema manager, Proc. of the Second IEEE Intl. Conf. on Data Engineering, February 1986, pp. 151-164.
3. D. Bryce and R. Hull, A Conceptual Basis for Graphics Based Engineering Data Management, Tech. Report TR-84-316, Computer Science Dept., Univ. of So. Calif., Los Angeles, 1984.
4. R. Catell, An Entity-Based Database User Interface, Proc. ACM SIGMOD Int. Conf. on the Management of Data, Santa Monica, California (1980).
5. P. P. Chen, "The Entity-Relationship Model - Toward a Unified View of Data," ACM Trans. on Database Systems 1, 1 (1976), 9-36.
6. J. B. Crampes, C. Y. Christment, and G. Zurfluch, The Big Project, ICOD-2 Second International Conference on Databases, The British Computer Society, Churchill College, Cambridge, September, 1983, pp. 259-286.
7. C. J. Date, An Introduction to Database Systems, Vol. I, Addison-Wesley, Reading, MA (1981).
8. R. Peter Dube and H. Randall Johnson, Computer-Assisted Engineering Data Base, The American Society of Mechanical Engineers, ASME 83-WA/Aero-11 (1983).
9. R. Peter Dube and Marcia Rivers Smith, Managing Geometric Information with a Database Management System, IEEE Computer Graphics and Applications, October, 1983, pp. 57-62.
10. Steven Evans, Daniel Lipkie, John Newlin, and Robert Weissman, "Star Graphics: An Object-Oriented Implementation," Computer Graphics 16, 3 (July 1982), pp. 115-124.
11. D. Fogg, Lessons from a "Living In a Database" Graphical Query Interface, Proc. ACM SIGMOD Int. Conf. on the Management of Data, 1984, pp. 100-106.
12. M. Hammer and D. McLeod, "Database Description with SDM: A Semantic Database Model," ACM Trans. on Database Systems 6, 3 (1981), 351-386.
13. C. F. Herot, "Spatial Management of Data," ACM Trans. on Database Systems 5, 4 (1980), 493-514.
14. R. Hull and C. K. Yap, "The Format Model: A Theory of Database Organization," Journal of ACM 31, 3 (July 1984), pp. 518-537.
15. H. R. Johnson, J. E. Schweitzer, and E. R. Warktine, A DBMS Facility for Handling Structured Engineering Entities, SIGMOD '83 Engineering Applications, May, 1983, pp. 3-11.
16. W. Kent, "Limitations of Record-Based Information Models," ACM Trans. on Database Systems 4, 1 (1979), pp. 107-131.
17. L. Kerschberg and J. E. S. Pacheco, A Functional Data Base Model, Pontificia Universidade Catolica do Rio de Janeiro, Rio de Janeiro, Brazil, February (1976).
18. R. King, Sembase: A Semantic DBMS, Proc. of the First Intl. Workshop on Expert Database Systems, October, 1984, pp. 151-171.
19. R. King and D. McLeod, "A Methodology and Tool for Designing Office Information Systems," ACM Trans. on Office Information Systems, 1985.
20. R. King and D. McLeod, Semantic Database Models, in: Database Design, S. B. Yao, ed., Springer-Verlag, New York (1984).
21. Mustapho Koriba, "Database Systems: Their Applications to CAD Software Design," Computer Aided Design 15, 5 (September 1983), pp. 277-287.

22. A. Malhortra, H. M. Markowitz and D. P. Pazel, "EAS–E: An Integrated Approach to Application Development," ACM Trans. Database System 8, 4 (December 1983), pp. 515-542.

23. Frank Manola and Alain Pirotte, An Approach to Multi-Model Database Systems, ICOD-2 Second International Conference on Databases, The British Computer Society, Churchill College, Cambridge, September, 1983, pp. 53-75.

24. N. McDonald and M. Stonebraker, CUPID: A User Friendly Graphics Query Language, Proc. ACM-Pacific Conference, San Francisco, California, April, 1975, pp. 127-131.

25. D. McLeod, K. Narayanaswamy, and K. V. Bapa Rao, An Approach to Information Management for CAD/VLSI Applications, SIGMOD '83 Engineering Applications, May, 1983, pp. 39-50.

26. Amihai Motro, "BAROQUE: A Browser for Relational Databases," ACM Trans. on Office Information Systems 4:2 (April 1986), 164-181.

27. Nils J. Nilsson, Problem Solving Methods in Artifical Intelligence, McGraw-Hill (1971).

28. Marc J. Rochkind, "The Source Code Control System," IEEE Transactions on Software Engineering, SE-1, 4 (December 1975), 364-370.

29. R. S. Shenoy and L. M. Paitnaik, "Data Definition and Manipulation Langugages for CAD Databases," IEEE Computer Aided Design 15, 3 (May 1983), pp. 131-134.

30. D. Shipman, "The Functional Data Model and the Data Language DAPLEX," ACM Trans. on Database Systems 6, 1 (1981), pp. 140-173.

31. J. M. Smith and D. C. P. Smith, "Database Abstractions: Aggregation and Generalization," ACM Trans. on Database Systems 2, 2 (1977), pp 105-133

32. Richard Snodgrass, The Temporal Query Language TQuel, Proc. ACM SIGACT-SIGMOD Symp. on Principles of Database Systems (1984), pp. 204-213.

33. M. Stonebraker and J. Kalash, TIMBER: A Sophisticated Relation Browser, Proc. 8th Int. Conf. Very Large Data Bases (1982), pp. 1-10.

34. Walter Tichy, A Data Model for Programming Support Environments and Its Application, Automated Tools for Information System Design, January, 1982, pp. 31-48.

35. Walter Tichy, Design, Implementation, and Evaluation of a Revision Control System, Sixth International Software Engineering Conference, September, 1982, pp. 58-67.

36. D. C. Tsichritzis and F. H. Lochovsky, Data Models, Prentice-Hall, Englewood Cliffs, New Jersey (1982).

37. Stig Ulfsby, Steinar Meen, and Jorn Oian, "TORNADO: A DBMS for CAD/CAM Systems," Computer Aided Design 13, 4 (July 1981), pp. 193-197.

38. J. D. Ullman, Principles of Database Systems, 2nd ed., Computer Science Press, Potomac, Maryland (1982).

39. H. K. T. Wong and I. Kuo, GUIDE: A Graphical User Interface for Database Exploration, Proc. 8th Int. Conf. Very Large Data Bases (1982), pp. 22-32.

SEMANTIC DATA ORGANIZATION ON A GENERALIZED

DATA MANAGEMENT SYSTEM

G. T. Nguyen and J. Olivares

IMAG
Université de Grenoble
Laboratoire de Génie Informatique
BP 68
38402 St.-Martin-d'Heres
France

ABSTRACT

A set of definition, manipulation, and control facilities is proposed
to handle the semantics of a generalized data model. A meta-model is specified which includes the basic concepts it captures, as well as its evolution rules. It not only concerns the data types but also the logical data
structures, their relationships, and the integrity constraints relevant to
the applications. We thus provide a semantic validation subsystem which
permits control of the design of database schemas, of the operations on the
data, and the modification of both the data structure and integrity constraints. The validation subsystem uses a meta-knowledge base to store,
retrieve, and update the information pertaining to the semantics of the
data model. It is implemented in first-order logic. A brief description
of a CAD/VLSI application is given.

1 INTRODUCTION

Semantic integrity in the database field usually amounts to tight correspondence between the data definitions and the data values. Since the
definitions are given in some data definition language and storage is implemented with specific database software, semantic integrity is controlled
by matching the data values against the database schema. From this are derived the so-called consistent databases or consistent data.

Inconsistent data can arise in case of bad modification issued by the
user, of uncontrolled access by simultaneous users, or hardware and software crashes. Maintaining semantic integrity thus amounts to

- controlling the consistency of user requests,

- a proper ordering of concurrent data accesses,

- the implementation of efficient restart facilities.

In the following, we are only interested in the first aspect. Consistent requests are controlled by the correctness of the items involved against the data definition statements. This is usually performed when the request is analyzed and compiled.

Similarly, another level of semantic integrity may be defined: the consistency of data definitions. This means the correctness of the data definitions with respect to the data model at hand. Here it is necessary to control the definition statements and the inter-data relationships to conform them with the semantics of the data model.

The focus is here on the control of consistent data definitions and on the maintenance of correct schemas and integrity contraints for the data model of TIGRE, a generalized database system [7].

Section II is a short introduction to TIGRE. Section III is a summary of our objectives and solutions Section IV describes the architecture of the validation subsystem. Section V explains its operations concerning semantic integrity controls. Section VI briefly sketches an application to CAD/VLSI. Section VII is a conclusion.

2. THE GENERALIZED DATABASE SYSTEM TIGRE

TIGRE was developed at IMAG (University of Grenoble) in cooperation with the BULL Corporate Research Center. It is intended to explore new storage and manipulation capabilities for multimedia objects, including graphics, vocal and documentation items. A full-scale prototype is being implemented on a BULL SPS 7 machine running Unix V5 (Fig. 1).

In the following, it is assumed that the database server exists and that it allows for the modeling of sophisticated applications such as CAD. The data model is an extension of the entity-relationship approach to data modeling. It provides strong data typing facilities, including the notions of entity, relationship, class, aggregation, specialization, and generalization.

On the one hand, TIGRE provides the data management software for large and complex objects, e.g., documents and VLSI circuits. On the other hand, the validation subsystem provides semantic representation, control, and manipulation capabilities for the applications using these objects.

3. OBJECTIVES AND SOLUTIONS

Consistency of databases has been formalized previously on the basis of first-order logic [5, 8, 9]. Semantic constraints are considered as axioms in a first-order theory. Facts, i.e., elementary data, must satisfy the constraints to provide consistent database sates. Rules form a knowledge-base (KB), and each database state is a model, in the sense of logic, of this KB, i.e., an interpretation in which all the formulae are satisfied.

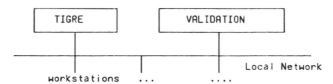

Fig. 1. Environment of the research.

MKB   KB

- definition of data
  model concepts
- semantic integrity
  in data model

$\longrightarrow$

- definition of
  application objects
- application integrity
  constraints

Fig. 2. MKB and KB interactions.

We consider here a meta-knowledge base (MKB) to control and maintain the consistency of the data definitions in the KB (Fig. 2).

3.1. The Meta-Knowledge Base

The meta-knowledge base is a symbolic representation of the generalized data model of TIGRE. It is implemented in first-order logic, precisely by Horn-clauses. It is used to define formulae, e.g., MKB requests, and to prove theorems, e.g., validate semantic integrity in the KB.

From a logic point of view, data definition statements are first-order logic axioms. Interpretation of these axioms will provide consistent data definitions in the KB for specific applications.

3.2. The Knowledge Base

The axioms in the MKB may have several logic models. A particular KB can also be considered as a first-order theory in which the axioms are derived from the data definition statements proved consistent with respect to the MKB.

Stated otherwise, the MKB is used to derive application-specific knowledge, and to control its consistency. Each interpretation of the theory in the MKB produces a KB state. The KB can evolve. It models therefore successive interpretations of the theory. It must also evolve from one consistent state to another.

For CAD/VLSI applications, the KBs are associated for instance to specific representations of a circuit, e.g., logic, symbolic or layout. They

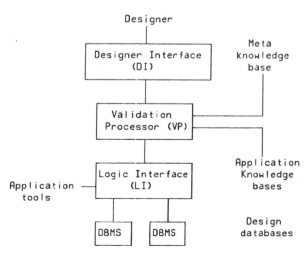

Fig. 3. Validation subsystem.

473

contain the design rules related to the particular technologies used in the design.

3.3. The Database

The application database is an interpretation of the axioms in the KB [6]. It includes for instance the cell library in CAD/VLSI applications. It also contains the information related to the objects being designed.

4. THE VALIDATION SUBSYSTEM

The semantic validation subsystem provides integrity controls for sophisticated applications developed on TIGRE, e.g., Office Automation and CAD. It is built with three components: the designer interface (DI), the database logic interface (LI), and the validation processor (VP).

4.1. Architecture

4.1.1. The designer interface

The designer interface is devoted to the control of the semantics as defined in the MKB. It is used for the correct definition and modification of the application KB. It concerns both the database schema and the integrity constraints. It is implemented in Prolog. The designer addresses queries and updates to the DI by commands that allow him to insert, delete, and update data definitions (Fig. 3).

4.1.2. The database logic interface

The purpose of LI is to control the consistency as defined in the KB. It is triggered by the application programs.

When TIGRE receives a request R, it transfers the control to the valida-tion subsystem after syntactic analysis of R. LI checks its semantic cor-rectness as follows:

- R is translated using a rewriting system,

- R is checked for semantic correctness by controlling the data types involved, the relationship usage, and the functional and multivalued dependencies,

- modification of R by addition of explicit side-effects detected.

4.2. The Meta-Knowledge Base

The MKB contains general knowledge about the data model and meta-rules which define the correct model usage, e.g., permitted modifications on the schema and on the constraints. The definition of the model includes the following specifications:

- basic data type definitions (integer, character strings, etc.),

- compound type definitions (records, arrays, etc.),

- class definitions (entity, relationships, etc.),

- aggregation definitions (union, intersection of class types),

- generalizations and specializations of class types.

For instance, if the user wishes to insert in the KB the explicit definition of the EMPLOYEE class, including their number, name, salary, address, and employment contract, he will submit the following request:

insert(class(EMPLOYEE)entity, attributes(integer(NUMBER), string(NAME), amount(SALARY), string(ADDRESS), contract(DOCUMENT)))))

where predefined data types are in lower-case letters.

Upon receiving the request, DI checks if a corresponding meta-rule exists in the MKB for the insertion of the new class type.

Similarly, the MKB includes:

- derivation rules,

- dynamic constraints definition,

- definition of functional, multivalued, and inclusion dependencies,

- constraint validation strategies.

If the user wants to define an implicit class type HEAD-DEPT, such that a department D is head of a project P if and only if its manager E is attached to D, the following request must be issued:

HEAD-DEPT(D,P): -employee(E,_), project(P,_), manages(P,E), dept(D,_), attached(E,D)

The MKB then invokes the meta-rule defining the valid operations on the KB. There, the following rule is triggered:

implicit(insert(class(HEAD-DEPT(D,P), relationships(employees(E,_), attached(D,E)))))

If the HEAD-DEPT rule does not contradict the existing rules, it is inserted in the KB. This approach allows for the implementation of a validation subsystem combining

- the definition of the concepts in the data model at hand,

- the implementation of semantic integrity controls.

The MKB and the DI allow the user to define and dynamically update a KB, with respect to the specifications of the meta-model. He can thereby define new entities, relationships, and constraints. He can also delete and modify the existing ones in a consistent way.

4.3. The Knowledge Base

An application knowledge base contains a schema and a set of integrity constraints. The schema is obtained by translating the application data structures by ad hoc rewriting systems [3]. The constraints are specific to the application and must be explicitly written by the application designer [1].

The KB enforces a strong (i.e., immediate) consistency in the application data base using deduction.

Specific inconsistencies are allowed in a generation mode. This is of particular importance in CAD applications, because the designers usually want to interact throughout the definition and the instantiation of an object. They are also allowed to propagate their updates through composition relationships [2]. This is detailed in Section VI

5. SEMANTICS CONTROL

The semantics controls are implemented at two different levels: modification of the KB and the data instance level.

5.1. KB Semantics Control

It is implemented by procedures in charge of the processing of the updates. For instance, if new derivation rules are to be inserted, they are checked for contradiction with existing ones. This is done using a database *sample*. It contains a set of logical object *prototypes* which are representatives of object equivalence classes [1, 2]. Two objects are *equivalent* if they satisfy the same set of constraints.

The new rules are tested by applying them to the database schema and the database sample. This guarantees that if the new rules are consistent, they are derivable on the sample.

The validation processor is invoked by the DI to check whether new definitions are derivable on the sample. If not, they are rejected. If they are, they are *certified*.

The sample is maintained automatically along the evolution of the KB. It contains representatives which are consistent with the semantic properties defined in the schema and the integrity constraints. Some of these are the functional and multivalued dependencies, and the cardinalities of roles in relationships relating entities.

5.2. Data Semantics Control

The semantics controls on the data instances are implemented by checking the update requests directed to the KB. They include the control of the data types, the attribute values, and so forth.

6. APPLICATIONS TO CAD/VLSI

An application of the above proposal is being implemented for CAD/VLSI in cooperation with CNET, the French National Research Center for Telecommunications [1, 2].

The definition of the objects are contained in the application KB, including:

- their structures (nets, ports, components),

- the design rules under consideration,

- the equivalence among different object representations,

- the equivalence among different versions of the same object.

An explicit representation of the semantics links between objects is used. It includes *weak* and *strong* relationships that model the equivalence and component relationships. They are used to propagate the updates performed on the object instances. Weakly related objects are independent. Modification of one of them is not visible from another. In contrast, strongly dependent objects are implicitly modified when updating one of them.

7. CONCLUSION

The integration in the same formalism of such different concepts as:

- the semantics of a generalized data model,

- the processing strategies for integrity enforcement,

- the definition of structural relationships among database objects

points to promising steps in the development of Expert Database Systems for sophisticated applications such as CAD [3]. This proposal is relevant to the specific area of CAD. Nonetheless, the theoretical foundations it assumes make it very flexible. Its application to other data models requires the formalization of the concepts they capture in first-order logic. The translation of the data definition statements and of the requests can then be performed by rewriting systems.

A first prototype of the validation subsystem has been implemented on a VAX 11/780 running Unix 4.2 BSD. It includes over 15,000 Prolog clauses. It is now being ported on APOLLO workstations. Current extensions include the definition of the environment, i.e., the sharing of multiple users, and the notions of incompleteness and dynamically evolving data structuers [4].

ACKNOWLEDGEMENTS

The authors are greatly indebted to Dominique Rieu and Marie-Christine Fauvet from IMAG for their suggestions to this proposal.

REFERENCES

1. G. T. Nguyen, Semantic data engineering for generalized databases, Proc. 2nd International Conf. on Data Engineering, Los Angeles, California, February 1986.
2. G. T. Nguyen, Object prototypes and database samples for expert database systems, Proc. 1st International Conf. on Expert Database Systems, Charleston, S. Carolina, April 1986.
3. D. Rieu and G. T. Nguyen, Semantics of CAD objects for generalized databases, Proc. 23rd ACM/IEEE Design Automation Conference, Las Vegas, Nevada, June 1986.
4. G. T. Nguyen and D. Rieu, Expert database concepts for engineering design, Submitted for publication.
5. R. Fagin, Horn clauses and database dependencies, Journal of the ACM, Vol. 29, No. 4, October 1982.
6. H. Gallaire et al., Logic and databases: a deductive approach. ACM Computing Surveys, Vol. 16, No. 2, June 1984.
7. M. Lopez et al., The TIGRE data model, Research Report TIGRE No. 2, IMAG, November 1983.
8. J. M. Nicolas, Logic for improving integrity checking in relational databases, Acta Informatica, No. 18 (1982).
9. R. Reiter, Towards a logical reconstruction of relational database theory. On conceptual modelling, Springer-Verlag (1984).

FREQUENCY SEPARATION ANALYSIS FOR OBJECT

ORIENTED DATABASES

Alex Kostovetsky

Digital Equipment Corporation
MR02
200 Forest St.
Marlboro, MA 01752, U.S.A.

ABSTRACT

The paper proposes a formalism necessary to apply the frequency separation technique to object oriented script driven structures. The frequency separation technique provides a natural and efficient way of decomposing a large database system into a set of smaller manageable subsystems with significant benefits in terms of system performance, availability and reliability.

Object oriented structures, however, pose specific problems for the frequency separation due to intricate and tangled interrelations between the objects. As shown in this paper, this difficulty can be resolved by applying a special linear transformation to the database original update frequency spectrum. The transformation matrix is obtained based on a notion of an event script which is described in detail. The proposed transformation algorithm is illustrated by an example taken from the field of plant design engineering.

INTRODUCTION

In recent years large databases have appeared as a vital instrument in storing and accessing massive arrays of data, in both engineering and transaction management environments. The required augmentation of the database size and growing sophistication of its structure normally create specific problems uncharacteristic for standard size databases [1]. In large scale engineering applications of computers, such as CAD/CAM projects, very large data volumes, 50-100 gigabytes, are not uncommon and have to be maintained in a database for a considerable time. The object oriented approach has been suggested and extensively researched in order to address the problem of economical and efficient storage and greater database responsiveness [2, 3]. The object oriented approach is particularly valuable in engineering and scientific databases as a way to preserve the natural composition of a complex object allowing at the same time a greater degree of evolutionary flexibility.

The frequency separation technique first suggested for large databases in [4] is based on the assumption that the database records are easily separable based on the frequency criterion. This assumption holds very well for network type databases as well as for many relational structures. However, the object oriented database, as a rule, contains so many semantic links among object instances that the frequency separation becomes a tricky issue. This paper suggests a formal algorithm to establish the frequency separability criterion for both transactions and engineering object oriented databases.

1. FREQUENCY SEPARATION APPROACH

The frequency separation approach was suggested as a way to improve various operational characteristics of a large database by dynamically decomposing it into a collection of smaller more manageable subsystems [4]. The technique separates parts of the original database according to their inherent frequencies of access. The separation process produces a hierarchical structure which ranks subsystems with respect to their position in the frequency spectrum. The separation can lead to considerable saving in both access cost and performance as compared to the original unseparated structure [4].

The idea of frequency decomposition can be applied to an object oriented database so that objects are grouped in the database according to their characteristic frequencies. Each group of objects can independently reside on its own dedicated DBMS. In this architecture the task of communicating messages between the frequency separated subsystems should be given to a pivot node. If properly applied, the frequency separation would lead to the following advantages:

- operational reliability of the entire system increased because each subsystem is independently supported;

- transactions with frequency separated group are faster due to the reduced search space;

- recovery and archiving can be handled easily for each subsystem;

- the entire system has a better overall performance and availability since a subsystem's failure will not affect other subsystems.

The obvious price to pay for the frequency decomposed structure is the increased complexity of the database organization. Each updatable unit, i.e., object instance, now has its specific residence in the system, and its address should be included in the software running the database. It is noteworthy that the frequency decomposed structure in the presented form distributes data between independent DBMSs without any replication and, therefore, the decomposition process does not create any concurrency or data communication problem.

In order to manage the more complex database organization brought about by the frquency separation process, one can suggest an additional pivot database. Its functions include:

- object allocation according to its characteristic frequency;

- routing messages between the subsystems;

- interface with the system database administrator (DBA);

- periodic restructuring of the system.

In principle, there is no limit on the database size when the frequency separation technique is used. The decomposition process can be applied as many times as necessary to achieve sufficiently small subsystems.

2 OBJECT ORIENTED DATABASE STRUCTURE

The object oriented database structure provides a way to directly manipulate with meaningful engineering or business entities called objects. Each generic object type defines an object frame. The frame is a semantic template which contains all generic object data inheritable by the object instances. Both the frame and the instances are stored in relational tables, so that each object instance corresponds to a tuple. The tuple carries the instance identifier and all data attributes necessary to describe the object completely.

In addition to storing object frames and instances the database should allow groupings of objects into various supertype objects and associations of object instances. This is achieved by either storing pointers in the attribute fields or by common attributes, such as project name.

Furthermore, in both design and production databases there is a need to preserve consistency of objects and notify appropriate users when an object changes its state. For example, when an object 'door' gets deleted, it should cause an automatic deletion of object instances of a keyhole and a handle that belong to the deleted door. A single change in an object state may trigger multiple consequences through the entire database, and therefore a mechanism is needed to carry out the derivative actions. An action performed on the object database is called an event. An antecedent event performed by the user triggers a sequence of consequent events. This sequence is called an event script. A script may consist of events like deletions, notifications, additions, or displays. The script events are executed sequentially and apply only to appropriate specified object instances in the object database. The scripts should, of course, be user definable and accessible for modification. Obviously, the execution of a script may cause changes in the objects which are not explicitly updated by the user. In other words, the frequency of antecedent update cannot be an accurate measure of an object update frequency and, therefore, the antecedent frequency cannot be used in the frequency decomposition procedure [4]. A prior transformation of the frequency spectrum is needed in order to produce a correct decomposition pattern for object oriented databases.

3. FREQUENCY SPECTRUM TRANSFORMATION

The scripts in the object database are composed and verified by the user; therefore, the exact effect of the scripts is assumed to be known. Consider an object oriented database which consists of N distinct data types

$$O_1, O_2, \ldots O_N.$$

Since each object instance has a script attached to it, one can examine the script and list all the objects whose instances are impacted by the script. For a more formal analysis it is convenient to introduce an $N \times N$ impact matrix S. The elements of this matrix are defined as follows:

$$S_{ij} = \begin{cases} 1, & \text{if a script attached to } O_j \text{ triggers an impact on } O_i \\ 0, & \text{otherwise.} \end{cases} \tag{1}$$

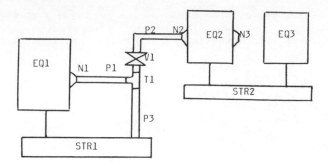

Fig. 1. Example of plant CAD.

Obviously, the diagonal elements S_{ii} of the matrix should be set equal to 1.   Suppose that the antecedent, i.e., user originated event frequency, is w_i and the total frequency resulting from both the user and the scripts is f_i.   Then, f_i can be computed as a sum of its own antecedent frequency w_i and those w_j which impact 0_i in their scripts.   In other words

$$f_i = \sum_{j=1}^{N} S_{ij} w_j \qquad\qquad (2)$$

This formula defines transformation of the antecedent frequency spectrum to the update frequency spectrum for a script driven object database.   This linear transformation should be performed before the standard frequency separation technique is applied to the object database.   The actual frequency separation can now be carried out based on the f_i values [4].

The second order impacts caused by indirect cascading triggering of scripts are not considered in this analysis.

EXAMPLE

As an example of a script driven object oriented CAD database consider the diagram shown in Fig. 1.

The following notations are used:

EQ - equipment
SRT - supporting structure
N   - nozzle
P   - pipe segment
V   - valve
T   - branch

Assume the following values for antecedent (user originated) frequencies in relative units:

EQ: $w_1 = 2$
STR: $w_2 = .5$
N: $w_3 = 4$
T: $w_4 = 6$
V: $w_5 = 6$
P: $w_6 = 10$

By analyzing the object database scripts dictated by the physical interrelationships between the design elements, one can come up with the following impact matrix S:

$$S = \begin{pmatrix} 1 & 1 & 1 & 0 & 0 & 0 \\ 0 & 1 & 0 & 0 & 0 & 0 \\ 0 & 0 & 1 & 0 & 0 & 1 \\ 0 & 0 & 0 & 1 & 0 & 1 \\ 0 & 0 & 0 & 0 & 1 & 1 \\ 0 & 0 & 1 & 1 & 1 & 1 \end{pmatrix} \qquad (4)$$

The frequency spectrum transformation formula (2) yields the following values for the object update frequencies:

$$
\begin{array}{lll}
\text{EQ:} & f_1 = 2 & \\
\text{STR:} & f_2 = 2 + .5 = 2.5 & \\
\text{N:} & f_3 = 2 + 4 = 6 & \\
\text{T:} & f_4 = 6 + 10 = 16 & (5) \\
\text{V:} & f_5 = 6 + 10 = 16 & \\
\text{P:} & f_6 = 4 + 6 + 6 + 10 = 26 &
\end{array}
$$

As one can see from the example, the spectrum transformation can significantly rearrange the object update frequencies, and therefore application of the frequency separation technique would result in a different decomposition pattern as compared to the original spectrum w_i.

CONCLUSION

An implementation of the object oriented script driven database is being currently considered at Computervision. As one can expect in the case of an industrial CAD database, the total number of objects and their attributes will soon reach a point where some kind of data distribution will be necessary. The frequency separation analysis suggested for object oriented databases provides a formalism to address the decomposition issue. The exact software configuration required to implement the decomposition is yet to be determined.

REFERENCES

1. F. Codd, "Extending the Database Relational Model to Capture More Meaning," ACM TODS, 4-4 (December 1979).
2. P. Lyngbaek and D. McLeod, "Object Management in Distributed Information Systems," ACM Trans on Office Info Systems (April 1984).
3. M. Minsky, "A Framework for Representing Knowledge," in: The Psychology of Computer Vision, McGraw-Hill, New York (1975).
4. A. Kostovetsky, "Time Separation Technique for Large Databases," Int. Journal of Computer and Info. Sciences, Vol. 12, No. 3 (1983).

Query Processing
and
Physical Structures
for Relational Databases

INDEX SELECTION IN RELATIONAL DATABASES

Kyu-Young Whang

IBM T.J.Watson Research Center
P.O.Box 218
Yorktown Heights, NY 10598

ABSTRACT

An index selection algorithm for relational databases is presented. The problem concerns finding an optimal set of indexes that minimizes the average cost of processing transactions. This cost is measured in terms of the number of I/O accesses. The algorithm presented employs an approach called *DROP heuristic*. In an extensive test to determine the optimality of the algorithm, it found optimal solutions in all cases. The time complexity of the algorithm shows a substantial improvement when compared with the approach of exhaustively searching through all possible alternatives. This algorithm is further extended to incorporate the clustering property (the relation is stored in a sorted order) and also is extended for application to multiple-file databases.

1. INTRODUCTION

We consider the problem of selecting a set of indexes that minimizes the transaction-processing cost in relational databases. The cost of a transaction[†] is measured in terms of the number of I/O accesses.

The index selection problem has been studied extensively by many researchers. A pioneering work based on a simple cost model appeared in [1]. A more detailed model incorporating index storage cost as well as retrieval and index maintenance cost was developed in [2]. Some approaches [3], [4] attempted to formalize the problem to obtain analytic results in certain restricted cases. In a more theoretical approach, Comer [5] proved that a simplified version of the index selection problem is NP-complete. Thus, the best existing algorithm for finding an optimal solution would have an exponential time complexity. In an effort to devise a more efficient algorithm, Schkolnick [6] discovered that, if the cost function satisfies a property called *regularity,* the complexity of the optimal index-selection algorithm can be reduced to one that is less than exponential. Hammer and Chan [7] took a somewhat different approach and developed a heuristic algorithm that drastically reduced the time complexity. However, no attempt has yet been made to establish the validity of this algorithm.

Although considerable effort has been devoted to developing algorithms for index selection, most past research has concentrated on single-file cases. Furthermore, the problem of how to

---

† The term *transaction* used here should not be confused with the atomic transaction as a unit of consistency and recovery. Here, it is used as a generic term for both queries and update activities against the database.

incorporate the primary structure (the clustering property) of a file is still awaiting a solution. The purpose of this paper is to introduce an index selection algorithm with a reasonable efficiency and accuracy, which is subsequently extended to include multiple-file cases as well as to incorporate the clustering property.

The approach presented in this paper bears some resemblance to the one introduced by Hammer and Chan [7]; but there is one major modification: the DROP heuristic [8] is employed instead of the ADD heuristic [9]. The DROP heuristic attempts to obtain an optimal solution by dropping indexes incrementally, starting from a full index set. On the other hand, the ADD heuristic adds indexes incrementally, starting from an initial configuration without any index, to reach an optimal solution.

Since we are pursuing a heuristic approach for index selection, the actual result is suboptimal. However, in an extensive test performed for validation, the DROP heuristic found optimal solutions in all cases. (On the other hand, the ADD heuristic found suboptimal solutions on several occasions.)

In determining optimality, we consider only the cost of accessing and maintaining the database and indexes; i.e., the cost of storing indexes in the storage medium is not considered. If desired, however, the index storage cost can be incorporated by making it part of the index maintenance cost (or, equivalently, the two costs can be combined to constitute the *overhead cost* as defined in [2]), so that it can fit in the framework of the algorithms to be presented. Thus, the general validity of our approach should remain intact.

We first present the index selection algorithm for single-file databases without the clustering property. This algorithm is tested for validation with 24 randomly generated input situations, and the result compared with the optimal solutions generated by searching exhaustively through all possible index sets. The algorithm is then extended to incorporate the clustering property. Extension to multiple-file cases is considered subsequently.

Section 2 introduces assumptions regarding the cost model and a simple index structure, while Section 3 describes the classes of transactions we consider and their cost formulas. Next, in Section 4, we present the index selection algorithm and its time complexity. In Section 5, we discuss the result of the test performed for the validation of the algorithm. Extension of the algorithm to incorporate the clustering property is the principal topic of Section 6. Finally, in Section 7, we discuss an extension of the algorithm for application to multiple-file databases.

2. ASSUMPTIONS

We assume that the relation is stored in a secondary storage medium, which is divided into fixed-size units called blocks [10]. For simplicity, we assume that a relation is mapped into a single file, an attribute to a column, and a tuple to a record. Accordingly, we shall use the terms *file* and *relation* interchangeably.

In processing a transaction the number of I/O accesses necessary to bring the blocks into the main memory depends on the specific buffer strategy. We assume, however, the following simple strategy: no block access will be necessary if the next tuple (or index entry) to be accessed resides the the same block as that of the current tuple (or index entry); otherwise, a new block access is necessary. We also assume that all TID (tuple identifier) manipulations can be performed in the main memory without any need for I/O accesses.

We assume that a B^+-tree index [12] can be defined for a column of a relation. The leaf level of the index consists of <key, TID-list> pairs for every unique value in that column. Each TID list contains the list of TIDs (or tuple identifiers) of tuples having the same column value. The leaf-level blocks are chained so that the index can be scanned without traversing the index tree. When index entries are inserted or deleted, we assume that splits or concatenations of index blocks are rather infrequent, so that modifications are done mainly on leaf-level blocks.

3. TRANSACTION MODEL

We consider four types of transactions: query, update, deletion, and insertion. The classes subsumed under these types are shown in Figures 1 to 4.

SELECT   <list of columns>

FROM   R

WHERE   P

Figure 1. General class of queries considered.

UPDATE R

SET   R.A = < new value$_A$ >,

SET   R.B = < new value$_B$ >,

.

.

.

WHERE   P

Figure 2. General class of update transactions considered.

DELETE R

WHERE   P

Figure 3. General class of deletion transactions considered.

INSERT   INTO R: <list of column values>

Figure 4. General class of insertion transactions considered.

In figures 1 to 4, "P" stands for the restriction predicate that selects the relevant tuples. We call the columns appearing in P *restriction columns*.

Cost formulas for those transactions are now introduced in the form of functions. In calculating the cost of a query we do not include the cost of writing the result, since that cost is independent of the index set and accordingly irrelevant for optimization purposes.

We define the following notation:

C	A column.
n	Number of tuples in the relation (cardinality).
p	Blocking factor of the relation.
L_C	Blocking factor of the index for column C.
F_C	Selectivity of Column C or of its index.
m	Number of blocks in the relation, which is equal to n/p.

489

t	A transaction
Restricted Set	Set of tuples that satisfy all the restriction predicates. Equivalent to $\left(\prod\limits_{C\epsilon\{\text{all restriction columns}\}} F_C\right) \times n.$
Partially Restricted Set	Set of tuples that satisfy the restriction predicates that can be resolved via indexes. Equivalent to $\left(\prod\limits_{C\epsilon\{\text{all restriction columns having indexes}\}} F_C\right) \times n.$

- **function b(m,p,k):** cost of accessing k randomly selected tuples in TID order

$$b(m,p,k) = m\left[1-\binom{n-p}{k}/\binom{n}{k}\right] \qquad (1)$$
$$= m\left[1-((n-p)!)/((n-p-k)!n!)\right]$$
$$= m\left[1-\prod_{i=1}^{k}(n-p-i+1)/(n-i+1)\right] \qquad \text{when } k \leq n-p, \text{ and}$$

$$b(m,p,k) = m \qquad\qquad\qquad\qquad \text{when } k > n-p.$$

The function is approximately linear when $k \ll n$ and approaches m as k becomes large. Equation 1 is an exact formula derived by Yao [13]. Variations of this function and approximation formulas for faster evaluation are summarized in [14].

- **function IA(C,mode):** cost of accessing a B^+-tree index from the root

A. mode = Query mode
$$IA = \lceil \log_{L_C} n\rceil + (\lceil F_C \times n/L_C \rceil - 1) \qquad (2)$$

B. mode = Insertion mode
$$IA = \lceil \log_{L_C} n\rceil$$

C. mode = Update mode
$$IA = \lceil \log_{L_C} n\rceil + (\lceil 0.5 \times F_C \times n/L_C \rceil - 1)$$

The function IA has three modes, depending on the purpose of accessing the index. In query mode, all the index entries with the same key value are retrieved. The first term in Equation 2 is the height of the index tree, while the second is the number of leaf-level index blocks accessed. One block access is subtracted from the second term since the cost of accessing the first leaf node is already included in the first term. In insertion mode, an index entry corresponding to the inserted tuple is placed after the last entry that has the same key value. Thus, only one leaf-level block will be accessed. This cost, however, is included in the first term. In update mode, the index entries containing the old value have to be searched to find the one with the TID of the updated tuple; thus, on the average, about half of the index entries will be searched.

- **function Query(t):** cost of processing a query

$$Query = b(m,p,|\text{partially restricted set}|) + \sum_{C\epsilon\{\text{all restriction columns having indexes}\}} IA(C,\text{query mode}) \qquad (3)$$

Queries are processed as follows. Indexes of all restriction columns are accessed in query mode to obtain the sets of TIDs that satisfy the corresponding simple restriction predicates. The intersection of these TID sets is formed subsequently to locate tuples in the partially restricted set. These tuples are retrieved and produced as output after the remaining restriction predicates are resolved. The first term in Equation 3 represents the cost of accessing data tuples, the second the cost of accessing indexes.

- **function Update(t):** cost of processing an update transaction

$$\text{Update} = \text{Query}(t) \tag{4}$$

$$+ \ b(m,p, |\text{restricted set}|)$$

$$+ \ |\text{restricted set}| \times 2 \times \sum_{C \in \{\text{all updated columns having indexes}\}} [IA(C, \text{update mode}) + 1]$$

The update cost consists of three parts: the first term of Equation 4 represents the cost of reading in blocks containing the tuples to be deleted, the second term the cost of writing out modified blocks, and the third term the cost of updating corresponding indexes. The third term is again divided into two parts: the cost of deleting index entries for old values and that of inserting index entries for new values. Since these two parts have the same value, a factor of two is introduced. In either part, one block access is added for each index entry modified to account for writing out the updated block. Let us note that, even for insertion of new index entries, the update mode is specified for function IA because index entries having the same key value must be ordered according to their TIDs.

- function Delete(t): cost of processing a deletion transaction

$$\text{Delete} = \text{Query}(t)$$

$$+ \ b(m,p, |\text{restricted set}|)$$

$$+ \ |\text{restricted set}| \times \sum_{C \in \{\text{all columns having indexes}\}} [IA(C, \text{update mode}) + 1]$$

The deletion cost is the same as the update cost except that the third term of the cost function represents the cost of deleting index entries for all existing indexes.

- function Insert(t,Ntuples__inserted): cost of processing an insertion transaction

$$\text{Insert} = \text{Ntuples\_\_inserted}$$

$$\times (1 + 1 + \sum_{C \in \{\text{all columns having indexes}\}} [IA(C, \text{insertion mode}) + 1])$$

Three different mechanisms contribute to the insertion cost: locating the place to insert a new tuple (one I/O access); writing out the modified block access (one I/O access); and modifying all existing indexes accordingly. In the third, function IA is called in insertion mode since the new index entry is always added at the end of the list of index entries that have the same key value.

4. INDEX SELECTION ALGORITHM (DROP HEURISTIC)

Input:

- Usage information: A set of various query, update, insertion, and deletion transactions with their relative frequencies.

- Data characteristics: Relation cardinality, blocking factor, selectivities, and index blocking factors of all columns.

Output:

- A near-optimal index set.

Algorithm 1:

1. Start with a full index set.

2. Try to drop one index at a time and, applying the cost functions, obtain the total transaction-processing cost to find the index that yields the maximum cost benefit when dropped.

3. Drop that index.

4. Repeat Steps 2 and 3 until there is no further reduction in the cost.

5. Try to drop two indexes at a time and, applying the cost functions, obtain the total transaction-processing cost to find the index pair that yields the maximum cost benefit when dropped.

6. Drop that pair.

7. Repeat Steps 5 and 6 until there is no further reduction in the cost.

8. Repeat Steps 5, 6, and 7 with three indexes, four indexes, ..., until there is no further improvement. The number of indexes considered together at the termination of the algorithm is denoted as k.

The variable k is the maximum number of indexes that produce incremental cost benefits when dropped together at a time. We need to consider dropping more than one index together because the presence of an index may have influence on the benefit of having other indexes. This interaction occurs when more than one column appears in the predicate specified in a transaction. For example, in processing a query having a predicate, (Column A = 'a' AND Column B = 'b'), the selectivity of either conjunct may not significantly reduce the number of blocks to be accessed; nevertheless, the joint selectivity of the two might. Let us note that this situation can occur because of nonlinearity of function $b(m,p,k)$, which returns the number of blocks to be accessed to retrieve a given number of tuples. When this happens, dropping either index alone may cause a heavy penalty in cost because the other index alone is not very useful, whereas dropping both together may produce benefit because maintenance costs of both indexes are eliminated for the same amount of loss in benefit.

The time complexity of the algorithm is $O(g \times v^{k+1})$, where g is the number of transactions specified in the usage information, v the number of columns in the relation, and k the maximum number of columns considered together in the algorithm when it terminates. The time complexity is estimated in terms of the number of calls to the cost evaluator, which is the costliest operation in the design process. In the algorithm, the cost evaluator is called for every k-combination of columns of the relation and for every transaction in the usage information. This contributes the order of $g \times v^k$. The procedure is repeated until there is no further reduction in the cost. Since the number of iterations is proportional to v, the overall time complexity is $O(g \times v^{k+1})$.

5. VALIDATION OF THE ALGORITHM

An important task in developing heuristic algorithms is their validation. In this section the result of an extensive test performed to validate the index selection algorithm (DROP heuristic) will be presented. In particular, we try to measure the deviations of the heuristic solutions from the optimal ones for various input situations generated by using different parameters. (These parameters are chosen from the ranges that are important in practical applications.) Optimal solutions are obtained by searching exhaustively through all possible alternatives (2^v combinations). We use v=10 for all test cases.

The input situations are generated as follows:

1) Two sets of the relation cardinality and column cardinalities are used: in the first set the relation cardinality is 1000; in the second it is 100,000. The column cardinalities are randomly generated between 1 and the relation cardinality, with a logarithmically uniform distribution.

2) Two sets of blocking factor and index blocking factor pairs are used: 1) 10 and 100; 2) 100 and 1000.

3) An input situation includes 30 transactions and their relative frequencies. Among them are 21 queries, 4 to 5 update transactions, 3 to 4 deletion transactions, and 1 insertion transaction. Using this template, three different sets of transactions are

created to provide different mixture of transactions. For each set, transactions are randomly generated as follows: for queries and deletion transactions, 1 to 3 (this number is randomly selected for each transaction) columns are randomly selected as restriction columns; for update transactions, 1 to 3 (this number is also randomly selected) columns are randomly selected as updated columns, and another randomly selected set of columns as restriction columns.

4) Two sets of relative frequencies are used. In the first set all transactions initially have identical frequencies. Later, the frequencies of deletion and insertion transactions are multiplied by an adjustment factor to keep the number of indexes in the result between 3 and 7. This adjustment is made to avoid extreme cases in which a full index set or an empty index set is the optimal solution. For the second, the relative frequencies of transactions are randomly generated between 100 and 500, with an interval of 50 between adjacent values.

The scheme described above generates 24 different input situations with varying statistics of the database and usage information as well as random mixtures of transactions. An example situation is shown in Figure 5 in an abbreviated form. The test results for both DROP and ADD heuristics are summarized in Table 1. In the first column of Table 1, the first digit of the input situation number represents the set of the relational cardinality, the second the set of the main-file blocking factor and the index blocking factor, the third the set of transactions, and the last the set of relative frequencies of transactions. The second column of the table shows the number of indexes present in the optimal solution. The CPU time shows the performance of the algorithms when run on a DECSYSTEM-2060. Percentage deviations are shown for the situations in which any deviation occurred. Marked by "opt" are the situations in which optimal solutions were found.

In all situations tested, the DROP heuristic found optimal solutions. Although the test is by no means exhaustive, the result is a good indication that the DROP heuristic will perform well in many practical situations. In comparison, the ADD heuristic produced suboptimal solutions in six cases; the maximum deviation encountered was 21.17%.

The reason why the ADD heuristic does not perform as well as the DROP heuristic is the following. With the ADD heuristic, a potentially most important index is selected first. Since the presence of an index affects selection of other indexes in such a way as to maximize its benefit, selection of the first index is tantamount to dictating the solution finally to be generated. Thus, a possible suboptimal decision made at the beginning may well persist in the result. On the other hand, in the DROP heuristic, the least significant index is dropped first. Accordingly, even if the decision made at the beginning is not optimal, it is unlikely that the resulting solution is significantly affected. Example 1 illustrates this point further.

Example 1: Figure 6 shows intermediate index sets at each step of the ADD heuristic and the DROP heuristic. For convenience, we chose a relation with only four columns. The symbol '1' in the figure indicates the presence of an index, and 'X' its absence. Only the first two iterations of the design process are shown since there is no more improvement in the third.

In this example, the DROP heuristic found the optimal solution. The ADD heuristic, however, resulted in a slight deviation from the optimal. Compared with the optimal solution, the solution that the ADD heuristic produced has an index on Column 1, but lacks one on column 4. Column 1 was assigned an index during the first iteration because the index set (1 X X X) was less costly than (X X X 1); this index subsequently stayed until the algorithm terminated. The index on Column 1 is absent in the optimal solution, however, since the presence of indexes on columns 3 and 4 renders it less significant than it would be without them. □

This example shows the error caused by selecting an insignificant index that looks as if it were important at the initial stage of the design using the ADD heuristic because the interaction among indexes on different columns has not been well established. In contrast, since all indexes are initially present in the DROP heuristic, the influence of an index on others is taken into account from the beginning, thereby reducing the probability of reaching an incorrect solution.

```
!Input Situation 2132!
Schema
     Relations
          Relation                    R
               Relcard               100000
               Nblocks               10000
               Blkfac                10

               Column                C1
                    Colcard          409
                    Niblk            1000
                    Iblkfac          100

               Column                C2
                    Colcard          1333
                    Niblk            1000
                    Iblkfac          100

                    .
                    .
                    .

               Column                C10
                    Colcard          328
                    Niblk            1000
                    Iblkfac          100

Usage
     Transaction 1
          Type      SQ        FREQ    500
          Select    R.C1
          From      R
          Where     R.C7 ="a"         AND
                    R.C10="b"

     Transaction 2
          Type      SQ        FREQ    100
          Select    R.C1
          From      R
          Where     R.C6 ="a"         AND
                    R.C8 ="b"         AND
                    R.C9 ="c"

     Transaction 3
          Type      SQ        FREQ    200
          Select    R.C1
          From      R
          Where     R.C3 ="a"         AND
                    R.C4 ="b"         AND
                    R.C9 ="c"

     Transaction 4
          Type      SQ        FREQ    100
          Select    R.C1
          From      R
          Where     R.C6 ="a"

               .
               .
               .

     Transaction 24
          Type      SU        FREQ    300
          Update    R
          Set       R.C3 ="a",
          Set       R.C6 ="b"
          Where     R.C8 ="c"         AND
                    R.C5 ="d"

               .
               .
               .

   · Transaction 29
          Type      SD        FREQ    200
          Delete    R
          Where     R.C7 ="f"         AND
                    R.C4 ="g"

     Transaction 30
          Type      INS       FREQ    150
          Insert    INTO     R:
                    <"a1","a2","a3","a4","a5","a6","a7","a8","a9","a10">
```

Figure 5. An input situation.

Table 1. Accuracy and Performance of the Index Selection Algorithm.

Input Situation	Number of Indexes	CPU time(seconds) / Deviation(%)					
		DROP Heuristic		ADD Heuristic		Ex. Search
1111	7	2.3	opt	2.0	0.21	36
1112	6	2.2	opt	2.1	opt	36
1121	6	2.4	opt	2.0	1.23	37
1122	6	2.3	opt	2.1	1.17	37
1131	6	2.5	opt	2.1	opt	39
1132	7	2.5	opt	2.1	1.17	39
1211	5	3.1	opt	1.7	opt	32
1212	3	1.9	opt	1.6	opt	31
1221	4	2.1	opt	1.7	opt	32
1222	5	2.0	opt	1.7	opt	32
1231	4	2.3	opt	1.7	opt	35
1232	5	2.2	opt	1.7	opt	35
2111	4	2.4	opt	2.1	16.71	39
2112	5	2.5	opt	2.1	21.17	40
2121	6	2.3	opt	2.0	opt	38
2122	7	2.5	opt	2.0	opt	37
2131	6	2.6	opt	2.2	opt	40
2132	6	2.7	opt	2.2	opt	40
2211	6	2.6	opt	2.0	opt	36
2212	4	2.4	opt	1.9	opt	36
2221	6	2.4	opt	1.9	opt	34
2222	6	2.3	opt	1.9	opt	34
2231	5	2.4	opt	2.0	opt	38
2232	5	2.4	opt	2.0	opt	38

Algorithm	ADD heuristic				DROP heuristic			
Column	1	2	3	4	1	2	3	4
Initial	X	X	X	X	1	1	1	1
Iteration 1	1	X	X	X	1	X	1	1
Iteration 2	1	X	1	X	X	X	1	1

Figure 6. Intermediate index sets during a design process.

As we can see in Table 1, an exhaustive search takes excessive computation time; in comparison, the DROP heuristic is far more efficient without significant loss of accuracy. Obviously, for larger input situations, the exhaustive-search method will become prohibitively time-consuming. In these cases, heuristic algorithms such as the DROP heuristic may be the only ones applicable.

6. INDEX SELECTION WHEN THERE IS A CLUSTERING COLUMN

In this section we extend the index selection algorithm to incorporate the clustering property. We present two algorithms for this extension.

Algorithm 2:
1. For each possible clustering column in the relation, perform index selection.
2. Save the best configuration.

Algorithm 3:
1. Perform index selection with the clustering column determined in Step 2 of the last iteration. (During the first iteration it is assumed that there is no clustering column.)
2. Perform clustering design with the index set determined in Step 1. The clustering property is assigned to each column in turn, and then the best clustering column is selected.
3. Steps 1 and 2 are iterated until the improvement in cost through one loop cycle is less than a predefined value (e.g., 1%).

Algorithm 2 is a pseudoenumeration since index selection is repeated for every possible clustering-column position. Naturally, Algorithm 2 has a higher time complexity than Algorithm 3, but has a better chance of finding an optimal solution. Both algorithms have been implemented and tested as a part of Physical Database Design Optimizer–an experimental system for developing various heuristics for the multiple-file physical database design [15]. In most cases tested the algorithms found optimal solutions. (The validation of these algorithms is combined with those of Algorithms 4, 5, and 6 in Section 7.) Let us note that the cost formulas have to be modified to take the clustering column into account. A complete set of cost formulas for multiple-file relational databases with the clustering property can be found in [16].

7. INDEX SELECTION FOR MULTIPLE-FILE DATABASES

We present, in this section, an extension of the index selection algorithm for application to multiple-file databases. The extended algorithm (Algorithm 4) is almost identical to Algorithm 1 except for the following:

1) The entire database is designed all together. This is done by treating all columns in the database uniformly, as if they were all in a single relation.

2) Clustering columns are incorporated by a technique similar to the one employed in Algorithm 3. In addition, multiple clustering columns are allowed with the restriction that at most one can be assigned to a relation. Accordingly, clustering design is repeated until as many clustering columns as are beneficial in reducing the overall cost are assigned to the database.

Let us note that, when considering a transaction involving more than one relation, the optimizer [17],[18] has to be invoked to find the optimal sequence of access operations as well as to determine the cost of evaluating the transaction.

Algorithm 4 has also been implemented and tested as a part of the Physical Database Design Optimizer. For the purpose of comparison, we now briefly introduce other multiple-file physical database design algorithms (Algorithms 5 and 6), which are based on the property of separability. The results of the tests for their validation are then compared with those for Algorithm 4.

The separability approach was proposed by Whang, Wiederhold, and Sagalowicz in [19] and, subsequently, in [20] and [21]. Other work based on this approach appeared in [22]. The separability-based approach enables the physical design of the entire database to be performed relation by relation independently of one another (we call this phase of design *Phase 1*)—if certain conditions are met. Features that violate these conditions can be incorporated by adding an adjustment step (which we call *Phase 2*) during each iteration. Thus, Phase 1 of Algorithm 5 is identical to Algorithm 2, and that of Algorithm 6 to Algorithm 3. The description of Phase 2, however, is beyond the scope of this paper and will not be discussed further. Interested readers are referred to the reference [15].

The three multiple-file design algorithms were tested extensively. The input situations tested consisted of seven schemas, each of which was accompanied by three variations of usage specification generated as follows. First, the transactions and their frequencies were defined in such a way that by intuition they looked most "natural". Second, according to the test results from the first usage specification, the frequencies were modified so that the costs of individual transactions were of the same order. This modification prevented a few most costly transactions from dominating the results of the design. Third, all the queries were eliminated from the usage specification leaving only update, insertion, and deletion transactions. This modification simulated a situation in which there was a high frequency of updates.

The test schemas were selected with various statistics. Among them, four were arbitrarily chosen, while the remaining three were extracted from the Ships-Monitoring-Database–a research vehicle for the Knowledge-Base Management Systems (KBMS) Project [23], [24] at Stanford University. Two schemas were defined as small subsets with the third encompassing the entire KBMS database. The skeleton of the KBMS schema is shown in Figure 7, using the notation defined in the Structural Model [25]. In Figure 7, the symbol --> represents a many-to-one relationship between relations, and --* a one-to-many relationship with different structural constraints.

The results of the tests thus obtained are summarized in Table 2. In the first column the first digit of the input situation number represents the schema, and the second the usage input. In the description, r stands for the number of relations, c the number of columns in the database, and t the number of transactions in the usage input. The CPU time shows the performance of the algorithms when run in a DECSYSTEM-2060. Marked by "*" are the situations in which any deviation occurred.

Optimal solutions were obtained by running the exhaustive-search algorithm. For Situations 70, 71, and 72, where exhaustive search was nearly impossible, however, the results of three design algorithms were compared; if they produced the same result, it was considered to be optimal. According to this criterion, in most situations tested, all three algorithms produced optimal solutions. Even in the situations that produced suboptimal solutions, the deviations (max. 6.6%) were far from being significant.

As we can see in Table 2, Algorithm 4 performs well with reasonable efficiency. Compared with the exhaustive search algorithm, it takes a negligible amount of time to complete the design without significant loss of accuracy. For a very large database (for example, one consisting of 250 relations and 5000 columns), however, even Algorithm 4 can become intolerably time-consuming. In these cases, Algorithms 5 and 6, which are based on the separability property, are the only algorithms applicable. Indeed, when a very large database is involved, the entire database design somehow has to be partitioned to achieve a reasonable performance in the design process.

Nevertheless, Algorithm 4 has its own advantages. Because it does not require that the database management system satisfy the conditions for separability, it can be easily implemented on top of any relational system that supports indexes and clustering columns, although, for the other algorithms, the system should satisfy these conditions as closely as possible to achieve better accuracy, and any violations of the conditions should be explicitly identified through analysis. Besides, it is worth mentioning once again that the performance of Algorithm 4 falls into a practically feasible range, especially when small to moderate-sized databases are considered.

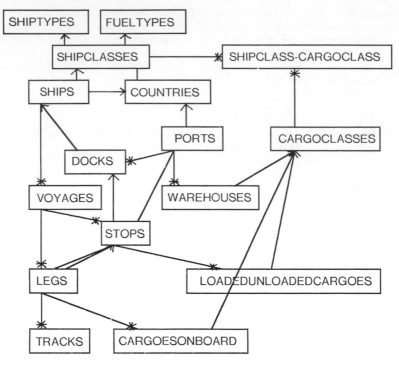

Figure 7. The KBMS schema.

8. SUMMARY AND CONCLUSION

Algorithms for the optimal index selection in relational databases have been presented. Algorithm 1, which employs the DROP heuristic, has been introduced for single-file databases and compared with the ADD heuristic. Test results show that the DROP heuristic produces more accurate results than the ADD heuristic.

The index selection algorithm using the DROP heuristic has been extended to incorporate the clustering property and to include cases for multiple-file databases (Algorithm 4). Algorithm 4 has subsequently been tested for validation and compared with other algorithms for multiple file databases including the exhaustive-search. The result shows that Algorithm 4 takes a negligible amount of time with no significant loss of accuracy in comparison with exhaustive search; furthermore, although it is not as fast as the other two algorithms, its performance is in a practically usable range.

Index selection has long been a subject of intensive research. Nevertheless, no successfully validated and reasonably efficient algorithm has been reported. We believe that our approach provides useful, easy-to-implement, and reliable algorithms for practical applications.

ACKNOWLEDGMENTS

The author wishes to thank Ravi Krishnamurthy and Steve Morgan for reading earlier versions of this paper and providing many thoughtful comments.

Table 2. Accuracy and Performance of Multiple-File Design Algorithms

Input Situation	Description	CPU time(s:seconds;m:minutes;h:hours;y:years)			
		Algorithm4	Algorithm5	Algorithm6	Ex.Search
10	2r, 6c, 7t	1.83s	1.25s	0.86s	26.91s
20	4r, 9c, 10t	5.41s	1.48s	1.23s	36.75m
30	4r, 12c, 12t	10.51s	3.44s	2.09s	13.93h
40	4r, 11c, 13t	6.62s	2.73s	2.04s	3.65h
50	5r, 12c, 12t	12.51s	4.89s	2.32s	* 25.85h
60	4r, 11c, 15t	13.93s	3.54s	2.63s	8.52h†
70	16r,110c, 81t	2.00h	4.80m	1.63m	10^{35} y†
11	2r, 6c, 7t	1.81s	1.26s	0.84s	26.46s
21	4r, 9c, 10t	5.91s	1.67s	1.35s	42.83m
31	4r, 12c, 12t	10.67s	3.43s	2.17s	14.00h
41	4r, 11c, 13t	9.90s	1.88s	1.42s	3.62h
51	5r, 12c, 12t	13.13s	5.00s	3.54s	26.63h
61	4r, 11c, 15t	21.51s	3.74s	2.71s	8.04h†
71	16r,110c, 81t	2.02h	4.60m	2.13m	10^{35} y†
12	2r, 6c, 5t	1.23s	0.86s	0.57s	17.23s
22	4r, 9c, 5t	1.50s	0.55s	0.43s	10.43m
32	4r, 12c, 6t	4.65s	1.73s	1.08s	5.95h
42	4r, 11c, 6t	0.95s *	0.43s	0.25s	* 29.95m
52	5r, 12c, 8t	5.04s	2.41s	1.49s	9.95h
62	4r, 11c, 6t	3.72s	1.81s	1.23s	2.12h†
72	16r,110c, 38t	24.40m	1.77m	21.76s	10^{35} y†

† Values are estimated.
* Situations that produced nonoptimal solutions.

REFERENCES

[1]   Lum, V.Y. and Ling, H., "An optimization problem of the selection of secondary keys," in *ACM Natl. Conf.,* ACM, 1971, pp.349-356.

[2]   Anderson, H.D. and Berra, P.B., "Minimum cost selection of secondary indexes for formatted files," *ACM Trans. Database Systems,* Vol. 2, No. 1, March 1977, pp. 68-90.

[3]   King, W.F., "On the selection of indices for a file," IBM Research Report RJ1341, IBM, San Jose, Calif., 1974.

[4]   Stonebraker, M., "The choice of partial inversions and combined indices," *Intl. Journal of Computer Information Sciences,* Vol. 3, No. 2, 1974, pp.167-188.

[5]   Comer, D., "The difficulty of optimum index selection," *ACM Trans. Database Systems,* Vol. 3, No. 4, Dec. 1978, pp.440-445.

[6]   Schkolnick, M., "The optimal selection of secondary indices for files," *Information Systems,* Vol. 1, March 1975, pp.141-146.

[7]   Hammer, M. and Chan, A., "Index selection in a self-adaptive database management system," *Proc. Intl. Conf. on Management of Data,* Washington, D.C., ACM SIGMOD, June 1976, pp. 1-8.

[8]   Feldman, E., et al., "Warehouse location under continuous economies of scale," *Management Science,* Vol. 12, No. 9, July 1966, pp.670-684.

[9]   Kuehn, A.A. and Hamburger, M.J., "A heuristic program for locating warehouses," *Management Science,* Vol. 10, July 1963, pp. 643-657.

[10]   Wiederhold, G., *Database Design,* McGraw-Hill Book Company, New York, 1983.

[11] Demolombe, R., "Estimation of the number of tuples satisfying a query expressed in predicate calculus language," in *Proc. Intl. Conf on Very Large Databases,* Montreal, Canada, 1980, pp. 55-63.

[12] Comer, D., "The ubiquitous B-tree," *ACM Trans. Database Systems,* Vol. 11, No. 2, June 1979, pp.121-137.

[13] Yao, S.B., "Approximating block accesses in database organizations," *Commun. ACM,* Vol. 20, No. 4, 1977, pp. 260-261.

[14] Whang, K., Wiederhold, G., and Sagalowicz, D., "Estimating block accesses in database organizations–a closed noniterative formula," *Commun. ACM,* Vol. 26, No. 11, Nov. 1983, pp. ⁻940-944.

[15] Whang, K., "A physical database design methodology using the property of separability," Ph.D. dissertation, Stanford University, Stanford, Calif., 1983, Rep. No. STAN-CS-83-968.

[16] Whang, K., "Transaction-processing costs in relational database systems," IBM Res. Rep. RC10952, Jan. 1985.

[17] Selinger, P.G. et al., "Access path selection in a relational database management system," in *Proc. Intl. Conf. on Management of Data,* Boston, Mass., May 1979, pp.23-34.

[18] Kooi, R. and Frankforth, D., "Query optimization in INGRES," *IEEE Database Engineering Bulletin,* Vol. 5, No. 3, Sept. 1982, pp. 2-5.

[19] Whang, K., Wiederhold, G., and Sagalowicz, D., "Separability—an approach to physical database design," in *Proc. Intl. Conf. on Very Large Data Bases,* Cannes, France, IEEE, Sept. 1981, pp. 320-332.

[20] Whang, K., Wiederhold, G., and Sagalowicz, D., "Separability–an approach to physical database design," *IEEE Transactions on Computers,* Vol. C-33, No. 3, Mar. 1984, pp. 209-222.

[21] Whang, K., Wiederhold, G., and Sagalowicz, D., "The property of separability in physical design of network model databases," *Information Systems,* Vol. 10, No. 1, 1985, pp. 57-63.

[22] Bonfatti, F., Maio, D., and Tiberio, P., "A separability-based method for secondary index selection in physical database design," in *Methodology and tools for data base design,* Chapter 6, North-Holland, 1983.

[23] Wiederhold, G., Kaplan, S.J., Sagalowicz, D., "Physical database research at Stanford," *IEEE Database Engineering Bulletin,* Vol. 5, No. 1, Mar. 1982.

[24] Wiederhold, G. and Milton, J., "Knowledge-based management dystems project," *IEEE Database Engineering Bulletin,* Vol. 6, No. 4, Dec. 1983.

[25] Wiederhold, G. and El-Masri, R., "The Structural Model for database design," In *Proc. Intl. Conf. on Entity Relationship Approach,* Los Angeles, Calif., Dec. 1979, pp. 247-267.

A PHYSICAL STRUCTURE FOR EFFICIENT PROCESSING OF RELATIONAL QUERIES

Elisabetta Grazzini   Renzo Pinzani   Fabio Pippolini

University of   University of   I.A.M.I.-C.N.R.
Florence   Florence   Florence
Italy   Italy   Italy

INTRODUCTION

The query processing subsystem of a relational Data Base Management System (DBMS) is often called the query optimizer. The degree of sophistication of a query optimizer design critically affects the performance of the DBMS. This is true even if the query optimizer is probably the most difficult subsystem to implement because each of the high-level database operations that the relational model of data allows may be implemented in more than one way; the choice of optimal algorithms depends on the characteristics of both the stored data and of the query itself.

During the past decade, a variety of algorithms for query optimization were proposed. Most of the studies regarding relational query optimization are concerned exclusively with minimizing the processing cost of a single part of each query.

Transforming algebraic expressions is the base of most strategies for query processing. These strategies aim at the reordering of the query tree by moving selections as far down the parse tree as possible, see Palermo (1974).

A variety of other useful manipulations are suggested such as alternative ways of computing joins. For example, the approach to rearranging the operations, proposed in Dayal (1983), defines the class of conjunctive generalization queries and describes four tactics for processing these queries. Other tree-manipulation rules for queries including outerjoins are described in Reiner and Rosenthal (1984).

Another approach consists of designing efficient algorithms for relational algebra operators. Particular attention has been devoted to joins as, for example, in Gotlieb (1975). In a more recent paper algorithms for relational algebra and set operations based on hashing are presented, see Bratbegsengen (1984). Krishnamurthy and Morgan (1984) present a two-step processing strategy that requires a single sequential scan of the database. In the first step a reduction scheme is used to find a subset of the database for each query which can fit into the main memory; in the second step the answer to the query is computed without further access to secondary storage. The optimizer developed in Yao (1979) for the selection of access strategies takes into account detailed database storage structures. The result is used to drive a query subsystem which generates the required access strategy from a small set of access operators provided by the model itself. In particular whenever possible the optimizer uses indices to access data and therefore produces a set of pointers. The records located by the pointer are read into the main memory and then processed. As a result some records that are not used in constructing the output for the query are also read into the main memory.

Another important algorithm is the QUEL decomposition algorithm, proposed by Wong and Youssefi (1976). In addition to using the idea that selections should be performed as early as possible, the QUEL algorithm uses an appropriate decomposition of Cartesian products and joins, and it takes advantage of the associativity and commutativity of the product for choosing a good way of ordering these products.

In this paper, we describe a new approach to query processing. First of all, we define a new operation on relations, called "Universal operation", and we prove that each relational operation can be expressed in terms of this new operation and all queries can be reduced to a sequence of Universal operations. We then propose a new storage structure for indices which allows us to perform a Universal operation efficiently. By using the Universal operation and the storage structure for indices, we can answer a query simply by reading and processing some sets of pointers to the tuples. The tuples are read into the main memory only at the final stage, i.e., when the resulting relation is produced.

In section 2, we introduce some definitions, the properties of the Universal operation and the method to be used in order to transform a query in a sequence of Universal operations. Section 3 describes a possible physical structure for the efficient implementation of the Universal operation; the execution cost of the Universal operation is defined as a function of some parameters. The actual values of parameters proposed depend on applications and the hardware architecture involved. Section 4 is devoted to comparing our procedure with a traditional one.

THE UNIVERSAL OPERATION AND ITS PROPERTIES

A relational data base is made up of a series of relations, represented by R, S, T,.... . R(A,B, ...) is the relation scheme for R. R.A is the set of the values associated with A in the tuples of R. IRI is the cardinality of R, that is, the number of tuples in R. IR.AI is the size of R.A, that is the number of different values in R.A. The k elements of a finite set X are here denoted by x_1, x_2, ..., x_k.

Let Q be a query on the sub-database made up of relations R_1, $R_2,R_3,....,R_n$. The tuples in the answer to Q are the tuples of R_1 x R_2 x x R_3 x .... x R_n satisfying a given predicate P. For sake of simplicity we assume P=p_1 and p_2 and ... and p_m, where and stands for the logical and operator. The simple predicates p_i may be unary or binary predicates. An unary predicate is of the R.A=v form where v is a constant value, a binary predicate is of the R.A=S.B form. Typically, unary predicates require select operations and binary predicates require join operations; R[R.A=x] means to select from R the tuples having the given value x in the attribute A; analogously (RxS)[R.A=S.B] means to select from Cartesian product RxS the tuples having the same value in the attributes A and B.

In the sequel we will show that the hypothesis about the predicate P and the equality operator are not a limit when our approach is used for evaluating relational queries. An example of query may be:

(Q)   (R[R.A=x and R.B=y]xS[S.C=z])[R.D=S.E]

which corresponds to select from R the tuples satisfying both predicates R.A=x and R.B=y, to select from S the tuples satisfying predicate S.C=z, then to join the so obtained subrelations by using the R.D=S.E predicate.

We assume that:
- for any relation R there exists a biunivocal function f:R--->I where I is the set of integers 1,2,..,q where q≥IRI. Number i such as f(r)=i is the tuple-identifier (TID) of tuple r belonging to R; f(R) is the set of TIDs associated with all the R tuples;
- relation R is partitioned according to its attribute A. A partition set

is represented as P<R.A=v> and is made up of the tuples having the same value v in the attribute A.

If R[p] is the set made up of the tuples in R satisfying predicate p and if X=R.A ∩ S.B, we have:

(1) P<R.A=x>=R[R.A=x]

(2) $(RxS)[R.A=S.B] = \bigcup_{x \varepsilon X} R[R.A=x]xS[S.B=x] = \bigcup_{x \varepsilon X} P<R.A=x>xP<S.B=x>$

We define the Universal operation. Let be:

$X=X_1xX_2x....xX_s \subseteq R_1xR_2x....xR_s$ and $Y=Y_1xY_2x....xY_k \subseteq \underline{R_1}xR_2x....xR_k$
where s≥1 and k≥1; if there exists h≥0 so that $R_i=\underline{R_i}$ for each i=1,2,..,h, the Universal operation between X and Y, represented as X&Y, is so defined

$X\&Y=$ $\{<x_1,x_2,....,x_h,x_{h+1},.....,x_s,y_{h+1},.....,y_k>$ where
$<x_1,x_2,....,x_h,x_{h+1},....,x_s>\varepsilon X,$
$<y_1,y_2,....,y_h,y_{h+1},.....,y_k>\varepsilon Y,$
$x_i=y_i$ for each i=1,2,...,h\}

We note that when h=0 there is no relation common to $R_1xR_2x......xR_s$ and to $\underline{R_1}xR_2x.......x\underline{R_k}$. In such a case we set X&Y equal to XxY. In particular cases, we have also:

(a) R[R.A=x and R.B=z] = R[R.A=x]&R[R.B=z]

(b) (R[R.A=x]xS)[R.B=S.C] = R[R.A=x]&(RxS)[R.B=S.C]

(c) ((RxS)[R.A=S.B]xT)[R.C=T.D] = (RxS)[R.A=S.B]&(RxT)[R.C=T.D]

By using (a), (b) and (c) we can represent a query as a sequence of &-operations applied to subrelations of either the R[R.A=x] or the (RxS)[R.A=S.B] form. For example query (Q) becomes:

R[R.A=x]&R[R.B=y]&S[S.C=z]&(RxS)[R.D=S.E].

The &-operation possesses the following properties:

(3) R[R.A=x and R.B=y] = R[R.A=x]&R[R.B=y] = P<R.A=x>&P<R.B=y>

(4) $(R[R.A=z]xS)[R.B=S.C]=R[R.A=z]&(RxS)[R.B=S.C] =$
$=P<R.A=z>\&\bigcup_{x \varepsilon X}(P<R.B=x>xP<S.C=x>)=\bigcup_{x \varepsilon X}(P<R.B=x>\&P<R.A=z>)xP<S.C=x>$
where X=R.B ∩ S.C

(5) $((RxS)[R.A=S.B]xT)[R.C=T.D]=(RxS)[R.A=S.B]\&(RxT)[R.C=T.D]=$
$=\bigcup_{x \varepsilon X}(P<R.A=x>xP<S.B=x>)\&\bigcup_{z \varepsilon Z}(P<R.C=z>xP<T.D=z>)$
$=\bigcup_{x \varepsilon X}\bigcup_{z \varepsilon Z}(P<R.A=x>\&P<R.C=z>)xP<S.B=x>xP<T.D=z>$
where X=R.A ∩ S.B and Z=R.C ∩ T.D.

If l<R.A=x> is the TID list of tuples in R satisfying predicate R.A=x, we have:

(6) f(P<R.A=x>) = l<R.A=x>

(7) f(P<R.A=x>)xf(P<S.B=z>) = l<R.A=x>xl<S.B=z>

(8) f(P<R.A=x>&P<R.C=z>) = l<R.A=x> ∩ l<R.C=z>

The method we propose for answering relational queries is the following:

(i)   for each unary predicate of an R.A=a type, we must determine the list of TIDs l<R.A=a> (see properties (1) and (6));

(ii) for each binary predicate of an R.A=S.B type, we must determine the set of TID list pairs l<R.A=x> and l<S.B=x> for each x∈R.A ∩ S.B (see properties (2) and (6));

(iii) for each combination of predicates of an R.A=a and an R.B=b type, we must calculate the intersection of TID lists l<R.A=a> and l<R.B=b> (see properties (3) and (8));

(iv) for each combination of two predicates, the one of R.A=a type and the other of R.B=S.C type, we must calculate the intersections between l<R.A=a> and l<R.B=x> for each x∈R.B ∩ S.C (see properties (4) and (8));

(v)   for each combination of two predicates, the one of R.A=S.B type and the other of S.C=T.D type, we must calculate the intersection between l<R.A=x> and l<R.B=z> for each x∈R.A ∩ S.B and for each z∈S.C ∩ T.D (see properties (5) and (8));

(vi)   the final result of steps (i)-(v) is a set of n-tuples of TID lists.
        Each list l_i is associated with the R_i relation in the result.
        Each n-tuple of lists is the image of a subrelation in the result.
        Namely, $f^{-1}(l_1) \times f^{-1}(l_2) \times \ldots \times f^{-1}(l_n)$ is the subrelation obtained
        from lists l_1, l_2,...,l_n. The union of all the subrelations so
        obtained is the answer to the query.

Let a temporay relation be the f-image of a subrelation of either the
R[R.A] or the (RxS)[R.A=S.B] form. The basic operation to be performed in
order to apply the preceding procedure are:
U-operation
        Compute the temporary relation associated to an unary predicate, i.e.
compute one TID list.
B-operation
        Compute the temporary relation associated to a binary predicate, i.e.
compute a set of pairs of TID lists.
UU-operation
        Compute the &-operation between two temporary relations, both of them
are associated to an unary predicate; i.e. intersect two TID lists.
UB-peration
        Compute the &-operation between two temporary relations, the one
associated to an unary predicate and the other associated to a binary
predicate, i.e. intersect one TID list with a set of TID lists.
BB-operation
        Compute the &-operation between two temporary relations associated to
two binary predicates; given two sets of TID lists, this means to
intersect each list coming from one set with each list coming from the
other one.

A PHYSICAL MODEL

        The following is a kind of physical data organization that permits us
to implement our procedure in an efficient way. We assume that:
- an index I(R.A) exists for the attribute R.A. An index is a   set of
  access-keys. An access-key is a pair (v,p) where v is a value appearing
  in at least one tuple of R in the attribute A   and   p is the pointer to
  l<R.A=v>;
- a shared index SI(R.A,S.B) exists for the attributes R.A and   S.B when
  R.A and S.B are joinable, i.e, when the values in R.A   may be compared
  with the values in S.B. A shared access-key is a triple (v,p,p') where p
  and p' are the pointers to l<R.A=v> and l<S.B=v>, respectively.
        If an attribute R.A is joinable to the attribute S.B, the indices for
R.A and S.B are represented in the shared index. In the same way, if the
attributes R.A, S.B, T.C, etc. can be joined together, the shared indices
can be merged.
        The structure of an access-key is shown in figure 1, where v is a
value appearing at least once in R.A, S.B, T.C, or others, and F<R.A> and
P<R.A> represent,respectively, the frequency of v in the attribute R.A and
the pointer to l<R.A=v>. If v does not appear in the attribute R.A, a null
value is coded in the corresponding pair. The other pairs are maintained
for the attributes S.B, T.C, etc., in the same way.
        When the index I(R.A) is involved, only the corresponding pair is
referred to and the same applies to the pairs R.A and S.B when the shared
index SI(R.A,S.B) is involved.
        The number of different attributes associated with the values is the
sharing degree of the access-keys.
        The physical structure of the inverted file whose access-keys are
like those shown in figure 1 can be chosen according to the
characteristics of the most probable queries and of relations in the data

| v | F<R.A> | P<R.A> | F<S.B> | P<S.B> | F<T.C> | P<T.C> | .... |

Fig. 1. Structure of an access-key

base. For example, B-trees or hash tables can be used. In the sequel we suppose that indices are structured as B^+-trees (see Wedekind (1974); B^+-trees permit us to take both equality and inequality predicates into account.

As far as representing the image $f(R[p])$ of a set $R[p]$ is concerned, we have:
- if p is a unary predicate, such as R.A=x, $f(R[p])$ is represented by the $l<R.A=x>$ list and by the $r<x>$ pointer to it (see figure 2a);
- if p is a predicate of an R.A=S.B type and x_1, x_2,...,x_k are the values in R.A $\cap$ S.B, $f((RxS)[R.A=S.B])$ is made up of pairs of TID lists and pairs of pointers. Two TID lists, $l<R.A=x_i>$ and $l<S.B=x_i>$, and the pair of the $r<x_i>$ and $s<x_i>$ pointers are associated with each value x_i. The $r<x_i>$ pointer refers to $l<R.A=x_i>$ and the $s<x_i>$ pointer refers to $l<S.B=x_i>$ (see figure 2b).

Now we are going to deal with the costs of the basic operations. The costs are defined by using some parameters. A set of them concerns with capabilities of the hardware architecture; another set of parameters depends on the data base which is referred to.

The hardware architecture parameters taken into account are:
Ca, the cost of an assign operation;
Cv, the time to move a single word;
Cc, the cost of a byte comparison operation;
Cr, the time to move a page from a disk device into the main memory;
Cw, the time to transfer a page from the main memory to a disk device.

The page is the unit of data transfer occurring between all the possible levels of the memory hierarchy.

The variable data base parameters are:
P, the page size;
s, the sharing degree of one index;
b, the byte length of each value in the access-key;
N, the number of distinct access-keys in one index.

The following data base parameters are fixed:
f, the number of bytes used for representing a F<R.A> value. It is two, hence a value may appear up to 65536 times in any attribute;
p, the number of bytes used for representing a P<R.A> value. It is four; the first two bytes are used for the page number and the other two for the displacement inside the page;
t, the number of bytes used for representing a TID. The TID is assumed to be the address of the tuple itself. Four bytes are used and the meaning is the same as in P<R.A>.

If the pages of a B^+-tree are filled to a degree of 80%, we can assume that the height H and number D of the leaves in the index I are $H=\log_{0.8n}((N+1)/2)+1$ and $D=5N\cdot(b+6s)/4P$ where $n=P/(b+6s)$ is the maximum number of entries in one B^+-tree node. The costs of the basic operations when our physical structure is used are the following.

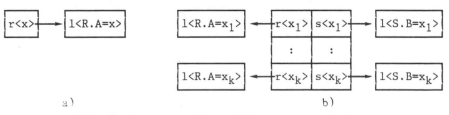

a)   b)

Fig. 2. Representation of temporary relations

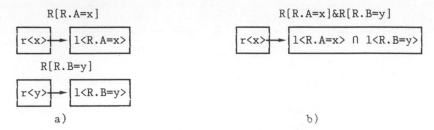

Fig. 3. The UU-operation

U-operation
    Given the predicate R.A=x, we only have to search the index I(R.A) for x and record the fields F<R.A> and P<R.A> of the access-key corresponding to x. The cost is:
    $$C(U)=H \cdot Cr+(H-1) \cdot [(1/3) \cdot b \cdot (n/2)] \cdot Cc+[(1/3) \cdot b \cdot ((n/2)-1)+b] \cdot Cc+3Cv$$
where we assume that the root page of the B^+-tree is always in the main memory and that 1/3 of two values must be compared before deciding that they are not equal.
B-operation
    Given the predicate R.A=S.B, the fields F<R.A> and F<S.B> of each access-key in SI(R.A,S.B) must be tested in order to verify that they are not non-null. Fields F<R.A>, P<R.A>, F<S.B> and P<S.B> are recorded when the test is satisfied. The cost is:
    $$C(B) = D \cdot Cr + 2N \cdot (2Cc + 3\mu Cv)$$
where μ=IR.A ∩ S.BI/N is the percentage of the values appearing in both R.A and S.B.
UU-operation
Given two unary predicates, namely R.A=x and R.B=y, the two TID lists, 1<R.A=x> and 1<R.B=y>, must be accessed and intersected. The result of the intersection is written in the mass memory. One of the pointers records the address of the intersection list (see figure 3). The cost is:

$$C(UU) = \left\lceil \frac{4v}{P} \right\rceil \cdot Cr + \left\lceil \frac{4w}{P} \right\rceil \cdot Cr + \frac{v+2w}{2} \cdot Cc + w \cdot Cv + \left\lceil \frac{\tau \cdot w}{P} \right\rceil \cdot Cw + 2Ca$$

where v and w are the number of TIDs in 1<R.A=x> and 1<R.B=y>, respectively, and τ, $0 \le \tau \le 1$, is the percentage value which indicates the

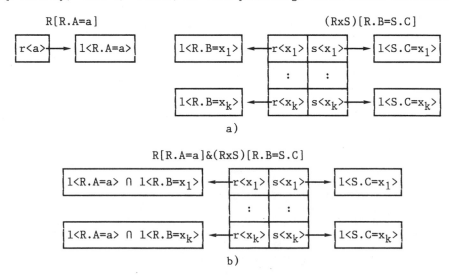

Fig. 4. The UB-operation

number of TIDs in l<R.A=x> which also belong to l<R.B=y>. We assume that (v+2w)/2 comparisons have to be made when two lists of length v and w are intersected. τ=δ if δ is the selectivity factor of the select operator using R.B=y as the select predicate.

UB-operation

Given an unary predicate R.A=a and a binary predicate R.B=S.C (see figure 4), lists l<R.A=a> and the set of k lists l<R.B=x_i> must be transferred from the mass memory to the main memory. Then l<R.A=a> must be intersected with each list l<R.B=x_i>, the results of the intersections are written in the mass memory and the pointers r<x_i> record the addresses of the intersection lists. Some intersections may result empty. Lists l<S.B=x_i> remain the same. The cost is:

$$C(UB) = \left[\frac{4v}{P}\right] \cdot Cr + \left[\frac{4}{P} \cdot \sum_{1}^{k}{}_i \ w_i\right] \cdot Cr + 2Cc \cdot \sum_{1}^{k}{}_i \ (v+2w_i) +$$

$$+ \ 2v \cdot Cv \cdot \sum_{1}^{k}{}_i \ \tau_i + \left[\frac{4v}{P} \cdot \sum_{1}^{k}{}_i \ \tau_i\right] + 2k \cdot Ca$$

In this case we have $\sum_{1}^{k}\tau_i=\delta$.

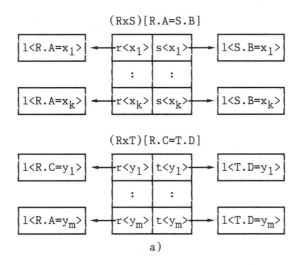

(RxS)[R.A=S.B]

l<R.A=x_1>	◄── r<x_1>	s<x_1> ──►	l<S.B=x_1>
:	:		
l<R.A=x_k>	◄── r<x_k>	s<x_k> ──►	l<S.B=x_k>

(RxT)[R.C=T.D]

l<R.C=y_1>	◄── r<y_1>	t<y_1> ──►	l<T.D=y_1>
:	:		
l<R.A=y_m>	◄── r<y_m>	t<y_m> ──►	l<T.D=y_m>

a)

(RxS)[R.A=S.B]&(RxT)[R.C=T.D]

l<R.A=x_1> ∩ l<R.C=y_1>	◄── r<x_{11}>	s<x_1>	t<y_1>
:	:	:	:
l<R.A=x_1> ∩ l<R.C=y_m>	◄── r<x_{1m}>	s<x_1>	t<y_m>
:	:	:	:
l<R.A=x_k> ∩ l<R.C=y_1>	◄── r<x_{k1}>	s<x_k>	t<y_1>
:	:	:	:
l<R.A=x_k> ∩ l<R.C=y_m>	◄── r<x_{km}>	s<x_k>	t<y_m>

b)

Fig. 5. The BB-operation

Given two binary predicates R.A=S.B and R.C=T.D, lists l<R.A=x$_i$> and l<R.C=y$_i$>, where x$_i$'s belong to R.A ∩ S.B and y$_i$'s to R.C ∩ T.D, are read from the mass memory (see figure 5a). Then each list l<R.A=x$_i$> is intersected with all the lists l<R.B=y$_i$>. Hence m·k is the total number of intersections. The results are written in the mass memory. Lists l<S.B=x$_i$> and l<T.D=y$_i$> remain unchanged. A new set of pointers must be obtained. Each element is made up of a triple of pointers (see figure 5b). These pointers are associated with R, S and T, respectively. The pointers in the same triple refer to the tuples of R, S and T connected to each other by the join predicates. The cost is:

$$C(BB) = \sum_1^k {}_i \left\lceil \frac{4v_i}{P} \right\rceil \cdot Cr + k \cdot \sum_1^m {}_i \left\lceil \frac{4w_i}{P} \right\rceil \cdot Cr + 2Cc \cdot \sum_1^k {}_i \sum_1^m {}_j (v_i + 2w_j) +$$

$$+ 2Cv \cdot \sum_1^k {}_i \sum_1^m {}_j \tau_{ij} \cdot v_i + \left\lceil \frac{4}{P} \cdot \sum_1^k {}_i \sum_1^m {}_j \tau_{ij} \cdot v_i \right\rceil \cdot Cw$$

If σ_1 and σ_2 are the join selectivity factors of the two joins involved, we have:
- $\sigma_1 \cdot \sigma_2 \cdot IRI \cdot ISI \cdot ITI$ is the number of tuples in the relation resulting from the two joins;
- $\sigma_1 \cdot IRI \cdot ISI$ is the total number of TIDs in the relation resulting from (RxS)[R.A=S.B];
- $\sigma_2 \cdot IRI \cdot ITI$ is the total number of TIDs in the relation resulting from (RxT)[R.C=T.D];

If τ is the reduction factor in the universal operation, we can obtain $\tau = 1/IRI$.

AN EXAMPLE

The performance of the procedure we propose is evaluated by taking a given query Q into account and by comparing our procedure with a

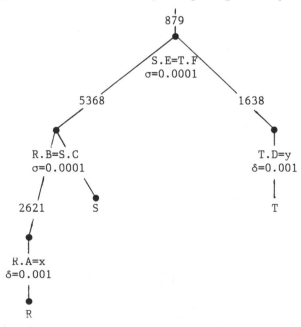

Fig. 6. The parse of example query

traditional one. The parse tree of Q is shown in figure 6. There are three operand relations, namely R, S and T. Their sizes, expressed in number of tuples, are 2621440, 20480 and 163840 for R, S and T. The predicates in nodes of the parse tree describe the relational algebra operators to be performed. Also the selectivity factors, i.e., fractions which describe the number of tuples in each operator's result are represented in the nodes of the parse tree. The numbers in the branches are the number of tuples resulting from the operation.

Query Q is equivalent to the following sequence of universal operations: R[R.A=x]&T[T.D=y]&(RxS)[R.B=S.C]&(SxT)[S.E=T.F].

The tuples in the answer to Q consist of values coming from R, S and T. When the I/O operations are taken into account during the performance evaluation phase, we refer to a three-level memory hierarchy. The three levels consist in the main memory, a cache memory and the disk devices. The cache is used as a transit memory. The data are moved in the hierarchy and are processed by page units. A page is the quantity of data transferred by a single I/O operation from a level to an adjacent one of the memory hierarchy.

Different pages sizes were taken into account in order to obtain the most significant evaluation possible. Only the results related to 4K and 16K byte pages are reported. Other page sizes give similar results.

The tuples were assumed to be 100-byte long. A page contains 40 or 160 tuples when the page size is 4K or 16K bytes long. The sizes of relation R, S and T, expressed in number of pages, are 65536, 512 and 4096, when the page size is 4K, and 16384, 128 lnd 1024, when the page size is 16K.

As far as the indices are concerned, we assume that the sharing degree is 2, the b byte length is 20. For each page size, different values for the total number of access-keys in the indices are taken into account when possible. The number of access-keys per page is 512 (128) when pages of size 16K (4K) are involved. For the sake of simplicity, we assume that the values are uniformly distributed among the tuples of a relation.

The basic costs we use in our performance evaluation are $C_a=C_c=1\mu s$ and $C_v=1.5\mu s$.

When computing C_r and C_w, we assume that a read operation moves a page into the main memory from the cache, fetching it first from a disk device when necessary. A write operation always moves a page residing in the main memory in the cache. If there is no free page frame in the cache, one page in the cache is written into the disk. Therefore, if R_m is the cost of a disk to the cache transfer, R_c the cost of a cache to the main memory transfer, H and H' are the cache hit ratio and the probability that there be a free frame in the cache during a write operation, we have $C_r=H\cdot R_c+(1-H)(R_c+R_m)$ and $C_w=H'\cdot R_c+(1-H')(R_c+R_m)$.

The actual values based on a bus bandwidth of 1 Mbyte per second are: when the page size is 4K we have $R_m=28msecs$ and $R_c=4msecs$ and when the page size is 16K we have $R_m=40msecs$ and $R_c=16msecs$.

The cache hit ratios are H=0.8 and H'=0.3 hence the values for C_r and C_w are, respectively, 9.6 msecs and 23.6 msecs for 4K page size, 24 msecs and 44 msecs for 16K page size.

The &-operation method

The steps involved in processing query Q according to the procedure we propose are the following:

Temporary relation R[R.A=x]

Since the selectivity factor is $\delta=0.001$, number N of access-keys in the I(R.A) index is 1000, provided that the values are uniformly distributed. The temporary relation is a list of 2621 TIDs.

Temporary relation T[T.D=y]

Here we have $\delta=0.01$; therefore N is 100 when I(T.D) is taken into consideration. The resulting list is made up of 1638 TIDs.

Temporary relation (RxS)[R.B=S.C]

Different values for number N1 of access-keys in SI(R.B,S.C) are taken into account. The possible range is 100≤N1≤5368709. The minimum value accords to the chosen value of μ, i.e., 0.01; the maximum value is equal to (IRI·ISI)/10000. The values we use in the experiments are chosen between 1000 and 256000. The temporary relation is made up of 5368709 tuples of RxS in accordance to the join selectivity factor. The number of pairs of lists is k1=μN1. For the hypothesis of uniform distribution of values among the attributes we set w1=ISI·(σ/k1)$^{\frac{1}{2}}$ as the length of lists made up of TIDs of the S relation and v1=IRI·(σ/k1)$^{\frac{1}{2}}$ as the length of lists associated to relation R. In fact, we can set v1/w1 = IRI/ISI; moreover we have k1·v1·w1=IRI·ISI; we solve the system and we set w1 equal to the floor of the solution in order to be sure that w1≥1.

Temporary relation (SxT)[S.E=T.F]

100≤N2≤(IRI·ITI)/10000 is the range of values for number N2 of access keys in SI(S.E,T.F); μ is 0.01 in this case, too. The values in the experiments are between 1000 and 8000. The temporary relation is made up of 335544 tuples of SxT. The number of lists is k2=μN2; w2=ISI·(σ/k2)$^{\frac{1}{2}}$ is the length of lists for S and v2=ITI·(σ/k2)$^{\frac{1}{2}}$ is the length of lists for T.

&-operation between R[R.A=x] and (RxS)[R.B=S.C]

One list made up of 2621 TIDs has to be intersected with k1 lists; each of them is formed by v1 TIDs. Different values of k1 and v1 are taken into account according with the values of N1 for temporary relation (RxS)[R.B=S.C]. All the lists refer to the R relation. As far the result is concerned, we assume that the temporary relation is made up of 5368 tuples from RxS in accordance with the selectivity factor of R.A=x. We set k1'=k1/10 the number of pairs of lists in the result; the length of lists for S remains the same as before the computation of Universal operation, i.e., w1. The length of lists for R becomes v1'=v1/100.

&-operation between T[T.D=y] and (SxT)[S.E=T.F]

Here exists one list made up of 1638 TIDs for the T relation and one set of lists; the lists refer to relation T and they have length v2. The resulting temporary relation is made up of 3355 tuples from SxT. The selectivity factor is now δ=0.01. The lists for S have always length w2 while the lists for T have length v2'=0.2·v2; 2/10 of the intersection are not empty; k2'=0.2·k2 is the number of pairs of lists in the result.

&-operation between (T[T.D=y]xS)[S.E=T.F] and (R[R.A=x]xS)[R.B=S.C]

Two sets of lists for the S relation have to be intersected. One set is made up of 5368 tuples from RxS, the other one of 3355 tuples from SxT. All the possible pairs of above defined values k1' and k2' are taken into account as number of lists in the sets. The lists for S have length w1 or w2 as they refer to RxS or SxT. The lists for R and T are v1' and v2' long, respectively. The result of the &-operation is always made up of 879 tuples from RxSxT and it is made up of a set of triples of lists. We assume that the number k of triples is equal to the ceiling of (k1'·k2')/100. The lengths of lists are v1' and v2' for R and T as they are not affected by the Universal operation. The lists for S are 879/(k·v1'·v2') long.

Access to tuples

The final step of our procedure consists in accessing tuples whose TIDs appear in the temporary relation resulting from the above Universal operation. For each triple of lists of TIDs, three accesses to the disks have to be performed in order to read the lists; then the pages containing the tuples are read from the mass storage. We assume that the number of pages to be read is determined by using the formula in Cardenas (1974) as the lists are sorted. Hence the average number of pages to be read when searching for K tuples tuples uniformly distributed among M pages is M(1-(1-1/M)^K). Here M is the number of pages for the relation taken into account and K is the number of TIDs in one lists for the same relation. Finally, the Cartesian product of the selected tuples is performed by making some projection, when it is necessary. We assume that the projected tuples are 100-byte long. If F is the value from Cardenas' formula, the cost for obtaining the final tuples is $\underline{k}\cdot(3+F)\cdot Cr+87900\cdot Cv/2$.

The traditional algorithm

We use the approach described in Bitton at al. (1983) in order to evaluate the execution time for query Q. This means that a certain number of basic database tasks is defined; a cost is given for each task. The execution cost of a relational algebra operation is obtained by determining which tasks must be executed and by counting how many times each of them must be performed. Finally the answer to a query requires a time equal to the sum of execution costs of the involved relational operations.

The basic tasks may be I/O or processing tasks. The I/O tasks are the read and write tasks as already introduced and refer to the page transfer between the memory hierarchy. The processing tasks concern the computations made by the processor after data are transferred to the main memory. The costs are defined by taking into account the time to carry out comparison and move operations. In particular we assume that 1/3 of two attribute values must be compared before deciding that they are not equal. We set that the attribute values are b-byte long and that a tuple is q-byte long. The basic tasks are the following.

Scan task

The scan task sequentially scans the tuples in a page in order to search for those satisfying a given predicate. Let δ be a fraction describing the portion of tuples satisfying the predicate and k be the number of tuples in a page. Thus we can calculate Csc, the scan cost, as $Csc = \delta\cdot k\cdot b\cdot Cc + (1-\delta)\cdot k\cdot(b/3)\cdot Cc + \delta\cdot k\cdot(q/2)\cdot Cv$.

One page sort task

If a page containing k tuples has to be internally sorted, $k\cdot\log k$ attribute comparisons and tuple moves are required. Thus Cso, the average cost of internally sorting a page, is $Cso = ((b\cdot Cc + (q/2)Cv)\cdot k\cdot\log k$.

Merge task

This task merges the tuples in one page from relation X with the tuples in one page from relation Y. The pages are internally sorted and contain k1 and k2 tuples, respectively. The tuples are q1- or q2-byte long as they are in X or in Y. The tuples are merged by taking a given attribute into account. Let $\sigma\cdot(k1+k2)$ be the average number of tuples satisfying the merge predicate; hence $\sigma\cdot(k1+k2)\cdot b$ byte must be compared for the tuples satisfying the predicate, $(1-\sigma)\cdot(k1+k2)\cdot(b/3)$ bytes are compared for the tuples not satisfying it. Thus the cost Cm of merging two pages is $\sigma\cdot(k1+k2)\cdot b\cdot Cc+(1-\sigma)\cdot(k1+k2)\cdot(b/3)\cdot Cc+\sigma\cdot(k1+k2)\cdot((q1+q2)/2)\cdot Cv$.

As far as the relational algebra operations are concerned, we assume that no index exists and that the pages of relations are internally sorted with respect to a prespecified relation key.

Select

A select operation includes the following actions:
- to read and to scan all the pages of the involved relations;
- to internally sort and to output all the pages containing the resulting relation; the tuples in page are sorted with respect to the attribute involved in the next join operation, if it exists.

The cost of a selection is $SEL=N\cdot(Cr+Csc) + \delta\cdot N\cdot([Cso]+Cw)$ where N is the size, expressed in number of pages, of the operand relation and δ is the select selectivity factor. The sort cost is square-bracketed as the sort tasks may not be necessary.

Join

The nested loops join algorithm is used. Given two relations X and Y, the smaller relation is chosen as the inner relation and the longer (say X) becomes the outer relation. One page of X is read and sorted on the join attribute, if it is necessary, then all pages of Y are sequentially read from the mass memory. As each page of Y is in main memory, it is

Table 1

relation	number of pages		tuples per page	
R	16384	65536	160	40
S	128	512	160	40
T	1024	4096	160	40
U	17	66	160	40
V	11	41	160	40
Z	68	269	80	20

probably sorted and then joined by merging it and the page of X which is present in the main memory. This procedure is repeated for each page of X. When an output page is full it is sorted on the relevant attribute of the subsequent relational operation, if it exists; finally the output page is written out. The cost of the join is then:

$$\text{JOIN} = N \cdot (C_r + [C_{so}] + M \cdot (C_r + [C_{so}'] + C_m)) + \Phi \cdot N \cdot M \cdot ([C_{so}''] + C_w)$$

where N and M are the sizes, expressed in number of pages, of X and Y; Φ is the join selectivity factor with respect to page sizes. The sort costs may be different as it may be different the number of tuples in one page.

As far as the execution time of computing the answer to Q is concerned, we assume that the following sequence of relational operations is performed.
- Perform a select on R by using the predicate R.A=x. U is the resulting relation; the pages of U are internally sorted with respect to R.B.
- Perform a select on T by using the predicate T.D=y. V is the resulting relation; the pages of V are sorted with respect to T.F.
- Perform a join on U and S by using the predicate R.B=S.C. Z is the resulting relation; the pages of Z are internally sorted with respect to S.E.
- Perform a join on V and Z by using the predicate S.E=T.F. The resulting relation is the answer to Q.

The sizes, in pages, of the relations as well as the number of tuples in a page are shown in table 1, where two values for both the size and the number of tuples are recorded for each relation. The first value always concerns the page size of 16K bytes and the second one concerns the 4K-byte size.

The two procedures for processing query Q were evaluated by taking the described values for parameters into account. The experimental runs were performed by varying the number N1 of distinct values of the join attribute for R and S and the number N2 of distinct values of the join attribute for S and T. The traditional algorithm is not affected by the different values of N1 and N2. The values of N1 are between 1000 and 256000. A new value is obtained by doubling the old one. The values of N2 are obtained in the same way between 1000 and 16000.

The comparison between the performance of the two procedures is illustrated in table 2. We compared our method with the traditional one by calculating the percentage of profit of Universal operation procedure. In table 2 each column corresponds to a value of N2 and each row to a value of N1; the first value in a column deals with the 4K byte page size while the second value is concerned with the 16K byte page size.

As it is shown in table 2, the percentage of profit decreases as numbers N1 and N2 increase. This seems to be due to two main factors: the number of pages in a shared index is proportional to the number of distinct values in the attribute and as a result the index scanning time increases; the number of TID lists increases as number N1 or N2 increases; therefore a greater number of read tasks is required in order to obtain lists which can be intersected even if the total number of TIDs remains the same.

Furthermore, when the values of N1 and N2 are given, the percentage

Table 2

N1	N2									
	1000		2000		4000		8000		16000	
1000	99.7	98.9	99.6	98.9	99.6	98.9	99.6	98.8	99.4	98.5
2000	99.6	98.9	99.6	98.9	99.6	98.9	99.5	98.8	99.4	98.5
4000	99.6	98.9	99.5	98.9	99.5	98.9	99.5	98.8	99.3	98.2
8000	99.4	98.7	99.4	98.7	99.4	98.7	99.3	98.3	99.1	97.8
16000	99.2	98.4	99.2	98.4	99.1	98.1	99.0	97.8	98.8	97.0
32000	98.9	97.8	98.8	97.5	98.7	97.3	98.6	96.6	98.3	95.7
64000	98.1	96.3	98.1	96.2	98.0	95.6	97.8	94.7	97.4	93.2
128000	96.9	94.1	96.8	93.4	96.6	92.6	96.3	91.2	95.8	88.7
256000	94.6	89.4	94.4	88.4	94.1	86.9	93.6	84.4	92.7	80.1

of profit decreases as the page size increases because the number of I/O
tasks the traditional algorithm requires decreases as the page size
increases. However, it should be pointed out that the percentage of profit
is always greater than 92% when the page size is 4K, and 80% when the page
size is 16K. Judging from our experimental results, we may conclude that
the procedure we propose is efficient. Moreover, it is particularly useful
when the page size determined by the application decreases.

CONCLUSIONS

A procedure for the execution of relational queries is presented in
this paper. It is based on three main factors: the presence of indices for
the attributes involved in the query; the representation of temporary
relations as lists of TIDs; the definition of the &-operation (which
corresponds to the intersection of the TID lists at a physical level).
A particular physical structure for indices is introduced. It
requires the same algorithms for its maintenance as traditional structures
do. A relevant degree of space is saved and there is an important decrease
in join performance time.
The results of some experiments show that this procedure has a better
performance level than traditional ones do. As an example, we report a
given query's percentage profit. The values are always between 99% and
80%. The percentage of profit increases as the page size decreases.
The experiments illustrated assume that indices are present for all
the simple predicates in the query. If this is not true for some query, we
can use the procedure we propose for the predicates referring to existing
indices and then proceed by using traditional algorithms.
A limit to this paper consists in our only taking equality predicates
into account. All the same, non-equality predicates can be managed in
the same way. For example, when performing a Universal operation, in which
a relation like R[R.A≠x] is one operand relation, we can search for the
list corresponding to R.A=x and put the sets' difference operation in the
place of the sets' intersection. Joins like (RxS)[R.A≤S.B] can be managed
by scanning the shared index and by taking out pairs of lists having at
least one non-null pointer. Each pointer for S must be assumed to be
associated with all the pointers for R which come before it. Some
particular tricks have to be used when a Universal operation between
temporary relations resulting from joins is performed in order to keep up
the correspondence between a list for one relation and a set of lists for
another one. As far as the management of the or operator between select
predicates is concerned, these predicates are treated in the proposed way
and the set union of the resulting temporary relations is obtained. This

union can be performed by referring to the pointers in the temporary relations alone.

The approach we propose seems to be attractive even if multiprocessor database machine are taken into account, see, for example, DeWitt (1979). For example, each of unary predicates in a query can be evaluated by one processor while sets of processors can be assigned to the evaluation of binary predicates. Moreover, the execution of Universal operations can be performed by several processors. The parallel evaluation of Universal operations is under investigation.

REFERENCES

Bitton, D., Boral, H., DeWitt, D.J. and Wilkinson, W.K., 1983, Parallel algorithms for the execution of relational database operations", ACM TODS, 324:353.

Bratbergsengen, K, 1984, Hashing methods and relational algebra operations, in "VLDB84", Singapore.

Cardenas, A.F., 1975, Analysis and performance of inverted data base structures, CACM, 253:263.

Dayal, U., 1983, Processing queries over generalization hierarchies in a multidatabases system, in "VLDB83", M. Schkolnick and C. Thanos, eds., Florence.

DeWitt, D.J., 1979, DIRECT - A multiprocessor organization for supporting relational database management systems", IEEE Trans. on Comp., 395:406.

Gotlieb, L.R., 1975, Computing joins of relations", in "ACM SIGMOD-International Symposium on Management of Data".

Krishnamurthy, R. and Morgan, S.T., 1984, Query processing on personal computers: a pragmatic approach, in "VLDB84", Singapore.

Palermo F.P., 1974, A database search problem, in "Information Systems COINS IV", J.T. Tou, ed., Plenum Press, N.Y.

Reiner, D. and Rosenthal A., 1984, Extending the algebraic framework of query processing to handle outerjoins, in "VLDB84", Singapore.

Wedekind, M., 1974, On the selection of access paths in a data base system, in "Data Base Management", North-Holland Pub.Co., Amsterdam.

Wong, E. and Youssefi, K., 1976, Decomposition-A strategy for query processing, ACM TODS, 223:241.

Yao, S.B., 1979, Optimization of query evaluation algorithms, ACM TODS, 133:155.

A HASH JOIN TECHNIQUE FOR RELATIONAL DATABASE SYSTEMS

Yasuo Yamane

Software Laboratory
Fujitsu Laboratories Ltd.
Kawasaki, Japan

In this paper, we will formulate and discuss an efficient method of equijoin in relational database systems. In this method, if either of two relations can be loaded into internal memory, the equijoin can be processed quickly using a hashing technique. Otherwise, the relations are recursively partitioned into subrelations using hashing to be loaded. We also analyze the method and evaluate the results of our experiments. We will concentrate on how we should partition relations.

1. INTRODUCTION

In relational database systems, equijoin (Ullman, 1980) is one of the most frequently used operations (hereafter referred to as 'join'). In many systems, such as RDB/V1 (Makinouchi et al., 1981), System R (Blasgen and Eswaran, 1977) join by the nested scan method (Gotlieb, 1975; Kim, 1980) and by the merge scan method (Blasgen and Eswaran, 1977; Kim, 1980) have been used. The order of time complexity of the nested scan is $O(mn)$ and that of the merge scan is $O(m \log m + n \log n)$, where m, n are the number of tuples of relations A and B, respectively.

In this paper, we will formulate and discuss a join method based on hashing (hereafter called 'hash join'). In this method, if either of two relations can be loaded into internal memory, the join can be quickly processed, in $O(m+n)$ time, using hashing (internal hash join). If neither relation can be loaded entirely, both are recursively partitioned into subrelations to be loaded; when either of the two subrelations is loaded, each pair of subrelations is joined by the above internal hash join (external hash join). The external hash join can also be processed quickly and with a small number of I/O accesses. In our experiment, the external hash join was generally processed faster and with fewer I/O accesses than the merge scan. In addition, this method is simple and can be easily implemented.

Several join methods using hashing have been proposed and discussed (Babb, 1979; Bratbergsengen, 1984; Dewitt et al., 1984). The hash join is not new; a join method using internal memory was discussed in detail (Dewitt et al., 1984), and recursive partitioning was discussed (Bratbergsengen, 1984). In this paper, we will formulate a method (hash join), analyze it, and concentrate on how we should partition relations.

In Chapter 2, we will explain our assumptions and the notation used throughout this paper. The basic concept of the hash join will also be discussed. In Chapter 3, we will discuss the internal hash join, in which only internal memory is used during processing. We will also analyze the expected execution time and the optimal partition. Then we will compare the results of our experiments with the internal hash join against the merge scan where the quick sort is used. In Chapter 4, we will discuss the external hash join, in which external memory is used during processing. We will analyze the expected number of I/O accesses and compare the hash join with the merge scan where the multi-way merge sort is used.

2. FOUNDATION

In this chapter, we will explain our assumptions and the notation used throughout this paper. The basic concept of the hash join is also discussed.

2.1 ASSUMPTIONS

A subset of a relation is called a 'subrelation'. The field on which the join is performed is called the join field. The two relations to be joined, called 'input relations', are represented by $A=\{a_1, a_2, ..., a_m\}$ and $B=\{b_1, b_2, ..., b_n\}$, where a_i and b_j are tuples of relations A and B, respectively. Thus A has m tuples and B has n tuples. A relation has a single join field. Join on multiple join fields is possible by regarding them as a single field, so the above condition doesn't lose generality. The value of the join field of a_i is represented by $\bar{a}_i$, and that of b_j by $\bar{b}_j$. The physical byte length of fields and tuples is fixed. We call the relation that is obtained as the result of join an 'output relation', represented by C. The number of tuples of C is represented by g. Throughout this paper, we assume the following unless otherwise specified:

Uniform distribution assumption

Input relations A and B have a common join field domain, which is a set of integers $R=\{0, 1, ..., r-1\}$; r is the domain size. The value of the join field of A, $\bar{a}_i$ ($1<=i<=m$) follows the uniform distribution on R; likewise $\bar{b}_j$ ($1<=j<=n$), and $\bar{a}_i$'s and $\bar{b}_j$'s are all mutually independent.

External memory is a directly accessible medium, such as disk storage. The unit for I/O is page; page size is fixed. A tuple is entirely included in a page. Pages contain tuples only. Therefore the maximum number of tuples included in a page is $\lfloor$(page size)/(tuple length)$\rfloor$ ($\lfloor x \rfloor$ indicates the greatest integer that is not more than x, and $\lceil x \rceil$ the smallest integer that is not less than x). We represent the number of pages of relations A, B and C by M, N and G, respectively.

2.2 BASIC CONCEPT OF THE HASH JOIN

First, relation A is partitioned into p subrelations $A_i=\{a \in A | h(\bar{a})=i\}$ ($0<=i<p$) by applying hash function h to the join field. In the same way, relation B is partitioned into p subrelations $B_i=\{b \in B | h(\bar{b})=i\}$ ($0<=i<p$). We call p the 'partition number'. Next, for each i ($0<=i<p$), A_i and B_i are joined. Here, it is not necessary to join A_i and B_j where $i \neq j$, because for any $a \in A$ and any $b \in B$, if $h(\bar{a}) \neq h(\bar{b})$, then $\bar{a} \neq \bar{b}$.

The idea is based on the 'divide and conquer' (Aho et al., 1974); we decompose the problem, "join two relations A and B" into p subproblems, "join two subrelations A_i and B_i" ($0<=i<p$), then solve each subproblem. This approach is applied recursively. That is, in solving a partitioned subproblem, we can partition the subproblem still further. When we join subrelations A_i and B_i, we can partition A_i into q subrelations A_{i0}, A_{i1},

..., $A_{i,q-1}$, and B_i into B_{i0}, B_{i1}, ..., $B_{i,q-1}$, and join A_{ij} and B_{ij} ($0<=j<q$). When we join A_{ij} and B_{ij}, we can partition A_{ij} into A_{ij0}, A_{ij1}, ..., $A_{ij,r-1}$ and so on.

This recursiveness is very important and useful especially in the hash join using external memory, discussed in Chapter 4. The tree diagram in Fig.2.1 shows the 'partition tree of relation A'. Partition trees of A and B are topologically quite similar. The root of the partition tree corresponds to an input relation. So, we call the input relation a 'root subrelation' (we regard the input relation as a subrelation). Each subrelation corresponding to a leaf node of a partition tree is called a 'leaf subrelation'. Subrelations other than root and leaf subrelations are called 'internal subrelations'. We give 'level' numbers to subrelations as shown in Fig.2.1. The stage at which level (i-1) subrelations are partitioned into level i subrelations is called 'stage i'. We call the maximum level 'the depth of the partition tree' and represent it by d. In Fig.2.1, d is 3. Below is the general algorithm for the hash join written in a Pascal-like language. The procedure is recursive.

Algorithm H   (general algorithm for the hash join)
    procedure   HASHJOIN (A, B, C);
    begin
        if termination condition holds true
            then LASTJOIN(A, B, C)
            else begin
                determine hash function h and partition number p;
                partition A into p subrelations $A_i=\{a \in A | h(\bar{a})=i\}$ ($0<=i<p$), and
                partition B into p subrelations $B_i=\{b \in B | h(\bar{b})=i\}$ ($0<=i<p$);
                for i:=0 to p-1 do HASHJOIN (A_i, B_i, C)
            end
    end

LASTJOIN is a procedure for final joins when the termination condition holds true. The candidate method for LASTJOIN is either the nested scan, the merge scan, or the hash join without recursion mentioned in Chapter 3.

3. INTERNAL HASH JOIN

In this chapter, we will discuss the internal hash join. It can be used when either of two input relations can be loaded into internal memory. We show the algorithm of the internal hash join and the analysis of its

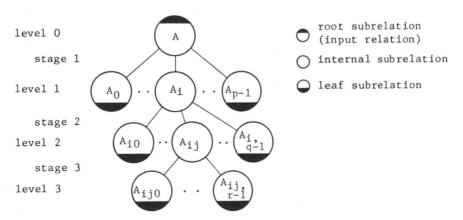

Fig. 2.1 Partition tree of relation A

expected execution time. The optimal partition number is also discussed. The internal hash join is also used, in the external hash join, discussed in Chapter 4.

3.1 ALGORITHM

The internal hash join algorithm is similar to Algorithm H except in the following two respects.

Optimization 1 : Recursion is not used.
For efficiency it is desirable for the recursion to be as shallow as possible. In the internal hash join, recursion is not absolutely necessary.

Optimization 2 : Only one relation is partitioned.
Partition only one relation into subrelations. The other relation is only scanned tuple by tuple. This optimization makes the use of memory space more efficient; it becomes unnecessary to load both relations entirely.

We show the internal hash join algorithm below. The execution frequency of each statement is written on its right. Value s is the number of tuple comparisons between relation A and B.

Algorithm I (internal hash join algorithm)
```
  procedure INTERNAL(A, B, C);              execution frequency
  begin
     determine which input relation is
        partitioned (let it be A);               1
     determine partition number p and
        hash function h;                         1
     for i:=0 to p-1 do A_i:={ };               p
     for i:=1 to m do begin                      m
        get a tuple a_i from A;                  m
        k:=h(ā_i);                               m
        A_k:=A_k U {a_i}                         m
     end;
     for j:=1 to n do begin                      n
        get a tuple b_j from B;                  n
        k:=h(b̄_j);                              n
        for each a∈A_k do                        s
           if ā=b̄_j                             s
              then generate the resulting
                   tuple and output it to C      g
  end
end
```

3.2 ANALYSIS

In this section, we will analyze the expected execution time for the internal hash join algorithm under our uniform distribution assumption mentioned in Section 2.1; we don't take I/O accesses into account. Here, as hash function $h(x)$ we select $h(x)=x \bmod p$, where $x \bmod p$ indicates $x-p\lfloor x/p \rfloor$. From the above execution frequency, the time complexity

$$T = c_1 s + c_2 p + c_3 g + c_4 m + c_5 n + c_6, \tag{3.1}$$

where c_i ($1<=i<=6$) is a machine-dependent constant. The expected number of tuples of the output relation C

$$E(g) = mn/r, \tag{3.2}$$

where $E(X)$ indicates the expected value of a random variable X. For the proof of (3.2), refer to the paper (Merrett and Otoo, 1979).

Lemma 1

Under the uniform distribution assumption, the expected value of s

$$E(s) = mn\{1/p + (p/r^2)\langle r/p\rangle(1-\langle r/p\rangle)\},$$
where $\langle x\rangle = x - \lfloor x\rfloor$, that is, the fractional part of x.
Proof. Subrelations A_i and B_i $(0\mathord{<}\mathord{=}i\mathord{<}p)$ are expressed as $A_i = \{a\epsilon A|\bar{a} \bmod p=i\}$
and $B_i = \{b\epsilon B|\bar{b} \bmod p=i\}$, respectively. First, $s = \sum\limits_{i=0}^{p-1}|A_i||B_i|$. Let $q=\lfloor r/p\rfloor$ and
$t=r \bmod p$, then for any $a\epsilon A$
$$P(a\epsilon A_i) = \begin{cases} (q+1)/r & (0\mathord{<}\mathord{=}i\mathord{<}t) \\ q/r & (t\mathord{<}\mathord{=}i\mathord{<}p), \end{cases}$$
where $P(E)$ represents the probability of event E. $P(b\epsilon B_i)$ is the same. $|A_i|$
and $|B_i|$ have binomial distribution, so
$$E(|A_i|) = \begin{cases} m(q+1)/r & (0\mathord{<}\mathord{=}i\mathord{<}t) \\ mq/r & (t\mathord{<}\mathord{=}i\mathord{<}p) \end{cases}$$
$$E(|B_i|) = \begin{cases} n(q+1)/r & (0\mathord{<}\mathord{=}i\mathord{<}t) \\ nq/r & (t\mathord{<}\mathord{=}i\mathord{<}p) \end{cases}$$
Since $|A_i|$ and $|B_i|$ are mutually independent random variables,
$$E(s) = \sum_{i=0}^{p-1} E(|A_i|)E(|B_i|)$$
$$= mn\{t(q+1)^2/r^2 + (p-t)q^2/r^2\}$$
$$= mn\{1/p + (p/r^2)(t/p)(1-t/p)\}$$
$$= mn\{1/p + (p/r^2)\langle r/p\rangle(1-\langle r/p\rangle)\} \qquad \square$$

From (3.2) and Lemma 1, the expected time complexity
$$E(T) = c_1 mn\{1/p + (p/r^2)\langle r/p\rangle(1-\langle r/p\rangle)\} + c_2 p + c_3 mn/r + c_4 m + c_5 n + c_6. \qquad (3.3)$$

Theorem 1

Let $m\mathord{<}\mathord{=}n$. Under the uniform distribution assumption, if $\sqrt{mn}\mathord{<}\mathord{=}r$ and
$m\mathord{<}\mathord{=}p\mathord{<}\mathord{=}n$, the expected time complexity $E(T)$ is $O(m+n)$.
Proof. From (3.2) and $\sqrt{mn}\mathord{<}\mathord{=}r$, $E(g)\mathord{<}\mathord{=}\sqrt{mn}$ $(\mathord{<}\mathord{=}(m+n)/2)$. Since $\langle r/p\rangle(1-\langle r/p\rangle)\mathord{<}\mathord{=}1/4$ because $0\mathord{<}\mathord{=}\langle r/p\rangle\mathord{<}1$, from Lemma 1, $\sqrt{mn}\mathord{<}\mathord{=}r$ and $m\mathord{<}\mathord{=}p\mathord{<}\mathord{=}n$, $E(s)\mathord{<}\mathord{=}5n/4$. Since $p\mathord{<}\mathord{=}n$, from (3.1), the order of $E(T)$ is $O(m+n)$. $\square$

Optimal partition number

Now, let us consider the optimal partition number p_{min} which minimizes the expected time. Extracting the terms dependent on p from (3.3), and dividing their sum by $c_1 mn$, we get the following function:
$$f(p) = e(p) + c_2 p/(c_1 mn), \qquad (3.4)$$
where $e(p) = (1/p) + (p/r^2)\langle r/p\rangle(1-\langle r/p\rangle)$. $\qquad (3.5)$
The integer which minimizes $f(p)$ is p_{min}.

Lemma 2

$e(p)$ is expressed as follows.
$$e(p) = \begin{cases} 1/r & \text{(if } r\mathord{<}p) \\ -\{k(k+1)/r^2\}p + (2k+1)/r & \text{(if } r/(k+1)\mathord{<}p\mathord{<}\mathord{=}r/k, \text{ for some integer } k) \end{cases}$$
Proof. If $r\mathord{<}p$, $\langle r/p\rangle = r/p$. So, $e(p) = 1/r$, based on (3.5).
If $0\mathord{<}p\mathord{<}\mathord{=}r$, let $k=\lfloor r/p\rfloor$, then $r/(k+1)\mathord{<}p\mathord{<}\mathord{=}r/k$ and $0\mathord{<}\mathord{=}r/p-k\mathord{<}1$. So, $\langle r/p\rangle = (r-kp)/p$. From this and (3.5), $e(p) = -\{k(k+1)/r^2\}p + (2k+1)/r$. $\square$

Theorem 2

Let $p_1 = r/\lfloor\sqrt{c_2 r^2/(c_1 mn) + 1/4} + 1/2\rfloor$. Then, the optimal partition number $p_{min} = \lfloor p_1\rfloor$ or $\lceil p_1\rceil$. $\qquad (3.6)$

519

Proof: From Lemma 2, the graph of $e(p)$ consists of the segments connecting the points $(r/k, k/r)$ $(k=1,2,\ldots)$ and a half line $y=1/r$ $(r<=p)$. Furthermore, it is convex. The second term of $f(p)$ in (3.4) is linear, so the value of p_1 that minimizes $f(p)$ is one (or either of two) of the values r/k $(k=1,2,\ldots)$; if two exist, we select the smaller value for memory efficiency. Let $p_1 = r/k_1$. Then, including the case where $k_1=1$,

$$k_1(k_1+1)/r^2 > c_2/(c_1 mn) \qquad \text{and} \qquad k_1(k_1-1)/r^2 <= c_2/(c_1 mn).$$

So,

$$k_1 = \lfloor \sqrt{c_2 r^2/(c_1 mn)+1/4} +1/2 \rfloor \quad \text{and} \quad p_1 = r/k_1 \quad = r/\lfloor \sqrt{c_2 r^2/(c_1 mn)+1/4} +1/2 \rfloor.$$

Since p_{min} is an integer, $p_{min} = \lfloor p_1 \rfloor$ or $\lceil p_1 \rceil$. $\square$

Now we have found the optimal partition number p_{min}, but, unfortunately, (3.6) includes the domain size r. In practical database

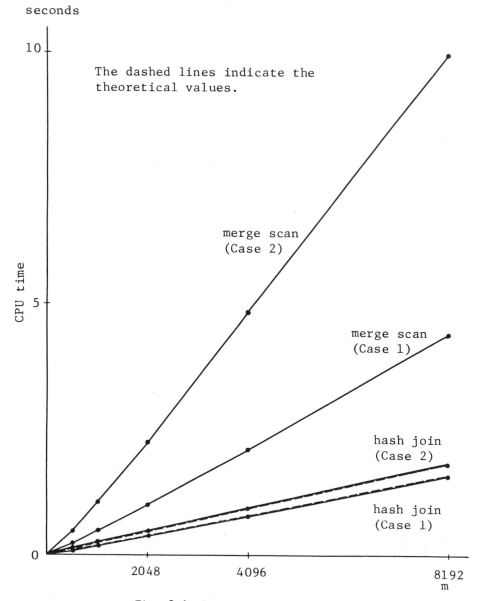

Fig. 3.1   Result of Experiment 1

systems, the domain size is often unknown. Here we encounter the problem of
what to do when we don't know the domain size. A nearly optimal solution is
to use p between m and n (for   example, $\lceil \sqrt{mn} \rceil$),   as   a   partition   number,
because   from Theorem 1, O(m+n) expected execution time is guaranteed if $\sqrt{mn}$
<=r (otherwise E(g) > $\sqrt{mn}$).

3.3 EXPERIMENT

We implemented the internal hash join method and the merge scan method
whose internal sort method is quick sort (Knuth, 1973). In this section, we
will evaluate the results of this experiment in terms of CPU time   and   show

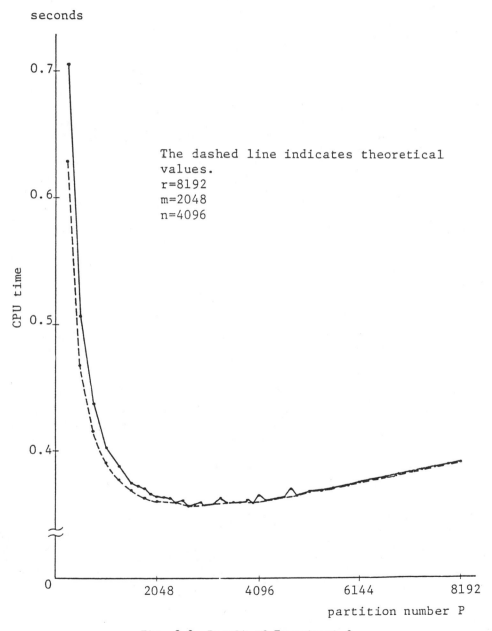

Fig. 3.2 Result of Experiment 2

the validity of the theoretical expression (3.3). I/O accesses are not taken into account in evaluating execution time. Relation A consists of two columns, A.id and A.jfld, and, similarly, relation B of B.id and B.jfld. Each field has a 4-byte integer value. Therefore, the tuple length is 8 bytes. A.jfld and B.jfld are join fields and the values are uniform pseudo-random numbers generated on the domain $R=\{0, 1, \ldots, r-1\}$. The output relation consists of three columns, C.id1, C.jfld and C.id2, corresponding to A.id, A.jfld and B.id, respectively. The experiment was done using a FACOM M-160 computer, as described below.

Experiment 1 : We measured the CPU time required for the hash join and that for the merge scan letting the parameters r, m, n be as follows:
Case 1 r=m=n
Case 2 r=4n=16m.
The partition number p was $\lceil \sqrt{mn} \rceil$. The result is shown in Fig.3.1.
Experiment 2 : We measured the CPU time required for the hash join, changing the partition number p. Other parameters were as follows:
r=8192, m=2048, n=4096.
The partition number p was moved in the range, 256<=p<=8192. The result is shown in Fig.3.2.

From Experiment 1, we see that the hash join outperforms the merge scan in terms of CPU time. And, from Experiment 2, we see that the CPU time required for the hash join is kept short in the range, m<=p<=n. We could get values of c_i's in (3.3) by solving the simultaneous equations which were obtained by substituting various real values gathered during the experiment. The result was as follows:
$c_1=0.01$, $c_2=0.01$, $c_3=0.11$, $c_4=0.03$, $c_5=0.03$, $c_6=1.8$.
In (3.3) time is given in milliseconds. The time theoretically computed using (3.3) is represented by dashed lines in Fig.3.1 and Fig.3.2. It closely approximates to the actual time. And, in Experiment 2, the optimal partition number computed using (3.6) is 2730 or 2731.

4. EXTERNAL HASH JOIN

In this chapter, we will discuss the external hash join. It is used when neither of two input relations can be loaded into internal memory.

4.1 ALGORITHM

The essential difference between this and the internal hash join is the use of recursiveness, as mentioned in Section 2.2, and the partitioning of both of the two input relations. We show the external hash join algorithm below; the INTERNAL procedure is shown in Algorithm I.

Algorithm E   (external hash join algorithm)
  procedure EXTERNAL(A, B, C);
  begin
    if either A or B is loaded into internal memory
      then INTERNAL(A, B, C)
      else begin
        determine hash function h and partition number p;
        partition A into p subrelations $A_i=\{a\epsilon A|h(\bar{a})=i\}$ (0<=i<p) and
        partition B into p subrelations $B_i=\{b\epsilon B|h(\bar{b})=i\}$ (0<=i<p)
          using external memory;
        for i:=0 to p-1 do EXTERNAL(A_i, B_i, C)
      end
  end

How to allocate buffers

We assume that there are $Z+2$ buffers BUF(0), BUF(1), ..., BUF($Z+1$) in internal memory and that partition number p is less than or equal to Z. BUF($Z+1$) is always used for storing the resulting tuples into the output relation. Relation A is partitioned into p subrelations as follows. BUF(Z) is used to read relation A, and p buffers BUF(0), ..., BUF($p-1$) of Z buffers are used to output tuples for subrelations A_0, ..., A_{p-1}. These Z buffers are called 'partition buffers'. Relation A is read page by page into BUF(Z), and for each tuple a in the buffer, compute $k=h(\bar{a})$ ($0<=k<p$) and put a in BUF(k). The content of BUF(i) ($0<=i<p$) is written to external memory when BUF(i) becomes full or when the partitioning of A is complete. Relation B is partitioned in the same way. For the internal hash join, we use Z buffers BUF(0), BUF(1), ..., BUF($Z-1$) to load one subrelation and buffer BUF(Z) to read the other subrelation.

On worst-case and hybrid method

For example, in the extreme case of all the tuples having the same join field values, we can't partition the relation at all. A simple way to avoid such a situation is to add the condition "and the current level is the maximum level" to the termination condition in Algorithm E and to use other join methods. When neither A nor B can be loaded under the above condition, we must use other methods, say the nested scan or the merge scan.

4.2 PARTITION SCHEME

In this section, we will discuss partition schemes; how to determine partition number p and hash function h. Partition schemes are grouped into two classes. One class consists of those in which we partition relations into subrelations before executing the internal hash join; we call these 'external partition schemes'. The other class consists of partition schemes concerned with the internal hash join; we call these 'internal partition schemes'. The difference between the two classes is that, in external partition schemes, the partition number is limited to the number of partition buffers, Z; this is not so in internal partition schemes. First, we discuss external partition schemes (simply called 'partition schemes' below as long as there is no ambiguity).

External partition schemes

Partition schemes are very important, for they determine the number of I/O accesses in the external hash join. Partition schemes are concerned with the construction of partition trees, which were mentioned in Section 2.2. The most important criterion in selecting the best partition scheme is the minimization of I/O accesses. We can break this criterion down into the following three sub-criteria.

Criterion 1 : The partition tree should be as balanced as possible. In other words, we should select a set of hash functions that distribute tuples as equally as possible.

Criterion 2 : The partition tree should be as shallow as possible. We consider the partition of relation A, which has M pages. Even if the partition tree is balanced, each stage requires 2M or more I/O accesses. This means that it is better to have as few stages as possible.

Criterion 3 : The partition tree should have as few nodes as possible. That is, there should be as few subrelations as possible. Let us assume that a subrelation A_i has M_i pages. The subrelation A_i is made so that M_i-1 pages are full of tuples, but the last one page may not be full. If the number of subrelations increases, the number of not-full pages may increase. Consequently, an increase in subrelations may cause an increase in I/O accesses.

P-ary partition scheme

Here, we will discuss a partition scheme in which the same partition number p is used in all the stages; this is called a 'p-ary partition scheme'. Let the partition number and hash function used in stage i be p_i and h_i, respectively. In this scheme, $p_i=p$ and $h_i(x)=\lfloor x/p^{i-1}\rfloor \bmod p$ $(p<=Z)$. When we represent x by $x=\sum_{i=0}^{\infty} x_i p^i$, $h_i(x)=x_{i-1}$. The partition tree for this scheme is a p-ary tree. Clearly, this scheme satisfies Criterion 1. And, if we let p be Z, Criterion 2 is also satisfied clearly. However, this scheme doesn't always satisfy Criterion 3, because balanced p-ary trees are not always optimal, as shown in Fig.4.1. The trade-off between Criterion 2 and Criterion 3 is discussed in Section 4.3.

On internal partition schemes

The problem is how to determine the partition number p' and hash function h' to join leaf subrelations using the internal hash join. When we use the p-ary partition scheme for the external partition, it is desirable that p' and p are relatively prime.

4.3 ANALYSIS

In this section, we will analyze the depth of the partition tree and the number of I/O accesses required by the external hash join. We use the p-ary partition scheme. First, we consider the partition of relation A. We represent the level i subrelation by the following expression:

$$A_{k_1 k_2 \cdots k_i} = \{a\in A | h_1(\bar{a})=k_1,\ h_2(\bar{a})=k_2,\ \ldots,\ h_i(\bar{a})=k_i\}. \tag{4.1}$$

The following proposition concerns the distribution of tuples to each subrelation.

Proposition

It is assumed that the probability of a tuple being distributed to a subrelation $A_{k_1 k_2 \cdots k_i}$ is the same for any $k_1, k_2, \ldots, k_i$, where $0<=k_j<p$ $(1<=j<=i)$, that is, for any $a\in A$,

$$P(a\in A_{k_1 k_2 \cdots k_i})=1/p^i. \tag{4.2}$$

Let $q=1/p^i$. The expected value e, and the variance s^2, of the number of tuples distributed to $A_{k_1 k_2 \cdots k_i}$, $|A_{k_1 k_2 \cdots k_i}|$, are as follows.

$$e=mq,\ s^2=mq(1-q). \tag{4.3}$$

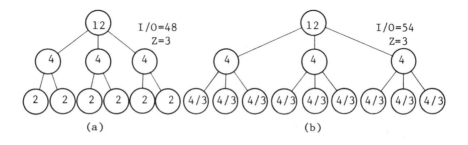

The 3-ary partition scheme corresponds to (b). 'I/O' indicates the number of I/O accesses. 'Z' indicates the number of partition buffers.

Fig. 4.1 Examples of partition trees

Accordingly, $s < \sqrt{e}$ (4.4)

Proof. From (4.2), $|A_{k_1 k_2 \cdots k_i}|$ is a random variable following the binomial distribution, so (4.3) holds true. And, $s^2 = mq(1-q) < mq = e$, so (4.4) holds true. □

Since $s < \sqrt{e}$, it can be said that the standard deviation s is comparatively small in comparison with e. Taking the above into account, to simplify the analysis, we assume the following.

Equal distribution assumption

In any level i, the number of tuples distributed to each subrelation is the same. Accordingly, the partition tree is balanced.

Let the number of pages of relations A and B be M and N, respectively, and $M \leq N$. Let the partition number $p \leq Z$ and the number of buffers used to entirely load one relation be Z. Let the depth of the partition tree be d. In level d, input relation A is partitioned into p^d leaf subrelations and the number of pages of each leaf subrelation is $\lceil M/p^d \rceil$. Therefore, $\lceil M/p^d \rceil \leq Z$ and $Z < \lceil M/p^{d-1} \rceil$. That is,

$$d = \lceil \log_p M/Z \rceil. \tag{4.5}$$

In (4.5), note that the depth of the partition tree of relation B is also d; therefore, the depth of the partition tree depends on the smaller relation.

Next, we consider the number of I/O accesses. Firstly, we consider the number of I/O accesses required to partition A into p^d leaf subrelations; we represent this by part(A). Then,

part(A) = (the number of I/O accesses to read A)
+ (the number of I/O accesses to write
and read internal subrelations)
+ (the number of I/O accesses to write leaf subrelations).

Since, in level i, the number of pages of each subrelation is $\lceil M/p^i \rceil$,

$$\text{part(A)} = M + 2 \sum_{i=1}^{d-1} \lceil M/p^i \rceil p^i + \lceil M/p^d \rceil p^d.$$

Similarly,

$$\text{part(B)} = N + 2 \sum_{i=1}^{d-1} \lceil N/p^i \rceil p^i + \lceil N/p^d \rceil p^d.$$

Secondly, we consider the number of I/O accesses required for internal hash joins; we represent this by 'internal'. Let the number of pages of the output relation be G. To read a leaf subrelation of A and one of B takes $(\lceil M/p^d \rceil + \lceil N/p^d \rceil)$ I/O accesses. So,

internal = $(\lceil M/p^d \rceil + \lceil N/p^d \rceil) p^d + G.$

From the above, the number of I/O accesses required in the external hash join

external = part(A) + part(B) + internal

$$= (M+N) + 2 \sum_{i=1}^{d} (\lceil M/p^i \rceil + \lceil N/p^i \rceil) p^i + G. \tag{4.6}$$

How to determine p in a p-ary partition scheme

Here, we will discuss how we should determine partition number p in a p-ary partition scheme. This problem is concerned with the trade-off between Criterion 2 and Criterion 3. If $p = Z$, then Criterion 2 is satisfied, but generally, Criterion 3 is not. Conversely, if p is too small, Criterion 2 is not satisfied. Let $d = \lceil \log_Z M/Z \rceil$. Then d represents the minimum depth of partition trees in all partition schemes. So, the trade-off is to minimize p, while holding the depth of partition tree to d.

4.4 EXPERIMENT

We implemented the external hash join and the merge scan using external sort. In this section, we will compare the results of the experiment, in terms of CPU time, elapsed time and the number of I/O accesses, and show how the theoretical expression (4.6) approximates the practical results. We used relations of the same type as used in Section 3.3. Both buffer and page were 4096 bytes. In the implementation of the external hash join, the buffer allocation method was the same as that discussed in Section 4.1. We used Z+2 buffers. We used the p-ary partition scheme. In the implementation of the merge scan, we used the (Z+1)-way merge (Knuth, 1973) as the external sort, and the quick sort as the internal sort to make initial runs. The experiments were done using a FACOM M-160 computer, as follows.

Experiment 1 : The sizes of two input relations were the same.
M=N and r=m=n
The depth of the hash join partition tree was held to 1, letting $Z=\lceil \sqrt{2M} \rceil -1$. The result is shown in Fig.4.2.

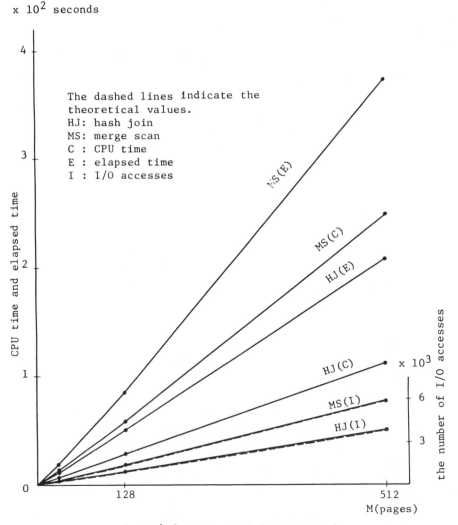

Fig. 4.2   Result of Experiment 1

526

Experiment 2 : The product of the sizes of A and B was kept constant; M was decreased and N was increased.

$MN=64^2$ r=32768 and Z=8
The result is shown in Fig.4.3.

The number of I/O accesses required in the merge scan is expressed by
$$mergescan= 3(M+N) + 2M\lceil\log_{Z+1}(M/(Z+1))\rceil$$
$$+ 2N\lceil\log_{Z+1}(N/(Z+1))\rceil +G. \qquad (4.7)$$
The number of I/O accesses predicted by (4.6) and (4.7) is shown in Figs.4.2, 4.3. The theoretical expectations closely approximate the practical values. From (4.6) and (4.7), in the case of Experiment 1, the number of hash join I/O accesses is about 2(M+N) less than that of the merge scan. In Experiment 1, this is the reason the hash join requires fewer I/O accesses than the merge scan, as mentioned in the paper (Bratbergsengen, 1984).

As mentioned in the paper (Bratbergsengen, 1984), from Experiment 2, we see that the greater the size difference of the two input relations, the more the hash join outperforms the merge scan in terms of CPU time, elapsed time and the number of I/O accesses. The reason is that, in the hash join,

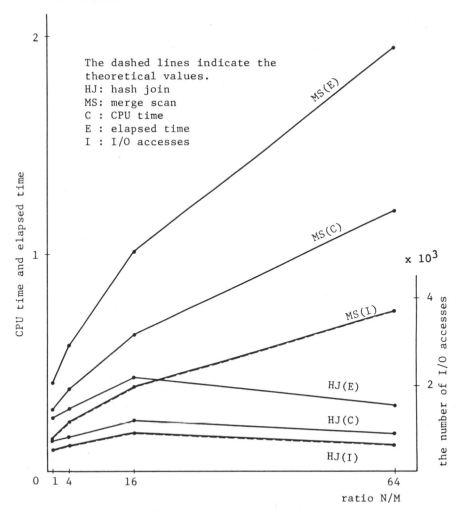

Fig. 4.3 Result of Experiment 2

527

the depth of the partition tree is determined by the smaller relation (not both), from (4.5), under the equal distribution assumption.

In the above experiments, we got the almost same results as predicted in (4.6). However, when M/p^d is close to Z, the leaf subrelations may range over two levels, causing unmatch between prediction and actual results.

5. CONCLUSION

As for the internal hash join, under our uniform distribution assumption, the internal hash join is superior to the merge scan in terms of CPU time. In the future, the larger internal memory becomes, the more powerful the internal hash join will be.

The advantages of the external hash join are that the depth of the partition tree depends on the smaller relation only (not both) and the processing speed can be fast and the number of I/O accesses small. In our experiments, CPU time, elapsed time and the number of I/O accesses required for the external hash join were generally less than those for the merge scan. However, we can't conclude that the external hash join is superior to the merge scan because we can't say our sort method is the best.

Other relational algebra operations, such as duplicate elimination in projection, intersection, union, difference and so on, are based on equality. To implement these operations using hashing will lead to efficient relational database systems. The idea of recursive partition is applicable to these operations.

Acknowledgements

I am indebted to M.Tezuka, S.Adachi, T.Nakada, R.Take and T.Okazaki for their helpful discussion. I would like to thank A.Makinouchi, who gave me the chance to explore the hash join technique and a lot of helpful advice.

References

Aho, A. V., Hopcroft, J. E., and Ullman, J. D., 1974, The Design and Analysis of Computer Algorithms, Addison Wesley.

Babb, E., 1979, Implementing a Relational Database by Means of Specialized Hardware, ACM Trans. Database Syst., Vol.4, No.1, pp.1-29.

Blasgen, M. W., and Eswaran, K. P., 1977, Storage and access in relational data bases, IBM Systems Journal, Vol.16, No.4, pp.363-378.

Bratbergsengen, K., 1984, HASHING METHODS AND RELATIONAL ALGEBRA OPERATIONS, Proc. Int. Conf. on VLDB, pp.323-333.

DeWitt, D. J., Katz, R. H., Olken, F., Shapiro, L. D., Stonebraker, M. R., and Wood, D., 1984, IMPLEMENTATION TECHNIQUES FOR MAIN MEMORY DATABASE SYSTEMS, SIGMOD Record Vol.14, No.2, pp.1-8.

Gotlieb, L. R., 1975, Computing joins of relations, Proc. International Conf. on Management of Data (ACM), pp.55-63.

Kim, W., 1980, QUERY OPTIMIZATION FOR RELATIONAL DATABASE SYSTEMS, Ph.D. Thesis, Report No. UIUCDCS-R-80-1034, Dept. of Computer Science, Univ. of Illinois.

Knuth, D. E., 1973, Sorting and Searching, The Art of Computer Programming Vol.3, Addison Wesley.

Makinouchi, A., Tezuka, M., Kitakami, H., and Adachi, S., 1981, THE OPTIMIZATION STRATEGY FOR QUERY EVALUATION IN RDB/V1, Proc.Int.Conf. on VLDB, pp.518-529.

Merrett, T. H., and Otoo, E., 1979, DISTRIBUTION MODELS OF RELATIONS, Proc. Int. Conf. on VLDB, pp.418-425.

Ullman, J. D., 1980, Principles of Database Systems, Pitman.

DATA ORGANIZATION METHOD FOR THE PARALLEL EXECUTION

OF RELATIONAL OPERATIONS

C. Thomas Wu

Northwestern University
Department of Electrical Engineering
and Computer Science
Evanston, Illinois 60201

ABSTRACT

We propose the use of an indexing technique called M-cycle hash file as a data organization method for storing relations in the database machine. The advantage of using the M-cycle hash file is the effective parallel execution of selection, projection, and join operations without requiring any specialized hardware. With no specialized hardware, our database machine has a much simpler design than other database machines. Moreover, the use of M-cycle hash file reduces the complexity of software, because there is no directory for maintaining the indices. In this paper, we review the M-cycle hash file, present the architecture of our database machine, and describe the high-level algorithms for the parallel execution of relational operations.

1. INTRODUCTION

In this paper, we describe how the indexing technique called M-cycle hash file [17, 18] can be used in the database machine architecture to support the parallel execution of selection, projection, and join operations of the relational database system.

Many database machine architectures have been proposed in the last decade to improve the performance of relational database management systems. More recently proposed are the architectures described in [9, 11]. The architectures that utilize the processor-per-track approach [14, 12, 16] or processor-per-head approach [2, 1] are considered obsolete today because the processor-per-track approach cannot store a very large database due to hardware limitations and because the processor-per-head approach cannot process the complex queries any better than a conventional processor (see [3]). As prescribed in [3, 15], we advocate the use of "off-the-shelf" conventional processors and moving-head disk drives to construct the cost-effective, efficient relational database machine. The relational database machine that we propose in this paper utilizes only conventional hardware.

The author's current address is Naval Postgraduate School, Department of Computer Science Code 52, Monterey, CA 93943.

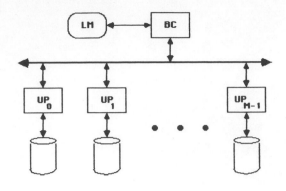

BC – Backend Controller
LM – Local Memory for BC
UP – Unit Processor
⊖ – Storage Device

Fig. 1. A simplified diagram of our proposed machine.

A simplified diagram of our proposed database machine is shown in
Fig. 1. BC is the back-end controller which communicates with the host and
controls the unit processors UP. Each unit processor UP_i is assigned to a
storage device; in other words, each storage device has a dedicated unit
processor. The architecture will be discussed more fully in Sec. 4. Since
a unit processor is dedicated to a single storage device, there is no costly
and complex interconnection network (e.g, crossbar switch) typically employed in the off-the-disk approach such as the one proposed in [6]. Our
database machine without any specialized processor, storage device, or interconnection network has a much simpler design than other database machines.

Since no specialized hardware is employed in our database machine,
the parallel execution of relational operations is achieved by an effective
data organization method. The indexing technique called M-cycle hash file
is the data organization method we use to execute the relational operations
in parallel. There are three significant characteristics of M-cycle hash
file that make the technique an ideal data organization method for the relational database machine. First, M-cycle hash file distributes the tuples
of relation evenly among the storage devices so that each unit processor
performs an equal amount of work in processing relational operations. Note
that this is not just dividing the relation randomly into equal pieces. Let
us call the part of the relation stored in a storage device a segment.
M-cycle hash file distributes the tuples so that when the selection operation, for example, is executed, each unit processor need only to access a
portion of the segment, and the size of the portion that each unit processor
accesses is the same no matter what selection criteria are applied. Second,
M-cycle hash file requires no directory (such as the one employed in a B^+-
tree) in maintaining indices. This will reduce the complexity of software
drastically. And third, M-cycle hash file preserves the order of tuples.
When the order of tuples is maintained, the relational operations such as
join and projection require no sorting, and the range query is efficiently
supported in the selection operation.

This paper is organized as follows. We first define the relational
operations in the next section. Then, we informally describe the M-cycle
hash file in Sec. 3 with an illustrative example. The emphasis of our de-

scription is on "what" it does, rather than "how" it does. In Sec. 4, we describe the architecture of our database machine and how the relational operations are carried out in it. We conclude the paper in Sec. 5.

2. TERMINOLOGY

A relation scheme R is a finite, ordered set of attribute names $\{A_1, \ldots, A_n\}$. To each attribute name A_i there corresponds a finite set of values D_i called a <u>domain</u>. An element of $D_1 X \cdots X D_n$ is called a <u>tuple</u>. A relation R on a relation scheme $\underline{R}$ is a finite set of tuples $\{t_1, t_2, \ldots, t_p\}$. Let t be a tuple and X the subset of attributes, then $t(A_i)$ denotes the value of the ith attribute of tuple t, and $t(X)$ denotes the values of X attributes of tuple t.

We define the selection operation as

$$\sigma_{COND}(R) = \{t \in R | t \text{ satisfies COND}\}$$

where $COND = C_1 \wedge C_2 \wedge \cdots \wedge C_m$ and each C_i is an expression to specify the selection condition for attribute A_i. The expression C_i can be (i) v, an exact domain value; (ii) *, a don't care condition; or (iii) [l, u], a range of domain values. If $C_i = *$, then every tuple t in R satisfies C_i. If $C_i = v$, then every tuple in R with $t(A_i) = v$ satisfies C_i. If $C_i = [l, u]$, then every tuple in R with $l \leq t(A_i) < u$ satisfies C_i. In the literature on file organization, the selection is called (i) range query if all C_i's are ranges of domain values; (ii) partial-match query if C_i's are either * or v; and (iii) exact-match query if all C_i's are v. Note that the partial-match and exact-match queries are special cases of range query, because [l, u] is equal to v if l = u and to * if l and u are possible minimum and maximum domain values.

We define the projection operation as

$$\pi_X(R) = \{t(X) | t \in R\}.$$

We define the join operation $|X|$ as

$$R|X|S = \{t \in T | t(R) = r \in R, \ t(S) = s \in S, \text{ and}$$

$$r(R \cap S) = s(R \cap S)\}.$$

The attributes that are common in $\underline{R}$ and $\underline{S}$ are called <u>joined attributes</u>.

3. M-CYCLE HASH FILE

M-cycle hash file is originally proposed as an efficient file organization technique capable of indexing multidimensional key space records. The technique is one of dynamic hash files, where the number of buckets varies with the insertion and deletion operations. It is directory-free, i.e., it maintains no directory for the indexing unlike other dynamic hash files such as extendible hashing [7], trie hashing [13], and dynamic hashing [10]. It adapts the interpolation hashing scheme [4, 5] to the independently accessible M storage devices (e.g., disk drives), and by allocating the records evenly among the M storage devices, it reduces the data retrieval time proportional to the number of storage devices utilized. This time reduction in data retrieval is achieved without an increase in data maintenance (insertion and deletion) operations.

In this section, we describe the M-cycle hash file informally by using an illustrative example. We will use the words "relation" and "tuple" in-

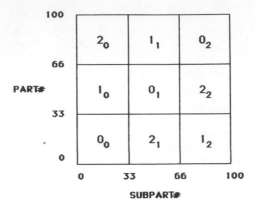

Fig. 2.   Initial configuration.

stead of "file" and "record" used in the original paper. The description
is essentially "what" it does, for this is all that is required to understand this paper. Readers interested in "how" the scheme carries out the
data maintenance and retrieval operations are referred to [18].

Let us use the relation COMPONENT with attributes PART# and SUBPART#
as our example. We assume here that the ranges of both PART# and SUBPART#
are between 0 and 100 (normally we can expect that the lower and upper
bounds of the key space are known a priori, but if not, then we would use
the technique described in [5]). We set the number of UPs equal to three,
i.e., M = 3. We use the expression t(PART#) and t(SUBPART#) to denote
the respective attribute values of tuple t.   Figure 2 shows the initial
configuration of how the relation is divided into nine equal regions. For
each region, there is a bucket in a storage device to store tuples that
belong to the region. Since every tuple is stored in exactly one bucket,
there is no duplication of data. A region is identified as U_P, where U is
the unit processor number and P the bucket address within the storage device attached to the unit processor U. For example, all tuples in region
1_0, i.e., all tuples t such that $33 \leq t(PART\#) < 66$ and $0 \leq t(SUBPART\#) < 33$,
are stored in bucket #0 of unit processor #1's storage device. Note that
the nine regions are distributed evenly among the three storage devices.
A bucket is physically implemented as a chain of pages, with each page having a capacity of c tuples. The capacity of a page is set equal to the
track, cylinder, or some other convenient size depending on the physical

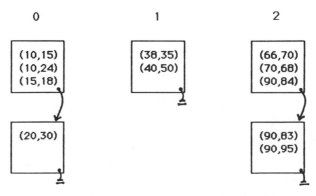

Fig. 3.   Content of storage device #0.

532

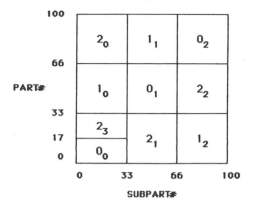

Fig. 4. After the region 0_0 has split.

property of the storage device. Figure 3 shows the content of unit processor #0's storage device.

The increase or decrease of the number of regions is determined by the storage utilization factor, which is defined to be the ratio of the number of tuples to the number of available slots. In Fig. 3, the current storage utilization factor is 11/15.* The lower and upper bounds of the storage utilization factor are set by the user (most probably the database administrator). Let α be the current storage utilization factor and α_L and α_U, respectively, the lower and upper bounds. If α exceeds α_U, then the region is split into two regions and all tuples that belong to the unsplit region are distributed between two split regions. Let us suppose that α_U = .70 and the tuple (40,50) was the last one inserted. Before (40,50) was inserted, α was 10/15 < .70. After (40,50) is inserted, the state which is depicted in Fig. 3, α becomes 11/15 > .70, and therefore, the region is split.

The next region to split is 0_0 in Fig. 3, and Fig. 4 is the state after region 0_0 is split. The region to split next is not necessarily

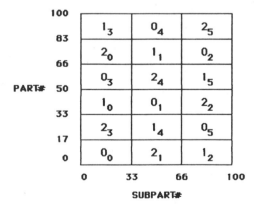

Fig. 5. After all regions of Fig. 3 have split.

---

*This is an example, and for the sake of simplicity, we define the storage utilization factor for a single storage device. In the actual implementation, there is only one storage utilization factor, which is globally defined over the M storage devices.

the region where the last record is inserted. The next region to split
follows the predetermined order of splitting, which is defined to be 0_0,
1_0, ..., 0_x, 1_x, ..., $(M-1)_x$, where the term x is the current largest
bucket address. When the last region $(M-1)_x$ is split, the number of regions
has doubled and the splitting will start all over from 0_0. Note that the
new largest bucket address will increase to $x + 2^1 M$, where i is the number
of times the state has doubled from the initial configuration. After all
the regions in Fig. 3 split, the number of regions has doubled, as shown
in Fig. 5, and the largest bucket address has increased from 2 to 2 +
$2^0 * 3 = 5$.

When the region 0_0 splits, the tuple (20,30) now moves to the newly
created bucket 2_3; all other tuples remain in bucket 0_0. Whether a tuple
remains in the same bucket or moves to the newly created bucket is determined by the hash function employed by the scheme. The address of the
newly created region is computed so that after all regions are split, there
will again be a cyclic permutation of storage device numbers. Figure 5 is
the state where all regions in Fig. 3 have split. Along the vertical coordinate, we have a cyclic permutation of (i) 0,2,1,0,2,1; (ii) 2,1,0,2,1,0;
(iii) 1,0,2,1,0,2, and along the horizontal coordinate, we have a cyclic
permutation of (i) 0,2,1; (ii) 2,1,0; and (iii) 1,0,2. Having a cyclic
permutation of unit processor numbers along all coordinates guarantees the
run time performance improvement that we desire, i.e., M times faster on
the average with M storage devices. The proof of this claim can be found
in [18].

The merge operation is an inverse of the split operation. When α becomes smaller than α_L the two regions that were split last will be merged
and all tuples in those two regions are collected into one bucket.

By preserving the order of tuples, M-cycle hash file efficiently supports the processing of a range query. And because the buckets are distributed among the independently accessible storage devices, the range query
can be processed in a parallel fashion. Moreover, the process time is
M times faster on the average by utilizing M unit processors than just
utilizing a single unit processor because the buckets are evenly distributed among the M storage devices. Consider, for example, the range query
($17.5 \leq$ part $3 < 66$, $33 \leq$ SUBPART# < 100). Buckets for regions # 1_4, 0_5,
0_1, 2_2, 2_4, and 1_5 must be retrieved; there are exactly two buckets per
storage device, and hence each unit processor performs the same amount of
work. Of course, the number of pages allocated for each bucket differs,
but the variance is very small.

The performance improvement of M-cycle hash file can be stated formally as

$$A^1(N)/M \leq A^M(N) \leq A^1/M + O(1) \tag{1}$$

where $A^M(N)$ is an average processing time for a range query on M storage
devices containing N tuples. The proof can be found in [18].

Parallel sequential access is also supported by M-cycle hash file.
It is very flexible in that the tuples can be sequentially accessed in parallel in the order of any attribute values or combination of attribute
values. For the sample relation COMPONENT, the parallel sequential access
can be done either in the order of PART# attribute values (i.e., row-major
order) or in the order of SUBPART# attribute values (i.e., column-major
order). Such flexibility is critical in performing projection and join
operations. If we wish to project COMPONENT relation on PART#, we sequentially access tuples in the row-major order. Duplicate elimination is
simplified because the tuples are accessed in the order of PART# values.

Moreover, the operation is carried out in parallel with each unit processor retrieving the same number of buckets (not exactly the same, but the difference is $O(1)$, as stated in Eq. (1)).

In summary, the outstanding features of M-cycle hash file are:

(1) there is no redundancy of storage space;

(2) the split and merge operations (to increase or decrease the number of regions) are localized to two unit processors;

(3) there is no limit on the number of storage devices used;

(4) the address computations for the query processings are very simple;

(5) the run time performance is very good.

By utilizing this M-cycle hash file organization technique for storing relations, the efficient parallel execution of relational operations is possible without any specialized hardware.

4. ARCHITECTURE

There are three major functions performed by the back-end controller BC. The first function is communication with the host. BC receives the user's transaction* request and sends the result back to the host. The second function is the creation of a process for each transaction request. For each process created, BC assigns a unique identification number. The third function is the management of the local memory LM. Each process is allocated a segment of LM by BC to store the intermediate and final results of operations. The actual allocation scheme is peripheral to the main subject and, therefore, will not be discussed here. There is a result buffer RB for each unit processor. The RBs are used to save the tuples sent back from the unit processors. The RB is implemented as a queue and the front of the RB can be read simultaneously by more than one process.

The process will send the control message to the appropriate unit processors to execute the operations. The format for the control message is

<PID, RID, ADDR, OP, EXPR>

where the terms stand for process id number, relation id number, bucket addresses, operation code, and expression, respectively. The PID identifies the process that is sending this control message. The RID identifies the relation on which the operation must be executed. The ADDR is the set of bucket addresses where the tuples of the relation that needs to be accessed are stored. The OP identifies the operation, and EXPR provides the necessary information to carry out the operation. The following table shows the value of EXPR for the corresponding operation OP:

OP	EXPR
SELECT	COND
PROJECT	attribute names
JOIN	joined attribute names

---

*In this paper, a <u>transaction</u> is a collection of relational operations.

We shall now describe the high-level algorithms for the selection, projection, and join operations. Update operations (insertion, deletion, and modification of a record) are carried out similarly to the selection operation, and they will not be discussed here. In the following, operand denotes the tuples of relation that need to be processed. The operand may not always be in the storage devices. It may also be in the local memory LM. For instance, if we desire to perform selection on the result of a join operation, the join operation is first executed and the intermediate result saved in LM. The selection is then performed on the intermediate result stored in LM.

The selection operation is carried out as follows. If the operand is in LM, then for each tuple in LM, selection criteria are applied and the tuples satisfying the criteria are saved in LM. If the operand is in the storage devices, the process will compute the addresses of buckets to be retrieved and send the control messages to UPs. Then the result is collected from the result buffers RB and saved in LM.

The resulting tuples from the unit processors are sent to BC as a packet which has the format <PID, TUPLES> (or <PID, RID, TUPLES>, depending on the operation), where TUPLES is a collection of tuples. The size of the packet is determined by the size of i/o bandwidth. The process can acknowledge the result of its request by checking the PID. A unit processor will send the terminating message <PID, UPID, 'finish'> to BC after all the tuples are returned. The process will know that the operation has been completed by reading the terminating messages from the UPs.

Let us assume that the selection criteria are $17.5 \leq t(PART\#) < 66$ and $33 \leq t(SUBPART\#) < 100$. Buckets to be retrieved are buckets #1 and #5 in unit processor #0's storage device, buckets #4 and #5 in unit processor #1's storage device, and buckets #2 and #4 in unit processor #2's storage device. If the PID is 0899, the control message

<0899, COMPONENT, (1,5), SELECT,

$(17.5 \leq t(PART\#) < 66$ and $33 \leq t(SUBPART\#) < 100)$ >

will be sent to unit processor #0. Similar control messages will also be sent to unit processors #1 and #2. The tuples satisfying the selection criteria are returned to the corresponding result buffers RB by the UPs. The process collects the tuples and saves the result in the local memory LM. After reading <PID, 0, 'finish'>, <PID, 1, 'finish'>, and <PID, 2, 'finish'>, the process acknowledges the completion of the selection operation and moves on to the next operation in the transaction.

The projection operation will proceed as follows. If the operand is in LM, then the projection operation is performed on the tuples in LM and the result saved in LM. If the operand is in the storage devices, the process first computes the bucket addresses in the order of projected attributes.* Then, it sends the control messages to the UPs, collects the result from the RBs eliminating the duplicates, and saves the result in LM.

Again using Fig. 5, let us suppose that we would like to project COMPONENT onto PART#. The control message <PID, COMPONENT, (0,5,1,3,2,4),

---

*We say "bucket addresses b_1, b_2, ..., b_q are computed (retrieved) in the order of projected attributes" if a relation is projected on X and for all tuples $t_i \in b_i$, $1 \leq i \leq q-1$, $t_i(X) < t_{i+1}(X)$. Analogously, we say "bucket addresses b_1, b_2, ..., b_q are computed (retrieved) in the order of joined attributes."

PROJECT, PART#> is sent to unit processor #0, <PID, COMPONENT, (2,4,0,5,1,3),
PROJECT, PART#> to unit processor #1, and <PID, COMPONENT (1,3,2,4,0,5),
PROJECT, PART#> to unit processor #2. Each unit processor retrieves the
buckets in the order specified, eliminates the duplicates, and sends the
result back to BC. The process terminates the operation when the terminating messages are received from all unit processors.

The join operation will proceed as follows. Let R1 and R2 be the two
operands involved in the join operation. If one operand R1 is in LM and
another operand R2 is in the storage devices, then the process computes
bucket addresses of R2 in the order of joined attributes and sends the control messages. Unit processors retrieve the tuples of R2, and the results
are collected in RBs. The tuples of R1 stored in LM must be sorted in the
order of joined attributes. This is the only exception which requires a
sorting routine. Tuples of R1 stored in LM and R2 stored in RBs are now
joined and the result is saved in LM. Since tuples of R1 and R2 are retrieved in the order of joined attributes, the merging is simple. If both
operands are in the storage devices, then the process computes bucket addresses of R1 and R2 in the order of joined attributes. The control messages are sent to UPs and results from UPs are collected in RBs. The
process then joins the tuples of R1 and R2 now stored in RBs and saves
the result in LM. The joining of tuples of R1 and R2 (both stored in RBs)
proceeds as follows. The join operation needs two segments in LM, and we
denote them as LM_1 and LM_2. If a tuple read from RB is of R1, then it is
moved to LM_1, otherwise the tuples of R2 will be joined with appropriate
tuples of R1 in LM_1 and the result saved in LM_2. Note that the process
must retrieve the tuples from RBs in a synchronized manner because the
unit processors return the tuples independently of each other. For example,
it is possible to have all resulting tuples in one RB_i while none is available in another RB_j. In this case, the process must wait until some results move into RB_j critical, because the work load of each unit processor
is well balanced (due to the equal distribution of tuples among the storage
devices), and thus we can expect an approximately equal rate of flow of
tuples into the Rbs.

CONCLUSION

We have shown in this paper that parallelism can be obtained in a relational database machine by the use of effective data organization methods
instead of specialized hardware. By using the M-cycle hash file organization technique, the parallel execution of relational operations is realized.
This approach results in simpler hardware and software. The hardware of
our database machine is simpler because there are no specialized processors,
storage devices, or interconnection network. The software of our database
machine is simpler because there is no need to maintain directories. Our
proposed database machine can be summarized as a software back-end, page
indexing, MIMD (Multiple Instruction Multiple Data stream), and off-the-
disk processing relational database machine.

Our database machine is similar to the MDBS [8] in that both do not
use any specialized processors, storage devices, or interconnection network.
Both distribute data among the storage devices without any duplication. A
major difference between the two proposals is that MDBS requires directories
while ours does not. Without directory maintenance, the software of our
database machine should be simpler. The possible tradeoff is in performance.
With directories, the user may adjust the data allocation among the storage
devices so that the most frequently executed operations can be processed
most efficiently. The analysis of the differences between the two is yet
to be investigated.

REFERENCES

1. E. Babb, "Implementing a Relational Database by Means of Specialized Hardware," ACM TODS, Vol. 4, No. 1 (1979).

2. J. Banarjee, R. I. Baum, and D. K. Hsiao, "Concepts and Capabilities of a Database Computer," ACM TODS, Vol. 3, No. 4 (1978).

3. H. Boral and D. J. DeWitt, "Database Machines: An Idea Whose Time Has Passed? A Critique of the Future of Database Machines," Database Machines (O. L. Leilich and M. Missihoff, eds.) (1983).

4. W. A. Burkhard, "Interpolation-based Index Maintenance," BIT, Vol. 3, No. 3 (1983).

5. W. A. Burkhard, "Index Maintenance for Non-form Record Distributions," Proceedings of 1984 Principles of Database Systems (1984).

6. D. J. DeWitt, "DIRECT-A Multiprocessor Organization for Supporting Relational Database Management Systems," IEEE Transactions on Computers, Vol. C-28, No. 6 (1979).

7. R. Fagin, J. Nievergelt, N. Pippenger, and J. R. Strong, "Extendible Hashing - A Fast Access Method for Dynamic Files," ACM TODS, Vol. 4, No. 3 (1979).

8. X. He, M. Higashida, D. S. Kerr, A. Orooji, Z. Shi, P. R. Strawser, and D. K. Hsiao, "The Implementation of a Multibackend Database System (MDBS): Part II - The Design of a Prototype MDBS," Advanced Database Machine Architecture (D. K. Hsiao, ed.) (1983).

9. D. K. Hsiao, Advanced Database Machine Architecture, Prentice-Hall, Inc. (1983).

10. P. A. Larson, "Dynamic Hashing," BIT, Vol. 18, No. 2 (1978).

11. O. L. Leilich and M. Missikoff, Database Machines, Springer-Verlag (1983).

12. S. C. Lin, D. C. P. Smith, and J. M. Smith, "The Design of a Rotating Associative Memory for Relational Database Applications," ACM TODS, Vol. 1, No. 1 (1976).

13. W. Litwin, "Trie Hashing," Proceedings of the 1981 ACM SIGMOD Conference (1981).

14. E. A. Ozkarahan, S. A. Schuster, and K. C. Smith, "Rap - Associative Processor for Database Management," AFIPS Conference Proceedings, Vol. 44 (1975).

15. G. Z. Qadah, "A Relational Database Machine: Analysis and Design," Ph.D. Thesis, University of Michigan (1983).

16. S. Y. W. Su and G. J. Lipovski, "CASSM: A Cellular System for Very Large Data Base," Proceedings of Very Large Data Bases Conference (1975).

17. C. T. Wu and W. A. Burkhard, "Multiple Storage Units for Dynamic File Organization," Proceedings of 1983 Conference on Information Sciences and Systems (1983).

18. C. T. Wu and W. A. Burkhard, "Multiple Storage Units for Efficient Associative Retrieval," Computer Science Technical Report No. CS-064, Department of EECS, University of California, San Diego (1983). Extended version to appear in ACM Transactions on Database Systems.

IMPLEMENTATION OF INFERENTIAL RELATIONAL DATABASE SYSTEM

Toshihisa Takagi[1], Fumihiro Matsuo[2], Shooichi Futamura[2],
and Kazuo Ushijima[1]

Department of Computer Science and Communication Engineering
Kyushu University 36, Hakozaki, Fukuoka 812, Japan[1]

Computer Center, Kyushu University 91, Hakozaki, Fukuoka 812,
Japan[2]

ABSTRACT

Tsuno has been developed to retrieve the relational database created by
Adbis which was developed for supporting the construction of scientific
database systems at the Computer Center, Kyushu University. Tsuno is an
interactive retrieval system having the inference facilities based on Horn
set refutation procedures and is now used as the query system for two
databases. This paper describes the implementation of Tsuno.

INTRODUCTION

We developed a database integration support system called Adbis[1] in
1981 to aid in building scientific database systems at universities. Adbis
is a set of subroutines for constructing an inferential relational database
system[2] which has inference functions based on Horn set refutation
procedures[3]. In 1982 we constructed the crystallographic database and
developed an interactive system, called Tsuno, for retrieving this database,
using Adbis at the Computer Center of Kyushu University.

Tsuno has the following features:

(1) provides users with the facility to define virtual relations by use
    of Horn clauses. This facility produces the disk space efficiency,
    since the facility reduces the redundancies of relations.

(2) has a traditional information retrieval command interface for end-
    users unfamiliar with descriptions of Horn set.

(3) can call the programs written in high level procedural languages
    such as Fortran77, and invoke TSS commands. Using this function,
    Tsuno links easily the retrieved data to application programs.

(4) can be used as a logic programming language like Prolog[4] by
    interpreting Horn set procedurally[5].

In this paper we describe the implementation of Tsuno. In particular, we explain internal data structures and inference mechanisms to retrieve relational databases.

OVERVIEW OF TSUNO

System Configuration

The inferential relational database system using Tsuno consists of three files: database, primitives, and Horn set. Figure 1 shows the system configuration.

database: The database is a relational database which consists of a set of relations. The database physically consists of a master file and some data files. In the master file, the definitions of the database are stored: database name, passwords, relation names, types and lengths of domains and so on. The master file is created by the database administrator using Adbis subroutines in the process of the database creation. The administrator gives the definitions of the database by the use of the database definition language. In a data file, a relation is stored. The data file consists of the tabular form and the inverted form. The tabular form expresses physically a set of tuples. The inverted form is a set of indexes for retrieving the tuples from the domain values quickly. We call the database the extensional data. We also call the relations stored in the extensional data the base relations.

primitives: The primitives consist of executable programs written in high level procedural languages such as Fortran77 and the definitions of these programs. We call the executable programs procedures. Each definition includes the information concerning a procedure: a number of arguments, types and lengths of arguments, information about which argument is an input and/or output parameter. This definition is described by the use of the procedure definition language.

Horn set: A Horn clause is a clause containing at most one positive literal. A Horn set is a set of Horn clauses. Following Gentzen[6], we express a Horn clause by the form

$$A_1, \ldots, A_m \rightarrow B_1, \ldots, B_n \qquad (n \leqq 1)$$

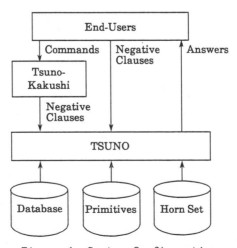

Figure 1. System Configuration

where $A_1,...,A_m$, $B_1,...,B_n$ are atoms. We call a clause a positive clause if both m=0 and n=1; a negative clause if both m>0 and n=0; a mixed clause if both m>0 and n=1; an empty clause if both m=0 and n=0. If all the terms in the clause are constants, we call the clause a ground clause. A regular clause is either a positive clause or a mixed clause. We call a set of regular clauses a regular set. The refutation is to obtain the empty clause form R∪{Q} by the resolution operation[7]. Where R is a given regular set, Q is a given negative clause, and {Q} means a set which has one element Q. A Horn set in Fig.1 means a regular set and physically is an ordinary text file written in the clausal form. We call the relation defined by Horn set a virtual relation. We also call the regular set the intensional data.

Tsuno: Tsuno is an interactive database retrieving system, which receives a negative clause as a query from users, performs refutations, and returns answers to users. We call a negative clause a closed query if it has no variables, and an open query otherwise. To a closed query, Tsuno answers true(=T) or false(=F) according as the refutation derives the empty clause or not. To an open query, Tsuno returns a set of tuples of constant values substituted for variables of the query in the refutations when the empty clause is derived. Otherwise, Tsuno answers false(=F). Regarding the base relations in the database and the procedures in the primitives as a set of positive ground clauses, Tsuno retrieves the database and calls the primitives in the refutation process.

Tsuno-Kakushi: Tsuno-Kakushi was developed for the end users unfamiliar with predicate logic. This interface receives traditional information retrieval system commands such as FIND, DISPLAY from users, translates their commands into negative clauses, and transfers them as queries to Tsuno. Tsuno-Kakushi draws the syntax from DIALOG and ORBIT[8]. But the use of Tsuno-Kakushi is limited only for the bibliographic data.

Examples

We show the notation of Tsuno and Tsuno-Kakushi, and the formats of answers by examples. The following examples are the retrieval of the crystallographic database[9] by Tsuno or Tsuno-Kakushi. The underlined parts of Figure 2 and Figure 3 mean the inputs from users; the others except "READY" are outputs from Tsuno.

Example.1. Figure 2 shows how to invoke Tsuno system and select the database, and presents the retrieval of chemical compounds by keywords and authors. "READY" is a TSS message, which indicates that commands can be received. "TSUNO" is a command for invoking Tsuno system. Tsuno returns "." as a prompting message. The negative clause "$DB('CXDB')->" means the selection and opening of the database named CXDB. "$DB" is the predicate symbol prepared by Tsuno and we call such a predicate a system predicate. In Tsuno, every input is in the clausal form except END command. "=T" implies that the opening of the database is successful. The negative clause in the next line denotes the search of the compounds of which keyword(KY) is "TRIAZOLE" and of which the author(AUTH) is "WERNER,P.-E.", and indicates the display of the compound name(CN), the molecular form(MF), and the bibliography(BI) of retrieved compounds. KY, AUTH, and BIB are base relations that are defined in the extensional data. "$A" is also the system predicate and "$A(CN,MF,BI)" indicates the display of the all constants substituted for variables CN, MF, and BI. The terms which are not enclosed by single quotes are variables. Variables ID, CO, and YR mean a sequential number attached to a compound, a journal coden, and publication year respectively. Tsuno finds three compounds that satisfy the condition represented by the above negative clause. The outputs, below the line "=1", imply the answers to the query from Tsuno. "=1, "=2", and "=3" correspond to the first, second, and third compound found by Tsuno respectively.

541

```
READY
TSUNO
.$DB('CXDB')->;
=T
.KY(ID,'TRIAZOLE'),AUTH(ID,'WERNER,P.-E.'),BIB(ID,CN,MF,BI,CO,YR)->$A(CN,MF,BI);
=1
CN=4-AMINO-3-(2,6-DICHLOROBENZYLIDENE-HYDRAZINO)-1,2,4-TRIAZOLE
HYDROCHLORIDE TRIHYDRATE
MF=C9 H9 CL2 N6 +, CL1 -, 3(H2 O1)
BI=L.GOTHE,S.SALOME,P.-E.WERNER:CRYST.STRUCT.COMMUN.,8,1,1979,11.32.10
=2
CN=4-AMINO-3-(2,6-DICHLOROBENZYLIDENEHYDRAZINO)-5-METHYL-1,2,4-TRIAZOLE
HYDROCHLORIDE
MF=C10 H11 CL2 N6 +, CL1 -
BI=P.-E.WERNER:  CRYST.STRUCT.COMMUN.,6,69,1977,9.32.33
=3
   (omitted)
```

figure 2. Retrieval using Tsuno

Example.2. Figure 3 is an example of Tsuno-Kakushi commands. FIND
command denotes the search of the compounds of which keyword(KW) is
"TRIAZOLE" and of which the author(AU) is "WERNER,P.-E." "=3" in the next
line means that Tsuno-Kakushi finds three compounds. KW is the second domain
name of the base relation KY. AU is also the second domain name of the base
relation AUTH. DISPLAY command indicates the output of the compound
name(CN), the molecular form(MF), and the bibliography(BI) of retrieved
compounds. CN, MF, BI are domain names of BIB. The results of two commands
are also identical to those of example 1.

INTERNAL DATA STRUCTURES

Inner Forms

Tsuno deals with two different forms of Horn clauses: clausal forms and
inner forms. Clausal forms are used for communicating with users. Inner
forms were designed to improve the time and space efficiencies in the
refutation process. Tsuno translates the clausal forms from users into the
inner forms and then executes the refutation on them. The clausal form

$$P_1,\ldots,P_{m-1} \rightarrow P_m$$

is translated into the inner form

$$m\ p_1\ldots p_m\ n_1\ldots n_m\ [t_1]\ldots[t_m]$$

where m is a number of literals in the clause, p_i is a coded number of
predicate symbol P_i, n_i is a number of terms of P_i, $[t_i]$ is a sequence of
coded terms $t_{i1},\ldots,t_{in_i}$ of P_i. All coded symbols m, p_i, n_i, t_{ij} are bit
strings of two bytes length in the inner form.

Tsuno can perform the refutation procedure efficiently on inner forms
by the following reasons:

   (1) since Horn clauses are represented in Polish notation, unification
       procedures[3] can be executed quickly.

542

.FIND KW=TRIAZOLE   AU=WERNER,P.-E.
=3
.DISPLAY CN MF BI
=1
CN=4-AMINO-3-(2,6-DICHLOROBENZYLIDENE
-HYDRAZINO)-1,2,4-TRIAZOLE HYDROCHLORIDE
TRIHYDRATE
MF=C9 H9 CL2 N6 +, CL1 -, 3(H2 O1)
   (omitted)

.END
READY

Figure 3. Retrieval using Tsuno-Kakushi

(2) the renaming process[3] of variables before the resolution can be
    carried out by simple bit operations, because variables are encoded
    into bit strings of two bytes length.

(3) the memory space for the refutation is reduced by the use of inner
    forms.

Management Tables

Tsuno performs the refutation, retrieves the database, and invokes the
primitives using the following management tables.

database definition table: Each entry of this table corresponds to one
relation and includes the definitions of the relation: a relation name, a
number of domains constituting the relation, relationships among domains,
types and lengths of domains and so on. When the user indicates for Tsuno to
open the database, Tsuno reads the definitions from the master file and puts
them into this table. In example.1, when the user inputs the clause
"$DB('CXDB')->", the definitions of the base relations KY, AUTH, and BIB are
registered.

procedure definition table: Each entry of this table corresponds to one
procedure and contains a number of arguments in the procedure, types and
lengths of arguments, information about which argument is an input and/or
output parameter. When the user indicates for Tsuno to use procedures, Tsuno
reads the definitions of procedures from the primitives and puts them on
this table.

regular set table: This table contains a set of regular clauses in the
inner form. Since a regular set defines virtual relations, this table may be
called a virtual relation definition table.

relation management table: This table is created by linking the
relations having the identical predicate symbols registered in the above
three table. (As mentioned at the system configuration, Tsuno regards base
relations in the database and procedures in primitives as positive ground
clause.) Each entry contains a relation name, a number of terms, a pointer,
called a clause pointer, to one of the above three tables in which this
entry is defined. If the relation of this entry is a virtual relation, that
is, the relation is defined by a regular clause, the pointer of this entry
indicates the address of the clause in the regular set table. If the
relation is a base relation, the pointer indicates the address where the
definitions of the relation are stored.

constant table: each entry contains constant itself and type and length of it.

domain management table: each entry contains type and length of this entry and the correspondence between the domain name and the relation name including this domain for the translation of Tsuno-Kakushi commands.

INFERENCE MECHANISMS

Tsuno retrieves the database through the following steps.

opening databases: When the user indicates for Tsuno to open the database, Tsuno opens the database and creates the database definition table. The relation management table is also created at this time. In example.1 the input of the clause "$DB('CXDB')->' corresponds to this step and the definitions of the base relations KY, AUTH, and BIB are stored in the database definition table.

receiving queries: Tsuno receives a negative clause in the clausal form as a query directly from users or from Tsuno-Kakushi, links the relations in the clause with the relations in the relation definition table, and translates the clause into an inner form. If the query is not a clausal form, Tsuno-Kakushi translates the query into the clausal form using the domain management table and transfers it to Tsuno.

selection of a literal and a clause: Tsuno selects one literal from the given negative clause and selects one clause by looking at the clause pointer of this literal in the relation definition table. Tsuno selects automatically the literal and the clause using the following strategy[10] based on the theoretical consideration concerning the effect of backtracking during deductions on computability of number theoretic functions[11].

[step1] Select a positive clause in preference to mixed clauses. If it is impossible, go to [step2].

[step2] Select a mixed clause so that a positive clause may be selected in the next selection. If it is impossible, select any mixed clause.

This is a strategy that adds "lookahead" to the unit preference strategy which is one of the best refinement on the basic resolution in the field of automated theorem-proving. The information necessary for applying [step2] has been beforehand embedded in the relation management table at the creation stage of this table.

database retrieval: If the selected literal is a base relation, Tsuno takes the definitions of the relation from database definition table, retrieves the database by using these definitions, translates the results into the set of the positive ground clause in the inner form, and then puts them into the regular set table.

calling primitives: If the selected literal is a procedure, Tsuno takes the definitions of the procedure from the procedure definition table, calls the procedure by using the definition, translates the results into the positive ground clause in the inner form, and puts it into the regular set table.

refutations: After the selection, Tsuno performs the resolution operation between the selected clause and the negative clause given as a query. If the result clause of the resolution is empty, Tsuno stops the refutation process. Otherwise, Tsuno repeats the selection and resolution by regarding the result clause as a new query in turn, until the empty clause is derived or it is known that the empty clause can not be derived from the negative clause given by the user. Tsuno carries out these refutations on inner forms in top-down and depth-first manner. Tsuno has also the inference mechanism by the negation as failure[12].

IMPLEMENTATION AND APPLICATIONS

Tsuno is implemented using Adbis on the operating system FACOM OS IV/F4 at the Computer Center of Kyushu University. FACOM OS IV/F4 is basically equivalent to IBM MVS. Tsuno system is written in Fortran 77 and the program size is about 1 mega bytes. Both the tabular forms and the inverted forms in extensional data are implemented by indexing techniques of B-tree[13]. B-trees are created by the direct access method of Fortran 77.

Two scientific database systems using Tsuno are now in practical service. One is XDT(CRYStallographic DaTa system) whose extensional data is created from the Cambridge Crystallographic Database compiled and distributed by the Cambridge Crystallographic Data Center[9]. This database consists of 11 relations, and 49 domains and includes 33,794 compounds. XDT has eight application programs which analyze and draw crystal structures. The second database system is GENAS(GENe Analyzing System)[14] whose extensional data is created from both EMBL Nucleotide Sequence Data Library compiled by the European Molecular Biology Laboratory of West Germany and Atlas of Protein Sequence and Structures compiled by National Biomedical Research Foundation, Georgetown University Medical Center. This database consists of 17 relations, 27 domains and includes 1,481 nucleotide sequences and 2,784 proteins. GENAS has 31 application programs which analyze the nucleotide sequences.

CONCLUSION

Although the selection strategy of Tsuno is founded upon theoretical consideration, the question whether inference mechanisms of Tsuno is sufficient and efficient in practical settings is not yet completely evident, since it does not always perform extremely complicated jobs in the database systems.

To retrieve the database quickly, we are investigating whether to use hashing instead of B-tree, and we intend to improve the performance of refutation procedures.

Tsuno may be regarded as a knowledge base management system, since it can store a number of facts and knowledge in the form of Horn set and has Horn set refuter as an inferential engine. To aid in constructing a knowledge base system, we developed a question-answering interface, called Waku, to provide end-users with frame-based interface. We also constructed as an application of Waku a question answering system.

REFERENCES

1. F. Matsuo, S. Futamura, and T. Takagi, Adbis - An Inferential Relational Database Management System, Trans. Inform. Process. Soc. of Japan, 24, 2, 249-255 (1983)(in Japanese).
2. J. Minker, Search Strategy and Selection Function for an Inferential Relational System, ACM Trans. Database Syst., 3, 1, 1-31 (1978).

3. L. Henshen, and L. Wos, Unit Refutations and Horn Sets, <u>J.ACM</u>, 21, 4, 590-605 (1974).
4. W. F. Clocksin, and C. S. Mellish, "Programming in Prolog," Springer-Verlag, Berlin Heidelberg (1981).
5. R. Kowalski, Predicate Logic as Programming Language, <u>Information Processing 74</u>, North-Holland, Amsterdam, 569-574 (1974).
6. G. Gentzen, Untersuchungen über das Logische Schliessen, <u>Math. Z.</u>, 39, 176-210, 405-431 (1935).
7. D. W. Loveland, "Automated Theorem Proving: A Logical Basis," North-Holland, Amsterdam (1978).
8. G. Salton, and M.McGill, "Introduction to Modern Information Retrieval," Mcgraw-Hill, New York (1983).
9. F. H. Allen, S. Bellard, M. D. Brice, B. A. Cartwright,, A. Doubleday, H. Higgs, T. Hummelink, B. G. Hummclink-Pcters, O. Kennard, W. S. Motherwell, J. R. Rodgers, and D. G. Watson, The Cambridge Crystallographic Data Center: Computer-Based Search, Retrieval, Analysis and Display of Information, <u>Acta Cryst.</u>, B-35, 10, 2331-2339 (1979).
10. F. Matsuo, and T.Takagi, A Horn Set Refuter for Database Manipulation, <u>Trans. Inform. Process. Soc. of Japan</u>, 25, 3, 458-464 (1984) (in Japanese).
11. F.Matsuo, Number Theoretic Functions Calculated by Deterministic Refutation of Horn Sets, ibid., 25, 3, 1984, 437-442 (in Japanese).
12. K. L. Clark, Negation as Failure, in "Logic and Data Bases," H. Gallaire and J. Minker, ed., Plenum, New York (1978).
13. D. Comer, The Ubiquitous B-tree, <u>Comput. Surv.</u>, 11, 2, 121-137 (1979).
14. S. Kuhara, F. Matsuo, S. Futamura, A. Fujita, T. Shinohara, T. Takagi, and Y. Sakaki, GENAS: a Database System for Nucleic Acid Sequence Analysis, <u>Nucleic Acids Res.</u>, 12, 1, 88-89 (1984).

Database Theory

TOWARDS A BASIC RELATIONAL NF^2 ALGEBRA PROCESSOR

H.-J. Schek

Technical University of Darmstadt
Alexanderstr. 24
D-6100 Darmstadt
West Germany

Abstract

Relations with relation-valued attributes (NF^2 relations) are proposed to serve as a model for internal data structures of a data base system. A related NF^2 relational algebra is described and a suitable subset of it is selected to define the interface of a basic NF^2 algebra processor. It will be shown that specific projection-selection-join-sequences expressed in 1NF relations at the conceptual level are mapped to nested projection-selection sequences of the internal of NF^2 relations. Costly joins therefore can be eliminated and substituted by equivalent cheaper linear queries. The selection of the subset of NF^2 algebra operations is motivated by this join elimination and by the restriction to single-scan operations.

1. INTRODUCTION

The relational model of data with relation-valued attributes, also called NF^2 model, is characterized by allowing relations as attribute values, i.e. by dropping the 1NF condition (1NF = First-Normal-Form, NF^2 = NFNF = Non-First-Normal-Form). This model is being discussed in the context of new data base applications e.g. in office systems (as forms in e.g. /SLTC82/), or in engineering applications (as model for complex objects /HL82/), or for textual data /SP82/. Advantages for the representation of dependencies have been seen already in /Ma77/ and /Ko81/ and a complete logical data base design method using non-normalized relations has been developed in /KTT83/. Operations on NF^2 relations and their properties have been discussed in /JS82, FT83, AMM83, AB84/ and a complete relational NF^2 algebra has been defined in /Ja84, SS84/.

In this paper we discuss the aspect of using the NF^2 relational model for a formal description of internal data structures of a data base system. The motivation for this - possibly unexpected - direction is that internal data structures often are not considered with the same theoretical foundation as the conceptual ones. Therefore, the mapping from the conceptual data model to the internal data structures and operations is often left open to the specific implementation and solved on an ad-hoc basis.

The idea here is to provide the NF^2 data structures and operations - or a suitable subset of them - at the interface of a powerful kernel system. Then the mapping from data models at the conceptual level to the kernel system can be described by well defined operations. Especially, this paper considers the design of a basic kernel system interface which is characterized by a subset of the whole NF^2 algebra. The selection of the subset is based on the consideration of the possibilities to map important expressions of a classical relational model at the conceptual level to internal NF^2 representations. Without considering the NF^2 relational algebra the idea of using "compacted" relations in order to better utilize hardware search facilities has already been proposed in /BRS82/.

The paper is structured as follows: First, the data structure of relations with relation-valued attributes is introduced formally in chapter 2. The building blocks of a related relational algebra are explained by examples in chapter 3 and a first justification for a nested projection-selection interface is given in chapter 4 by considering simple expressions. By looking into the transformation of a 1NF interface into possible NF^2 interfaces in chapter 5 we will further justify a subset of the NF^2 algebra, the nested projection-selection-projection expressions, to become the interface of a basic relational NF^2 algebra processor.

2. RELATIONS WITH RELATION-VALUED ATTRIBUTES (NF^2 RELATIONS)

2.1 The Data Structure

A short formal definition of the data structure is given which preserves the usual notions of schema, value and attributes of a relation known from the conventional 1NF model. In the 1NF case δ sets $D_j, i = 1, 2, \ldots, \delta$ are given which serve as "basic domains". Their elements will not be regarded as decomposable by the DBMS. For our purpose it is sufficient to introduce one basic domain $D := D_1 \cup \ldots \cup D_\delta$. This setting simplifies the discussion without loosing generality. The restriction of attribute values to be in certain subsets of D (e.g. D_1) can be regarded as a special kind of integrity constraints which are not of interest in this paper. We further assume a given set **NAME** the elements of which will be used for naming purposes.

As abbreviation we denote with $\mathbf{P}(S)$ the powerset of a set S and with $S_1 \times S_2$ the Cartesian product of two sets S_1 and S_2. $\emptyset$ is the empty set. For an element $s \in S_1 \times S_2$ of a Cartesian product we write $s = \langle s_1, s_2 \rangle$ with $s_1 \in S_1$ and $s_2 \in S_2$.

The domain of 1NF relations consists of powersets of Cartesian products of D. If we allow relations as attribute values, we need the following

Definition 1 "Complex Domains C": The set **C** of "complex domains" is defined by
(1) $D \in \mathbf{C}$
(2) $C_1, C_2 \ldots C_k \in \mathbf{C} \Rightarrow \mathbf{P}(C_1 \times \ldots \times C_k) \in \mathbf{C}$
(3) nothing else is in **C**

Definition 1 simply allows powersets of Cartesian products of the (basic or complex) domains to be also a complex domain. In the 1NF case the schema of a relation indicates the structure of the relation and serves for naming. These two roles will be visible by the next

Definition 2 "Schema S and Description B":
The set of allowed schemata **S** and descriptions **B** is defined by
(1) $X \in \mathbf{NAME} \Rightarrow \langle X, \emptyset \rangle \in \mathbf{B}$
(2) $b_i := \langle X_i, S_i \rangle \in \mathbf{B}, i = 1, \ldots, k, X_i \neq X_j \text{ for } i \neq j \Rightarrow \{b_1, b_2, \ldots, b_k\} \in \mathbf{S}$
(3) $X \in \mathbf{NAME} \wedge S \in \mathbf{S} \Rightarrow \langle X, S \rangle \in \mathbf{B}$

Obviously, we have a pair as a description consisting of a name as first component and a schema as second. The names which appear in the descriptions b_i of a single schema must be unique. They will play the role of the 1NF attributes. As abbreviation we introduce $sch(b)$ as the schema component of a description b.

As we assume one basic domain D only the domain of a given description is uniquely determined by its schema, according to

Definition 3 "Description-Related Domain": The function $dom : \mathbf{B} \rightarrow \mathbf{C}$ is defined by

$$\langle N, S \rangle \in \mathbf{B} \Rightarrow dom(\langle N, S \rangle) = \begin{cases} D & \text{for } S = \emptyset \\ \mathbf{P}(\underset{y \in S}{\times} dom(y)) & \text{else} \end{cases}$$

Obviously, the values of this function are in **C**.

Definition 4 "NF^2 Relation": A pair $R = \langle b, v \rangle$ with $b \in B$ is an NF^2 relation,

$$iff \quad sch(b) \neq \emptyset \text{ and } v \in dom(b)$$

In the usual terminology v is called "instance" or "value" of the schema defined by b. We do not consider further restrictions for v, e.g. satisfaction of integrity constraints, throughout this paper. Furtheron we use the following notational conventions: For the two components of a relation R we use $des(R)$ for the first component of R, $val(R)$ for the second component of R. We extend the function sch to be directly applicable to relations by the following shorthand notation: $sch(R) := sch(des(R))$. An element $y \in sch(R)$ is called an "attribute description" and its first component abbreviated by $nam(y)$ is called an "attribute" of relation R. This corresponds to the usual terminology. As the schema of a relation is the set of attribute descriptions and not only the set of attributes (i.e. names) as in the flat relational model we introduce $attr(R)$ to be the set of attributes according to

Definition 5 "Attributes":

$$attr(R) := attr(des(R)) := \{X \mid (\exists y \in sch(r) \land X = nam(y)\}$$

As in the 1NF case an element of a relation value is called a tuple. We can take over the set-of-mapping notion of a relation: A tuple t of a relation R is a mapping of the form

$$t : attr(R) \rightarrow \bigcup_{y \in sch(R)} dom(y)$$

The value $t(A)$ of the function t is called "attribute value" of A. For a set of attributes $A = \{A_1, \ldots, A_n\} \subseteq attr(R), t \in val(R)$, we also apply the usual shorthand notation $t := t(A_1, A_2, \ldots, A_n) := \langle t(A_1), t(A_2), \ldots, t(A_n) \rangle$

Now we want to construct again a relational data structure using the attribute value $t(A)$ to an attribute A of a relation R. As the elements of a schema have different names (see def. 2) we are able to assign the description to the attribute A, abbreviated by $d(A)$. Using $d(A)$ and $t(A)$ we obtain again a (sub-) relation according to the following

Lemma 1: Let R be a (NF^2) relation, $t \in val(R)$ and $A \in attr(R)$ with description $d(A)$. The pair $\langle d(A), t(A) \rangle$ is a relation iff $sch(d(A)) \neq \emptyset$

Proof: Obviously, the first component $d(A)$ is in $\mathbf{B}$ since $A \in attr(R)$. For the second component we have $t(A) \in dom(d(A)$ by definition.

Lemma 1 indicates that an attribute value together with its attribute description may again be a (1NF or NF^2) relation supposed it has a non-empty schema. If the schema of an attribute description is empty the attribute value is from the atomic domain D (see definition 3). Therefore we can characterize a 1NF relation as a special case of a relation by

Lemma 2: A relation R is a 1NF relation $\Leftrightarrow \forall y \in sch(R) : sch(y) = \emptyset$ Otherwise it is a "real" NF^2 relation.

The main consequence from these definitions is the observation that the operations of the relational algebra may be applied in a nested manner, i.e. repeatedly to a given relation R but also to relations which are constructed from its relation-valued attributes. This is the reason why we do not allow other data structures like lists or tuples and combinations of these with sets.

2.2. A Sample Relation and its Different Representations

Let us consider a relation IDEPT containing information about departments with their administrational and technical staff including sets of courses the technical employees attended. The abbreviations we use are D for department number, DN for department name, IAE for administral employee with name (AN) and job description (AJD). With ITE we denote the technical employees with name (TN), job description (TJD) and courses (IC) with CN as course number and Y as year. The schema of the IDEPT relation in its linear form looks like the following:

IDEPT (D,DN,IAE (AN,AJD),ITE (TN,TJD,IC (CN,Y)))

This schema can be represented by the tree of figure 1:

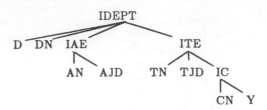

Figure 1: tree representation of the IDEPT relation

The tabular representation of the schema and a sample value of IDEPT is shown in figure 2.

IDEPT								
		IAE		ITE				
D	DN	AN	AJD	TN	TJD	IC		
						CN	Y	
1	RESEARCH	121 122 130 140 169	LIBRARY SECRETARY TRANSLATION PATENTS SECRETARY	511	SOFTWARE ENGINEERING	1 2 5	75 76 79	
				552	BASIC RESEARCH	1 2	82 79	
				678	PLANNING	2 4	76 82	
2	DEVELOPMENT	119 125 135	SECRETARY TRANSLATION PROCUREMENT	650	DESIGN	1 2	75 77	
				780	MAINTENANCE	3	82	
				981	PLANNING	2 3	81 82	

Figure 2: tabular representation of a sample IDEPT relation

For further considerations we need also a 1NF representation of the information in the IDEPT table, shown in fig. 3 below. Here we use the same abbreviations for attributes but without the prefix "I" (for internal) for the administration employees (AE), technical employees (TE) with their courses (C), and DEPT for the department data.

3. BUILDING BLOCKS OF THE NF^2 RELATIONAL ALGEBRA

As shown in section 2.1 the pair consisting of the description of an attribute and of the value of a tuple at this attribute form either again a relation or an atomic attribute value pair. The first case is important since we are allowed to apply any relational algebra operation defined for NF^2 relations in a repeated manner also for these subrelations. This means that we do not extend the number of different types of operations but stay with the few concepts projection (π), selection (σ), Cartesian product ($\times$), natural join ($\bowtie$), together with the set operations union ($\cup$), intersection ( ), and difference ($-$). The only new operators are nest (ν) and unnest (μ).

552

DEPT	
D	DN
1	RESEARCH
2	DEVELOPMENT

AE		
D	AN	AJD
1	121	LIBRARY
1	122	SECRETARY
1	130	TRANSLATION
1	140	PATENTS
1	169	SECRETARY
2	119	SECRETARY
2	125	TRANSLATION
2	135	PROCUREMENT

C		
TN	CN	Y
511	1	75
511	2	76
511	5	79
552	1	82
552	2	79
678	2	76
678	4	82
650	1	75
650	2	77
780	3	82
981	2	81
981	3	82

TE		
D	TN	TJD
1	511	SOFTWARE ENGINEERING
1	552	BASIC RESEARCH
1	678	PLANNING
2	650	DESIGN
2	780	MAINTENANCE
2	981	PLANNING

Figure 3: 1NF representation of the IDEPT example

Rather than giving formal definitions for the building blocks of the NF^2 relational algebra as done in /SS84/ or differently (without building blocks) in /Ja84/ we will give an impression of the expressive power by a series of suitable examples. In its elementary form, i.e. without recursivity, the formal definitions known from the 1NF case of $\pi, \sigma, \times, \cup, -$ are unchanged and are taken over for the NF^2 case.

3.1 Nest and Unnest

The nest operation $R' := \nu[A1, A2, \ldots, Ak : B](R)$ applied on $R(A1, \ldots, Ak, \ldots, An)$ "along $A1, A2, \ldots, Ak$" produces a schema $R'(B(A1, \ldots, Ak), Ak + 1, \ldots, An)$ with a new relation-valued attribute B. The tuples of R' are obtained by grouping together all $(A1, \ldots, Ak)$-subtuples which have identical values on the remaining attributes $Ak + 1, \ldots, An$. As an example we use

Q1: Determine the set of technical employee jobs to every department on the TE relation

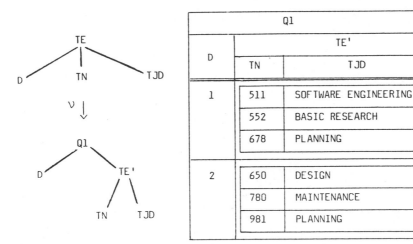

Q1		
	TE'	
D	TN	TJD
1	511	SOFTWARE ENGINEERING
	552	BASIC RESEARCH
	678	PLANNING
2	650	DESIGN
	780	MAINTENANCE
	981	PLANNING

Figure 4: Result of the nest operation (Q1) on the TE relation of fig. 3

The corresponding algebra operation is

$$Q1 = \nu[TN, TJD : TE'](TE)$$

The schema tree and the resulting tuples are shown in figure 4. A new attribute TE' is generated which is relation-valued with attributes TN and TJD.

The unnest operator μ is inverse to nest in the sense that the effect of a previous nest is undone. In our example we have

$$TE = \mu[TE' : TN, TJD](Q1)$$

Obviously, the attribute which is unnested, in our case TE', must be relation-valued. Attributes for the resulting schema may be necessary to guarantee different names (see def. 2,(2)). In our example we may use the attributes TN and TJD of TE'.

3.2 Nested Projection ($\pi - \pi$-Type)

In the usual format of the projection $\pi[L](R)$ the list L of attributes from R specifies which attributes have to be taken over for the resulting relation. We apply this setting also for relation-valued attributes without change. E.g. the projection $\pi[D, DN, ITE](IDEPT)$ generates a result table with the atomic attributes D, DN and the complete relation-valued attribute ITE. However, a generalization is necessary if we want to generate Q2:

Q2: Establish a table which contains the department data together with the technical employee data but without their courses.

We find a solution by applying the projection operation twice and use the following nested projection expression

$$Q2 = \pi[D, DN, \pi[TN, TJD](ITE) : TE'](IDEPT)$$

Figure 5 shows the schema transformation by this expression on the resulting table. Notice that a new attribute TE' is generated.

Q2			
D	DN	TE'	
		TN	TJD
1	RESEARCH	511	SOFTWARE ENGINEERING
		552	BASIC RESEARCH
		678	PLANNING
2	DEVELOPMENT	650	DESIGN
		780	MAINTENANCE
		981	PLANNING

Figure 5: Result of a nested
projection on IDEPT

Q3					
D	DN	DTE			
		TN	TJD	C	
				CN	Y
1	RESEARCH	Ø			
2	DEVELOPMENT	650	DESIGN	1	75
				2	77

Figure 6: Result of a $\pi - \sigma$-operation
on IDEPT

3.3 The Projection-Selection Combination ($\pi - \sigma$-Type)

The previous example showed how we are able to select columns of subrelations. Now we want to select rows of subrelations. The obvious way to perform this is to apply the relational selection within a projection:

Q3: Establish a table which contains department data together with technical employee data but only for design people.

$$Q3 = \pi[DN, \sigma[TJD = \ 'DESIGN'](ITE) : DTE](IDEPT)$$

The result is shown in figure 6. As the selection operator does not influence the schema, the new name DTE is only optional here. Notice that the DTE value in the first department is the empty set since there are no design people.

3.4 Nested Selection ($\sigma - \sigma$ and $\sigma - \pi$)

In the selection formula F of a selection $\sigma[F](R)$ on a relation R we have formula atoms of the form $A \ominus c$ or $A \ominus B$ where A, B are attributes of R and c denotes a constant. The comparison operator $\ominus$ may also be a set comparison from $\{=, \neq, \subset, \subseteq, \supset, \supseteq, \epsilon, \notin\}$ in addition to the arithmetic ones $\{=, \neq, <, \leq, >, \geq\}$. This allows to compare a **whole** RV-attribute value with a constant (set) or with another attribute. With the same argument we had before with the projection we may want to perform a selection or a projection on a subrelation before we do the comparison within the selection formula. An example demonstrates this:

Q4: Determine those research departments which have a library person among their administration staff.

$$Q4 = \sigma[DN = \ 'RESEARCH' \land (\sigma[AJD =' LIBRARY'](IAE) \neq \emptyset](IDEPT)$$

The first department is a match since both conditions contained in the (outer) selection formula are satisfied. The selection within the selection together with the comparison with $\emptyset$ realizes an exists condition. For the schema of any selection on R we have $sch(\sigma[\ldots](R)) = sch(R)$. The next example shows the application of a π within a σ in combination with a set comparison.

Q5: Determine departments with their employees supposed that there is a technical employee who has attended courses 1 and 2.

$$Q5 = \sigma[\sigma[\pi[CN](IC) \supseteq \{1,2\}](ITE) \neq \emptyset](IDEPT)$$

The innermost projection determines the set of courses to a single technical employee. The next selection selects employees with the desired education and the outmost selection checks whether there is at least one in a department.

3.5 Cartesian Product within the Projection ($\pi - \times$-Type)

The definition of the Cartesian product $R \times S$ of two relations known for the 1NF case holds also for the NF^2 relations. Under the assumption $attr(R) \cap attr(S) = \emptyset$ (as in the 1NF case) the schema of $R \times S$ is given by $sch(R \times S) = sch(R) \cup sch(S)$. This operation connects two relations at their roots. If we want to connect a relation which appears as A-attribute value of a relation R(A(A1,A2),B, ...) with another relation S(S1,S2) we generate a table by a $\pi - \times$ building block according to $R' = \pi[(A \times S) : AS, B, \ldots,](R)$. The schema is shown in figure 7.

3.6 Other Operators within the Projection

For completeness we allow every relational algebra operation within the π-list, i.e. also the building blocks $\pi - \nu$, $-\mu$, $\pi - \cup$, and $\pi - -$. However, they will not be required for a description of the interface of our basic algebra processor. Definitions and examples can be found in /SS84/.

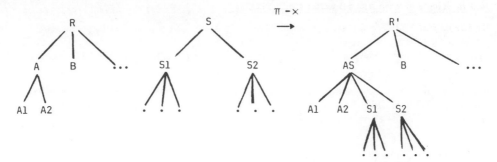

Figure 7: Schema transformation of a $\pi - \times$ building block

4. EXPRESSIONS FORMED WITH BUILDING BLOCKS

As the result of any operation introduced so far is again a relation we may form expressions in the usual way. But, in addition, we want to allow expressions also for relation-valued attributes as the following considerations show.

4.1 Natural Join within a Projection

The following example shows that the natural join is desirable within the projection

Q6: Establish a table which shows the information of Q1 but additionally the courses attended.
We find a solution by the formulation

$$Q6 = \pi[D, (TE' \bowtie C) : TEC](Q1)$$

As any natural join $A \bowtie B$ between two relations is only an abbreviation for a sequence of operations of the form $A \bowtie B := \pi[L](\sigma[F](A \times B))$ we also find a sequence of building block operations for Q6

$C' = \pi[TN : TN', CN, Y](C) \dots$ for renaming
$X1 = \pi[D, (TE' \times C') : TEC'](IDEPT) \dots$ for Cartesian Product
$X2 = \pi[D, \sigma[TN = TN'](TEC')](X1) \dots$ "Natural Join" selection
$Q6 = \pi[D, \pi[TN, TJD, CN, Y](TEC') : TEC](X2) \dots$ duplicate column elimination

4.2 Projection-Selection Expressions within the Projection

A slight generalization of Q6 would be to ask for

Q7: Establish a table which shows the technical employees supposed they have attended the course with number 1.

$$Q7 = \pi[D, \pi[TN, TJD](\sigma[CN = \ '1'](TEC) : TE1](Q6)$$

This result could have been obtained also by the sequence of building blocks

$$X3 = \pi[D, \sigma[CN =' 1'](TEC)](Q6)$$
$$Q7 = \pi[D, \pi[TN, TJD](TEC) : TE1](X3)$$

The point is that, from an implementation and performance point of view, the execution of the sequence of building blocks will be less efficient due to the many intermediate results than the execution of the projection-selection expression within the projection. As projection-selection combinations are most frequent we consider these as candidates for allowed expressions within the π-list to be executed by the basic NF^2 algebra processor.

4.3 Is Selection within a Selection Redundant ?

In the example Q4 we showed the application of a $\sigma - \sigma$-formulation. In pursuing the objective of defining a basic NF^2 algebra interface we are interested whether nested selection also should be supported there. Simple considerations show that we are able to transform a nested selection into a nonnested selection with a nested projection. The Q4 example shows how

$$Q4 = \pi[D, DN, IAE, ITE]($$
$$\sigma[DN =' RESEARCH' \wedge X \neq \emptyset]($$
$$\pi[D, DN, IAE, ITE, \sigma[AJD =' LIBRARY'](IAE) : X](IDEPT)))$$

The "trick", obviously, consists in first taking over the whole attribute IAE and to generate a new attribute X which contains the subtuples with the desired property. The next step selects the whole tuple if it is a research department and if there is at least one library employee ($X \neq \emptyset$). The final projection drops the auxiliary attribute X. This consideration shows that general nested selections are not necessarily to be supported by the basic NF^2 processor since a simple equivalent execution can be found using a nested projection.

5. MAPPING CONSIDERATIONS

For further considerations on a basic NF^2 algebra interface we will now assume that we map a conventional 1NF relational (algebra) interface at the conceptual level to an internal NF^2 interface. Clearly, the reason for such a mapping would be to gain performance for the execution of the expected expressions while keeping the efficiency on a similar level as in a system without this mapping.

We note in passing that the above scenario is complementary to the problem of a "universal relation" interface /Ul82/. We may regard the 4NF interface as the general conceptual data base interface which is defined on the (physical) NF^2 level. A universal relation (as external interface) would be defined on top of the conceptual one. So, if the mapping problem between the universal (external) view and the conceptual 4NF view has been solved and if the mapping and optimization between this conceptual and the internal NF^2 level can be solved we have also a mapping from the universal relation view to the physical NF^2 interface by combination. It would be an interesting task to do this mapping in one step. For the following discussion, however, we restrict ourselves to the 4NF-NF^2 mapping problem which is simpler.

We observe that we are able to describe the mapping by NF^2 relational algebra expressions and therefore we may generate executable expressions for the internal interface by taking the user expressions at the conceptual level together with the mapping expressions. As an example we take the four relations of figure 3 to be offered at the conceptual level and map these on the NF^2 relation IDEPT at the internal level. Examples of mapping equations are then

$$DEPT = \pi[D, DN](IDEPT)$$
$$AE = \mu[IAE : AN, AJD](\pi[D, IAE](IDEPT))$$
$$\dots$$

The problem of equivalent expressions and lossless properties, e.g. that IDEPT can be reconstructed from relations DEPT,AE, ...is not the subject of this paper but must be analysed carefully (for more details see /SS83/). For the following discussion an inspection of a few basic query type examples is sufficient to motivate our basic algebra processor. As abbreviation we use C-expression and I-expression for an expression at the conceptual and internal level respectively.

5.1 Mapping of the Projection-Selection Expression

The most simple but frequent C-expression is a projection-selection combination on a single C-relation. We want to know how difficult an equivalent I-expression becomes when the C-relation appears as subrelation within the schema (-tree) of an I-relation. As an example we consider

Q8: Determine administration employee numbers of secretaries.

The related expression at the conceptual level is

$$C - Q8 = \pi[AN](\sigma[AJD =' SECRETARY'](AE))$$

A direct formulation against IDEPT would be

$$Q8' = \pi[\pi[AN](\sigma[AJD =' SECRETARY'](IAE)) : ANS](IDEPT)$$

The figure 8 shows that Q8' is not yet equivalent to C-Q8 since we still have a relation-valued attribute.

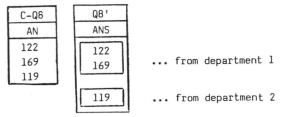

Figure 8: Result of Q8 and Q8'

The equivalent expression to C-Q8 to be executed at the internal level is obtained by a unnest

$$I - Q8 = \mu[ANS : AN](Q8')$$

The important point is that the C-expression $\pi\sigma$ appears as an expression **within** the projection list of the I-expression. As the $\pi\sigma$ sequences on a single C-relation are considered important we require that our basic NF^2 algebra processor must support nested $\pi - \pi - \ldots \pi - \sigma$ expressions.

5.2 Mapping of Projection-Selection-Join Sequences

An important C-expression is the natural join between two or more relations. Let us first consider, in a more general setting than in our example, which possibilities exist to represent internally two C-relations R and S assuming there is one common attribute A between R and S. As we want to support joins we concentrate on internal representations which "materialize" the join internally. For notational convenience we introduce $AS^- := attr(S) - \{A\}$. Then three principal possibilities exist which may be represented by

(a) $I1 = R \bowtie S$
(b) $I2 = R \bowtie (\nu[AS^- : IS](S))$
(c) $I3 = \pi[AS^-, \ (A' \bowtie R) : IR[(\nu[A : A'](S))$

The first possibility (a) is the obvious one: The join between R and S is stored. It corresponds to the approach by /SkS81/ called "denormalization". In order to reconstruct R and S from I1 one must take the outer natural join. In the next case (b) we first nest along the attributes of S without A. This is reasonable e.g. if A is a non-key attribute of S because then we factor out common information by nesting. If, in addition, A is key in R, I2 represents a DBTG-set with owner R and member S. The third case is complementary to (b) in that we nest along A of S and then bring the tuples of R "inside" of S. This case e.g. may be reasonable if A is a member of a key in S. A graphical representation of these three possibilties is given in figure 9. Note that I3 is not obtained by changing the roles of R and S. As in the first case we must take outer joins also for the case (b) and (c).

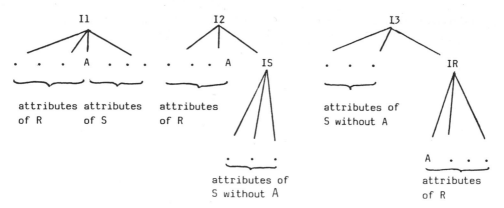

Figure 9: "Materialized" Joins between R and S

The interesting fact for our further considerations is that the join between R and S is executed simply by sequentially scanning through I2 or I3 and unnesting without any comparisons or sorting. This will be evident when we now consider the mapping of the general C-expression

$$CE := (\pi[_1]\sigma[F_1](R)) \bowtie (\sigma[L_2]\sigma[F_2](S))$$

L_1 and L_2 contain the attribute A and other attributes of R and S respectively and F_1, F_2 are usual selection formulae on R and S respectively which must not contain a condition on A.

In the first case (a) the equivalent I-expression obviously is

$$IE1 = \pi[L_1 \cup L_2]\sigma[F_1 \wedge F_2](I1)$$

The more interesting cases are (b) and (c) with the following I-expressions

$$IE2' = \pi[L_1 \cup IS']\sigma[F_1 \wedge IS' \neq \emptyset](\pi[attr(R), \pi[L_2]\sigma]F_2](IS) : IS'](I2))$$

$$IE3' = \pi[L_2 \cup IR']\sigma[F_2 \wedge IR' \neq \emptyset]((\pi[AS^-, \pi[L_1]\sigma[F_1](IR) : IR'](I3))$$

In both cases we observe a query of the type $(\pi\sigma\pi) - (\pi\sigma)$. The one $\pi\sigma$ is executed inside the π-list as it was specified at the C-interface, the remaining $\pi\sigma$ to be executed outside is slightly modified. The selection formula $F_1 \wedge IS' \neq \emptyset$ or $F_2 \wedge IR' \neq \emptyset$ guarantees that tuples with empty subrelations which may have been produced by the inner selection are discarded. The outmost projection must be augmented by the relation-valued attribute IS' or IR' respectively.

The I-expression IE2' and IE3' contain the desired information but still not in the 1NF format. In order to produce the result equivalent to CE we do a final unnest

$$CE \equiv \mu[IS' : L_2](IE2') \equiv \mu[IR' : L_1](IE3')$$

Notice that no empty sets to be unnested will appear. In the general case (not to be discussed further in this paper) unnest on empty relations will produce undefined attribute values.

The important observation here is, again, that whatever materialized join has been prepared, our basic NF^2 processor should be able to process nested $(\pi\sigma\pi) - (\pi\sigma\pi)\ldots - (\pi\sigma)$ sequences in order to utilize the join materialization. Notice that no nested selection is necessary. Only selection formula with atomic attributes occur augmented by a comparison with empty sets for relation-valued ones. This is not too astonishing as we assumed a conventional relational model on top. Nevertheless we see that this subset of the NF^2 relational algebra, characterized by nested $\pi\sigma\pi$, is a kind of a minimum requirement for our basic NF^2 processor.

In our previous example, IDEPT materializes the join between TE and C and the joins between DEPT and AE as well as DEPT and TE according to the second possibility. As an example we use

Q9: Which research department has a technical employee with a planning job who has visited the course 4.

The C-expression is

$$C - Q9 = (\pi[D]\sigma[DN =' RESEARCH'](DEPT) \bowtie ($$
$$\pi[D](\sigma[TJD =' PLANNING'](TE) \bowtie (\pi[TN]\sigma[CN = 4](C))))$$

The I-expression on IDEPT is

$$I - Q9 = \pi[D]\sigma[DN =' RESEARCH' \wedge TE' \neq \emptyset]($$
$$\pi[D, DN, (\pi[TN]\sigma[TJD =' PLANNING' \wedge C' \neq \emptyset]($$
$$\pi[TN, TJD, (\pi[CN]\sigma[CN =' 4'](IC)) : C'](ITE))) : TE'](IDEPT))$$

It can be seen that - instead of two natural joins we have a "simple" nested query of the type $\pi\sigma\pi - \pi\sigma\pi - \pi\sigma$.

Figure 10 shows a simple graphical representation for the query I-Q9. The nodes of the schema tree are marked with the conditions. A conjunction is formed with the predicates at the attributes of every relation. A circle around a node indicates that this attribute is projected to its father node.

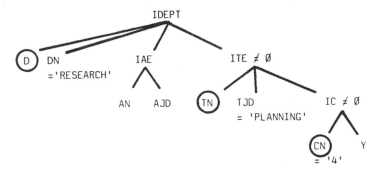

Figure 10: Graphical representation of I-Q9

6. REMARKS TO THE EXECUTION OF NESTED PROJECTION-SELECTION EXPRESSIONS

A few statements are in place here to comment on the execution of a query of the type I-Q9 and to show why it is simple. Without going into the details of an implementation (see /DOPSSW84/, /DGW85/) we assume the following main properties

1) A tuple of an I-relation is stored clustered in as few pages as possible. This set of pages is the address space of a tuple. If a tuple is smaller than a page it must not be fragmented. If a tuple is larger than a page it is not allowed to store other tuples into the address space of this tuple. One page must not contain tuples of different I-relations.

2) One page out of the address space of a tuple is the "root" page. It contains the page numbers of other address space pages and the tuple directory consisting of the subtuples. Once the tuple directory has been fetched the beginning of any subtuple is found with one additional page access in the average.

To execute now e.g. I-Q9 let us first assume that the only access strategy for IDEPT is a sequential scan. In this case for every tuple of IDEPT the root page is fetched. With the tuple directory there and with the specified projection part of I-Q9 we determine the numbers of further required pages to be fetched too. These pages contain the attribute values of D, DN, ITE including all IC values but not necessarily the IAE subtuples. Next we check the predicates in a single pass through the tuple in these pages starting with the conditions nearest to the root of the tuple, in our example with DN = 'RESEARCH'.

As it can be seen easily no tuple or subtuple is touched twice. We can implement the nested $\pi\sigma\pi - \pi\sigma\pi - \ldots - \pi\sigma$ expressions by nested loops which scan through the tuples only once.

If we have an access path e.g. with DN and TJD on IDEPT we may restrict the number of pages to be fetched and the number of predicate checks as usually. In our intended system architecture we assume that indices as access path are also stored and retrieved as NF^2 relations by the basic NF^2 processor. Therefore, if an access path exists we fetch the suitable addresses first by a (nested) $\pi\sigma\pi - \ldots - \pi\sigma$ expression and - possibly after some operations on addresses - we use the resulting addresses to restrict the tuples and subtupels of the target NF^2 relation. This so-called ψ-operation is a special address selection which is also offered at the basic NF^2 algebra processor interface. More details on the ψ-operation and on the retrieval function in general but also on update functions are contained in /DOPSSW84/. The representation of some access paths as NF^2 relations is described in /SS83/.

7. CONCLUSION

Starting with a formal definition of relations with attribute values which may again be relations we have introduced building blocks for a NF^2 relational algebra. It was obtained by a nested application of the standard relational algebra operations. The building blocks have been developed with the objective in mind to design the interface of a basic NF^2 algebra processor to be regarded as the next extension of a page oriented DB layer (buffer) in the architecture of our data base system.

The following observations led us to the decisions on the interface of a basic NF^2 algebra processor

(1) The building block $\sigma - \sigma$, i.e. the selection within the selection formula (generally the application of any other algebra operation "op" within the selection formula) can be replaced by a nested projection of the form $(\sigma\pi) - \sigma$ (generally $(\sigma\pi)$-op)

(2) The aspect of mapping a $\pi\sigma$ expression on a 1NF relation to an equivalent expression when the 1NF relation appears inside the schema tree of a NF^2 relation motivated us to support nested projections of the form $\pi - \pi - \ldots\pi\sigma$.

(3) The mapping of projection-selection-join sequences at a 1NF relational interface of the form $\pi\sigma \bowtie \pi\sigma \ldots \bowtie \pi\sigma$ led us allow nested projections of the form $\pi\sigma\pi - \pi\sigma\pi - \ldots - \pi\sigma$. It was shown that costly joins may be eliminated by these simple "linear" queries. The last format is the most general one which contains the previous ones as special cases.

Therefore we have designed the basic algebra processor such that, together with the ψ-operation for address selection and with update functions, the nested $\pi\sigma\pi - \ldots\pi\sigma$ expressions, will be available.

As a joint effort of the whole data base research group in Darmstadt we have started to implement a first prototype of this kind. A system which allows not only relation-valued attributes but also lists, tuples, sets, and all combinations between these by a SEQUEL-like interface language is described in /Lu85/.

Acknowledgement

I want to thank the members of the Darmstadt data base group Uwe Deppisch, Volker Obermeit, Bernd Paul, Marc Scholl, Gerhard Weikum for many fruitful constructive discussions and collaboration. Liz Klinger carefully put the manuscript into this form. First ideas to a kernel NF^2 algebra processor have aeready been discussed with V. Lum and my former colleagues at the IBM Heidelberg Scientific Centre. Thanks also to Liz Klinger who carefully put the manuscript into this form.

8. REFERENCES

/AB84/ S. Abiteboul, N. Bidoit: Non First Normal Form Relations to Represent Hierarchically Organized Data, Proc. 3rd ACM SIGACT/SIGMOD Symp. on Princ. of Database Systems, Waterloo, Ontario, Canada, 1984

/AMM83/ H. Arisawa, K. Moriya, T. Miura: Operations and Properties on Non-First-Normal-Form Relational Databases, Proc. VLDB Conf., Florence, Italy, 1983

/BRS82/ F. Bancilhon, P. Richard, M. Scholl: On Line Processing of Compacted Relations, Proc. VLDB Conf., Mexico City, 1982

/DGW85/ U. Deppisch, J. Guenauer, G. Walch: Storage Structures and Address Techniques for the Complex Objects of the NF^2 Relational Model, GI-Conference on "Data Base Systems for Office, Engineering, and Science Environments", Karlsruhe, March 1985, (in German)

/DOPSSW85/ U. Deppisch, V. Obermeit, B. Paul, H.-J. Schek, M.H. Scholl, G. Weikum: A Subsystem for the Stable Storage of Versioned Hierarchically Structured Tuples, GI-Conference on "Data Base Systems for Office, Engineering, and Science Environments", Karlsruhe, March 1985, (in German)

/FT83/ P.C. Fischer, S.J. Thomas: Operators for Non-First-Normal-Form Relations, Proc. IEEE COMPSAC, 1983

/HL82/ R. Haskin, R. Lorie: On Extending the Functions of a Relational Database System, Proc. ACM SIGMOD Conf., Orlando, Fl., 1982

/Ja84/ G. Jaeschke: Recursive Algebra for Relations with Relation-Valued Attributes, Techn. Report No. 84.01.003 IBM Heidelberg Scientific Centre, 1984

/JS82/ G. Jaeschke, H.-J. Schek: Remarks on the Algebra of Non-First-Normal-Form Relations, Proc. 1st. ACM SIGACT/SIGMOD Symp. on Princ. of Database Systems, Los Angeles, Ca. 1982

/Ko81/ J. Kobayashi: An Overview of Database Management Technology, TRCS-4, SANNO College, 1753 Kamikasuya, Isehara, Kanagawa 259-11, Japan

/KTT83/ Y. Kambayashi, K. Tanaka, K. Takeda: Synthesis of Unnormalized Relations Incorporating More Meaning, Int. Journal of Information Sciences, Special Issue on Databases, 1983

/Lu85/ V. Lum, P. Dadam, R. Erbe, J. Guenauer, P. Pistor, G. Walch, H. Werner, J. Woodfill: Design of an Integrated DBMS to Support Advanced Applications, GI-Conference on "Data Base Systems for Office, Engineering, and Science", Karlsruhe, March 1985

/Ma77/ A. Makinouchi: A Consideration on Normal Form of Not-Necessarily-Normalized Relations in the Relational Data Model, Proc. VLDB Conf., Tokyo, Japan, 1977

/SkS81/ M. Schkolnick, P. Sorensen: The Effects of Denormalization on Database Performance, Res. Rep. RJ3082 (38128). IBM Res. Lab., San Jose, Ca., 1981

/SLTC82/ N.C. Shu, V.Y. Lum, F.C. Tung, C.L. Chang: Specification of Forms Processing and Business Procedures for Office Automation, IEEE, Trans. on Softw. Eng., Vol. SE-8:5, 1982

/SP82/ H.-J. Schek, P. Pistor: Data Structures for an Integrated Database Management and Information Retrieval System, Proc. VLDB Conf., Mexico City, Mexico, 1982

/SS83/ H.-J. Schek, M.H. Scholl: The NF^2 Relational Algebra for a Uniform Manipulation of the External, Conceptual and Internal Data Models (in German), in: Sprachen fuer Datenbanken, J.W. Schmidt (ed.), Informatik Fachberichte Nr. 72, Springer, Berlin Heidelberg New York Tokyo, 1983

/SS84/ H.-J. Schek, M.H. Scholl: An Algebra for the Relational Model with Relation-Valued Attributes, Technical Report No. DVSI-1984-T1, Technical University of Darmstadt, 1984

/Ul82/ J.D. Ullman: Principles of Database Systems, 2nd edition, Pitman, London, 1982

Two Classes of Easily Tested Integrity Constraints:
Complacent and FD-complacent Integrity Constraints

Hirofumi Katsuno

Electrical Communication Laboratory, NTT
Musashino-shi, Tokyo, 180, Japan

Abstract

This paper presents a new syntactic definition of complacent integrity constraints which are easily tested by selections without applying joins. Complacent integrity constraints were originally defined by Bernstein and Blaustein, but their definition of complacency was inappropriate.

This paper proves that the new definition of complacency is appropriate, i.e., a syntactically complacent integrity constraint is semantically complacent, and if an integrity constraint, IC, is semantically complacent then there exists a syntactically complacent integrity constraint equivalent to IC.

The paper shows that the problem deciding whether a given integrity constraint is complacent is an intractable problem. Hence, the paper presents an efficient algorithm which detects a subset of complacent integrity constraints.

The paper presents another class of easily tested integrity constraints, called FD-complacent. The integrity constraints are tested like complacent integrity constraints by using functional dependencies.

1. Introduction

Bernstein and Blaustein[3] have proposed a class of integrity constraints, called complacent assertions, which can be easily tested. From the semantic point of view, complacent assertions are integrity constraints that can be tested without applying costly join operations. They attempted to characterize complacent assertions syntactically and to develop a partial test to decide whether a given assertion is complacent or not. However, their efforts fail since their syntactic definition of complacency is inappropriate[4,7].

This paper presents a new syntactic definition of complacent integrity constraints. It can be shown that this definition is appropriate, i.e., syntactically complacent integrity constraints are semantically complacent, and if an integrity constraint, IC, is semantically complacent then there exists a syntactically complacent integrity constraint equivalent to IC.

Concerning an algorithm deciding whether a given integrity constraint is complacent or not, this paper shows that the decision problem is co-NP-hard, i.e., it is plausible to believe that the problem is intractable. Hence, this paper presents an efficient algorithm such that (1) if the algorithm says that a given integrity constraint IC is complacent then IC really is complacent, and that (2) if the algorithm says that IC is not complacent then IC may or may not be complacent.

Do complacent integrity constraints cover all integrity constraints that can be easily tested? In order to answer the question, it is necessary to define the meaning of "easily tested". If this is assumed to mean "without applying joins", the answer is no. The paper presents another class of integrity constraints, called FD-complacent, that can be tested by using functional dependencies rather than joins.

Jarke and Koch[5] have developed a fast method for evaluating quantified queries. They introduced a class of quantified queries, called perfect nested expressions, that include complacent integrity constraints.* Roughly speaking, perfect nested expressions are evaluated by applying selections and semi-joins. Jarke and Koch showed several techniques for transforming quantified queries into perfect nested expressions. However, they did not use a technique, called "instantiation" in this paper. Integrity constraint that can not be transformed into perfect nested queries by the Jarke and Koch's methods can be transformed into complacent integrity constraints by using the method presented in this paper.

Basic concepts are given in Section 2. The semantic and syntactic definitions of complacency are given in Section 3, and an equivalence is shown between them. The computational complexity of the problem deciding whether a given integrity constraint is complacent or not is discussed in Section 4. An efficient algorithm for detecting a subset of complacent integrity constraints is presented in Section 5. Section 6 presents FD-complacent integrity constraints which are tested like complacent integrity constraints by using functional dependencies. Some concluding remarks are presented in Section 7.

2. Basic Concepts

This paper uses the relational database model. It is assumed that the reader is familiar with basic concepts of relational database used in the Ullman's book[10]. A database scheme is a finite set of relation schemes. Each relation scheme $\underline{R}$ consists of a relation name, R, and a finite set of attributes, $\{A_1, \ldots, A_n\}$. An example is shown in Figure 1.

Let $dom(A)$ be the domain of values for an attribute A. A relation $R(A_1, \ldots, A_n)$

Database Scheme $\underline{D} = \{\underline{PE}, \underline{PD}, \underline{DE}, \underline{PM}\}$

Relation Scheme   Intuitive Interpretation

$\underline{PE}(P\#, E\#)$ The scheme shows which project$(P\#)$ each employee $(E\#)$ participates in.

$\underline{PD}(P\#, D\#)$ The scheme shows which department $(D\#)$ each project belongs to. A functional dependency $P\# \rightarrow D\#$ holds.

$\underline{DE}(D\#, E\#)$ The scheme shows which department each employee belongs to. A functional dependency $E\# \rightarrow D\#$ holds.

$\underline{PM}(P\#, M\#)$ The scheme shows which machine $(M\#)$ is used in each project.

Figure 1. A sample database scheme.

---

* Strictly speaking, perfect nested expressions do not include integrity constraints (i.e., quantified queries having no free variables). However, it is not difficult to extend their definition of perfect nested expressions to include integrity constraints.

corresponding to the relation scheme $\underline{R}$ is a finite subset of $dom(A_1) \times \cdots \times dom(A_n)$. A database state of database scheme $\underline{D} = \{\underline{R}_1, \ldots, \underline{R}_n\}$ is a set of relations $\{R_1, \ldots, R_n\}$ where each relation R_i corresponds to relation scheme $\underline{R}_i$.

An integrity constraint is described as a well-formed formula of a language called tuple calculus. In the tuple calculus, the symbols $\underline{R}, \underline{S}, \ldots$ are used for relation schemes, and the symbols $r, s, \ldots$ are used for tuple variables. Each tuple variable has a corresponding relation scheme over which the tuple variable ranges. The symbols $A, B, \ldots$ are used for attributes. For a tuple variable r and an attribute A, $r.A$ is called an indexed tuple variable which denotes the attribute A of tuple r. A term is defined as either a constant or an indexed tuple variable.

An atomic formula is defined as either $t = t'$ or $t < t'$, where t and t' are terms. If both t and t' are indexed tuple variables, atomic formulas $t = t'$ and $t < t'$ are called join atoms. A well-formed formula (wff) is defined as follows.

a) An atomic formula is a wff.
b) If P and P' are wffs, $P \vee P'$, $P \wedge P'$, $P \supset P'$, $P \equiv P'$ and $\neg P$ are wffs.
c) If r is a tuple variable ranging over R and P is a well-formed formula, $(\forall r \in R)P$ and $(\exists r \in R)P$ are wffs.

An integrity constraint is written as a wff having no free tuple variable. A quantifier-free wff $P(r)$ is called a selection formula if $P(r)$ has neither join atoms nor free tuple variables except for r.

Example 1. The following wff is an integrity constraint IC_1 stating that if a project has some employees and it uses some machines, it must belong to some department.

$$(\forall e \in PE)(\forall m \in PM)(e.P\# = m.P\# \supset (\exists d \in PD)(d.P\# = e.P\#)). \quad \square$$

For a relation scheme $\underline{R}$, a functional dependency (FD) $X \to A$ denotes a wff,

$$(\forall r \in R)(\forall r' \in R)(r.X = r'.X \supset r.A = r'.A).$$

If $X \to A$ is imposed on $\underline{R}$ as an integrity constraint, it is said that $X \to A$ holds in $\underline{R}$. X is a key for a relation scheme $\underline{R}$, if for any $A \in U$ $X \to A$ holds in $\underline{R}$ and for any proper subset Y of X there is some A such that $X \to A$ does not hold in $\underline{R}$.

Wffs can be interpreted as true (T) or false (F) under each database state and value assignment to free tuple variables in a manner similar to the interpretation of first order predicate calculus, but the details are omitted here. If a wff P is true under a database state D and a value assignment φ, it is denoted by $\langle D, \varphi \rangle \models P$. In particular, since the interpretation of integrity constraints can be determined only by the database state, $D \models IC$ is used instead of $\langle D, \varphi \rangle \models IC$. If $D \models IC$, it is said that IC holds in D. If a wff is true under any database state and value assignment, the wff is called logically valid. Wffs P and P' are called equivalent if $P \equiv P'$ is logically valid.

It is known that the wffs

$$(\forall r \in R)(A \vee B(r)) \equiv A \vee (\forall r \in R)B(r), \tag{2.1}$$

$$(\exists r \in R)(A \wedge B(r)) \equiv A \wedge (\exists r \in R)B(r), \tag{2.2}$$

$$(\forall r \in R)\neg B(r) \equiv \neg(\exists r \in R)B(r), \tag{2.3}$$

$$(\exists r \in R)\neg B(r) \equiv \neg(\forall r \in R)B(r), \tag{2.4}$$

where A has no free occurrence of r, are logically valid, but the wffs

$$(\forall r \in R)(A \wedge B(r)) \equiv A \wedge (\forall r \in R)B(r), \tag{2.5}$$

$$(\exists r \in R)(A \vee B(r)) \equiv A \vee (\exists r \in R)B(r) \tag{2.6}$$

are not logically valid[5]. In Sections 5 and 6, (2.1)-(2.4) are used as transformation rules. For example, (2.1) is used as

$$(\forall r \in R)(A \vee B(r)) \Rightarrow A \vee (\forall r \in R)B(r).$$

Integrity constraints are often considered as wffs in prenex normal form[3,8]. However, it should be noted that in the tuple calculus defined above some wffs can not be transformed into wffs in prenex normal form, since the wffs (2.5) and (2.6) are not logically valid. Hence, this paper does not assume prenex normal form.

When a database state is changed, one has to ensure that the new database state obeys integrity constraints. However, it is very time-consuming to evaluate the integrity constraints on the new database state. Consequently, many researchers have proposed a method for simplifying the integrity constraints by using the fact that the old database state obeys the integrity constraints[2,8,9].

Example 2. Let us consider the situation where a tuple $\langle p_1, m_1 \rangle$ is inserted into the PM relation in a database state D of the sample database scheme shown in Fig. 1. Let D' be the new database state. Let IC_1' be

$$(\forall e \in PE)(e.P\# = p_1 \supset (\exists d \in PD)(d.P\# = e.P\#)).$$

Then, IC_1' is a simplified integrity constraint of IC_1 and $D \models IC_1'$ is equivalent to $D' \models IC_1$. Hence, it suffices to ensure that IC_1' holds in D. □

When simplified integrity constraints are used, what happens? In general, the number of quantifiers of a simplified integrity constraint is one less than that of the original integrity constraint, and some indexed tuple variables in the original integrity constraint become constants in the simplified one. In the following, it is assumed that all the integrity constraints are simplified ones. The methods making use of the constants which were originally indexed tuple variables are investigated.

3. Definition of Complacency

This section first considers a simple example, and a complacent integrity constraint is semantically defined based on the example. Next, a complacent integrity constraint is syntactically defined. Finally, an equivalence between a semantic definition and a syntactic definition is shown.

Example 3. Let us examine how the simplified integrity constraint IC_1' shown in Example 2 can be checked. Let IC_2' be

$$\neg(\exists e \in PE)(e.P\# = p_1) \vee (\exists d \in PD)(d.P\# = p_1).$$

Then, IC_1' is equivalent to IC_2'. Hence, IC_1' can be checked in the steps:
(Step 1) Perform selections.

$$P_e' = \{e \in PE \mid e.P\# = p_1\}.$$
$$P_d' = \{d \in PD \mid d.P\# = p_1\}.$$

(Step 2) Based on the set emptiness of the results of Step 1, decide whether $D \models IC_1'$.

$$D \models IC_1' \iff P_e' = \emptyset \quad \text{or} \quad P_d' \neq \emptyset. \qquad \square$$

As in this example, if an integrity constraint is checked by (1) selecting each relation independently of other relations, (2) deciding whether the results obtained from the selections are empty, and (3) making a Boolean combination of the true-false values obtained in the second step, the constraint is called a complacent integrity constraint. Since complacent integrity constraints can be checked without applying joins but by examining whether the results of some selections are empty or not, they can be easily checked. A formal definition is given below.

In what follows, to avoid double subscripts, we assume that each tuple variable s_i ranges over a relation scheme $\underline{S_i}$, and the relation scheme $\underline{S_i}$ is one of $\{\underline{R_1}, \ldots, \underline{R_n}\}$. In addition, we assume that $\underline{S_i} \neq \underline{S_j}$ if $i \neq j$.

An integrity constraint IC is semantically complacent, if there exist selection formulas $P_i(s_i)$ $(1 \leq i \leq m)$ and a Boolean formula $B(x_1, \ldots, x_m)$ such that (1) $B(x_1, \ldots, x_m)$ has no implication symbol* and that (2) for any database state D, $D \models IC$ if and only if $B(a_1, \ldots, a_m)$ is true, where a_i is true if $\{s_i \in S_i \mid D \models P_i(s_i)\}$ is not empty, and a_i is false otherwise.

It should be noted that since the assumption has been made that if $i \neq j$ then $\underline{S_i} \neq \underline{S_j}$, it is not allowed to use several selection formulas on a relation scheme $\underline{R}$ at the same time.

Example 4. Let $P_1(s_1)$ be $s_1.P\# = p_1$, $P_2(s_2)$ be $s_2.P\# = p_1$, $S_1 = PM$, $S_2 = PD$. Let $B(x_1, x_2)$ be $\neg x_1 \vee x_2$. Then, for any database state D, $D \models IC_1'$ if and only if $B(a_1, a_2)$ is true, where a_i is true if $\{s_i \in S_i \mid D \models P_i(s_i)\}$ is not empty, and a_i is false otherwise. Hence, IC_1' is semantically complacent. □

An integrity constraint IC is syntactically complacent if IC is

$$B((\exists s_1 \in S_1)P_1, \ldots, (\exists s_m \in S_m)P_m),$$

where $B(x_1, \ldots, x_m)$ is a Boolean formula having no implication symbol and $P_i(s_i)$ $(1 \leq i \leq m)$ is a selection formula.

If an integrity constraint is equivalent to a syntactically complacent integrity constraint, it is said that the constraint is reducible to a syntactically complacent integrity constraint.

Example 5. Let $B(x_1, x_2)$ and $P_i(s_i)$ be the same as those in Example 4. Since IC_2' shown in Example 3 is $B((\exists s_1 \in S_1)P_1(s_1), (\exists s_2 \in S_2)P_2(s_2))$, IC_2' is syntactically complacent.

The following theorem shows an equivalence between semantic and syntactic definitions.

Theorem 3.1. IC is semantically complacent if and only if IC is reducible to a syntactically complacent integrity constraint.

Proof

($\Rightarrow$) By the definition of semantic complacency, there exist selection formulas $P_i(S_i)$ $(1 \leq i \leq m)$ and a Boolean formula $B(x_1, \ldots, x_m)$ such that for any database state D, $D \models IC$ if and only if $B(a_1, \ldots, a_m)$ is true, where a_i is true if $\{s_i \in S_i \mid D \models P_i(s_i)\}$ is not empty, and a_i is false otherwise. The following equivalence holds.

a_i is true. $\iff$ $\{s_i \in S_i \mid D \models P_i(s_i)\}$ is not empty.
$\iff$ $D \models (\exists s_i \in S_i)P_i(s_i)$.

Hence, by using induction on the number of connectives in B, it can be proved that

$B(a_1, \ldots, a_m)$ is true $\iff$ $D \models B((\exists s_1 \in S_1)P_1(s_1), \ldots, (\exists s_m \in S_m)P_m(s_m))$.

Therefore, IC is equivalent to $B((\exists s_1 \in S_1)P_1(s_1), \ldots, (\exists s_m \in S_m)P_m(s_m))$, that is, IC is reducible to a syntactically complacent integrity constraint.

($\Leftarrow$) This part is straightforward. □

The last part of the above proof shows that for a given syntactically complacent integrity constraint IC it is straightforward to give a procedure for deciding $D \models IC$ by the three steps: (1) select, (2) test set emptiness, and (3) make a Boolean

---

* Condition (1) is imposed to simplify Algorithm 1 in Section 5 and it is easy to remove this condition.

combination. Hence, for a given integrity constraint IC, if a syntactically complacent integrity constraint equivalent to IC is found, then IC can be easily tested.

In what follows, an integrity constraint is called complacent if the constraint is semantically complacent.

Let $B(x_1, \ldots, x_m)$ be a Boolean formula having no implication symbol, and let P'_i be $(q_i s_i \in S_i) P_i(s_i)$, where q_i is either a universal or existential quantifier, and $P_i(s_i)$ is a selection formula. An integrity constraint $B(P'_1, \ldots, P'_m)$ is said to be in split form. If some q_i is a universal quantifier, $B(P'_1, \ldots, P'_m)$ is not syntactically complacent. However, by transforming $(\forall s_i \in S_i) P_i(s_i)$ into $\neg(\exists s_i \in S_i) \neg P_i(s_i)$, it is easy to transform $B(P'_1, \ldots, P'_m)$ into a syntactically complacent integrity constraint.

4. Intractability of the Complacency Decision Problem

This section examines the computational complexity of the problem of deciding whether a given integrity constraint is complacent. In what follows, the decision problem is called the complacency problem.

In order to decide whether a given integrity constraint IC is complacent, first, select an appropriate syntactically complacent integrity constraint IC' and, next, decide whether IC and IC' are equivalent. There are two problems. One problem is how a syntactically complacent integrity constraint is selected in the first step. If the number of integrity constraints to be selected is infinite, this procedure is undecidable. However, it can be shown that it suffices to consider a finite number of syntactically complacent integrity constraints.*

The other problem is the decidability of the second step. Since predicates are restricted to equality and inequality in the tuple calculus defined in this paper, the step is decidable[1]. Hence, the complacency problem is decidable. However, the following theorem shows that it is plausible to believe that the complacency problem is intractable.

Theorem 4.1. The complacency problem is co-NP-hard.
Proof

It is known that the problem of deciding, in propositional logic, whether a given conjunction of clauses is a tautology is co-NP-complete. We show that any instance of the tautology problem is polynomial-time reducible to an instance of the complacency problem.

Let $B(x_1, \ldots, x_k)$ be a conjunction of clauses. Let IC be

$$(\forall r_1 \in R_1)((r_1.A = c \vee \neg B((\exists r_3 \in R_3)r_3.A_1 = r_1.A_1, \ldots, (\exists r_3 \in R_3)r_3.A_k = r_1.A_k))$$
$$\supset (\exists r_2 \in R_2)r_2.B = r_1.A),$$

where $\underline{R_1}$ is a relation scheme having attributes, $A, A_1, \ldots, A_k$, $\underline{R_2}$ is a relation scheme having an attribute B, and $\underline{R_3}$ is a relation scheme having attributes, $A_1, \ldots, A_k$. Then, the following hold.

$\qquad B(x_1, \ldots, x_k)$ is a tautology.
$\quad \Longleftrightarrow \quad IC$ is equivalent to $\neg(\exists r_1 \in R_1)(r_1.A = c) \vee (\exists r_2 \in R_2)(r_2.B = c)$.
$\quad \Longrightarrow \quad IC$ is complacent.

Next, it is shown that if IC is complacent then $B(x_1, \ldots, x_n)$ is a tautology. Assume that IC is complacent and that $B(x_1, \ldots, x_n)$ is not a tautology. Then, there are selection formulas $P_i(R_i)$ $(1 \le i \le 3)$ and a Boolean formula $B'(x_1, x_2, x_3)$ such that for any database state D, $D \models IC$ if and only if $B'(a_1^D, a_2^D, a_3^D)$ is true, where a_i^D is true if and only if $D \models (\exists r_i \in R_i) P_i(r_i)$.

For each database state D, a new database state $\phi(D)$ is defined as follows.

$\quad$* Since this proof is rather lengthy, it is omitted here.

(Case 1) $D \models IC$

Since $B(x_1, \ldots, x_n)$ is not a tautology, there are Boolean values b_i $(1 \le i \le n)$ such that $B(b_1, \ldots, b_n)$ is false. It is possible to construct a tuple $\bar{r}_1$ such that

(1) $\bar{r}_1.A \ne c$

(2) for each i $(1 \le i \le n)$, if b_i is T then there is a tuple r_3 in R_3 such that $r_3.A_i = \bar{r}_1.A_i$, otherwise there is no tuple r_3 in R_3 such that $r_3.A_i = \bar{r}_1.A_i$

(3) there is no tuple r_2 in R_2 such that $r_2.B = \bar{r}_1.A$.

Let $\phi(D)$ be a database state obtained from D by replacing R_1 in D with $R_1 \cup \{\bar{r}_1\}$.

(Case 2) $D \not\models IC$

Let $\bar{r}_2$ be a tuple such that $\bar{r}_2.B = c$. Let $\phi(D)$ be a database state obtained from D by replacing R_2 with $R_2 \cup \{\bar{r}_2\}$.

It is easy to show that in Case 1 $D \models IC$ and $\phi(D) \not\models IC$, and that in Case 2 $D \not\models IC$ and $\phi(D) \models IC$. On the other hand, if there is a database state D such that for each i $(1 \le i \le 3)$ $a_i^D = a_i^{\phi(D)}$, then $D \models IC \iff \phi(D) \models IC$ follows from $D \models IC \iff B'(a_1^D, a_2^D, a_3^D)$ is true. Hence, this is a contradiction. Therefore, it suffices to show that there is a database state D such that for each i $(1 \le i \le 3)$ $a_i^D = a_i^{\phi(D)}$.

In a sequence of database states, $D, \phi(D), \phi^2(D), \ldots$, Cases 1 and 2 are applied interchangeably. Since R_2 has only one attribute B, once R_2 is replaced with $R_2 \cup \{\bar{r}_2\}$, afterward when applying Case 2, $D = \phi(D)$ holds. Hence, there is a database state D such that for each i $(1 \le i \le 3)$ $a_i^D = a_i^{\phi(D)}$. □

5. Algorithm for complacency

It is important from the practical point of view to obtain an efficient algorithm such that the algorithm transforms not all but many complacent integrity constraints into syntactically complacent integrity constraints. This section shows such an algorithm (more rigidly, an algorithm transforming many complacent integrity constraints into split forms).

Lemma 5.1. Let a wff P be

$$(\forall r \in R)Q(\neg r.A = a \vee P_1) \quad [\text{or} \quad (\exists r \in R)Q(r.A = a \wedge P_1)],$$

where Q is a sequence of quantifiers and P_1 is a wff. Let P_1' be a wff obtained from P_1 by replacing each $r.A$ in P_1 with a. Let P' be

$$(\forall r \in R)Q(\neg r.A = a \vee P_1') \quad [\text{or} \quad (\exists r \in R)Q(r.A = a \wedge P_1')].$$

Then, for any database state D and value assignment φ, $\langle D, \varphi \rangle \models P$ if and only if $\langle D, \varphi \rangle \models P'$.

Proof

Let R_a be a relation scheme whose relation in a database state is $\{r \mid r \in R$ and $r.A = a\}$, and let $R - R_a$ be a relation scheme whose relation in a database state is $R - R_a$. Then, for any database state D and value assignment φ,

$\langle D, \varphi \rangle \models P$
$\iff \langle D, \varphi \rangle \models (\forall r \in R_a)Q(\neg r.A = a \vee P_1) \wedge (\forall r \in R - R_a)Q(\neg r.A = a \vee P_1)$
$\iff \langle D, \varphi \rangle \models (\forall r \in R_a)Q(\neg r.A = a \vee P_1)$
$\iff \langle D, \varphi \rangle \models (\forall r \in R_a)Q(\neg r.A = a \vee P_1')$
$\iff \langle D, \varphi \rangle \models (\forall r \in R_a)Q(\neg r.A = a \vee P_1') \wedge (\forall r \in R - R_a)Q(\neg r.A = a \vee P_1')$
$\iff \langle D, \varphi \rangle \models P'$ □

If P' is different from P (i.e., P_1 includes at least one occurrence of $r.A$), P' is called an instantiation of P.

Corollary 5.2. Let IC be an integrity constraint which contains a wff P as a subformula, and let P be the form of Lemma 5.1. Let P' be an instantiation of P, and let IC' be the integrity constraint obtained from IC by replacing P with P'. Then, for any database state D, $D \models IC$ if and only if $D \models IC'$.

Algorithm 1.
Input. An integrity constraint IC.
Output. An integrity constraint IC' equivalent to IC, and the value of Boolean variable *complacent*. If *complacent* is equal to true then IC' is in split form; otherwise IC may or may not be complacent.
Procedures.
- *move-negation-innermost(P)*: replace $\equiv$, $\supset$ of P with $\vee$, $\wedge$, $\neg$, and bring negation symbols of P immediately before atomic formulas, and let the result be P again.
- *move-quantifier(P, change)*: if quantifiers of P can be moved to the interior by using (2.1)-(2.2) and

$$(\forall r_1 \in R_1)(\forall r_2 \in R_2)P' \Rightarrow (\forall r_2 \in R_2)(\forall r_1 \in R_1)P'$$
$$(\exists r_1 \in R_1)(\exists r_2 \in R_2)P' \Rightarrow (\exists r_2 \in R_2)(\exists r_1 \in R_1)P'$$

  then move them as inner as possible, let the result be P again, and let *change* be true; otherwise, return P and *change* without altering them.
- *instantiation(P, change)*: If P has a subformula to which an instantiation can be applied, let P' be the wff obtained from P by replacing the subformula with its instantiation. If an instantiation is applied to P, let *change* be true and return P' as P, otherwise, return P and *change* without altering their value.
- *simplify(P, change)*: if P can be simplified by some simple rules (the rules are shown in the below) then let the result be P again and let *change* be true, otherwise return P and *change* without altering their value.

Method.
(1) *complacent := false*; $P := IC$;
(2) *move-negation-innermost(P)*;
(3) **repeat**
(4)   *change := false*;
(5)   *move-quantifier(P, change)*;
(6)   *instantiation(P, change)*;
(7)   *simplify(P, change)*;
(8)   **if** P is in split form **then** *complacent := true*;
(9) **until** *complacent* **or** not(*change*);
(10) print P as IC';
(11) print the value of *complacent*.

It should be noted that the rules used in Step (7) are not explicitly stated in Algorithm 1. In general, the more complex the rules are, the wider class of complacent integrity constraints Algorithm 1 detects, but the more time it takes. For example, the following rules can be considered as candidates.

$$a = b \Rightarrow F, \quad a = a \Rightarrow T, \quad \neg a = b \Rightarrow T, \quad \neg a = a \Rightarrow F,$$

$$T \vee P \Rightarrow T, \quad T \wedge P \Rightarrow P, \quad F \vee P \Rightarrow P, \quad F \wedge P \Rightarrow F,$$

$$(\forall r \in R)T \Rightarrow T, \quad (\exists r \in R)F \Rightarrow F, \quad \neg T \Rightarrow F, \quad \neg F \Rightarrow T.$$

Like (2.5) and (2.6), it should be noted that $((\exists r \in R)T) \equiv T$ and $((\forall r \in R)F) \equiv F$ are not logically valid.

Theorem 5.1. Algorithm 1 works correctly, and it requires $O(n^3)$ time, if the above rules are used in Step (7), where n is the length of IC.

Outline of Proof

The correctness of Theorem 5.1 is obvious.

Before the time complexity is evaluated, the data structure representing an integrity constraint must be determined. For brevity, the data structure is omitted here. By using a kind of list structure, each Step (2), (6)-(8) can be executed in $O(n)$ time, and Step (5) can be executed in $O(n^2)$ time. By using Corollary 5.2, each atomic formula of IC' can be changed at most three times in the process of iterating the repeat-until loop (for example, $r.A = s.B \Rightarrow a = s.B \Rightarrow a = b \Rightarrow F$). Therefore, since there are at most n atomic formulas, Algorithm 1 iterates the repeat-until loop at most $O(n)$ times. Hence, the total time spent in Algorithm 1 is $O(n^3)$. $\qquad \square$

Note 5.1. By passing from *instantiation* and *simplify* to *move-quantifier* the information that which part of P is changed, it is possible to show that Algorithm 1 can be executed in $O(n^2)$ time. However, the detailed algorithm is omitted here.

6. FD-complacency

Complacent integrity constraints can be checked without applying costly join operations. Is this the only case where joins can be avoided? The answer is no. Another case is considered in this section: that is the case where a join atom can be transformed into a selection atom by using FDs. This case is a generalization of a technique for transforming cyclic queries into acyclic queries[6]. The following example is first considered.

Example 6. Let IC_3 be

$$(\forall d \in PD)(\forall e \in PE)(d.P\# = e.P\# \supset (\exists r \in DE)(r.E\# = e.E\# \land r.D\# = d.D\#)).$$

This integrity constraint represents that if a project is managed by a department, all the employees engaged in the project must be members of that department. When a tuple $\langle p_1, e_1 \rangle$ is inserted into the relation PE, a simplified form IC'_3 is

$$(\forall d \in PD)(d.P\# = p_1 \supset (\exists r \in DE)(r.E\# = e_1 \land r.D\# = d.D\#)).$$

IC'_3 is not complacent. However, the join atom $r.D\# = d.D\#$ can be transformed into a selection atom. Since an FD $P\# \rightarrow D\#$ holds in PD, let $f_{d,P\#,D\#}(p_1)$ be the value of $D\#$ corresponding to p_1 in the relation PD. Let IC''_3 be

$$\neg(\exists d \in PD)(d.P\# = p_1) \lor (\forall r \in DE)(r.E\# = e_1 \land r.D\# = f_{d,P\#,D\#}(p_1)).$$

Then, for any database state D in which $f_{d,P\#,D\#}(p_1)$ can be defined,

$$D \models IC'_3 \quad \Longleftrightarrow \quad D \models IC''_3.$$

If $(\exists d \in PD)(d.P\# = p_1)$ is first evaluated, the value $f_{d,P\#,D\#}(p_1)$ is determined in the process of the evaluation, and $D \models IC''_3$ can be tested without applying joins. For any database state D in which $f_{d,P\#,D\#}(p_1)$ is not defined (i.e., there is no tuple in PD which has the value p_1 at the attribute $P\#$), $D \models IC'_3$. $\qquad \square$

Integrity constraints like IC''_3 are called semantically FD-complacent. Semantically FD-complacent integrity constraints can be checked by (1) selecting each relation independently of other relations in some predetermined order, (2) deciding whether the results obtained from the selections are empty, and interpreting constant atoms,* and (3) making a Boolean combination of the true-false values obtained in the second

---

* Constant atoms are defined in the below, and a situation where constant atoms appear is shown in Example 8.

step. In Step (1), the order of selections must be determined so that the values of FD-functions used in each selection may be determined in the process of applying other selections preceding the selection.

Next, FD-complacent integrity constraints are syntactically defined. Based on the definition, this section presents an algorithm which detects a subset of FD-complacent integrity constraints. More rigidly, for a given integrity constraint IC the algorithm yields an integrity constraint IC' equivalent to IC, and if it says IC' is FD-complacent, IC' can be checked by the above steps (1)-(3).

It is necessary to generalize wffs so that functions induced by FDs can be described. Let $\underline{R}$ be a relation scheme in which an FD $X \to A$ holds, and let r be a tuple variable on $\underline{R}$. The function symbol $f_{r,X,A}$ is called an FD-function. Then, a constant term is defined as either a constant or $f_{r,X,A}(t)$, where $t = \langle t_1, \ldots, t_k \rangle$, $|X| = k$, and each t_i is a constant term. A variable term is defined as an indexed tuple variable. A term is redefined as either a constant term or a variable term. A constant atom is defined as either $t = t'$ or $t < t'$, where t and t' are constant terms. Wffs, join atoms and selection formulas are similarly defined as done in Section 2. A redefined wff is called a generalized wff.

A constant term t is interpreted under a database state D as follows. If t is a constant c, t is interpreted as c. Let t be $f_{r,X,A}(x)$, where $x = \langle t_1, \ldots, t_k \rangle$. If each t_i is interpreted as c_i and there is some a such that $\langle c_1, \ldots, c_k, a \rangle$ is included in the relation obtained from R by projection on XA, then t is interpreted as a, otherwise t is interpreted as undefined. Let the value "undefined" be denoted by U. The interpretation of $=$ and $<$ is extended so that $c = U$, $U = U$, $c < U$, $U < c$, and $U < U$ may be interpreted as F, where c is a constant.

Example 7. Let FDs $A \to B$ and $B \to C$ hold in a relation scheme $\underline{R}$. Let r be a tuple variable on $\underline{R}$. Then, $f_{r,B,C}(f_{r,A,B}(a_1))$ is a constant term, but $f_{r,A,B}(r.A)$ is not a term. If $\langle a_1, b_1, c_1 \rangle$ is included in a relation R in D, $f_{r,B,C}(f_{r,A,B}(a_1))$ is interpreted as c_1 under D. □

For a sequence of attributes $X = A_1. \ldots A_k$ and for a tuple $x = \langle a_1, \ldots, a_k \rangle$ where each a_i is a term, $r.X = x$ denotes $r.A_1 = a_1 \wedge \ldots \wedge r.A_k = a_k$ and $\neg r.X = x$ denotes $\neg r.A_1 = a_1 \vee \ldots \vee \neg r.A_k = a_k$.

Let $B(x_1, \ldots, x_{m+k})$ be a Boolean formula having no implication symbol. Let a generalized wff P_i be $(q_i s_i \in S_i) P_i'(s_i)$, where q_i is a quantifier and $P_i'(s_i)$ is a selection formula. A generalized wff P is in split form, if P is $B(P_1, \ldots, P_m, C_1, \ldots, C_k)$, where C_i is a constant atom.

For a generalized wff P, a directed graph $G_P = (N, E)$, called the evaluation order graph of P, is defined as $N = \{r \mid r \text{ is a tuple variable occurring in } P\}$ and as $(r, s) \in E$ if and only if either (1) an FD-function on r occurs in the scope of s in P or (2) $f_{s,X,B}(x)$ occurs in P, and $f_{r,Y,C}(y)$ occurs in the tuple x. Case (2) is called a nesting of $f_s(f_r(.))$.

For an acyclic evaluation order graph $G_P = (N, E)$, a sequence of all the nodes in N $\langle r_1, \ldots, r_n \rangle$ is called an evaluation order if it satisfies the condition that if $i < j$ then G has no arc from r_j to r_i. It is obvious that if G_P is acyclic then there is at least one evaluation order of G_P.

A closed generalized wff IC is syntactically FD-complacent if the conditions (1)-(3) are satisfied:
(1) IC is in split form $B(P_1, \ldots, P_m, C_1, \ldots, C_k)$.
(2) The evaluation order graph G_{IC} is acyclic.
(3) If an FD-function $f_{r,X,A}(x)$ occurs in IC then there is a tuple variable s_i such that $r = s_i$ and that $P_i(s_i)$ is $(\exists s_i \in S_i)(s_i.X = x \wedge P_i''(s_i))$, where $P_i''(s_i)$ is a selection formula.

Theorem 6.1. If a generalized wff IC is syntactically FD-complacent then IC is semantically FD-complacent. Conversely, if IC is semantically FD-complacent then there is a syntactically FD-complacent generalized wff IC' equivalent to IC.

Proof

Theorem 6.1 can be proved in a similar way to the proof of Theorem 3.1. $\quad\square$

Semantically FD-complacent generalized wffs are called FD-complacent in the following. An algorithm detecting a subset of FD-complacent integrity constraints is considered in the below.

Let an FD $X \to A$ hold in a relation scheme $\underline{R}$. Let a generalized wff P be $(\forall r \in R)Q(\neg r.X = x \vee P_1)$ [or $(\exists r \in R)Q(r.X = x \wedge P_1)$], where Q is a sequence of quantifiers, P_1 is a generalized wff, $x = \langle c_1, \ldots, c_k \rangle$, and each c_i is a constant term. Let P_1' be a generalized wff obtained from P_1 by replacing each $r.A$ with $f_{r,X,A}(x)$. Let P' be $(\forall r \in R)Q(\neg r.X = x \vee P_1')$ [or $(\exists r \in R)Q(r.X = x \wedge P_1')$]. P' is called an FD-instantiation of P.

Lemma 6.1. For any database state D and any value assignment φ, $\langle D, \varphi \rangle \models P$ if and only if $\langle D, \varphi \rangle \models P'$.

Proof

There are two cases:
(Case 1) $f_{r,X,A}(x)$ is defined under $\langle D, \varphi \rangle$.

This case is similar to Lemma 5.1.

(Case 2) $f_{r,X,A}(x)$ is not defined under $\langle D, \varphi \rangle$.

In this case, $r.X = x$ is false under $\langle D, \varphi \rangle$. Then, the interpretations of P and P' under $\langle D, \varphi \rangle$ are $(\forall r \in R)Q(T)$ [or $(\exists r \in R)Q(F)$]. Hence, $\langle D, \varphi \rangle \models P \iff \langle D, \varphi \rangle \models P'$. $\quad\square$

Example 8. Let an FD $A \to B$ hold in R, and let an FD $A \to C$ hold in S. Then, the following integrity constraint IC can be transformed by FD-instantiations:

$$IC = (\forall r \in R)(\forall s \in S)(\neg r.A = a \vee \neg s.A = a \vee r.B = s.B \vee r.C = s.C \vee c = s.C)$$
$$\Rightarrow \quad (\forall r \in R)(\forall s \in S)(\neg r.A = a \vee \neg s.A = a \vee f_{r,A,B}(a) = s.B \vee$$
$$r.C = s.C \vee c = s.C)$$
$$\Rightarrow \quad IC' = (\forall r \in R)(\neg r.A = a \vee r.C = f_{s,A,C}(a)) \vee$$
$$(\forall s \in S)(\neg s.A = a \vee f_{r,A,B}(a) = s.B) \vee c = f_{s,A,C}(a) \quad\square$$

It should be noted that IC' in Example 8 cannot be tested in the spirit of FD-complacency, i.e., whichever selections on R or S is first evaluated, the value of an FD-function is not determined yet. To guarantee that this kind of problem does not occur, FD-instantiations are restricted to those concerning keys.

Let X be a key for a relation scheme $\underline{R}$ whose attribute set is U. Let a generalized wff P be $(\forall r \in R)Q(\neg r.X = x \vee P_1)$ [or $(\exists r \in R)Q(r.X = x \wedge P_1)$]. Let P_1'' be a generalized wff obtained from P_1 for each $A \in U - X$ by replacing each $r.A$ with $f_{r,X,A}(x)$. Let P'' be $(\forall r \in R)Q(\neg r.X = x \vee P_1'')$ [or $(\exists r \in R)Q(r.X = x \wedge P_1'')$]. Then, if P'' is different from P (i.e., P_1 includes at least one occurrence of $r.A$), P'' is called a key-instantiation of P on r.

It is easy to prove the following lemmas.

Lemma 6.2. If P_1 has no occurrence of $r.A$, where $A \in X$, then P_1'' has no occurrence of the tuple variable r.

Lemma 6.3. Let IC be a generalized wff which contains P as a subformula. Let P' be a key-instantiation of P, and let IC' be a generalized wff obtained from IC by replacing P with P'. Then, for any database state D, $D \models IC$ if and only if $D \models IC'$.

The following algorithm detects a subset of FD-complacent integrity constraints.

Algorithm 2

Input. An integrity constraint IC (an ordinary wff having no free tuple variable).

Output. A generalized wff IC' and the values of Boolean variables, *complacent* and *FD-inst*.

Procedures.

- *move-negation-innermost, move-quantifier, instantiation* and *simplify* are the same as those in Algorithm 1 except that those procedures handle not only ordinary wffs but also generalized wffs.
- *key-instantiation($P, change$)*: If P has a subformula to which a key-instantiation can be applied, let P' be the generalized wff obtained from P by replacing the subformula with its key-instantiation. If a key-instantiation is applied to P, let *change* be true and return P' as P, otherwise let *change* be false.

Method.

   (1) *complacent := false*; $P := IC$; *FD-inst := false*;
   (2) *move-negation-innermost(P)*;
   (3) **repeat**
   (4)   **repeat**
   (5)   *change := false*;
   (6)   *move-quantifier($P, change$)*;
   (7)   *instantiation($P, change$)*;
   (8)   *simplify($P, change$)*;
   (9)   **if** P is in split form **then** *complacent := true*;
  (10)   **until** *complacent* **or** **not**(*change*);
  (11)   **if** **not**(*complacent*) **then** *key-instantiation($P, change$)*;
  (12)   **if** *change* **then** *FD-inst := true*;
  (13) **until** *complacent* **or** **not**(*change*);
  (14) print P as IC';
  (15) print the values of *complacent* and *FD-inst*.

Theorem 6.2

(1) IC' is equivalent to IC.

(2) If Algorithm 2 yields *complacent = true* and *FD-inst = false*, IC' is a complacent integrity constraint.

A lemma is needed to prove that if Algorithm 2 yields *complacent = true* then IC' is FD-complacent.

Lemma 6.4. Assume that Algorithm 2 yields *complacent = true* and *FD-inst = true*. Then, IC' is in split form $Q(P_1, \ldots, P_m, C_1, \ldots, C_k)$.

(1) Assume that a key-instantiation is applied to a tuple variable s_i in the repeat-until loop (3)-(12). Then, $P_i(s_i)$ is either $s_i.X = x$ or $\neg s_i.X = x$, where X is a key for S_i.

(2) $G_{IC'}$ is acyclic.

(3) If $f_{r,X,A}$ occurs in IC', there exists s_i such that $r = s_i$.

Proof

(1) Assume that a key-instantiation on s_i is applied to a generalized wff P in line (11). Then, P has a subformula $(\forall s_i \in S_i)Q(\neg s_i.X = x \vee P_1)$ (or $(\exists s_i \in S_i)Q(s_i.X = x \wedge P_1)$), where X is a key for S_i. Since all the possible instantiations have already applied in line (7), P_1 has no occurrence of $s_i.A$, where $A \in X$. Let P_1' be a generalized wff obtained from P_1 by the key-instantiation on s_i, and let P' be a generalized wff obtained from P by replacing P_1 with P_1'. It follows from Lemma 6.2 that P_1' has no occurrence of

the tuple variable s_i. Since the split form IC' is obtained from P' by successively applying (2.1)-(2.4), (key-)instantiation, and simplification rules, it is obvious that the generalized wff concerning s_i in the split form is either $(\forall s_i \in S_i)s_i.X = x$ or $(\exists s_i \in S_i)\neg s_i.X = x$.

(2) It is obvious that (i) if $G_{IC'}$ has an arc from r to s then a key-instantiation on r is applied in the repeat-until loop (3)-(12), and that (ii) a key-instantiation on r can be applied only one time in Algorithm 2. Hence, it suffices to show that if G_{IC} has an arc from r to s and a key-instantiation on s is applied in the repeat-until loop, then a key-instantiation on r is applied before a key-instantiation on s is done. Assume that $G_{IC'}$ has an arc from r to s and that a key-instantiation on s is applied before a key-instantiation on r is done. At the time immediately before the key-instantiation on r is applied to P, it follows from the discussion in (1) that P has no atomic formula on s except for $s.X = x$, where X is a key for S and x is a tuple consisting of constant terms. Since the key-instantiation on r does not change x and s_i occurs in IC' only as $P_i(s_i)$ (i.e., $s.X = x$), there is no occurrence of $f_{r,Y,A}$ in the scope of s in IC'. It is trivial that there is no nesting of $f_s(f_r(.))$. Hence, there is no arc from r to s. This contradicts the assumption that $G_{IC'}$ has an arc from r to s.

(3) If the quantifier of s disappears in the process of iterating the repeat-until loop, the disappearance is caused by applying the simplification rules. Then, it is easy to show by induction that if the quantifier of r disappears from P then FD-functions concerning r also disappear. Hence, if $f_{r,X,A}$ occurs in IC', there exists s_i such that $r = s_i$. □

Theorem 6.3. If Algorithm 2 yields *complacent=true* and *FD-inst=true*, $IC'(= Q(P_1, \ldots, P_m, C_1, \ldots, C_k))$ is FD-complacent, i.e., IC' can be interpreted under any database state D by the following steps.

(1) Interpret each P_i as p_i in an evaluation order of $G_{IC'}$. Then, the value of each FD-function is naturally determined in the process of successive selections before the value is used.

(2) Interpret C_i as c_i. Then, the values of each FD-function occurring in C_i are already determined in Step (1).

(3) Calculate $B(p_1, \ldots, p_m, c_1, \ldots, c_k)$.

Outline of Proof

Theorem 6.3 follows from Lemma 6.4.   □

Note 6.1. If Algorithm 2 yields *complacent=false*, $G_{IC'}$ may be cyclic. For example, let IC be

$$(\forall s \in S)(\exists t \in T)[\neg s.Y = y \vee (\forall r \in R)(\exists u \in U)(\neg r.X = x \vee (r.A = s.A \wedge t.B = u.B))].$$

If X is a key for R and Y is a key for S, then IC' obtained by Algorithm 2 is

$$(\forall s \in S)(\exists t \in T)[\neg s.Y = y \vee$$
$$(\forall r \in R)(\exists u \in U)(\neg r.X = x \vee (f_{r,X,A}(x) = f_{s,Y,A}(y) \wedge t.B = u.B))].$$

$G_{IC'}$ is shown in Figure 2.

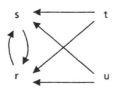

Figure 2. An evaluation order graph.

7. Concluding Remarks

This paper presents two classes of integrity constraints — complacent and FD-complacent. When those integrity constraints are defined, it is assumed that for all the database states the same evaluation strategy is applied. However, the database states in which a relation is empty have exceptional properties:

(1) If R is empty in a database state, for any wff P $(\forall r \in R)P$ [or $(\exists r \in R)P)$] is true [or false] under D and any value assignment.

(2) In practical databases, cases where a relation is empty seldom occur.

Therefore, it is possible to extend the definition of complacency and FD-complacency so that empty relations may be handled as exceptional cases.

An integrity constraint IC is extended complacent if IC is equivalent to a syntactically complacent integrity constraint under any database state such that all the relations occurring in IC are not empty. Let Algorithm 1' be the same as Algorithm 1 except that the procedure move-quantifier uses not only (2.1) and (2.2) but also (2.5) and (2.6). Then, it is easy to prove that if Algorithm 1' yields $complacent = true$, IC' is extended complacent.

Concerning FD-complacency, let Algorithm 2' be the algorithm obtained from Algorithm 2 by extending move-quantifier in the same way as the case of Algorithm 1'. An integrity constraint IC is extended FD-complacent if Algorithm 2' yields $complacent = true$ for the input IC. An extended FD-complacent integrity constraint IC can be tested in the same way as FD-complacent integrity constraints under any database state such that all the relations occurring in IC are not empty.

Algorithm 2' has a desirable property which Algorithm 2 does not have. That is, it is easy to prove that $G_{IC'}$ is acyclic regardless of the value of the Boolean variable $complacent$. Hence, there is no case corresponding to the example in Note 6.1.

References

1. W. Ackermann, Solvable Cases of the Decision Problem. North-Holland Publishing Company, 1968.
2. P. A. Bernstein and B. T. Blaustein, A Simplification Algorithm for Integrity Assertions and Concrete Views. Proc of COMPSAC, p.90-99, 1981.
3. P. A. Bernstein and B. T. Blaustein, Fast Methods for Testing Quantified Relational Calculus Assertions. Proc of ACM-SIGMOD, p.39-50, 1982
4. P. A. Bernstein, private communication, January 1983.
5. M. Jarke and J. Koch, Range Nesting: A Fast Method to Evaluate Quantified Queries. Proc. of ACM-SIGMOD, p.196-206, 1983.
6. Y. Kambayashi and M. Yoshikawa, Query Processing Utilizing Dependencies and Horizontal Decomposition. Proc. of ACM-SIGMOD, p.55-67, 1983.
7. H. Katsuno, A New Definition of Complacent Assertion. Internal Memorandum, 1982.
8. I. Kobayashi, Validating Database Updates. Inform. Systems, Vol.9 (1984) p.1-17.
9. J-M. Nicolas, Logic for Improving Integrity Checking in Relational Data Bases. Acta Informatica, Vol.18 (1982) p.227-253.
10. J. D. Ullman, Principles of database systems. Computer Science Press, 2nd edition, 1982.

SEMANTIC CONSTRAINTS OF NETWORK MODEL

(EXTENDED ABSTRACT)

Yahiko Kambayashi and Tetsuya Furukawa

Dept. of Computer Science and Communication Engineering
Kyushu University
Hakozaki, Higashi, Fukuoka 812, Japan

ABSTRACT

In this paper semantic constraints expressed by network database model
are summarized. In relational model functional dependencies and a join
dependency can be expressed by a set of relations.

Network model is different from relational model by the following
reasons; (1) The basic structure is a DBTG set not a relation, (2) ordered
set can be expressed directly, (3) to improve the effeciency of query
processing, redundancy is permitted, (4) multiple paths between two
attributes are permitted. Semantic constraints for one DBTG set correspond to
(1), (2) and (3), which are functional dependencies, existence constraints,
ordered sets, structual constraints and view constraints.

Semantic constraints for a set of DBTG set types correspond to
multivalued and join dependencies. Correspondence between Bachman diagrams
and such dependencies are given. Problems of Lien's definitions of
dependencies are pointed out and new definitions are given. Multiple paths
can be used to express constraints which are difficult to be handled by
relational model, such as cycles caused by functional dependencies and
constraints expressed by directed graphs. Multiple paths which are used to
improve effeciency of query processing are also summarized.

1. Introduction

Although there are many commercially available relational database
systems, network database systems are still very common because of their
efficiency. In order to design a network database, we need to know what kind
of semantic constraints can be expressed by network model.

One of advantages of relational model is the separation of the two
problems; (1) preservation of semantic constraints and (2) query processing.
(1) is performed by the relational database design theory and (2) is by query
optimization procedures. In network model, however, these problems are mixed
and thus both semantic constraints and representative queries must be
considered in database design stage. The separation of the two problems makes
it easy to solve the design problems. Recent research in the relational
database model, however, shows the following fact. By decomposing a relation
only a limited set of semantic constraints can be expressed. The set consists
of one acyclic join dependency and a set of functional dependencies
satisfying some conditions. Network model can keep ordered sets directly and
permit multiple paths between attribute sets. It can represent semantic
constraints which cannot be preserved by relational model efficient way. These

constraints include constraints on nulls, dependencies on nulls, frequently
used values and time versions. The purpose of this paper is to list such
constraints which can be used for network database design.

2. Basic Concepts

A network schema consists of record types and DBTG set types. The
occurrence of a record type is a set of records which have values for a same
set of data items (in this paper it is called attributes which are used in
relational model). R(X) represents record type R consisting of attribute set
X. The occurrence of a DBTG set type is a set of DBTG sets which are one to
many connections of records of the two record types. One DBTG set consists of
one owner record of the owner record type and any number of member records of
the member record type, represented as multilist structures. A multilist
structure may be implemented by only pointer links or pointer links with
direct pointers from member records to the owner record. It is sometimes
implemented as physically sequential location for VIA mode in CODASYL. S<R,Q>
represents DBTG set type S such that R is the owner record type and Q is the
member record type in S. Set type name S is omitted if it is not required.

A network schema is represented by a Bachman diagram, a directed graph
$B(V_B, A_B)$ where V_B is a set of vertices corresponding to record types and A_B
is a set of arcs corresponding to set types such that the direction of each
arc is from the owner record type to the member record type.

The key of a record type R(X) is a minimum set of attributes K, each
combination of whose values determines at most one record in R. In Bachman
diagrams, the key of a record type is represented as underlined attributes.
In CODASYL the key defined in this paper can be a CALC key. If there are no
CALC keys for a record type R(X), the key K for R is not included in X.

We will use the following assumptions for network model.

Assumption 1: In each DBTG set type, if there exists an attribute in both
owner record type and member record type, the attribute value of a member
record is identical to the value of the owner record.

Assumption 2: The set of the vertices satisfying the following condition is
connected in the Bachman diagram. Each vertex in the set corresponds to a
record type contains an identical attribute.

Assumption 1 is called the structural constraint in CODASYL. If a Bachman
diagram does not satisfy Assumption 2, it can be modified so as to satisfy
the assumption by renaming of attributes.

A relation is expressed as a form of a table. Each column of the table
is labeled by an attribute name and each row corresponds to a tuple. Each
field of the table has the value to which the corresponding attribute is
mapped by the corresponding tuple. t[X] denotes the part of tuple t
corresponding to the attribute set X. R[X] shows projection of R on attribute
set X, which is a set of t[X] where t is in R.

A functional dependency (FD): X->Y is satisfied in relation R, if and
only if for any pair of tuples t and s, t[X]=s[X] implies t[Y]=s[Y]. A join
dependency (JD) $*[X_1, X_2,..., X_m]$ is satisfied in R if and only if R is
always expressed by $R[X_1]*R[X_2]*...*R[X_m]$. When m=2, it is called an
multivalued dependency (MVD): X->->Y|Z where $X=X_1 \cap X_2$, $Y=X_1-X$ and $Z=X_2-X$.

A JD is represented by a hypergraph $D(V_D, E_D)$, where V_D is a set of
vertices corresponding to the attributes and E_D is a set of hyperedges
corresponding to relation schemes. Acyclicity of a JD is decided by the
acyclicity of the corresponding hypergraph [BEERF8105].

It is important to treat null values in databases. Existence constraints
are the constraints with respect to null values.

A network schema expresses these constraints. In order to design network
databases, we need to know what constraints network schemata express.
Existence constraints and FDs are two important constraints represented by a

DBTG set type, althogh there is some difference on the definition of functional dependencies. There are papers on network schema design using only FDs in conventional sense [KUCKS8203] [KUCKS8305] and using MVDs [LIEN8204].

3. Semantic Constraints for a Single DBTG Set

We can classify semantic constraints for a single DBTG set as follows.

(A) Constraints expressed by a DBTG set.
(B) Constraints introduced by redundancies of a DBTG set.

There are three types of constraints for (A) and two types for (B).

A.   There are the following constraints represented by a DBTG set in the CODASYL network model.

(1) Existence constraint: Each member record cannot exist without its owner record.
(2) Functional dependency: Each member record has at most one owner record and each owner record has any number of member records.
(3) Odered set: In a DBTG set type there is an order on the member records having the same owner record.

Existence constraints may not be required since a singular set (permitting SYSTEM as the owner) can be used. There are several variations of existence constraints, which are determined by combinations of two classes called set membership. The insertion class is AUTOMATIC or MANUAL. The retention class is FIXED, MANDATORY or OPTIONAL.

If the retention class of a DBTG set type is AUTOMATIC, the owner of the inserted record is determined by a predetermined rule. There is no such predetermined rule for MANUAL.

If the retention class is FIXED, DISCONNECT and RECONNECT operations cannot be applied for the occurences of the DBTG set type. A member record of one DBTG set cannot be transfered to another DBTG set. If the retention class is MANDATORY, a record once inserted cannot exist without its owner record while it can be transfered to another DBTG set. RECONNECT operation can be applied but DISCONNECT operation cannot. If it is OPTIONAL both of the operations can be applied.

AUTOMATIC insertion class means that the inserted record must have its owner record, and FIXED or MANDATORY retention class means that the deletion of an owner record cause deletions of its member records because member records cannot be disconnected. So existence constraints can be partitioned into insertion and deletion existence constraints in network model.

A functional dependency (FD) from the key of a member record type to its owner record type is satisfied since each member record determines its owner record. We need to handle the null value problem, because members without any owner and an owner without any member are permitted, which will be discussed in Section 4.

Clustering records is an example of FDs. For example, for attribute AGE we can introduce a new attribute AGE_INTERVALS of which values are 'under twenty', 'twenty to thirty', 'over thirty' etc. There is an FD: AGE-> AGE_INTERVAL, so by selecting AGE_INTERVAL as the owner record the clustering of data is realized.

The index structure is also an example of FDs. We can make a DBTG set type such that the owner record type is an index of the member record type. Index structures are effective for query processing.

There is an order of member records in a DBTG set because of its multilist structure. Unordered set can be expressed if the order is ignored. In relational model we need to introduce a new attribute to realize an ordered set and if records are not arranged by the order, processing the record by the order is very hard. In network model if member records are ordered by their usage frequency, retrieval from the owner record to its member records can be optimized.

B. If there are redundancies of data values, we need to check the consistency. Such a constraint is not a constraint expressed by a DBTG set but a constraint which should be maintained by a DBTG set. As in network database design, it is very important to consider the efficient processing of typical queries, we need to introduce redundancy to improve efficiency of query processing. Typical ones are as follows.

(4) Structural constraint: If an attribute is contained in both owner and member record types, its values are identical in each DBTG set (see Assumption 1 in Section 2).

We can always realize a non-redundant structure. The redundancy may help to reduce the access cost. If an attribute is contained in both owner and member record types, the attribute in the member record type can be eliminated. If such an attribute exists, we can retrieve the value without accessing the owner record. Since the access from member to owner requires more cost than the access in the opposite direction, such redundancy will help to reduce the access cost.

(5) View constraint: If a value of the owner record is calculated with values of member records, these values must satisfy the constraint induced by the calculation.

For example, owner record can have average salary of all salary values of member records. Another example is the current value. If the member records are ordered by the price history of one item, the first record value is the current price which can be also duplicated in the owner record to increase efficiency of query processing.

4. Global Constraints in Network Model

In this section, we discuss the global constraints in network model. There are two types of the constraints; objects and multivalued dependencies (MVDs) (equal to an acyclic join dependency (AJD)). Lien's definitions of an MVD and an FD with nulls are modified to satisfy these constraints.

One of the differences between network model and relational model is a representability of different relationships among attributes. In network model, they are represented by multi-paths between record types which will be discussed in Section 5. In this section, we assume that Bachman diagram is acyclic when directions of arcs are ignored.

Let C_1, C_2,..., C_n be n connected components obtained by removing a connected component C in a connected acyclic Bachman diagram. Since C and C_i ($1 \leq i \leq n$) are connected by one arc (DBTG set type S_i), by specifying the values of key attributes K of the record types in C, the set of values of attribute set Y_i contained in C_i are determined through the relationship of DBTG set type S_i. The relationship between Y_i and Y_j ($i \neq j$) is indirect, that is, they are related through the values of K. Thus each values of K determines the set of the values of Y_i (i=1, 2,..., n). By this fact there is an MVD: K->->Y_i for i=1, 2,..., n.

Example 1: Let a network schema be the Bachman diagram shown in Fig. 1. If we take $\{R_3, R_4, R_5\}$ as a connected component C. $C_1 = \{R_1, R_2\}$ and $C_2 = \{R_6\}$ are connected component obtained by removing C. So we realize that an MVD: BC->->AB|D (equal to BC->->A|D) expressed by this Bachman diagram. Similarly, MVDs: C->->AB|D and B->->A|CD are obtained from the Bachman diagram.

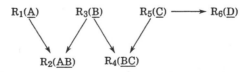

Figure 1   MVDs expressed by a Bachman diagram

Property 1: An acyclic Bachman diagram expresses an MVD: K->->Y_i where K is the key attributes of a connected component C of the given Bachman diagram and Y_i is an attribute set corresponding to one connected component produced by removing C from the diagram.

A set of MVDs which is not conflict-free causes either key-breaking anomaly or transitive anomaly [SCIO8104]. Since the set of MVDs expressed by an acyclic Bachman diagram does not cause such kinds of anomaly because of Assumption 2 in Section 2, the set is conflict-free.

As a conflict-free set of MVDs corresponds to an acyclic join dependency (AJD) [BEERF8105], an acyclic Bachman diagram expresses one AJD. The Bachman diagram in Fig. 1 expresses AJD *[AB,BC,CD].

Null values in network model are also restricted. Typical types of null values are non-existence (including temporally cases), unknown (the value exists but unknown) and no-information (completely unknown). The following two methods can be used to express nulls in network model.
(1) Data values expressing nulls
(2) Database structure

Usually unknown nulls are expressed by nulls with identifiers in method (1) since unknown nulls are used to represent real values and each different unknown null should be distinguished. In method (2) nulls are expressed by the DBTG sets which have no member records or records in a member record type which are not in any DBTG sets. The method is suitable to express the non-existence null values because non-existence values can be considered as non-existence of its owner (or member) record in a DBTG set type.

If nulls are expressed by the structure, objects (attribute sets whose values are not nulls in tuples) representable in network model are restricted. When there are correspondences of records between two record types R_1 and R_2, there must exist corresponding records in the record types on the path from R_1 to R_2. So the attribute values of the records are not nulls. This fact leads the following property.

Property 2: If an object can be expressed by network model, there exists the subgraph of the Bachman diagram such that the attribute set of the record types in the subgraph is identical to the object.

The constraints expressed by the global structure in network model are combined. Lien defined a multivalued dependency with nulls (NMVD) and showed that a conflict-free and contention-free (called CC-free) set of MVDs is equivalent to an acyclic Bachman diagram on constraints [LIEN8204]. There are, however, following problems in his results.
(1) Since FDs are not considered, Bachman diagrams are redundant.
(2) A set of MVDs which is not contention-free can be expressed by an acyclic Bachman diagram.
(3) The definition of NMVD is not safficient and a CC-free set of NMVDs of his definition is not equivalent to the constraints expressed by an acyclic Bachman diagram.
The definition of NMVD by Lien is as follows.

Multivalued dependency with nulls $X\text{->->}Y$ holds in relation $R(U)$ $(XY=U)$ if for every tuple t which is X-total it is true that

$$(R[XZ=t[XZ]])[Y]=(R[X=t[X]])[Y] \quad (Z=U-X-Y).$$

In relational model a set of MVDs which is not contention-free is not used for decomposition of a relation because the lossless join property is not preserved. In network model, however, such a set of MVDs can be expressed by the structure.

Example 2: MVD set {AB->-> D|CE, AC->->BD|E} is expressed by Bachman diagram shown in Fig. 2. The set is not contention-free.

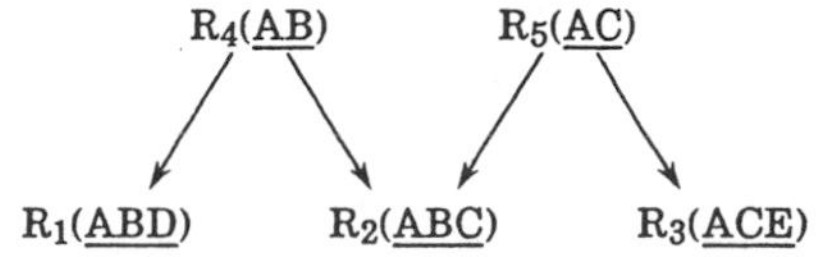

$R_4(\underline{AB})$ $R_5(\underline{AC})$

$R_1(\underline{ABD})$ $R_2(\underline{ABC})$ $R_3(\underline{ACE})$

Figure 2 Expression of a contention MVD

The third problem is caused by the fact that the tuples of which key values are not total are ignored in Lien's definition of NMVD. Property 2 shows that the null patterns are limited in network model.

Example 3: The relation shown in Fig. 3 (a) satisfies an NMVD: BD->->A|D by Lien's definition. One of the Bachman diagrams expressing the NMVD is shown in Fig. 3 (b). From the Bachman diagram, (-0-1) is not obtained for values of attributes ABCD but (10-1) is obtained which is not in the original relation, because value of BC (0-) is related to value of A (1) and values of D (1) and (0). Furthere more value of ABCD (0--1) is not obtained because the related record of R$_3$ does not exist. We realize by Property 2 that object AD can not represented by this Bachman diagram.

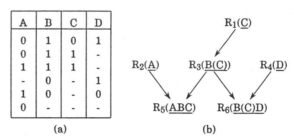

A	B	C	D
0	1	0	1
0	1	1	-
1	1	1	-
-	0	-	1
1	0	-	0
0	-	-	-

(a)   (b)

Figure 3 Modification of NMVD's definition

The definition of NMVD by Lien should be modified as follows.

Definition 1: Multivalued dependency with nulls
    Multivalued dependency with nulls X->->Y holds in a relation R(U), (XY=U) if for every tuple t whose X values are not nulls
        (R[XZ=t[XZ]])[Y]=(R[X=t[X]])[Y] (Z=U-X-Y)
is satisfied and when t[X] are nulls either of t[Y] or t[Z] are nulls.

    Lien also defined NFD (functional dependency with nulls).
    Functional dependency with nulls X->Y holds in a relation R(U) (XY=Z) if for every two tuples t$_1$ and t$_2$ which are X-total it is true that t$_1$[X]=t$_2$[X] implies t$_1$[Y]=t$_2$[Y].
    We also need to modify the definition considering nulls in the key attributes.

Definition 2: Functional dependency with nulls
    Functional dependency with nulls X->Y holds in a relation R(U), (XY=Z) if for every two tuples t$_1$ and t$_2$ whose X values are not nulls it is true that t$_1$[X]=t$_2$[X] implies t$_1$[Y]=t$_2$[Y].

5. Cycles and Semantic Constraints

    Since a cycle in a network database can express multiple relationships between attributes, cycles can be utilized to preserve semantic constraints which are not easily preserved by relational model. Cycles in network model can be classified into the following three types.
(A) Representation of a cycle caused by dependencies
    (A-1) Cyclic join dependency
    (A-2) Cycle caused by functional dependencies
(B) Representation of a graph structure
    (B-1) Bill of materials type, time versions
    (B-2) General graph
(C) Cycles generated by considering query processing
    (C-1) Index
    (C-2) Decomposition of multiple attribute joins
Cycles of types (B) and (C) are not considered in relational model.

A.   In Section 4, we showed that an acyclic Bachman diagram expresses an AJD. The result is easily extended to cyclic one. If the network schema has a cycle, it often expresses a cyclic JD. If there is a cyclic JD, the Bachman diagram representing the dependency have a cycle. Especially if there are more than one path between two attributes, the corespondence of attribute values at each path is different each other.

   There are two kinds of cycles caused by FDs.
   (A-2-1) Two or more paths of dependency chains
   (A-2-2) Key breaking condition
These kinds of cycles of FDs cause problems for relational databases.

Example 4: For the example of (A-2-1), we have three relations $R_1(BCD)$, $R_2(AB)$ and $R_3(AC)$ for FDs: D->BC, B->A and C->A. Each FD is kept in each relation, but the FD: D->A can be generated by joining R_1 and R_2 or R_1 and R_3. These two FDs should not be contradictory. Such a constraint cannot be easily kept by relational model. These FDs are kept by network model in the network schema shown in Fig. 4 (a). Because by structural constraint values of attribute A are identical in each DBTG set of any DBTG set types and every record of R_4 determines the same record of R_1 for each path $\langle R_1, R_2 \rangle$ $\langle R_2, R_4 \rangle$ or $\langle R_1, R_3 \rangle$ $\langle R_3, R_4 \rangle$.

   A FD set having the key breaking property is known to be very hard to be maintained. A set of FDs: AB->C and C->B is an example of such FD set. (A-2-2) shows that such a set is maintained by network model. The network schema shown in Fig. 4 (b) keeps this FD set because every record of R3 determines same record of R1 for either path $\langle R_1, R_2 \rangle$ $\langle R_2, R_3 \rangle$ or $\langle R_1, R_3 \rangle$ by structural constraints.

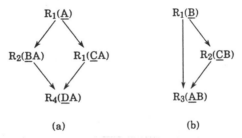

(a)   (b)

Figure 4   Maintenance of cycles caused by FDs

B.   In relational model bill of materials structure is not reflected in the logical schema design. It is treated as query processing. In network model, on the other hand, it is represented by two DBTG set types between two record types. One record type, which is an owner in both DBTG set types, keeps materials data. One DBTG set type expresses owner materials is composed of its members. The other DBTG set type is index for the member record type.

   Time versions are similar to bill of materials. For example, the history of reorganizations of departments by basic operations, merging and division is also expressed by the same structure. The owner record type keeps departments data. Two DBTG set types express "change to" and "change from" respectively. The member record type is intermediate data for reorganizations.

   The graphs for bill of materials problems and time version problems are directed acyclic. By using the same structure, two DBTG set types between two record types, we can .express unrestricted directed graphs. Such a graph is useful to express some kinds of constraints which cannot be expressed by other methods.

C.   In network model, navigation is normaly used for query processing. In order to improve effeciency of navigation we may need to modify network schemata, which sometimes introduces cycles of type (C).

Example 5: The network schema shown in Fig. 2 expresses JD *[ABD, ABC, ACE].
In relational model relations R_1(ABD), R_2(ABC), R_3(ACE) are formed by the JD.
By addition of index record type R_6 of attribute A for R_4 and R_5, we realize
a cycle R_6, R_4, R_2, R_5. The index will not change the constraints expressed
by the schema, but it will improve the effeciency of processing. As a result
a cycle is formed.

There are several classes of acyclicity of JDs [FAIG8307]. In general a
cycle (C-2) is generated by decomposing a multi-attributes join into more
than one joins when the network schema expresses an σ-acyclic but γ-cyclic
JD.

6. Concluding Remarks

For network database design there are typically the following three
approches.
(1) Design using 1-n relationships and n-m relationships
(2) Design using semantic constraints such as dependencies
(3) Design by typical queries
The first approach is commonly used. Interface to a well-known conseptual
model, E-R model, is very easy. Various semantic constraints handled by
network model discussed in this paper, however, are not utilized by this
approach. For (2) there are two kinds of papers, considering only FDs
[KUCKS8203] [KUCKS8305] and considering only MVDs (FDs are treated as special
case of MVDs) [LIEN8204].
We are currently developing design procedure considering all the
constraints discussed in this paper together with characteristics of typical
queries.

ACKNOWLEDGEMENTS

The authors are grateful to Professor Shuzo Yajima of Kyoto University
for his valuable discussion.

References

[BEERF8105] Beeri,C., Fagin,R., Maïer,D., Mendelzon,A., Ullman,J.D., and
             Yannakakis,M., "Properties of Acyclic Database Schemes", ACM
             SIGACT Symp. on the Theory of Computing, pp.355-362, May 1981.
[CHENK8406] Chen,H., Kuck,S.M., "Combining Relational and Network Retrieval
             Methods", Proc. ACM SIGMOD Int. Conf. on Management of Data, pp.
             131-141, June 1984.
[DAYAB8201] Dayal,U., Bernstein,P.A., "On the Updatability of Network View-
             Extending Relational View Theory to the Network Model",
             Information Systems, Vol.7, No.1, pp.29-46, Jan. 1982.
[DAYAG8206] Dayal,U., Goodman,N., "Query Optimization for CODASYL Database
             Systems", Proc. ACM SIGMOD Int. Conf. on Management of Data, pp.
             138-150, June 1982.
[GOLD8109]   Goldstein,B.S., "Constraints on Null Values in Relational
             Database", Proc. 7th Int. Conf. on VLDB, pp.101-110, Sept. 1981.
[KUCKS8203] Kuck,S.M., Sagiv,Y., "A Universal Relation Database System
             Impremented Via the Network Model", Proc. the 1st ACM Symp. on
             PODS, pp.147-157, March 1982.
[KUCKS8305] Kuck,S.M., Sagiv,Y., "Designing Globally Consistent Network
             Schemas", Proc. ACM SIGMOD Int. Conf. on Management of Data, pp.
             185-195, May 1983.
[LIEN8204]   Lien,Y.E., "On the Equivalence of Database Models", J. ACM, Vol.
             29, No.2, pp.333-362, April 1982.
[SCIO8104]   Sciore,E., "Real World MVD's", Proc. ACM SIGMOD Int. Conf. on
             Management of Data, pp.121-132, April 1981.
[ZANI7906]   Zaniolo,C., "Design of Relational Views over Network Schemas",
             Proc. ACM SIGMOD Int. Conf. on Management of Data, pp.179-190,
             June 1979.

JOIN-DECOMPOSITION OF MVD-CONSTRAINTS USING THE CHARACTERIZATION BY "BASES" - AN INTRODUCTIVE STUDY

Yoshito Hanatani

INRIA BP. 105, 78153 Le Chesnay Cédex, France
Lab. d'Informatique, 14032 Caen Cédex, France

(In memory of Professor Shikao Ikehara)

ABSTRACT

We call "MVD-constraints" the constraints which can be expressed by a set of MVDs (multi-valued-dependencies). There are two categories of MVD-constraints : cat 1) those which can be relaized when we impose a single JD (join-dependency), called join-realizable, and cat 2) the others. Cat 1 has two sub-categories, cat 1.1) those which can be equivalently represented by a single JD, called join-representable, and cat 1.2) the others.

Our objective is the decomposition of an MVD-constraint into two factors, the main factor which is join-representable (cat 1.1) and the complementary factor consisting of embeded MVD-constraints on every components of the main factor. Based on our filter-like representation of MVD-constraints and on our characterizations of these categories, introduced in our earlier works [H1] [H2] [HF], we propose a strategy to find such a decomposition.

1 - INTRODUCTION

1.1 Problems

Consider the MVD-constraints given by the following set M_0 of MVDs (multi-valued-depednencies) over the universe $U \overline{=} ABCDE$:

$$M_0 \equiv \{C \twoheadrightarrow D \mid ABE, \ D \twoheadrightarrow E \mid ABC\}$$

This MVD-constraint is an example of MVD-constraint which cannot be realized by imposing a single JD as the only constraint. That is, no JD can imply the same MVD-constraint as M_0 does. Although the form of M_0

is simple, it is difficult enough to understand the nature of this constraint.

Now consider the following constraint consisting of a JD on the same U and of sets of embeded MVDs on each component of the JD :

$$j \equiv \bowtie \{ABC, CDE\},$$

$$M_{ABC} \equiv \{\}, \quad M_{CDE} \equiv \{C \twoheadrightarrow D \mid E, \; D \twoheadrightarrow C \mid E\}.$$

This time we have a general view given by the JD and moreover the embeded MVD-constraints are considerably simple. Still, this constraints is in facts equivalent to the one given by M_0.

In the following, we denote by M, M' etc... and MVD-constraint and by j, j' etc... a JD-constraint. For any M and j, let us symbolize by M/j, the constraint given by the collection of the MVD-constraints induced by M on the components of j. We symbolize also by {j, M/j} the constraint given by the pair. We are not interested in {J, M/j} with a cyclic j in the sense of [BFMY]. We define therefore a <u>g-divisor</u> (g=global) of an M as an M' for which there is a j equivalent to M' such that {j,M/j} is equivent to M. M'$\equiv_\phi$ is a <u>g-divisor</u> of any M, because it is equivalent to $\bowtie\{U\}$. We call it the <u>trivial</u> g-divisor. Naturally we are interested in finding a maximal g-divisor.

Let M_I be the given constraint and M_T, the constraint to be found. M_T is invisible for us. We suppose that M_I contains the information concerning the form of M_T, and we expect to find M_T from M_I as the result of repeated pairs of operations : "generation of a new candidate and test of it". At every step the new candidate M_{k+1} must be found between M_k and M_T. This means that M_{k+1} must be among the nearest possibilities. The orientations of development of the chain is hoped to be inherited from M_k to M_{k+1}. for this reason the nearest possibility is hoped to be unique. One of criteria for the next candidate is the join representability defined in the next section. But this is not enough. In fact, our M_0 has two join representable sub-constraints which are the nearest to M_0, i.e. $M_1 \equiv \{C \twoheadrightarrow DE, D \twoheadrightarrow E\}$ and $M_2 \equiv \{C \twoheadrightarrow D, C \twoheadrightarrow E\}$. If we have to take one of these two as a candidate to be tested, then we cannot expect that the information concerning the orientation is inherited by it. Fortunately, we have some reason to reject them and take the next nearest as a candidate, i.e. $M'_0 \equiv \{C \twoheadrightarrow DE\}$. This permit us also to avoid the useless tests.

We have the following problems :

a) how to generate the MVD-sub-constraints of an MVD-constraint,
b) how to compare the constraint power of them,
c) how to recognize among them those which are join-representable,
d) how to obtain from them the forms of JDs which represent them,
e) how to orient the search of a candidate, and
f) how to avoid to   the maximum the useless tests.

1.2 Tools

The problems a), b), c), d) express the need of tools and the problems e), d) concern how to use the tools. We have to begin by supplying tools.

The tools we supply are :

1) simple representations of MVD-constraints with a simple characterization of their relative strength,

2) a characterization of the equivalence between MVD-constraints and JD-constraints using the representations of MVD-constraints.

Almost all of these theories are developed in the former works [H1], [H2], [H3] and [HF]. We restate them briefly in Section 2 and 3 for the coherency of terminology.

2 - REPRESENTATIONS OF CONSTRAINTS

2.1 MVD-and JD-constraints

Let $MVD(U)$ and $JD(U)$ be the set of all MVDs and all JDs, respectively on a fixed universe set U of attributes. Let $\Gamma^{MV} \equiv_{def}$ $P(MVD(U))$ and $\Gamma^{J} \equiv_{def} JD(U)$, where $P(MVD(U))$ denotes the power set of $MVDS(U)$. And let $\Gamma \equiv_{def} \Gamma^{MV} \cup \Gamma^{J}$. We mainly study on the constraints given by an element of P. A constraint given by an element of Γ^{MV} and of Γ^{J} is called a MVD-constraint and JD-constraints, respectively. For any subset C of Γ we denote by C^{MV} and C^{J} its MV-part and J-part, i.e. $C^{MV} \equiv_{def} C \cap \Gamma^{MV}$ and $C^{J} \equiv_{def} C \cap \Gamma^{J}$.

We denote by "$\geq$" the usual preordering of the constraint power on Γ instead of "$\models$" and we denote by "$=$" the associated equivalence relation.

The quotient sets $(\Gamma/=)$, $(\Gamma^{MV}/=)$ and $(\Gamma^{J}/=)$ according to this equivalence relation "$=$" are partially ordered by the order relations induced by "$\geq$". The partial orderings of these quotient sets are also denoted by "$\geq$".

For any element d of Γ the corresponding element of $(\Gamma/=)$ is denoted by $[d]_=$. The elements of its MV-part $[d]_=^{MV}$ and its J-part $[d]_=^{J}$ are called <u>MV-representations</u> and <u>join-representations</u> of d, respectively. For any M of Γ^{MV}, the constraint given by M is said to be <u>join-representable</u>, if $[M]_=^{J} \neq \phi$.

2.2 Perspective

We intend to give some simple representations of the elements of $(\Gamma^{MV}/=)$ and give a simple characterization of the partial ordering "$\geq$" using a preordering on the representations. The outline is the following : 1) we define a binary relation "$\Sigma \supset\sim M$" between two sets B and Γ^{MV}, where B is a set of objects called bases, 2) we define a preordering "$\leq_{Resp}$" on B by copying the relation "$\geq$" of Γ^{MV}, and show that $((P^{MV}/=), \geq)$ and $((B/=_{Resp}), \leq_{Resp})$ are isomorphic, 3) we define independently a preordering "$\leq$" on B and show that "$\leq$" is identical to "$\leq_{Resp}$".

2.3 Relation "accepts/respects"

For any subset S of U and for any MVD on U X $\twoheadrightarrow$ Y, let us define a relation "$\curlyvee$" as follows :

$$S \curlyvee \quad X \twoheadrightarrow Y \quad <=>_{def} \quad X \nsubseteq S \quad V \quad Y \subseteq S \quad V \quad \bar{Y} \subseteq S$$

This relation can be interpreted as "Under the truth-assignment S, the formula X $\twoheadrightarrow$ Y is true", because X $\twoheadrightarrow$ Y can be interpreted as a propositional formula in terms of elements of U.

On the basis of this relation, we define a relation between a family Σ of subsets of U and an element M of Γ^{MV} :

$$\Sigma \curlyvee M \quad <=>_{def} \quad \forall S \in \Sigma \quad \forall m \in M \; (S \curlyvee m)$$

We propose to read Σ M as "Σ accepts M" and as "M respects Σ" (the term "respects" is due to Dr. R. FAGIN). For example the empty family ϕ of subsets of U accepts every element M of Γ^{MV}. And any M respects every {S} with card(S) $\geq$ card(U)-1. Whereas, the singleton family {ϕ} does not accept M containing an MVD $\phi \twoheadrightarrow$ Y with Y $\neq \phi$, $\neq$ U. When there is no ambiguity, Σ or M of $\Sigma \curlyvee M$ will be replaced by S or m, if $\Sigma \equiv$ {S} or M $\equiv$ {m}, respectively.

2.4 Filters and bases

Let n = card(U). Then subsets S such that card(S) $\geq$ n-1, play any role in the relation "accepts/respects" as far as it concerns only MVD-constraints. We define therefore :

$$\Phi \equiv_{def} \{S \subset U \mid card(S) \leq n-2\},$$

and call the elements of Φ MVD-filters or simply, filters. We call a family of filters a basis. The set of all bases is

$$\mathbf{B} \equiv_{def} \quad P(\Phi).$$

2.5 Preordering "$\leq_{Resp}$" on $\mathbf{B}$

We intend to copy the preordering "$\geq$" on Γ^{MV} and define on $\mathbf{B}$ a preordering "$\leq_{Resp}$". We remark that the strength of an element M of Γ^{MV} can be evaluated by $M^+ \equiv_{def}$ {m $\in$ MVD(U) $\mid$ M $\geq$ {m}}, and that $M_1 \geq M_2$ is equivalent to $M_1^+ \supseteq M_2^+$. We first define a function from $\mathbf{B}$ into Γ^{MV} by :

$$\Sigma_1 \leq_{Resp} \Sigma_2 \quad <=>_{def} \quad Resp(\Sigma_1) \supseteq Resp(\Sigma_2).$$

Clearly "$\leq_{Resp}$" is a preordering of $\mathbf{B}$. Let us denote by "$=_{Resp}$" the associated equivalence relation. And let us denote also by "$\leq_{Resp}$" the partial ordering induced on the quotient set ($\mathbf{B}/=_{Resp}$).

2.6 Isomorphism between $(\Gamma^{MV}/=)$ and $(B/=_{Resp})$

We show : a) there is a homomorphism f from Γ^{MV} onto $(B/=_{Resp})$ and b) there is a homomorphism g from B onto $(\Gamma^{MV}/=)$.

We define a function "Base" from Γ^{MV} into B which gives the lower limit of over-estimation :

$$\text{Base(M)} \equiv_{def} \{S \in \Phi \mid S \succ M\}.$$

We can compare this to a DNF (disjunctive normal form) which evaluates the conjunction of the formulas associated to the members of M.

Evidently it characterizes the relative strength of the MVD-constraints :

Theorem 2.6.1. ([H1], [HF]).

$\forall M, M' \in \Gamma^{MV}$ ($M \geq M'$ <=> Base(M) $\subseteq$ Base(M')). $\square$

This function is conjugated to the function "Resp" as follows :

Lemma 2.6.2.

(1) $\forall \Sigma \in B$ (Resp (Base (Resp(Σ))) $\equiv$ Resp (Σ))

(2) $\forall M \in \Gamma^{MV}$ (Base (Resp (Base (M))) $\equiv$ Base (M)). $\square$

Now the properties a), b) above are clear. We see by 2.6.1 and by 2.6.2(1) that Base is a homomorphism from Γ^{MV} onto $(B/=_{Resp})$, and we see by 2.6.2 (2) that Resp is a homomorphism from B onto $(\Gamma^{MV}/=)$. Thus we have :

Theorem 2.6.3

$((\Gamma^{MV}/=), \geq)$ and $((B/=_{Resp}), \leq_{Resp})$ are isomorphical. $\square$

The association between Γ^{MV} and B corresponding to this isomorphism can be defined as follows :

$$\Sigma \text{ is } \underline{\text{a basis of}} \text{ M } <=>_{def} \Sigma =_{Resp} \text{Base(M)}$$

By the onto-ness of "Base", every element of B is a basis of some M.

2.7 Characterization of the order relation on B

If we have no simple characterization of the relation "$\leq_{Resp}$" on B, then representing an MVD-constraint by a basis is useless for our purpose. We need a definition which refers no objects of other domains.

In the following we define the notion "Σ forces S", from which we define a binary relation "$\leq$" on $\mathbf{B}$ which will be found to be equivalent to "$\leq_{Resp}$".

For any $\Sigma \in \mathbf{B}$, we consider the set $B(\Sigma)$ of all bi-partitionnings $\pi \equiv \{\Sigma_1, \Sigma_2\}$ of Σ, where the conditions are $\Sigma_1 \not\equiv \phi$, $\Sigma_2 \not\equiv \phi$, $\Sigma_1 \cup \Sigma_2 \equiv \Sigma$ and $\Sigma_1 \cap \Sigma_2 \equiv \phi$. For every $\pi \in B(\Sigma)$, we associate a subset of U called the scope of π, defined by $sc(\pi) \equiv_{def} (\cap \Sigma_1) \cup (\cap \Sigma_2)$. If $sc(\pi) \equiv U$, π is called a cut of Σ. If no element π of $B(\Sigma)$ is a cut of Σ, we say that Σ is convergent and that Σ converges to $\cap \Sigma$. For instance, any singleton $\{S\}$ converges to S, because $B(\{S\}) \equiv \phi$. And the empty element ϕ of $\mathbf{B}$ converges to U.

For any $\Sigma \in \mathbf{B}$ and any $S \in \Phi$, we say that Σ forces S and we denote it by $\Sigma \Vdash S$, if some subset Σ_0 of Σ converges to S.

Example Let $U \equiv ABCDE$, let $\Sigma_3 \equiv \{CDE, BDE, ADE, ABE, AB\}$ and let $\Sigma_4 \equiv \{CDE, BDE, ADE, ABE\}$. Then $\Sigma_3 \Vdash A$, because $\{ADE, AB\}$ converges to A. But $\Sigma_4 \not\Vdash A$.

Lemma 2.7.1

For any $\Sigma \in \mathbf{B}$ and for any $S \in \Phi$, we have :

Σ converges to S $<=>$ ($S \subseteq \cap \Sigma$ and $\{S\} \leq_{Resp} \Sigma$). □

Proof $<=$) Let $S_0 \equiv \cap \Sigma$ and let $m \equiv S \rightarrow S_0$. Clearly $m \in Resp(\Sigma)$. So by hypothesis, $m \in Resp(\{S\})$ i.e. $\{S\} \sim S \rightarrow S_0$ as $S_0 \cup S \equiv S_0 \not\equiv U$, it follows that $S_0 \subseteq S$ i.e. we have shown that $\cap \Sigma \equiv S$. It remains to show that Σ is convergent. Let $\pi \equiv \{\Sigma_1, \Sigma_2\}$ be an arbitrary element of $B(\Sigma)$. Let $S_1 \equiv \cap \Sigma_1$ and $S_2 \equiv \cap \Sigma_2$. We have to show that $S_1 \cup S_2 \equiv sc(\pi) \equiv U$. Case 1) $S_1 \equiv S_2 \equiv S$. Then $sc(\pi) \not\equiv U$ is trivial. Case 2) Otherwise. Without loss of generality, we may assume that $S \subsetneq S_1$. Let $m' \equiv S \rightarrow S_1$. Then $m' \notin Resp(\{S\})$. By hypothesis, it follows that $m' \notin Resp(\Sigma)$. On the other hand, $m' \in Resp(\Sigma_1)$. As $\Sigma_2 \equiv \Sigma - \Sigma_1$, it follows that $m' \notin Resp(\Sigma_2)$. This means that there exists at least one element S' in Σ_2 for which we have $S_1 \not\subseteq S'$ and $S_1 \cup S' \not\equiv U$. This implies that $S_1 \cup S_2 \not\equiv U$, because $S_2 \subseteq S'$. We have shown that $sc(\pi) \not\equiv U$ for all π in $B(\Sigma)$. That is, Σ is convergent.

$=>$) $S \subseteq \cap \Sigma$ is trivial. It remains to show that $\{S\} \leq_{Resp} \Sigma$. Let $m \equiv X \rightarrow Y$ be any MVD such that $m \notin Resp(\{S\})$. Then by $S \equiv \cap \Sigma$, we have $X \subseteq \cap \Sigma$, $Y \not\subseteq \cap \Sigma$, $\bar{Y} \not\subseteq \cap \Sigma$. We want to show that $m \notin Resp(\Sigma)$. Let $\Sigma_1 \equiv_{def} \{S' \in \Sigma \mid Y \subseteq S'\}$ and let $\Sigma_2 \equiv_{def} \{S' \in \Sigma \mid \bar{Y} \subseteq S'\}$. It suffices to show that $\Sigma_1 \cup \Sigma_2 \not\equiv \Sigma$. Suppose to the contrary that $\Sigma_1 \cup \Sigma_2 \equiv \Sigma$ were the case. Then $\bar{Y} \not\subseteq \cap \Sigma$ implies $\Sigma_2 \not\equiv \phi$, and $Y \not\subseteq \cap \Sigma$ implies $\Sigma_1 \not\equiv \phi$. But this means that $\pi \equiv \{\Sigma_1, \Sigma_2\}$ is in $B(\Sigma)$ and moreover $sc(\pi) \supseteq Y \cup \bar{Y} \equiv U$, contradicting to the hypothesis. That is, $\Sigma_1 \cup \Sigma_2 \equiv \Sigma$ is impossible. Hence we have $m \notin Resp(\Sigma)$ as wanted. □

Theorem 2.7.2

For any $\Sigma \in B$ and for any $S \in \Phi$, we have :

$$\Sigma \Vdash S \qquad <=> \qquad \{S\} \leq_{Resp} \Sigma.$$

Proof =>) By definition of "$\Vdash$", there is a subset Σ_0 of Σ which converges to S. By 2.7.2, it follows $\{S\} \leq_{Resp} \Sigma_0$. By the transitivity of "$\leq_{Resp}$", it suffices to show $\Sigma_0 \leq_{Resp} \Sigma$. But this follows immediately from $\Sigma_0 \subseteq \Sigma$, by definition of "$\leq_{Resp}$".

<=) We want to show this by 2.7.1. Let $\Sigma_0 \underset{def}{\equiv} \{S' \in \Sigma \mid S \subsetneq S'\}$. The condition $S \subseteq \cap \Sigma_0$ is trivial. It suffices to imply from the hypothesis that $\{S\} \leq_{Resp} \Sigma_0$. Suppose to the contrary that there exists an MVD $m \equiv X \twoheadrightarrow Y$ such that $m \notin Resp(\{S\})$ and $m \in Resp(\Sigma_0)$. Then from such an m we can find an MVD m' such that $m' \notin Resp(\{S\})$ and $m' \in Resp(\Sigma)$, which contradicts to our hypothesis. In fact, it suffices to put $m' \equiv S \twoheadrightarrow Y$. Because, $m \notin Resp(\{S\})$ implies $m' \notin Resp(\{S\})$, $m \in Resp(\Sigma_0)$ implies $m' \in Resp(\Sigma_0)$, and $m' \in Resp(\Sigma - \Sigma_0)$ by the fact that any S' in $\Sigma - \Sigma_0$ does not satisfy $S \subseteq S'$ by definition of Σ_0. We have shown that $\{S\} \leq_{Resp} \Sigma$ implies $\{S\} \leq_{Resp} \Sigma_0$. □

Our definition of the preordering "$\leq$" on B is the following. We may read it also as "forces".

$$\Sigma \leq \Sigma' \qquad <=> \qquad \forall S \in \Sigma (\Sigma' \Vdash S).$$

Following property is immediate :

Theorem 2.7.3

For any Σ, $\Sigma' \in B$, we have :

$$\Sigma \leq \Sigma' \qquad <=> \qquad \Sigma \leq_{Resp} \Sigma'. \quad □$$

Example

For our example M_0, $Base(M_0) \equiv \{CDE, BDE, ADE, ABE, AB, AE, BE, DE, A, B, E, \phi\}$. Let $\Sigma_0 \equiv Base(M_0)$, $\Sigma_1 \equiv \Sigma_0 - \{\phi\}$, $\Sigma_2 \equiv \Sigma_1 - \{A,B,E\}$, $\Sigma_3 \equiv \Sigma_2 - \{AE, BE, DE\}$. Then $\Sigma_0 = \Sigma_1 = \Sigma_2 = \Sigma_3.$ □

3. RELATION BETWEEN Γ^{MV} AND Γ^{J}

3.1 Perspective

We want to see the relation between Γ^{MV} and Γ^{J} by intermediate of B. The order structure of B is first introduced by copying the order relation "$\geq$" of Γ^{MV} using a binary relation "$\supset\sim$" between B and Γ^{MV}. If we introduce a new order structure on Γ by copying the order structure of Γ^{MV}, then the relation between B and Γ may become more apparent. In fact, it is the case. This time we use the binary relation "$\geq$" itself in place of "$\supset\sim$". The outline is :

1) we define a preordering "$\geq_{MDV}$" on Γ by copying the order structure of Γ^{MV},

2) we define a hierarchy in $\mathbf{B}$: $\mathbf{B} = \mathbf{B}_{-n} \supset \ldots \supset \mathbf{B}_{-2}$,

3) we give some results on the isomorphism between $(\mathbf{B}_{-2}, \leq)$ and $((\Gamma^J/=_{MVD}), \geq_{MVD})$,

4) we give a result on a characterizationn of the join-representability.

3.2 Preordering "$\geq_{MVD}$" on Γ

We define a new binary relation "$\geq_{MVD}$" on Γ by copying the preordering "$\geq$" of Γ^{MV}. The following property of "$\geq$" on Γ^{MV} is refered : $M_1 \geq M_2 \Longleftrightarrow M_1 \supseteq M_2$. For any $d \in \Gamma$, let $MV(d) \equiv_{def} \{m \in MDV(U) \mid d \geq m\}$. Then our definition of the new binary relation on Γ is

$$d_1 \geq_{MDV} d_2 \Longleftrightarrow_{def} MV(d_1) \supseteq MV(d_2).$$

Clearly the relation "$\geq_{MVD}$" is a preordering on Γ. We denote by "$=_{MDV}$" the associated equivalence relation. The followings are some notable properties of it :

3.2.1 $\forall d_1, d_2 \in \Gamma \ (d_1 \geq d_2 \Rightarrow d_1 \geq_{MVD} d_2)$,

3.2.2 $\forall d \in \Gamma, \forall M \in \Gamma^{MV} \ (d \geq_{MVD} M \Longleftrightarrow d \geq M)$,

3.2.3 $\forall j \in \Gamma^J \ M \in \Gamma^{MV} \ (j =_{MVD} M)$.

The quotient sets $(\Gamma/=_{MVD})$, $(\Gamma^{MV}/=_{MVD})$, $(\Gamma^J/=_{MVD})$ are partially ordered by the induced order relations which we shall denote also by "$\geq_{MVD}$". The quotient set $(\Gamma^{MV}/=_{MVD})$ is in fact identical to $(\Gamma^{MV}/=)$, because by 3.2.2, the relation "$\geq_{MVD}$" on Γ^{MV} is identical to "$\geq$". An element C of $(\Gamma/=_{MVD})$ is often denoted by $[d]_{MVD}$ using $d \in C$. By 3.2.3, its MV-part $[d]_{MVD}^{MV}$ is never empty. Whereas for an $M \in \Gamma^{MV}$, the J-part $[M]_{MVD}^J$ can be empty. If it is not empty, we say that the MVD-constraint M is join-realizable, and its elements are called join-realizations of the constraint M. By 3.2.2, all join-realizations j of M satisfy $j \geq M$. It may happen that no join-realization j of M satisfy $j = M$. That is, an M may will be join-realizable without being join-representable. On the contrary, if M is join-representable, then its join-representations j must be among its join-realizations and must be the weakests of them.

3.3 Hierarchy in $\mathbf{B}$

Let $card(U) = n$. For any $S \in \Phi$, we define the degree of S by $d(S) \equiv_{def} card(S) - n$. So, $-n \leq d(S) \leq -2$. We define the (-k)-segment of

Φ by $\Phi_{-k} \equiv_{def} \{S \in \Phi \mid -k \leq d(S)\}$, for $k = n,\ldots,2,1$. Clearly, $\Phi \equiv \Phi_{-n} \supset \cdots \supset \Phi_{-2} \supset \Phi_{-1} \equiv \phi$.

Now we define the set of the <u>bases of order $-k$</u> or <u>$(-k)$-bases</u> ($k = n,\ldots,2,1$) by $B_{-k} \equiv_{def} P(\Phi_{-k})$. Clearly, $B \equiv B_{-n} \supset \cdots \supset B_{-2} \supset B_{-1} \equiv \{\phi\}$. We define the <u>rank</u> of $\Sigma \in B$ by rank $(\Sigma) = -k \iff_{def} \Sigma \in B_{-k} - B_{-k+1}$, for $k = n,\ldots,2$.

For any $M \in \Gamma^{MV}$, let $B(M) \equiv_{def} [Base(M)]_{Resp}$ and let $B_{-k}(M) \equiv_{def} B(M) \cap B_{-k}.B_{-k}(M)$ is the set of the <u>$(-k)$-bases</u> of M. If $B_{-k}(M) \neq \emptyset$, we say that M is <u>$(-k)$- representable</u>. Note that $B_{-2}(M)$ contains <u>at most</u> one element. Because for any $\Sigma_1, \Sigma_2 \in B_{-2}$, $\Sigma_1 \leq \Sigma_2$ implies $\Sigma_1 \subseteq \Sigma_2$, by definition of "$\leq$" using the forcing. If M is (-2)-representable, the element of $B_{-2}(M)$ is called <u>the</u> (-2)-basis of M.

3.4 Relation between Γ^{MV} and Γ^{J}

We list the results given in [H2] and [H3].

Theorem 3.4.1

Let $M \in \Gamma^{MV}$. Then, M is join-realizable iff M is (-2)-representable. A join-realizable M has a unique (-2)-basis. □

This implies that $((\Gamma^{J}/=_{MVD}), \geq)$ and $(B_{-2}, \subseteq_J)$ are isomorphical. Now let $\Sigma \in B_{-2}$ and let $J(\Sigma) \equiv_{def} [Resp(\Sigma)]_{MVD}$. We have a way of generating $J(\Sigma)$ from Σ as follows : Let $\overline{\Sigma} \equiv_{def} \{\overline{S} \mid S \in \Sigma\}$ and let $I(\Sigma) \equiv_{def} \overline{\Sigma} \cup \{\{A\} \mid A \in \cap\Sigma\}$, which we call the <u>induced covering</u> of U given by Σ. Then :

Theorem 3.4.2

Let Σ and $J(\Sigma)$ be as above. Then, $\bowtie I(\Sigma) \in J(\Sigma)$ and $\forall j \in J(\Sigma)$ $(\bowtie I(\Sigma) \geq j)$. Moreover, any $j \in J(\Sigma)$ can be essentialy (modulo "=") obtained from $\bowtie I(\Sigma)$ by replacing simultaneously some of its complete subgraphs (i.e. cliques) by the corresponding hyper-edges.□

Example

Let $U \equiv ABCDE$ and $\Sigma_4 \equiv \{CDE, BDE, ADE, ABE\}$. Then, $I(\Sigma_4) \equiv \{AB, AC, BC, CD, E\}$. $J(\Sigma_4)$ consists essentially of two JDs : $\bowtie I(\Sigma_4)$ and $\bowtie \{ABC, CD, E\}$. The latter is equivalent to $\{C \twoheadrightarrow D, \phi \twoheadrightarrow E\}$. □

As to a characterization of the join-representability, we have the following one :

Let G be a graphe in the sense of a collection of edges. Let C be fa subset of G. We say that C is a <u>pure cycle</u> of G, if C forms a cycle with more than three edges and none of its chords (edges connecting pairs of non-adjacent vertices of the cycle) is among the elements of G.

<u>Theorem</u> 3.4.3

Let Σ be <u>the</u> (-2)-basis of M ∈ Γ^{MV}, then, M is join-representable
<=> I(Σ) has <u>no</u> pure cycle.

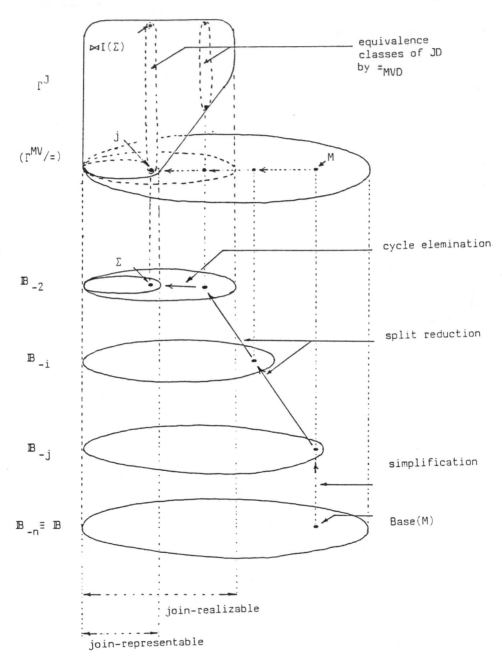

Figure 1. JD-constraints, MVD-constraints and reductions of bases.
The vertical lines represent the MVD-equivalence, =$_{MVD}$.

594

4. STRATEGY

4.1 Simplification of basis

Let $\Sigma \in B(M)$ for some $M \in \Gamma^{MV}$. Suppose there is an $S \quad \Sigma$ such that $\Sigma-\{S\} \;\|^- \; S$. Then $\Sigma-\{S\} = \Sigma$. We call such an S a <u>redundant</u> element of Σ. An elimination of a redundant element of Σ is called a <u>simplification</u> of Σ. When there is no redundant element in Σ, we say that Σ is <u>non-redundant</u>.

4.2 Intention

Let Σ_I be the given constraint. We assume that it is already in the form of a non-redundant basis. We symbolize by Σ_T the constraint to be found. It is therefore invisible for us. The condition for Σ_T is that it represents a maximal g-divisor of Σ_I. We intend to attain it by successive generations and tests of candidates for it. We are looking for a possibility of linear search producing form a candidate the next candidate. We are therefore looking for transformation rules of bases which, being applied to a basis Σ such that $\Sigma_I \leq \Sigma \leq \Sigma_T$ with $\Sigma \neq \Sigma_T$, produces another basis Σ' such that $\Sigma \leq \Sigma'^I \leq \Sigma_T$ with $\Sigma \neq \Sigma'$. The condition $\Sigma \leq \Sigma_T$ with $\Sigma \neq \Sigma_T$ and the condition demanded for Σ_T prescribe the rules. We suppose that Σ contains the information of the orientation of developement of the chain $\Sigma, \Sigma',...$ The successor Σ' of Σ must be chosen so that there may no loss in the inheritance. For this reason, Σ' must be chosen to be nearer possible to Σ. But when Σ has more than one potential successors, we need to avoid choosing one of them, because otherwise we certainly loose the information of the orientation and the search can be nomore linear. Fortunately, in many cases such successors has a unique common successor Σ' which verifies also the condition $\Sigma' \leq \Sigma_T$, and we can choose Σ' as the successor. For the help of infering $\Sigma' \leq \Sigma_T$, we use also a conjecture :

Conjecture 4.2.1.

Let Σ be maximal g-divisor of Σ_I. Then for any join-representable sub-constraint Σ' of Σ_I there exists a join-representable sub-constraint Σ'' of Σ_I such that $\Sigma'' \leq \Sigma$ and $\Sigma'' \leq \Sigma'$, where Σ'' can be identical with Σ and Σ'. In this conjecture, "Σ_1 is a sub-constraint of Σ_2" means that $\Sigma_2 \leq \Sigma_1$.

4.3 Transformation rules

Let Σ be a non-redundant basis such that $\Sigma_I \leq \Sigma \leq \Sigma_T$ and $\Sigma \neq \Sigma_T$. The result Σ' must satisfy $\Sigma \leq \Sigma' \leq \Sigma_T$. If $\Sigma \neq \Sigma_T$ is by the fact that Σ does not satisfy the condition of join-representability, we have the following general rules :

<u>Split reduction</u> : this rule can be applied when rank$(\Sigma) < -2$. It consists of replacing every S in Σ with $d(S) < -2$ by <u>all</u> its super sets of degree -2. □

<u>Cycle elemination</u> : this rule can be applied when rank$(\Sigma)=-2$ but $I(\Sigma)$ contains at least one pure cycle. Let $PC(\Sigma)$ be the collection of all the pure cycles found in $I(\Sigma)$. Then for every pure clycle C in $PC(\Sigma)$ and for every chord S of C, we add to Σ the complement set $U-S$. □

These rules are guided by the principle of "nearer is better" and also by our Conjecture 4.2.1. For each of them, the result Σ' seems to satisfy $\Sigma \leq \Sigma' \leq \Sigma_T$, but we have not yet its rigorous proof.

The remaining case is the case where Σ is join-representable. Σ is found to be $\neq \Sigma_T$ after the test. For this case, our conjecture does not work. Further generation consists of adding some elements of Φ_{-2}. The only criteria might be the acyclicity and the "distance" from Σ_I. That is, our strategy is imperfect. However as we see in the sequel, it is considerably powerful.

4.4 Example 1

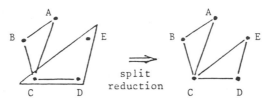

Figure 2. Example 1 : a successful case of the split-first strategy. The result of split reduction already represents a join-representable MVD-constraint. Every hyperedge represents the complement with respect to the universe U = ABCDE .

Consider our example M_0 over $U \overline{=} ABCDE$. We begin by $\Sigma_3 \overline{=} \{CDE,BDE,ADE,ABE,AB\}$. As rank $(\Sigma_0) < -2$, we apply a asplit reduction and we obtain $\Sigma' \overline{=} \{CDE,BDE,ADE,ABE,ABC,ABD\}$. $I\{\Sigma'\} \overline{=} \{AB,AC,BC,CD,DE,CE\}$ has no pure cycle. Hence, Σ' is a candidate. By elimination of cliques of $I(\Sigma')$, we get a join-representation $j' \overline{=}$ $\bowtie \{ABC,CDE\}$ of Σ'. We project Σ_3 on ABC and CDE, to get $\{A,B,C\}$ and $\{CDE,DE,,E,\phi\} = \{E,\phi\}$. M_0/j' is therefore, ϕ on ABC and $\{c \twoheadrightarrow D, D \twoheadrightarrow E\}$ on CDE. We verify by the chase method that $\{j', M_0/j'\}$ is equivalent to M_0.

4.5 Example 2

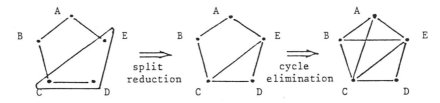

Figure 3. Example 2 : an unsuccessful case of the split-first strategy. The expected result is the clique K5. Every hyperedge represents the complement with respect to the universe U = ABCDE .

Consider another example over $U \overline{=} ABCDE$ given by : $M'_0 \overline{=} \{O \twoheadrightarrow D, AC \twoheadrightarrow B, AD \twoheadrightarrow E, BD \twoheadrightarrow C, BE \twoheadrightarrow A\}$. Base $(M'_0) \overline{=} \{CDE,ADE,ABE,BCD,AB,DE,$ $AE,CD,A,B,D,E,\phi\}$. The result of simplification is $\Sigma'_0 \overline{=} \{CDE,ADE,ABE,BCD,$ $AB\}$. The result of a split reduction is $\Sigma'_1 \overline{=} \{CDE,ADE,ABE,BCD,ABC,ABD\}$. $I(\Sigma'_1) \overline{=} \{AB,BC,CD,AE,DE,CE\}$. It has a pure cycle $\{AB,BC,CE,AE\}$. We apply a cycle elimination to Σ'_1 and get $\Sigma'_2 \overline{=} \{CDE,ADE,ABE,BCD,ABC,ABD,BDE,ACD\}$. $I(\Sigma'_2) \overline{=} \{AB,BC,CD,AE,DE,CE,AC,BE\}$. By elimination of cliques, we get a join-representation $j'_2 \overline{=} \bowtie \{ABCE,CDE\}$ of Σ'_2. M'_0 /j'_2 is given by

{CE,AE,BC,AB} on ABCE and {E,ϕ} on CDE. They are expressed by {AC$\twoheadrightarrow$B| E, BE$\twoheadrightarrow$A| C} and {C$\twoheadrightarrow$D| E, D$\twoheadrightarrow$C| E}. We want to know whether {j'_2, M'_0/j'_2} is equivalent to M'_0. The answer is unfortunately no, because the following relation instance r satisfies {j'_2, M'_0/j'_2} but not M'_0.

	A	B	C	D	E
r :	0	0	1	0	0
	1	0	0	0	1
	2	2	0	0	0
	3	3	1	0	1

We are in face of the situation that $\Sigma_I \leq \Sigma'_2 \leq \Sigma_T$ with $\Sigma'_2 \neq \Sigma_T$. We can verify that the only g-divisor of M'_0 is $\Sigma'_3 \equiv \phi_{-2}$. But we have no criteria to eliminate the two (-2) bases between Σ'_2 and Σ'_3.

4.6 Example 3

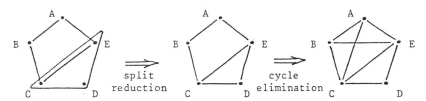

Figure 4. Example 3 : a successful case of the split-first strategy; to be compared with Example 2, Figure 3. Every hyperedge represents the complement with respect to the universe U = ABCDE .

Consider on the same U a third example given by $M''_0 \equiv$ {C$\twoheadrightarrow$D, E$\twoheadrightarrow$D, AC$\twoheadrightarrow$B, BE$\twoheadrightarrow$A}, whose non-redundant basis is $\Sigma'' \equiv$ {CDE,ADE,ABD,BCD,AB}. By splitting AB, we obtain $\Sigma''_1 \equiv$ {CDE,ADE,ABD,BCD,ABE,ABC}. $I(\Sigma''_1) \equiv$ {AB,BC,CE, AE,CD,DE} has a pure cycle {AB,BC,CE,AE}. By cycle elimination applied to Σ''_1 we obtain Σ''_2 having $j''_2 \equiv \bowtie$ {ABCE,CDE} as a join-representation. M''_0/j''_2 is then given by {CE,AE,AB,BC} and {D,ϕ}, which can be expressed by {AC$\twoheadrightarrow$B| E, BE$\twoheadrightarrow$A| C} and {C$\twoheadrightarrow$D| E, E$\twoheadrightarrow$C| D}. This time, {j''_2, M''_0/j''_2} and M''_0 are equivalent.

4.7. FURTHER PROBLEMS

The limit of our strategy, say "split first", shown by example 2, is certainly related to the fact that a pure cycle which is implicitly found in the initial basis, is reduced by the split reduction. We are obliged to take the "split-first" strategy because we have no definition of the cyclicity for the bases other than the (-2)-bases. A suitable extension of the notion of cyclity may enable us to inverse our reductions to obtain a more efficient strategy, say "cycle elimination first" strategy. Such an extension may also give us a clear sight on the problems about the conflict-freedom of MVD-constraints.

5. CONCLUSION

A relational database in stock represents a formula in a conjunctive normal form. Giving a database schema is giving the general form of the normal form, and essentially it is the only way to make the underlying semantics of the real workd reflect in the syntactical structure.

Expressing the relative independence, the MVD-constraint is the most important factor to determine the database schema. The equivalence M={j,M/j} that we studied on must be respected in the design of database schema, because otherwise the conflict between M and j will produce anormalies in the evolution of the database.

The real interest resides in the equivalent transformation from {j,M/j} to {j',M/j'}, which will provide a relational database sysstem with a considerable flexibility.

In this paper we are concentrated to the transformation from M to {j,M/j} with j maximal, because the rest is rather trivial. We have examined the possibility of applying the representations of MVD-constraints to this problem. The powerfullness of this tool was found to be evident and we got an imperfect but still powerful strategy of transformation.

This preliminary study raised the following open problems :

1) a suitable generalization of the notion of cyclicity which is actually defined only for the (-2)-bases, to all general bases,

2) a characterization of g-divisors in terms of bases and 3) a foundation of an algorithm which generates the g-divisor for any given MVD-constraint.

ACKNOWLEDGEMENT

The author would like to thank Professor Claude Delobel for a precious remark.

REFERENCES

[ABU]   A.V. Aho, C. Beeri and J.D. Ullman, "The Theory of Joins in Relational Database", ACM Transactions on Database Systems, Vol. 4, N° 3 Sept. 1979, pp. 297-314.

[A]   W.W. Armstrong, "Dependency structures of data base relationship", Proceedings IFIP 74, Norht-Holland Publi. Co. 1974.

[AD]   W.W. Armstrong, C. Delobel, "Decompositions and functional dependencies in relations", ACM Transaction on Database Systems, Vol. 5, N° 4, Dec. 1980.

[BFH]   C. Beeri, R. Fagin, J.H. Howard, "A Complete Axiomatization for Functional and Multivalued Dependencies in Database Relations", Proc. ACM SIGMOD Int. Conf. Management of Data, Toronto, 1977, pp. 47-61.

[BFMY]   C. Beeri, R. Fagin, D. Maier, M. Yannakakis, "On the Desirability of Acyclic Database Schemes", Journal of ACM Vol. 30, N° 3, July 1983, pp. 479-513.

[FMU]   R. Fagin, A.O. Mendelzon, J.D. Ullman, "A Simplified Universal Relation Assumption and its Properties", ACM Trans. on Database System, Vol. 7, N° 3, Sept. 1982 pp. 343-360.

[H1]   Y. Hanatani, "Bases for FD-MVD-Structures", Rapports de Recherche INRIA N° 236, 1983.

[H2]   Y. Hanatani, "Graphtheoretical Classification and Generation of JD's by means of Bases for MVD-Structures", Rapports de Recherche INRIA N° 170, 1982.

[H3]   Y. Hanatani, "Eliminating Cycles in Data Base Schemas", in "Advances in Data Base Theory", Vol. 2, eds. H. Gallaire, J.M. Nicolas, J. Minker, Plenum Press, 1984.

[HF]   Y. Hanatani, R. Fagin, "A Simple Characterization of Database Dependency Implication", Information Processing Letters, Vol. 22, N° 6, 1986, pp. 281-283.

[SDPF]   Y. Sagiv, C. Delobel, D.S.J. Parker, R. Fagin, "An Equivalence Between Relational Database Dependencies and a Fragment of Propositional Logic", Journal of ACM, Vol. 28, N° 3, 1981, pp. 435-453.

Database Research and Developments in the Pacific Area Countries

DATABASE RESEARCH AND DEVELOPMENT IN TAIWAN

C.C. Chang

Institute of Applied Mathematics
National Chung Hsing University
Taichung, Taiwan

The data base research and development in Taiwan can be illustrated in two parts: the theoretical part and the practical part.

So far as the research is concerned, there is an interesting phenomenon: We believe that data base storage problem is closely related to the data analysis problem.

For instance, when we store books in a library, we always perform a data analysis first. After this data analysis, which is called cataloging in library science, similar books will be stored together. Through this arrangement, when one wants to retrieve a batch of books related to the same topic, he does not have to go to different shelves.

Therefore, when one stores different records into a disk, one should do same thing, namely, putting similar records into adjacent areas such that the retrieval time can be reduced. We were very much inspired by Dr. Ghosh's pioneering work in this area.

It is also interesting to note that in case there are several disks and we have to distribute records onto these disks, we shall use just the opposite strategy: We should distribute similar records onto different disks. Later, when one retrieves similar records, this can be done in one small number of time slots as different disks can be accessed in parallel.

We even believe that this result will have a profound influence on the modern warehouse design. In the old days, in a warehouse, similar items are stored in the same physical location. Nowadays, if one employs several robots to retrieve items, one should distribute similar items onto different locations.

Hashing is considered an effective means to organize and retrieve data and is widely used in database management systems, operating systems and compilers. With a hashing function, a key word is converted into a near-random number (called address) and this number is used to determine where the key word is stored. If a hashing function can be found that is one-to-one from the set on n keys to the address space of n locations, it is a minimal perfect hashing function.

To design a minimal perfect hashing function for given set of keys is a fascinating research topic and has received increasing attention today. It is always possible to construct a minimal perfect hashing function for any arbitrary key set if the size of key set is small. To design a minimal perfect hashing scheme which can be applied to cases where the size of key set is large (For example, the key set of roughly five hundred COBOL reserved words) should be an even more challenging work.

We also believe that the result will have a huge influence on the problem of translating a higher-level language.

Consider the English-to-Spanish translation, we briefly describe the steps:

1. To divide the sequence of English characters into a sequence of tokens. That is, recognize the English words in the given text

2. Verify the grammatical structure of each sentence.

3. Using a preestablished English-to-Spanish dictionary, look up each English word obtained in "1" to its equivalent Spanish word. Then, applying certain rules, transform the grammatical structure found in "2" to its equivalent Spanish grammatical structure. Use the latter structure to produce the translated sentence in Spanish words.

Since the English- to- Spanish dictionary is searched very frequently, we may use hashing function to store and retrive data for this dictionary.

As to the practical aspect of data base research I like to point out that there was no commercially available data base management systems made in Taiwan. We had to totally rely on imported data base management systems.

I am proud to announce that we do have some data base management systems produced in Taiwan and going through the field tests at this moment. I myself developed a data base system based on the hierarchical model. There is a friendly user interface which is entirely menu driven and entirely interactive. It is now running on the CDC Cyber-170-720 under the NOS-2.1 operating system. The National Chung Hsing University and a military research organization have been using this system for about one year. This system can accept Chinese characters as input and output Chinese characters too.

Another data base management system developed in Taiwan is based upon the network model. However, it is a modified version. Only two levels are allowed. We made this two level decision after a careful performance analysis. Again, this system has a friendly user interface which is menu driven and interactive. The National Tsing Hua University is testing this system which is running on the VAX computer under the VMS operating system.

A symbolic solver for partial differential equations has been developed in the National Cheng Kung University. It is now running on the CDC Cyber 172 under the NOS-2.0 operating system. A particular set of two-variable second-order linear partial differential equations with constant coefficients and enough boundary conditions can be solved by using this package. Moreover, the symbolic solutions are obtained interactively by entering some given values and boundary conditions.

Another symbolic solver for linear difference equations had been developed in National Chung Hsing University. By using this package, a general expression that satisfied the given difference equation and boundary conditions can be determined. Again the symbolic solutions can be obtained interactively by entering some given difference equations and boundary conditions. It is now running on the VAX computer under VMS operating system.

This package will be further extended to solve more difference equations which have solutions in mathematical sense.

DATA BASE RESEARCH AND DEVELOPMENT IN KOREA

Sukho Lee

Department of Computer Engineering
Seoul National University
Seoul, Korea

ENVIROMENT OF DATA BASE R & D

Let us look at the environment of data base research and development in Korea.

Until 1980, the data base management systems has not been fully recognized and utilized in Korea. Some nonrelational, commercial DBMSs are kept at computer center as one of casual utility software. But some commercial information retrieval system packages are rather actively used in research center and companies.

Data base research and development activities in Korea are mainly dependent on two major institutions, Seoul National University(SNU) and Korea Advanced Institute of Science and Technology(KAIST) through computer science graduate students theses. Currently, the theses research topics span from data models, data language, concurrency and integrity to distributed data base and knowledge base management system. Some experimental relational DBMS are designed and implemented. Most underlying data models are relational, however, the industries and computer system suppliers are mainly interested in selling conventional, nonrelational DBMS. Thus data base design methodology, application system development based on their own DBMS and utilization of DBMS in their business computerization are major concerning matters.

It should be noted that one of common problems sharing among non-DBMS-producing countries, is interfacing input and output of their own characters with commercial DBMS. In Korea, efficient manipulation of Hangul(Korean Character) data is the most important problem in data processing. Some research activities of this area have been continued.

Databank services such as VIDEOTEX, TELETEX, are provided by DACOM (Data COMunication co.) through its nationwide communication network, DACOMNET. The various services are gradually expanded.

As an official data base research organization, SIGDB (Special Interest Group on DB) of KISS (Korean Information Science Society) becomes a good momentum for database people, now about 200 members, to share new ideas, experiences and technical problems in academy and industry. It holds a regular meeting semiannually, and some technical papers are presented.

RESEARCH ACTIVITIES

The major database research activities are categorized as follows:
The much efforts have been made to develop the system which has the Hangul data management capability and its own Hangul data query language.
The first result of these researches was HQL(Hangul Query Language)system
of SNU. The system includes the Hangul query, data definition and data
manipulation facilities. It was implemented on the existing lower level
system modules of INGRES.

As a successive development, the portable HQL system project is on
going based on IBM-PC/AT workstation which can interface with other
RDBMSs. This project is supported by national research project fund.

Another branch of Hangul research is the development of natural Hangul query language interface whose prototype system NHI (Natural Hangul
Interface) was implemented. Concurrently, the updating and natural language question answering capabilities is being extended. The system
takes advantages of the good experience of Korean-English machine translation system in developing NHI.

The research of Knowledge Base Management System(KBMS) is conducted
as a national research project, to prepare the Next Genetation Computing.
The KBMS is greatly influenced by Japan's Fifth Generation Computing, and
the major research topics are semantic data models, deductive database,
and AI programming paradigms, etc.

The Logic Programming Language, Prolog, is used as a machine language for implementing KBMS. The Sphinx is being developed in KAIST,
which provides procedure-oriented, object-orinted, and rule-oriented paradigms. There are also some efforts to merge inference engine and Database machine, and to build up a deductive database. A project of Database machine is currently supported by the Ministry of Science and Technology. The research of Engineering Database is at the very beginning,
and a research of Temporal(historical) Database has been studied as a
basic research for statistical database.

One of major researches is for DDBMS(Distributed DBMS). Aiming to
develop a working DDBMS, the major components of the DDBMS, namely concurrency control module and recovery module are investigated for design.
Other components such as task assignment, file allocation, and reliability mechanism are also designed.

In industry, the inclination of DBMS installations is popular in
these days. DBMSs are used as a core of office automation. As software
productions, they have adopted the commercial DBMS into Korean environment, and some OA software packages are available on IBM-PC level machines.

FUTURE PERSPECTIVES

Currently, national efforts are put into five nation-wide computer
network projects. These networks are Administration computer Network,
Education and Research Computer Network, Banking Computer Network, Defence Computer Network and Security Computer Network. These projects are
expected to trigger many distributed data base research activities.
Nevertheless, research and development activities will not be bound to
data base area.

In universities, relational data base system will be continuously

developed and knowledge processing systems will become also major research area.

DATABASE RESEARCH AND DEVELOPMENT IN AUSTRALIA

Leszek A. Maciaszek

Department of Computing Science
University of Wollongong
Wollongong, NSW 2500, Australia

Australia, the largest island and the smallest continent in the world, has the population of about 16 million. Being itself one of the world's most competitive computer markets, with representatives from every computer-manufacturing country, Australia cannot compete economically overseas in terms of the size of its population. With the smaller population, Australia can only win on the sophistication of research projects and the quality of manufacturing processes. While the efforts in this direction have been undertaken, the extra funds are needed to match spending levels in other developed countries. In particular, there is an urgent need for greater private sector investment in higher education and research.

There are two major research grant sources: the Australian Research Grant Scheme (ARGS) and the Australian Computer Research Board (ACRB). Other major sponsors in computer science are: the government departments (in particular Department of Science and Department of Defence), the Australian Coal Association, Telecom, the Australian Wool Corporation, and the Commonwealth Scientific and Industrial Research Association (CSIRO). CSIRO has also compiled a database called Australian Computing Research In Progress (ACRIP). It details nearly 400 research projects being undertaken in Australia.

Searching through the ACRIP database shows that Australian computing research is alive and well. Many projects are industry or application specific, many are in the domain of databases, and many are led by internationally recognized experts. By way of illustration, we list below some instances of database and closely related research.

One of the most interesting examples of longlasting and collective research efforts seems to be the development of educational database systems by Monash University (under the supervision of Dr K.McDonell). This software package, widely used in Australian universities for teaching and research, comprises three separate modules: (1) IQBE - Instructional Query-by-Example, (2) IRA - Instructional Relational Algebra, (3) MDBMS - Mini DBMS. It is an open system. All three modules have been written for Unix programming environment. The largest is MDBMS, which conforms to 1978 CODASYL specifications. The modules are still being extended but they have already proved sound and extremely useful for database education at both undergraduate and graduate level.

In the largest department of computer science among the Australian universities, i.e. in the University of Queensland, there is an influential group of database experts led by Prof. S.Nijssen and Dr E.Falkenberg. Their research concentrates on the 4th and 5th generation technology. As an example, IBM Australia Ltd released the microcomputer version of SQL (SQL/PC) that was developed at Queensland Information Technology by a team led by Professor S.Nijssen. The system is functionally equivalent to mainframe SQL and can be used for prototyping applications, before running them on the mainframe.

The sustained cooperation among researchers in Melbourne has resulted in inventing some of the best known dynamic hashing schemes. The main contributors are: Dr K.Ramanohanarao (University of Melbourne), Prof. K.Gupta (Monash University, currently - James Cook University of North Queensland), and Dr R.Sacks-Davis (Royal Melbourne Institute of Technology).

Dr I.Hawryszkiewycz from Canberra College of Advanced Education published recently an important monograph on database analysis and design (Science Research Assoc., Chicago, 1984). The book has already been adopted by many universities for teaching more advanced courses in databases.

This panelist has for many years been involved in developing an integrated database design methodology and implementing a pertinent computer-assisted design system. The methodology addresses the whole life-cycle of databases and is being developed for both network and relational environments. The CAD system is being implemented for Macintoshes and is partly operational (conceptual modelling and network logical design).

Research on knowledge bases, expert systems and logic for databases is also flourishing. Among the contributors are: Dr C.Sammut (University of New South Wales), Dr J.Lloyd (University of Melbourne), Dr J.Debenham and Dr J.R.Quinlan (NSW Institute of Technology). The Machine Intelligence Project is a major initiative which involves University of Melbourne and NSW Institute of Technology. The project looks at new architectures for plausible reasoning in expert systems. University of NSW is developing a micro-programmed Prolog machine for use in a workstation and a high-speed database access software for looking up Prolog clauses.

Apart from the aforementioned SQL/PC, many other interesting database software products have been Australian designed and implemented. The following are only few examples.

CDB - Portable Relational Database (first operational in 1982) is implemented in C for a wide range of equipment. ACCESS - network database management system has been operational since 1978. CO-CAM is a 4th generation DBMS and reporting system implemented as long ago as 1977. QUICKPICK (first operational in 1983) is an application generator for professional software developers using the PICK operating system. TODAY (developed in 1984) is a comprehensive 4th generation language which runs under UNIX.

For this panel, I have chosen to present some examples of database research and development in Australia rather than a comprehensive list of achievements. I believe that they correctly mirror the state of art and indicate the major research potentials. Is the Australian contribution to database theory and practice satisfactory in international terms? How does it compare to other developed countries with similar work force and technology? What can be the influence of major international computer science programs (such as the 5th Generation Project or the Strategic Defense Initiative program) on database research in Australia? Is the Australian research potential properly utilized internationally? The panel, I hope, will provide an opportunity to address those and other questions, to assess the current situation, and to indicate further goals.

DATABASE RESEARCH AND DEVELOPMENT IN JAPAN:

ITS PAST, PRESENT, AND FUTURE

Yoshifumi Masunaga

University of Library and
Information Science
Tsukuba Science-city, Ibaraki 305, Japan

Abstract This paper investigates the present, past, and future of database
research and development in Japan. The world-wide activities of database
research and development is first overviewed, and then the survey of database
activities in Japan is made extensively. As in USA and Europian countries,
Japan has also recognized that databases are essentially important to build
information systems for every type of applications such as business and
non-business data processing.

1. Introduction

ENIAC, the first electronic computer in the world, was developed by
Eckert and Mauchly in 1946. As it is widely accepted, the generation of
computers are divided into four; the first, second, third, and fourth
generation in which vacuum tubes, transistors, IC's, LSI and VLSI's, are used
as electronic components, respectively, recently, ICOT (Institute for new
generation COmputer Technology, Japan) is proposing the fifth generation
computers.

By the term "databases," we mean both databases and database management
systems. It is said that the necessity of database research and development
was strongly recognized by NASA in the late 1950's. CODASYL was founded in
1959, and a database management system IDS was developed by Backman in 1963.
In Japan, it is said that the database activity was first initiated around
1967 and IDS-like system was implemented on a Japanese computer. However,
the word database, or data base, or data bank was not recognized in Japan at
that time. Indeed, one member of the SIGSOFT (Special Interest Group on
Software) of IPSJ (the Information Processing Society of Japan) retrospected
and said that "the SIGSOFT had completely no idea on how to respond to the
request from CODASYL," when SIGSOFT first received the CODASYL draft proposal
in 1968. Then, couple of database research and development groups were
formed in around 1970, and SIGDB (SIG on Data Bases) of IPSJ was established
in 1974 with 163 members. The third international conference on VLDB (Very
Large Data Bases) was held in Tokyo in 1977, and the twelfth VLDB conference
will be held in Kyoto in 1986.

At present, there are six computer manufacturing companies in Japan:
Fujitsu, Hitachi, Mitsubishi, NEC, Oki, and Toshiba (in alphabetic order).
Of course, IBM, DEC, UNIVAC, NCR, and other foreign companies also have

strong bases for both marketing and manufacturing in Japan. Besides, many other Japanese companies exist for small computers and applications. All of them are now very active in database area. Ministry and academic circles are also active in research and development in building database based information systems.

In the following of this paper, we first overview world-wide activities of the research and development (R&D) of databases, and then report the Japanese activities in this area in detail.

2. A World-wide Overview of Database R&D

From database point of view, the last twenty years and so could be divided into the following five periods:
(1) The first period: -1969 (the ex-relational period).
(2) The second period: 1970-1974 (the period to substantiate the relational model of data).
(3) The third period: 1975-1979 (the period of prototyping of relational database management systems).
(4) The fourth period: 1980-1984 (the period of commercial release of relational database management systems).
(5) The fifth period: 1985-   (the period of the use of relational dbms's in every data processing applications).

These periods could be characterized in terms of the following activities:

(a) Announcement of commercial database management systems.
(b) Academic circle's activities.
(c) Other related events.

The first period (-1969):

IDS, the first dbms in the world, was marketed in 1963. TOTAL ('68), SYSTEM 2000 ('69), IMS ('69) were also on the market. The development of ADABAS was initiated, although it was not on the market until 1972. The 1960's was really a period for non-relational dbms's prosperity. However, as the term "software crisis" was first used by a NATO conference in 1969, there was a recognition that databases should be independent from application programs. It should be noted that the information algebra was proposed in 1962.

The second period (1970-1974):

The relational model of data was proposed in 1970. There were several institutional activities to design and implement relational data languages and database management systems. For example, SQUARE ('74) and Peterlee relational test vehicle. In this period, the University of California at Berkeley and the IBM Research Laboratory at San Jose, California, initiated INGRES project and the System R project, respectively.

Important conferences were first organized in this period. For example, the ACM SIGFIDET Conference was first held in 1971. SIGFIDET was changed into SIGMOD and the first SIGMOD Conference was held in 1973. Very hot discussion on the relational approach vs. the CODASYL approach was made in the 1974 SIGMOD Workshop. The call for papers for the first VLDB Conference appeared in 1974.

IBM System 370, which is known as the first product of the 3.5 generation (using IC's) computer was released in 1970. The first micro processor was delivered by Intel in 1971. The energy crisis began in 1974.

The third period (1975-1979):

In this period, a couple of full scale relational dbms's were institutionally developed. For example, INGRES prototype and System R were implemented in 1976 and 1979, respectively. QBE, a relational dbms with two dimensional interface, was marketed in 1978.

Database research became very popular in the world. For example, ACM TODS was first published in 1976, and the first VLDB Conference was held in 1975. Many other conferences such as the E-R Conference were first held in this period.

ANSI/X3/SPARC reported the final report in 1979.

The fourth period (1980-1984):

Many relational dbms's were announced as commercial products. DB2 for MVS operating system ('83), SQL/DS for DOS operating system ('81), INGRES for UNIX operating system by RTI ('80), and even a system for personal computers, say dBASEII for MS-DOS or CP/M-86 ('82)were available in the market of database systems.

The research and development of the distributed relational databases were in the final stage at this period. System R* project was almost completed in 1984. Great effort for developing DDM and Multibase was undertaken by CCA.

Applicability of relational databases to non-business data processing applications, such as engineering design databases, was first investigated extensively in this period. Development of office system applications were widely made and became popular. For example, OBE in 1980 and EMCOMPASS in 1981.

Of course database research was still flourishing in this period. The ACM PODS Conference was first held in 1982. Semantic data modelling became a hot issue.

IBM 308X, which is a 4G (VLSI) computer, was marketed around 1982.

The fifth period (1985-):

This period has just started. However, it seems clear that this will be a period for extensive research and development of how to organize non-business data processing dbms's based upon the relational approach. Presently, it has not yet been clear that the relational approach is the best one for such applications. But the very characteristics such as the high degree of data independence, the high degree of non-procedural property, the simplicity of the model, etc., which only the relational can provide seems essentially necessary to such applications.

Knowledge based approach for database management will be emphasized. Integration technique of databases and knowledge bases will also be extensively investigated.

3. Database Research and Development in Japan

3.1 An Overview in Computers in General

In order to understand what was done in Japan for the research and development of databases correctly, it seems quite reasonable to overview it from the following two areas:
(1) Role of industry under the (proper) guidance of the Ministry of International Trade and Industry (MITI) of Japan, and

(2) Role of Japanese academicians in it.

As it is well known, a country Japan needs natural resources from outside, the Japanese government took the following measures to promote computer manufacturing and its related activities:
(a) A special law for the promotion of machine industry was established in 1956, and was valid until 1970. Also a special law for the promotion of electronic industry was established from 1957 to 1970. These two laws were merged into one, called a special law for the promotion of specific electronic and machine industries in 1970, in order to cope with the situation caused by the liberalization of capital. This law has been valid since that time to now, although its name was changed in 1978.
(b) A special law for information technology promotion enterprise was established in 1970 in order to promote software development in Japan. This law is still valid.

Under these laws, IPA (Information-Technology Promotion Agency) was established, and various research unions were formed with the member of Japanese computer makers. Research and development funds were offered to them through IPA. That is why the Japanese 64kbit RAM technology became the highest in the world.

MITI has recognized that information is the third element following to the material, and energy. Also they understood the role of information industry and thus, the information industry is classified into two categories; computer industry, and information processing industry. More precisely, the former is further divided into computer manufacturing industry and VLSI industry, and the latter is classified into software industry, information processing services, and information supply services, i.e., database services.

3.2 An Overview of Databases R&D

Although MITI felt the importance of the promotion of database creation, processing, store, and its use, the governmental action on it was very slow. The reason seemed that the MITI was unable to understand that softwares will not be produced spontaneously even though good hardwares are provided.

From the administrative point of view, it is said that the first year of database activities in Japan is 1979, since the Machine Promotion Agency made a proposal for the promotion of database industry. Responding to this proposal, IPA started its measures to promote database softwares from 1980. Also a liason round-table conference for database service traders was formulated in 1979, and there were 29 members in 1982. It should be noted that a ledger for databases available in Japan was compiled in 1982. More than six hundred databases (duplication deleted) were registered in 1983. However 55% databases registered in 1983 are bibliographic database in Japan, while in case of Europe 65% databases are fact database. In 1984, Information Processing System Development Section was established under the Machine and Information Industry Bureau of MITI, which is the section for databases. Database Promotion Center was also established in 1984. If we compare Japanese actions with other countries, USA started to produce large databases such as NASA, CAS, and NTIS in 1964, and France and W. Germany begun database production promotion around 1973. In Japan, however, there are three large scaled originally developed databases, which are produced by JICST (Japan Information Center for Science and Technology), JAPATIC (Japan Patent Infromation Center), and Nikkei Shinbun-sha Co. Ltd.(the producer of NEED).

Little policy for the promotion of dbms development has been undertaken in Japan except in a few cases. ADABAS-like dbms named INQ was developed by NEC under the request of MITI, a relational database machine named EDC was developed by the Electro Technical Laboratory of MITI in 1980. Furthermore, a relational database machine named DELTA was developed by ICOT (Institute for new generation COmputer Technology) in 1984. The interoperable computer system project will be initiated by MITI in 1986 as a seven year plan under which multi-media dbms's will be developed.

Many CODASYL-like, or TOTAL-like, or IMS-like dbms's were developed for Japanese computers before 1980. The best seller in 1981 was the AIM by Fujitsu Co. Ltd., which is a CODASYL dbms.

Of course, full scale relational dbms's were implemented by Fujitsu Res. Lab., and Hitachi Co. Ltd. in 1979 and 1984, respectively. The former is named RDB/V1 and the latter is named RDB1, both of which are functionally compatible with SQL/DS. Fujitsu is now developing a distributed version of RDB/V1 named RDB/DV. NTT (Nippon Telegraph and Telephone corporation) has developed a distributed database management system named DEIMS-3 which integrates CODASYL databases. QBE-like interfaces were also available in certain systems.

The research and development of the dbms's for engineering databases, such as CAD/CAM (a typical example), has a relatively long history in Japan, particularly in NEC, Hitachi, and Mitsubishi. Office applications are extensively under development in many companies.

3.3 Japanese Academic Circles

The Ministry of Education, Culture, and Science founded the Grant-in-Aid for Special Project Research on the organization of scientific information for three years in 1977. This project provided a very good place for Japanese university database people to get together periodically and then present and discuss their ideas about databases very freely. DBMS architecture, database design, database machines, data models, and the relational database theory were widely discussed. Also ideas on knowledge bases and their management were made which became a seed for the fifth generation computer project of Japan. The Ministry is now conducting another special research project on knowledge information processing and integration since 1984 for the three year plan.

The Information Processing Society of Japan (IPSJ) was established in 1960, and is now one of the biggest societies in Japan. Total members in 1983 was accounted to 19,000. There are three main activities on databases that the IPSJ holds:
(a) Annual conferences (spring and fall).
(b) Special Interest Group on Database Systems.
(c) Database symposium (once a year).
In recent years, about 10% of the total of more than 1000 IPSJ annual conference papers are appearing in database systems. SIGDB was started in 1974. It holds a regular bi-monthly meeting, and several papers are presented regularly. A database symposium is held once a year which usually covers the burning topics. About one hundred people usually attend such symposium.

Japan is now one of the most popular "host" countries for international conferences. In 1977, the third VLDB Conference was held in Tokyo and had attended by 300 participants from different countries. Besides, IFIP 80 and ICSE were held in Tokyo in 1980 and in 1982, respectively. The International Conference on Foundation of Data Organization is going to be held in Kyoto in

1985, and it has also been decided that the twelfth VLDB conference will be held in Kyoto in 1986.

4. Conclusion

Japan has its own Kanji, Hiragana, and Katakana characters. In order to make dbms's widely used in Japanese enterprises, we should implement dbms's which can handle Japanese characters. INGRES, for example, has not yet been modified to support Kanji. Also Kanji causes troubles in producing proper database activities. For Japan, of course, databases should be in Kanji databases. However, they can not be offered to other countries, and vice versa. The European countries are facing to the similar problem.

At the tenth VLDB conference at Singapore, five out of 48 were Japanese papers. All Japanese papers were on theoretical works. However, the author knows that many excellent practical developments have been made in Japan in database industry, including distributed databases, CAD/CAM, OA and others. In many cases, they were not presented or distributed world-widely, mainly because our mother tongue is different from the Anglo-Saxon's mother tongue and Japan locates in Far-East. We hope that the fifth generation computer will contribute to overcome such barrier in near future so that Japan may take leadership in database activities in the world.

References

(1) Computer White Paper 1982, JIPDEC, Computer AG, Inc., 1982 (in Japanese).
(2) S. Uemura, "Database systems: Where we stand and where to go," Information Processing, Vol.23, No.10, pp.884-892, 1982 (in Japanese).

DATABASE RESEARCH AND DEVELOPMENT IN CHINA

Shu Gang Shi, Zhen Mei Zheng
Mong Chi Liu and Chun Ruan

Department of Computer Science
Wuhan University
Wuhan, China

ABSTRACT

This paper introduces the survey and organization for database research and development in China.

THE SURVEY FOR DATABASE RESEARCH AND DEVELOPMENT

We did not originate the research and development of database systems until the middle of 1970s. For some reason, we began to do it later than other countries. But being taken seriously by our goverment, we have made great progress in less ten years.

In November 1977, the first National Academic Conference on Database convoked at Yellow Mountain. In the conference, the fundamental problems were discussed, and the international research tendencies were introduced. This conference play an important role in promoting the research for database in China.

In September 1982, the second National Academic Conference convened at Nino Bo. The 142 deputies representing eighty-nine organizations, and coming from the sixteen provinces of the whole country, attented the conference, and 96 papers were received. During the time between this two conferences, the research on database in China had made great progress. We began to design and implement database systems. There were approximate 30 database systems beging developed, such as System DBS-130, System RTDB, System ACD-1, System DBMS-R and so on. The persons who are engaged in this work were more than 700.

In September 1984, the third National Academic Conference on Database, opened in Tain Tin. There were 177 deputies coming from all over the country attended the conference. During the conference, seven special themes were arranged to discussed and 77 papers were reported, which included database theory, database systems implemention, distributed database, database application and so on. This conference was at the time that the research on relational database had made great progress, the need for database in our society become more and more strong and the persons engaged in this area had increased greatly. Because of this, it has been decided on that conference the National Academic Conference will be held once a year.

RESEARCH INSTITUTION

In China, the research on database systems have been done in some specialized scientific research organizations. These organizations may be national, such as the Institute of Computational Technique in the Academy of Science of China, or they may be local or under particular ministry, like Huabei Institute of Computer which belongs Electronic Industry Ministry. Almost every province or city has established computer centers now.

In addition, many universities, like Beijing University, Qinghua University, Fudan University, Nanjing University, Jiaotong University, Renmin University, and Wuhan University are also working in the area of database. In China, we have about hundreads of universities and colleges. More than 150 of them have had computer science department or speciality. The course of "Introduction to database systems" is offered to undergraduate students while other advanced courses on database such as "The Theory of Database" are offered to graduate students. As a result, there are at least 10,000 persons who are educated in fundamental database knowledge every years, several dozens becoming Masters on it. But less persons have got ph.D degree of database. Nearly all kinds of subjects in database area have been discussed.

Another important aspect is that there are a great many users of database who are developers as well as users of it. Database systems are going into every field in our country.

In addition, with regard to academic activity, there is a annual conference as described above, which is "Database Academic Annual Conference." It is convoked by the specialized group of database.

THE PRESENT SITUATION

At present the main work on the database in China is to undertake the research on relational database theory, the implemention of relational database systems on mini- and micro- computers, and to introduce kinds of database systems in large computers, developing and applying.

Moreover, many persons are making researches and discuss theoretically and practically on distributed database systems on local area network for micro computers. Database machine as well as knowledge base is also attracting a lot of scholars and experts.

DATABASE RESEARCH AND DEVELOPMENT IN SINGAPORE

K.P. Tan* and R.A. Cook**

*Department of Information Systems and Computer Science
**Institute for Systems Science
National University of Singapore
Kent Ridge, Singapore 0511, Singapore

1. INTRODUCTION

This short paper reports on various current database research interests at the National University of Singapore and at the National Computer Board and the Systems and Computer Organisation of the Ministry of Defence. It also records some results of a recent study into database usage in Singapore, and mentions activities in data analysis and database design methodologies.

2. RESEARCH IN DATABASE AND RELATED TOPICS

2.1 Acyclic databases [1]

In relational database theory, database schemas may be partitioned into two classes: tree schemas and cyclic schemas. Recent work on the universal relation databases includes: Graham reductions of Graham, and Yu and Dszoyoglu, the canonical connection of Maier and Ullman, the tree and cyclic schemas of Beeri, the tree projections of Goodman and Shmueli, and Fagin's γ-acyclicity. The relationship among these concepts was examined in a study. In analysing lossless joins, a new characterisation for γ-acyclic databases is provided. If the reduction of relation R to schema D transforms it into a tree schema, then $R \supseteq GR(D)$. A separate work gives a sample characterisation of multivalued dependencies that are equivalent to a single join dependency.

2.2 Locking performance [2]

Locking is the most popular technique for concurrency control. A simple analytic model was proposed for studying its performance. This model is powerful enough to handle nonuniform data access, shared locks, multiple transaction classes, transactions of fixed or variable length, and static as well as dynamic locking. Analysis shows that blocking imposes an upper bound on throughout; this leads to a rule of thumb for limiting the workload on a system. This limit can be exceeded by using static locking, or by restarting transactions whenever there is a conflict. The model also makes possible a study of the interaction between resource and data contention, and the effect that changes in multiprogramming level, transaction length, and granularity of locks have on performance.

2.3   Improvement in third normal form [3]

Some Codd's third normal form relation schemas may contain removable superfluous attributes because the definitions of transitive dependency and prime attribute are inadequate when applied to sets of relations. To remove this discrepancy a new normal form is defined. A set of relations can be constructed from a given set of functional dependencies under the deletion normalization algorithm. An attribute that is essential is always prime.

2.4   Less costly checking for join dependency [4]

In relational database design, to check whether a relation instance is in fifth normal form, we need to check it against the join dependency. Some constraint equalities in the join dependency were found to be redundant. To remove this superfluity, the universe of attributes is partitioned into n disjoint sets and a new notation of join dependency is proposed. The checking time in each run of n tuples of a relation is reduced by a factor or n/2 when n>3.

2.5   Elimination of avoidable checking of integrity constraints [5]

The usage of semantic constraints to monitor the integrity of databases is under study. An integrity subsystem is more efficient, in that it can save program development effort and prevent errors of implementation in application, if certain constraints that are not affected by a specific update can be excluded automatically from the integrity check. An algorithm to eliminate such avoidable integrity checks is proposed. It is applicable to semantic assertions in any high level language based on relational algebra or calculus.

2.6   A logical database design tool -- datadict [6]

Based on IBM's Business System Planning, Yourdon's Structured Analysis and Structured Design, Infocom's Information Engineering and Jackson Structured Programming, DATADICT is designed and implemented as an automated data analysis and logical database design tool. It provides a data dictionary that supports data analysis and design software, complementing a system development methodology developed at the Ministry of Defence's Systems and Computer Organisation. DATADICT is implemented on a personal computer workstation for portability, and was developed using dBase II.

2.7   Non-destructive Sorting by a long key [7]

For a 16-bit computer, a long key of 5-9 digits can be regarded as a 2-component compound key: key1--key2. A sorting method (quicksort embedded within block sort) is introduced to sort a large volume of records in external storage by the long key, without destroying the original data file. Only 3 passes of the file are required.

2.8   Automatic schema navigation

Some work has been done to use expert systems to search relational databases for a set of attributes without requiring the enquirer to specify to which relations the attributes belong. This research has also led to work on database schemas for storing knowledge rules for expert systems.

3. DATABASE APPLICATION

3.1 A survey of database users

This survey was carried out by the Database Special Interest Group of the Singapore Computer Society by questionaire to owners of mainframe and minicomputers.

Results showed that the penetration of DBMS into installations has increased markedly over the last 3 years, and that nearly all installations are using their DBMS for a wide range of applications and are practising the sharing of data between applications. The majority of the respondants to the questionaire were from the public sector.

Amongst programming languages, productivity tools and COBOL are both in use by about 75% of the population. 88% are using query languages and 65% data dictionary tools. However, the level of satisfaction with query languages was low, averaging less than 'good' on almost all measures.

Owners reported satisfactory achievement with such purported advantages of DBMS as data sharing, improved security, integrity and standardisation, and the reduction of data redundancy and inconsistency. However they did not feel that DBMS help to reduce costs. Major frustrations were the lack of experience and experienced specialists in Singapore, and the time required to write programs, convert programs and create the data dictionary. Anticipation of future demands also ranked high as problems.

The leading DMBS products amongst the survey respondants were, in order:
Image (HP)
IDMS (Cullinet)
ADABAS (Software AG)
IMS (IBM)

It is interesting that some relational database products are beginning to feature in the list of those in commercial use, notably SQL/DS and INGRES.

3.2 Database methodologies

A considerable amount of work has been done at the National Computer Board, the Systems and Computer Organisation and the Institute of Systems Science to develop and teach a set of coherent methodologies for the analysis of data to produce a data model, and for the logical and physical design of databases starting from the same data model. The database design process is general but has been refined to suit each individual DBMS. These methodologies have been developed in Singapore for the local computer industry, by synthesising many existing published methods. The methods and models mesh with system design methodologies. The DATADICT tool mentioned above supports some of these methodologies.

Interest is now focussing on data administration techniques, the effective use of productivity tools for systems development, and on the commercial exploitation of relational DBMS.

4. REFERENCES

[1] N. Goodman, D. Shmueli and Y.C. Tay, GYO Reductions, Canonical Connections, Tree and Cyclic Schema, and Tree Projections, J. Computer and Sys. Sci. 29 3, December, 1984, 338-358.
[2] Y.C. Tay and R. Suri, Choice and Performance in Locking for Databases, Proc. 10th Int. Conf. on VLDB, Singapore (Aug. 1984), 119-128.

[3] T.W. Ling, F.W. Fompa and T. Kameda, An Improved Third Normal Form for Relational Databases, ACM TODS, 6 2, (June 1981), 329-346.

[4] K.P. Tan, A Less Costly Constraints Checking for Join Dependancy, Proc. 10th Int. Conf. on VLDB, Singapore (Aug. 1984), 63-68.

[5] T.W. Ling and P. Rajagopalan, A Method to Eliminate Avoidable Checking of Integrity Constraints, IEEE 1984 Proc. Trends and Applications, 60-68.

[6] T.J. Tan, K.P. Tan and A.M. Goh, A Data Analysis and Logical Database Design Tool, Proc. 10th Int. Conf. on VLDB, Singapore (Aug. 1084), 71-77.

[7] K.P. Tan and L.S. Hsu, Block Sorting of a Large File in External Storage by a 2-component Key, The Computer Journal, 25 3, (1982), 327-330.

Author index

Subject index

operations, 529
Parallel sequential access, 535
Parameterized version, 433
Parent table, 145
Parser, 40
Part-subpart hierarchy, 461
Partial correlation coefficient, 7
Partial directory, 403
Partial match query, 113, 159,
        177, 287, 288, 298, 391,
        409, 531
Partial orders, 134, 139
Partial randomness, 81
Partial range query, 287
Partition
  buffer, 523
  number, 516
  set, 503
  tree 517
Perfect hashing function, 91
Perfect nested expression, 564
Performance
  evaluation, 305
  of statistics metadata, 14
  prediction, 399
  related characteristics, 401
Permissible update, 327
Physical allocation structure, 378
Physical database
  design optimizer, 496
  organization, 293
Physical lock, 183
Physical locking scheme, 183
Physical page, 295
Physical storage structure, 378
Pipelining, 284, 294
Plug and socket diagram, 433
Point
  query, 448
  set query, 448
  -in-interval query, 448
Polish notation, 542
Polygon, 249
  tree, 243
Polygonal query, 238
Polyhedra, 444
Polyhedral range query, 237
Polyhedron, 444
  tree, 243
Positive clause, 541
Predicate
  locking, 184
  symbol, 53
Prenex normal form, 566
Primary object, 461
  rep, 462
Primary vertex, 324
Prime number, 93
Prime relation, 58
Procedure definition table, 543
Projection, 146, 345

-selection combination, 554
Prolog, 474, 539
  machine, 612
Prototype, 476
Pruned trie, 392, 397
Pseudo TRACER file, 129
Pseudo-quad tree, 239
QUEL decomposition algorithm, 502
QUICKPICK, 613
Quad tree, 239
Quantified query, 564
Query
  class, 400
  distribution function, 295
  processor, 40
  tree evaluator, 40, 41
  tree optimizer, 40, 41
Question-answering systems,
        108, 110
Quick sort, 521
R-relaxed CR (RCR) property,
        104, 105
RDB/DV, 619
RDB/V1, 515, 619
REBUILD, 147
RECONNECT, 579
RTDB, 621
Random variable, 4
Range
  query, 79, 80, 287, 288, 294,
        297, 531
  search query, 409
  -select query, 11
Rank, 335, 347
Rate-distortion
  function, 106
  theory, 104, 106
Realm, 378
Record
  array, 285
  type, 378, 578
  usage statistics, 377
  -based model, 466
  -query incidence matrix, 117
  -to-area mapping, 377, 382
Rectangle, 207
Recursive partitioning, 515
Recursive unsolvability, 358
Redo, 41
Reduction, 208
Redundancy of a filing scheme, 227
Refutation procedure, 539
Region, 208
  query, 441
Register structure, 268
Regression coefficient, 4, 7
Regular set table, 543
Regularity, 487
Relation, 128
  cardinality, 491
  management table, 543